Ecological Studies, Vol. 117

Analysis and Synthesis

Edited by

O.L. Lange, Würzburg, FRG
H.A. Mooney, Stanford, USA

Ecological Studies

Volumes published since 1989 are listed at the end of this book.

Springer
New York
Berlin
Heidelberg
Barcelona
Budapest
Hong Kong
London
Milan
Paris
Santa Clara
Singapore
Tokyo

José M. Moreno Walter C. Oechel
Editors

Global Change and Mediterranean-Type Ecosystems

With 144 illustrations

Springer

José M. Moreno
Departamento de Ecología
Facultad de Biología
Universidad Complutense
28040 Madrid
Spain

Walter C. Oechel
Department of Biology
San Diego State University
San Diego, CA 92182
USA

Cover illustration: Average results from general circulation model of summer equilibrium changes for air temperature (°C) after doubling of CO_2. (Redrawn from Warrick et al., 1990. See Chapter 19, page 388).

Library of Congress Cataloging-in-Publication Data
Global change and Mediterranean-type ecosystems/José M. Moreno, Walter C. Oechel, editors.
p. cm. – (Ecological studies; v. 117)
Includes bibliographical references and index.
ISBN 0-387-94352-8 (hardcover: alk. paper)
1. Mediterranean-type ecosystems. 2. Climatic changes – Environmental aspects. 3. Vegetation and climate. I. Moreno, José Manuel. II. Oechel, W.C. III. Series.
QH541.5.M44G58 1995
574.5′222 – dc20 95-5548

Printed on acid-free paper.

Production coordinated by Chernow Editorial Services, Inc., and managed by Francine McNeill; manufacturing supervised by Jeffrey Taub.
Typeset by Best-set Typesetter Ltd., Hong Kong.
Printed and bound by Braun-Brumfield, Inc., Ann Arbor, MI.
Printed in the United States of America.

9 8 7 6 5 4 3 2 1

ISBN 0-387-94352-8 Springer-Verlag New York Berlin Heidelberg
ISBN 3-540-94352-8 Springer-Verlag Berlin Heidelberg New York

Preface

Mediterranean-type ecosystems (MTEs) comprise a vegetation form with worldwide importance and distribution. MTEs are found in areas around the Mediterranean Basin, in North America (California), Chile, Australia, and South Africa. MTEs are usually associated with current or historic centers of prosperity and cultural development. With large permanent or transient (tourism) populations, they are among the most frequently visited areas of the world. In addition, the population in these areas is growing and is expected to continue doing so.

The contribution of MTEs to the balance of energy and materials of the earth system is probably small, because of their reduced area and the low-standing crop density of carbon in plant and soil systems. However, the impact of global change in MTEs is of greater global significance than one would anticipate given their human population and limited extension. In fact, MTEs are second only to the tropics in terrestrial biodiversity. Several areas of MTEs have been identified as among the first few plant biodiversity hot-spots of the world. Additionally, many of the world's agricultural, horticultural, and medical plants are of Mediterranean origin. Furthermore, MTEs are feeding and resting grounds for an important number of migratory birds. Changes in these grounds may have serious consequences for the bird populations of much larger areas of the world.

MTEs are transitional areas between the arid and mesic regions of the world. The contrasting temperature and precipitation regime between summer and winter places important restrictions on plant growth. These have important consequences for ecosystem functioning. It is becoming apparent that CO_2 effects on plant growth can be particularly important when water is limited, a situation that is present every year in MTEs. Further, the relatively small stature, but complex nature, of MTE shrublands makes them excellent models for understanding the impact of rising CO_2 on ecosystem functioning.

MTEs, particularly in the Mediterranean Basin, have been used since ancient times, producing complex landscapes patterned by the activity of people. Recent socioeconomic changes have brought important modifications in the use of the land, which may have resulted in large-scale modifications in ecosystem functioning, particularly the consequences of fire. The contrasting socioeconomic situations that still exist within these areas and the expected different future trends, which will further develop the current North-South contrasts with regard to economic development and population growth, also make MTEs excellent model systems to differentiate the various interactions of global change forcings.

Water is a limiting resource in MTEs. In some areas, current water demands cannot be satisfied from available resources. The increasing population, plus the elevated demands for water, make these areas most vulnerable to global change. Understanding how climate and other global changes can impact ecosystems to alter water availability is crucial in the area.

The above list of topics on the significance of global change for MTEs is not exhaustive, yet it signifies the relevance of these areas of the world under the current global change paradigm. To understand this, an international symposium met in Valencia (Spain) in September 1992, with representative scientists from all the Mediterranean-type areas of the world. The overall aim was to evaluate how global change could affect some of the main ecosystem processes in MTEs. The present volume contains the majority of the scientific presentations delivered at that meeting. Following the symposium, the participants gathered in a workshop to summarize the likely sensitivity of MTEs to global change; possible significant feedback on the global earth system; areas where MTEs could be particularly good model systems to study processes that are relevant for understanding global change impacts; and to recommend research activities for a better understanding of the response of MTEs to global change. The recommendations of the workshop are available from any of the editors.

We hope that this volume will contribute to the understanding of the problems of Mediterranean-type areas of the world and to raise awareness of their significance. While some MTEs are among the most developed areas of the world, important socioeconomic differences persist between them and across boundaries within a continent. Because the population in less-developed areas may be most vulnerable to threats of global change, it is imperative than we develop studies conducive to a better understanding of how global change can affect all areas and to help us implement actions to avoid or mitigate some of the most undesirable expected impacts.

The Valencia meeting was sponsored by Universidad Internacional Menéndez Pelayo in collaboration with Centro de Estudios Ambientales del Mediterráneo (CEAM), to which we express our gratitude. We thank the UIMP and CEAM staffs for their help in ensuring the success of the meeting. We also thank the following researchers who helped in reviewing the manuscripts: E. Allen, R. Askins, L. Ayerbe, J. Blondel, P. Bottner, F. Castro, D. Cayan, F. Chapin III, E. Chuvieco, S. Davis, J. Ehleringer, K. Everett, C. Field, P. Flanagan, T. Gilmanov, T. Halliday, P. Harley, D. Hilbert, J. Hom, P. Jordano, B. Lacaze, W. Lawrence, H. Le Houérou, G. Malanson, J. Meliá, R. Minnich, F. Mowry, E. O'Neill, J. Peñuelas, S. Rambal, P. Riggan, F. Rodá, J Roy, A. Ruiz de Elvira, J. Seeman, B. Strain, J. Tenhunen, N. van Bremen, J. Walsh, and P. Zedler.

José M. Moreno
Walter C. Oechel

Contents

Contributors

Pere Alberch	Departamento de Biodiversidad y Biología Evolutiva, Museo Nacional de Ciencias Naturales, José Gutierrez Abascal 2, 28006 Madrid, Spain
Edith B. Allen	Department of Botany and Plant Sciences, University of California, Riverside, CA 92521-0124, USA
Michael F. Allen	Department of Biology, Systems Ecology Research Group, San Diego State University, San Diego, CA 92182-0057, USA and Division of Environmental Biology, National Science Foundation, Arlington, VA 22230, USA
Anna Avila	Centre de Recerca Ecológica i Aplicacions Forestals, Universitat Autónoma de Barcelona, 08193 Bellaterra, Spain

M. Badiani	Department of Biochemistry, University of Viterbo, Via S. Camillo de Lellis, Viterbo, Italy
I. Bettarini	IATA-CNR, P. le delle Cascine, 18 50144 Firenze, Italy
W.J. Bond	Department of Botany, University of Cape Town, Rondebosch, 7700, South Africa
P. Bottner	C.E.F.E.-C.N.R.S., Route de Mende, 34033 Montpellier, France
Josep Canadell	Department of Integrative Biology, University of California, Berkeley, CA 94720, USA and Unitat d'Ecologia, Universitat Autònoma de Barcelona, 08193 Bellaterra, Spain
Manuel Castro	Departamento de Geofísica y Meteorología, Facultad de Física, Universidad Complutense, 28040 Madrid, Spain
M.M. Chaves	Instituto Superior de Agronomia, Tapada de Ajuda, 1399 Lisboa, Portugal
M.M. Coûteaux	C.E.F.E.-C.N.R.S., Route de Mende, 34033 Montpellier, France
Frank W. Davis	Center for Remote Sensing and Environmental Optics, Department of Geography, University of California, Santa Barbara, CA 93106, USA
G. Debussche	Centre Emberger – C.N.R.S., 34033 Montpellier, France

Fred Edwards	Department of Biology, Systems Ecology Research Group, San Diego State University, San Diego, CA 92182-0057, USA
Casimiro Fernández	Departamento de Geofísica y Meteorología, Facultad de Física, Universidad Complutense, 28040 Madrid, Spain
Miguel A. Gaertner	Departamento de Geofísica y Meteorología, Facultad de Física, Universidad Complutense, 28040 Madrid, Spain
Clemente Gallardo	Departamento de Geofísica y Meteorología, Facultad de Física, Universidad Complutense, 28040 Madrid, Spain
Diego García	Departamento de Biología Vegetal y Ecología, Universidad de Sevilla, 41080 Sevilla, Spain
Peter C. Harley	Atmospheric Chemistry Division, National Center for Atmospheric Research, Boulder, CO 80307, USA
Steven J. Hastings	Global Change Research Group and the Systems Ecology Research Group, San Diego State University, San Diego, CA 92182, USA
David W. Hilbert	Groupe de Recherche en Ecologie Forestière, Départment des sciences biologiques, Université du Québec à Montréal, Québec, Canada H3C 3P8
Cara L. Hinkson	Global Change Research Group and the Systems Ecology Research Group, San Diego State University, San Diego, CA 92182, USA

Allen S. Hope	Department of Geography, San Diego State University, San Diego, CA 92182, USA
Mitchell A. Jenkins	Global Change Research Group and the Systems Ecology Research Group, San Diego State University, San Diego, CA 92182, USA
Dimitris Kotzias	CES-Joint Research Centre, Environment Institute, Atmosphere-Biosphere Interactions Unit, 21020 Ispra (Va), Italy
George P. Malanson	Department of Geography, University of Iowa, Iowa City, IA 52242, USA
Rafael Márquez	Departamento de Ecología Evolutiva, Museo Nacional de Ciencias Naturales, José Gutiérrez Abascal 2, 28006 Madrid, Spain
A. Martín	Departamento de Biología Vegetal y Ecología, Universidad de Sevilla, Apdo 1095, 41080 Sevilla, Spain
J. Merino	Departamento de Biología Vegetal y Ecología, Universidad de Sevilla, 41080 Sevilla, Spain
O. Merino	Departamento de Biología Vegetal y Ecología, Universidad de Sevilla, 41080 Sevilla, Spain
Joel Michaelsen	Center for Remote Sensing and Environmental Optics, Department of Geography, University of California, Santa Barbara, CA 93106, USA
G.F. Midgley	National Botanical Institute, Stress Ecology Research Unit, Rondebosch, 7700, South Africa

F. Miglietta	IATA-CNR, P. le delle Cascine, 18 50144 Firenze, Italy
José M. Moreno	Departamento de Ecología, Facultad de Biología, Universidad Complutense, 28040 Madrid, Spain
Sherri J. Morris	Department of Biology, Systems Ecology Research Group, San Diego State University, San Diego, CA 92182-0057, USA and Department of Botany, Ohio State University, Columbus, OH 43214, USA
Zev Naveh	Faculty of Agriculture and Engineering, Technion, Israel Institute of Technology, 32000 Haifa, Israel
Walter C. Oechel	Department of Biology, San Diego State University, San Diego, CA 92182, USA
John F. O'Leary	Department of Geography, San Diego State University, San Diego, CA 92182-0381, USA
John S. Pate	Department of Botany, University of Western Australia, Nedlands, Western Australia 6009, Australia
J.S. Pereira	Instituto Superior de Agronomia, Tapada de Ajuda, 1399 Lisboa, Portugal
Josep Piñol	Centre de Recerca Ecológica i Aplicacions Forestals, Universitat Autónoma de Barcelona, 08193 Bellaterra, Spain

S. Rambal	Centre Emberger – C.N.R.S., 34033 Montpellier, France
A. Raschi	IATA-CNR, P. le delle Cascine, 18 50144 Firenze, Italy
Ferran Rodà	Centre de Recerca Ecológica i Aplicacions Forestals, Universitat Autónoma de Barcelona, 08193 Bellaterra, Spain
Tomás Santos	Departamento de Biología Animal I (Vertebrados), Facultad de Biología, Universidad Complutense, 28040 Madrid, Spain
F. Selvi	Department of Botany, University of Firenze, Via La Pira, 4, 50121 Firenze, Italy
Günther Seufert	CEC-Joint Research Centre, Environment Institute, Atmosphere-Biosphere Interactions Unit, 21020 Ispra (Va), Italy
Carmelina Spartà	CEC-Joint Research Centre, Environment Institute, Atmosphere-Biosphere Interactions Unit, 21020 Ispra (Va), Italy
W.D. Stock	Department of Botany, University of Cape Town, Rondebosch, 7700, South Africa
Douglas A. Stow	Department of Geography, San Diego State University, San Diego, CA 92182-0381, USA
Boyd R. Strain	Department of Botany, Duke University, Durham, NC 27706, USA

José Luis Tellería — Departamento de Biología Animal I (Vertebrados), Facultad de Biología, Universidad Complutense, 28040 Madrid, Spain

Jaume Terradas — Centre de Recerca Ecológica i Aplicacions Forestals, Universitat Autónoma de Barcelona, 08193 Bellaterra, Spain

Richard B. Thomas — Department of Botany, Duke University, Durham, NC, 27706, USA

V.R. Vallejo — Departamento de Biología Vegetal, Universitat de Barcelona, 08028, Barcelona, Spain

P. van Gardingen — Institute of Ecology and Resource Management, University of Edinburgh, Edinburgh EH9 3JG, Scotland

Antonio Vázquez — Departamento de Ecología, Facultad de Biología, Universidad Complutense, 28040 Madrid, Spain

Bruno Versino — CEC-Joint Research Centre, Environment Institute, Atmosphere-Biosphere Interactions Unit, 21020 Ispra (Va), Italy

R. Villar — Departamento de Biología Vegetal y Ecología, Universidad de Sevilla, 41080 Sevilla, Spain

George L. Vourlitis — Global Change Research Group and the Systems Ecology Research Group, San Diego State University, San Diego, CA 92182, USA

1. Relevance of Regional Models for Analyzing Future Climate Change in the Iberian Peninsula

Manuel Castro, Casimiro Fernández, Clemente Gallardo, and Miguel A. Gaertner

This chapter is devoted to showing that the particular characteristics of the Iberian Peninsula make necessary the application of atmospheric numerical models for analyzing the possible consequences of a global change in such a Southwest European area. After a short description of a research project currently in progress on regional climate modeling, some preliminary results of a month-long present climate simulation for the Iberian Peninsula are included.

For the analysis of future climate changes due to the enhanced levels in the atmosphere of the so-called greenhouse gases of direct or indirect anthropogenic origin, two approaches are usually taken: the reconstruction of the climatic past by means of paleoclimatic data (analogue method) and the application of General Circulation Models (GCMs) similar to those currently used in medium-range weather forecasting. The paleoanalogue method comprises two independent parts: (a) the estimation of global temperature sensitivity to atmospheric CO_2 levels based on supposed concentration of this gas at various periods in the past in relation to the corresponding global temperatures, and (b) the reconstruction of regional climate patterns in selected past epochs supposedly similar to that expected for the next century. The application of these methods entails a number of problems related to the extrapolation on a global scale of data collected in highly specific locations, the lack of sufficient knowledge of the interrelationships among the factors determining climatic changes in the past, changes in continental extension and location with respect to the present, and, finally, differences in the heating rate. For these reasons, the reconstruction of the climates of the past is not considered suitable for prediction (Crowley, 1990; Mitchell, 1990). In spite of these shortcomings for prediction of the consequences of warming as a result of the greenhouse effect, if the historical reconstructions were sufficiently accurate, they might provide reliable estimates of the spatial distribution of climatic changes, as well as data for calibration of the GCMs under climatic conditions different from those existing currently. The most powerful tools available to atmospheric physics for research into the possible climatic effects of the increase in greenhouse gases are the GCMs. These models are based on the numerical resolution of the equations of conservation of mass, momentum, heat, and water vapor describing the behavior of air parcels on the rotating planet Earth. This complex system of nonlinear differential equations (called primitive equations) has to be solved by numerical methods, in such a manner that the spatial variations of the predicted variables (wind, temperature, surface pressure, humidity, cloud water, rain water, etc.) are determined at grid points horizontally and vertically distributed throughout the Earth's atmosphere (finite difference models) or by a finite number of prescribed mathematical equations (spectral models). The variations in time of these variables are calculated in discrete time intervals on the basis of given initial conditions (e.g., Washington and Parkinson, 1986).

The GCMs are verified by simulating current climate using observed conditions as input. Typical integration times of the models vary from 5 to 100 years, depending on the research objective, the availability of adequate initial data, and the computing resources. On completion of the verification, the model is executed once more with higher concentrations of greenhouse gases, and the differences between the two simulations are examined by means of statistical tests of the significance of the response with respect to the natural variability of the model. There are two major groups of GCMs: equilibrium and time dependent. The first type is executed using a fixed quantity of greenhouse gases (for example, double concentration of CO_2), whereas the second, more complex, type allows this quantity to be gradually changed with time. The GCMs are also used for the analysis of the climatic effects due to alterations in the land-use distributions and anomalies in sea surface temperatures. The most sophisticated among them are able to represent the feedback between the atmosphere and the Earth's surface. These are the so-called coupled atmosphere–ocean GCMs (AOCGCMs) in which the temperature and dynamics of the ocean are taken into account, as well as the distributions of sea ice, snow, and vegetation cover, together with their mutual interactions.

The GCM's applicability is limited by the memory capacity and speed of the computer used to perform the numerical integrations of the primitive equations. This is essentially due to the limitation in the integration time step length imposed to avoid the solutions becoming numerically unstable, which in turn depends on the grid size (or smallest resolved wavelength), the propagation speed of the fastest moving perturbation (or wave), and the integration method chosen. Thus, an increment in the spatial resolution (grid size) would require a considerable increase in computer time, not only for the higher number of grid points (or mathematical functions) but also for the considerable reduction in the time step. As an example, Table 1-1 presents the computer time (CPU-time) needed to run a GCM for 24-hour simulation at different horizontal resolutions by using a supercomputer (Cray X-MP).

Current GCMs have a horizontal resolution in the range of 1° to 10° for both latitude and longitude and 2–19 vertical layers. The AOCGCMs are normally run at resolutions coarser than 3° × 3°, since they need a more extended simulation period to allow the system to reach equilibrium. As will be seen, although these spatial resolutions are sufficient to represent large-scale climatic forcings, they are not able to simulate climatic features on a regional or local scale. For this, horizontal resolutions of the order of 30 × 30 km (or higher in certain regions) and more vertical layers are needed, but current supercomputer resources are unable to run GCMs under such conditions. Another problem associated with atmospheric numerical modeling is that the necessary spatial discretization does not allow for important physical processes with scales smaller or similar to model grid size to be directly resolved by the equations. The only way to incorporate such subgrid processes into the model is by means of semi-empirical expressions relating their statistical effects with the variables calculated in each grid point. This procedure is called parameterization of physical processes. The physical processes parameterized in a completed GCM and their interactions are shown in Figure 1-1.

In spite of these limitations, current GCMs are able to simulate acceptably many aspects of the global climate system. However, there are important differences between observed and simulated fields of the atmo-

Table 1-1. Computer Power Needed for Running CCM1-NCAR Model (19 Vertical Layers) on a Cray X-MP

Horizontal Grid Size	Number of Horizontal Grid Points	Time Step (s)	CPU Min/Day Simulation
≈ 4.5° × 7.5°	1,920	1,800	1
≈ 2.9° × 2.9°	8,192	900	5
≈ 1.8° × 1.8°	18,432	600	15
≈ 1.1° × 1.1°	51,200	300	72
≈ 0.3° × 0.3°	583,200	90	3,000
≈ 0.1° × 0.1°	5,120,000	30	78,000

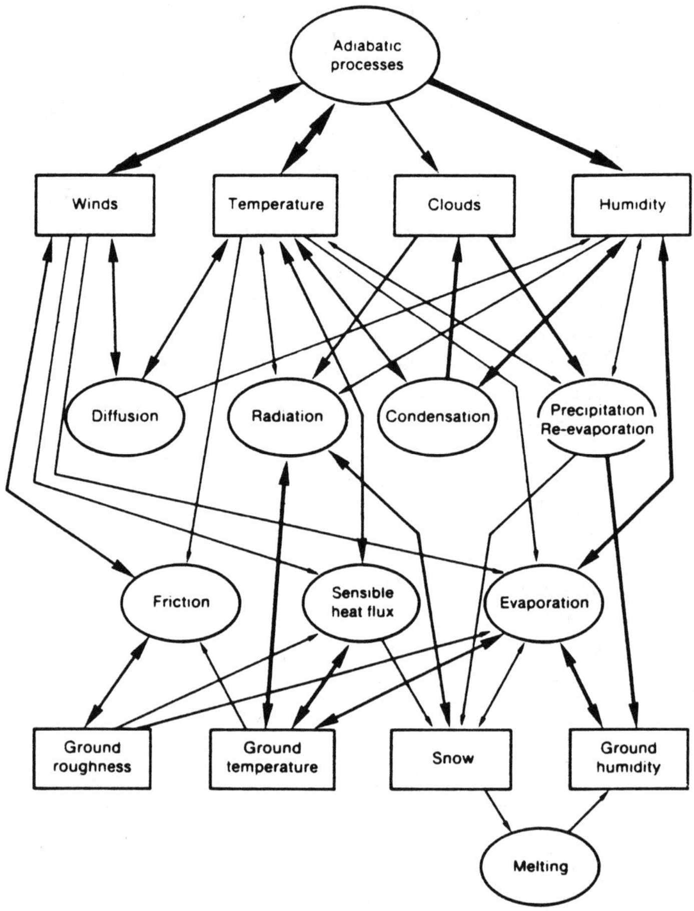

Figure 1-1. Physical processes parameterized in a GCM (ECMWF) and their interactions. (From Houghton, 1984.)

spheric variables that are most influential on climate. As an example, Figure 1-2 shows some results obtained from one of the more complete existing GCMs (UKHI model developed at the UK Meteorological Office's Hadley Center) for current climate simulation. As can be seen, the model acceptably simulates sea-level pressure patterns, which are associated with atmospheric circulation near the surface. However, in Figure 1-3 shows a poorer agreement in the precipitation fields. These significant errors at the subcontinental or regional scale are also observed in other sophisticated GCMs (e.g., IPCC, 1990).

Concerning double CO_2 concentration climate simulations, Figure 1-4 shows some results obtained from the above-mentioned UKHI model. Substantial changes in temperature, precipitation, and soil moisture content fields are observed. Major surface warming is seen at higher latitudes during late autumn and winter, and less in the summer months. In the tropics, warming is less than global, with smaller variations through the seasons. At mid-latitudes, summer warming is higher than global warming. Precipitation increases at high latitudes and in the tropics throughout the year, and in mid-latitudes in winter. Finally, soil moisture increases in northern high-latitude continents in winter and decreases in mid-latitudes during summer.

In summary, as the Intergovernmental Panel on Climate Change Report states (IPCC, 1990), though the amount of qualitative large-scale agreement between models, both temperature and precipitation distributions show strong discrepancies when regional scale variations are considered. In a broad sense, this lack of reliability is fundamentally due to two reasons: poor spatial resolution and bulk, and inadequate parameterization of physical processes. Thus the GCM's coarse resolution itself gives rise to a double effect:

1. It prevents realistic reproduction of atmospheric processes with a size smaller than or similar to the model-mesh which determines regional climate.
2. It distorts the continent-ocean lines and orographic features, which largely contributes to incorrect simulations.

An example of the poor reliability of the results produced by a GCM in the Mediterranean basin region is shown in Figure 1-5. The unreliability is due mostly to the conjunction of both the effects mentioned. The important cyclonic activity in this region, which determines decisively its characteristic climate, is ignored in the simulation as a consequence of the relatively small size of these cyclones, and particularly the drastic smoothing of the orography, which is the principal factor responsible for such an intense cyclogenetic activity.

On the other hand, the parameterization schemes included in most GCMs to describe surface physics, turbulent diffusion, clouds, and precipitation have been developed and tuned for coarse resolution models, and therefore the smaller the grid size, the less suitable are the results. Thus, increasing resolution does not by itself improve forecasting, unless it is accompanied by a more adequate description of the physical processes involved (Boer and Lazare, 1988; Boville, 1991). But introducing more detailed physical parameterizations into the model in turn implies a considerable growth of computer time. For this reason, current GCMs cannot adequately simulate the atmospheric perturbations generated, or strongly influenced, by the parameterized physical processes that may become decisive in the climate of certain regions. Therefore, it must be borne in mind that the adequacy of these model results is seriously affected when they are used to analyze impacts on vegetation, hydric resources, etc. For example, typical climate impact models are applied to

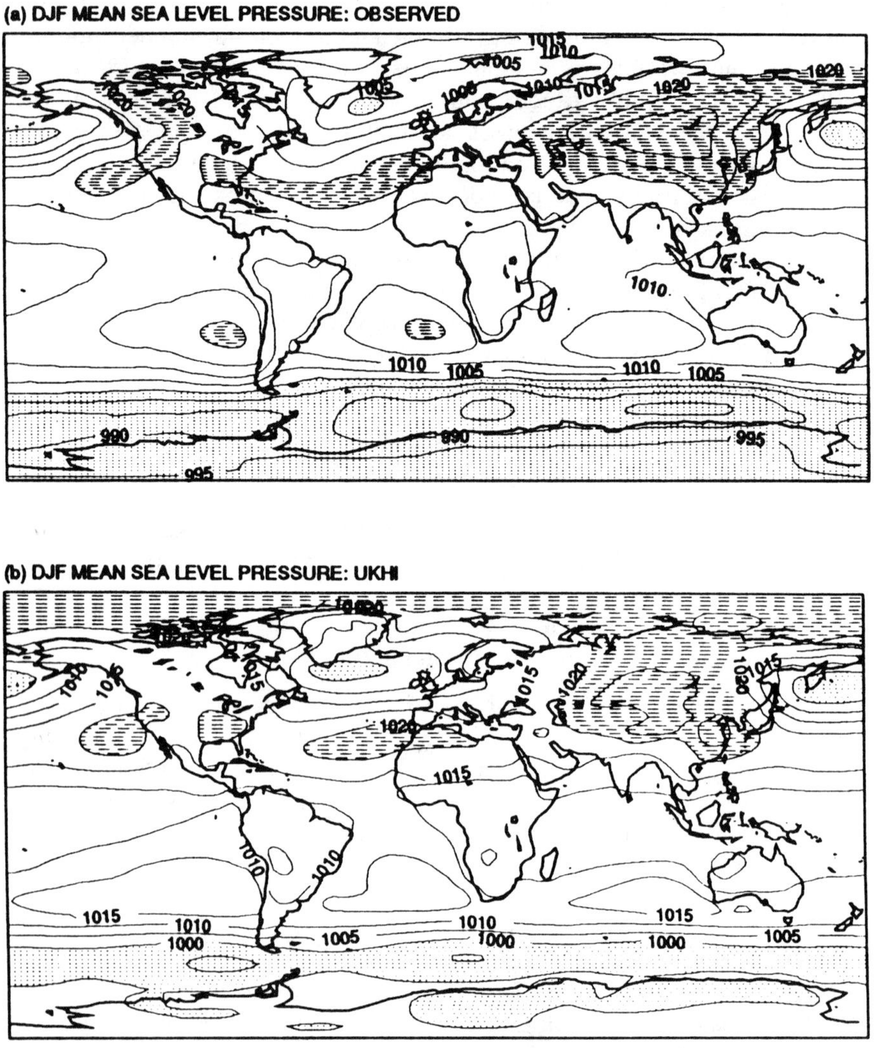

Figure 1-2. Sea-level pressure (hPa) for December–January–February (a, b) and June–July–August (c, d) for (a, c) observed and (b, d) simulated by the UKHI model. (From IPCC, 1990.)

areas of 10 × 10 km, while the GCMs grid-sizes are of the order of 10^2 km.

As can be seen later, in the particular case of the Iberian Peninsula, such specific shortcomings of current GCMs prevent their results being directly applied to analyze the impact of global change on regional climate, as von Storch et al. (1993) pointed out. There are some specific factors that reduce the reliability of GCMs, such as:

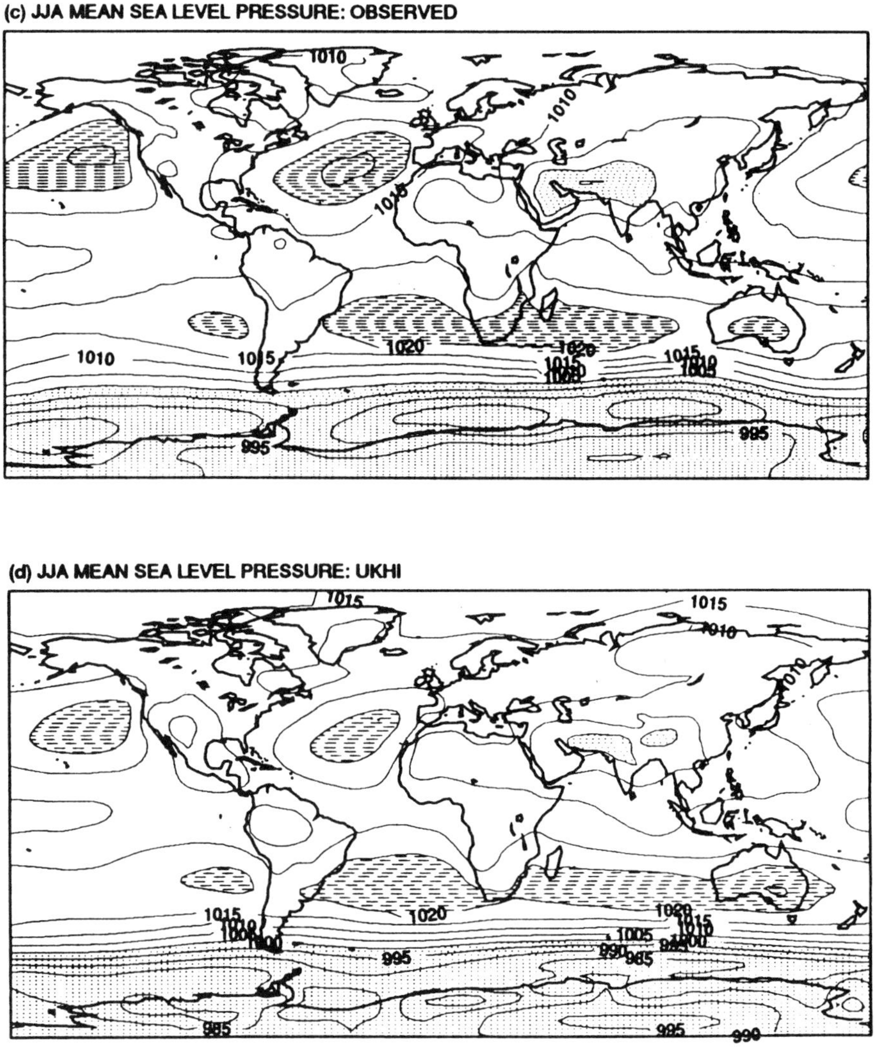

Figure 1-2. *Continued*

1. *Its critical geographic location*. As this region lies in the transition zone between middle and subtropical latitudes, a relatively small error in the latitudinal configuration of the general circulation could distort the associated meteorological processes that determine the climate of this region.
2. *Its small area*. Though models offer similar results, on the average, for large-scale features climatic along meridians, there is little agreement among them on the zonal variation along parallels.

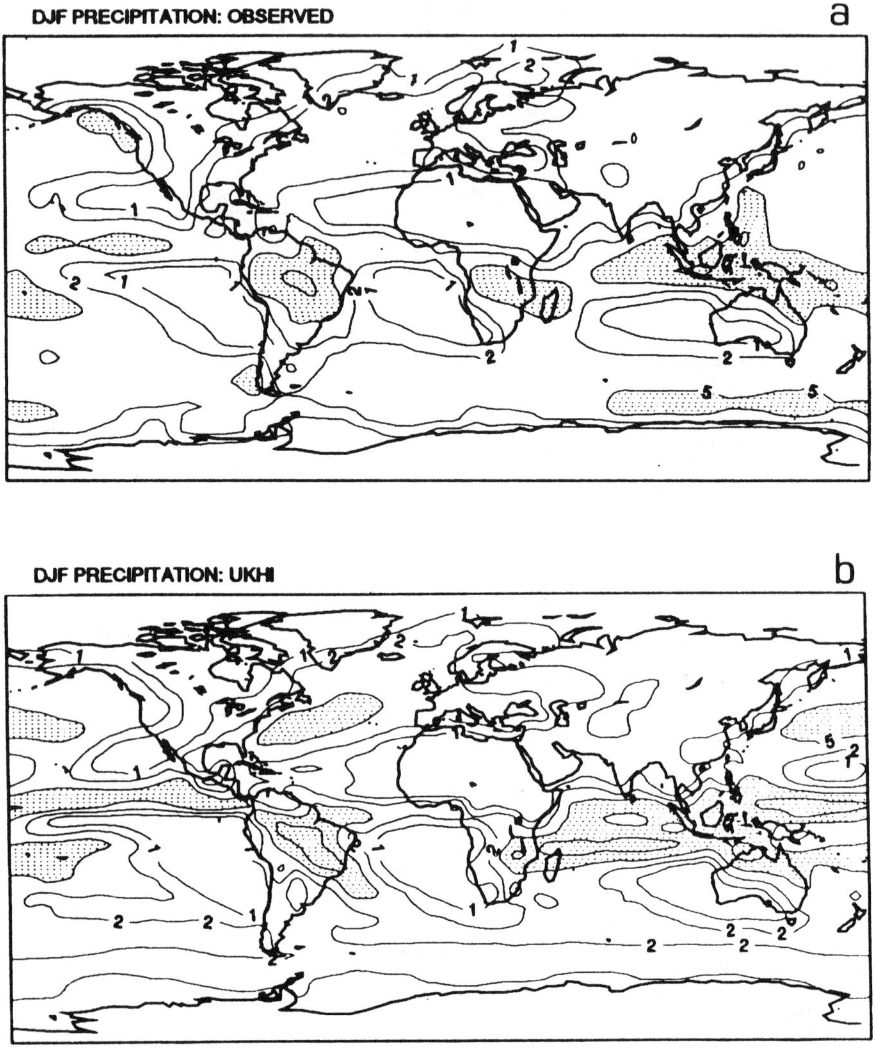

Figure 1-3. Same as Figure 1-2, except for precipitation (mm/day). (From IPCC, 1990.)

3. *Its complex orography*. The aforementioned artificial topographic smoothing involved in the coarse resolutions of the GCMs prevents a correct simulation of the important processes linked or influenced by the abundant mountain ranges crossing the Iberian Peninsula.
4. *Its great variety of vegetation (from humid forests to deserts)*.

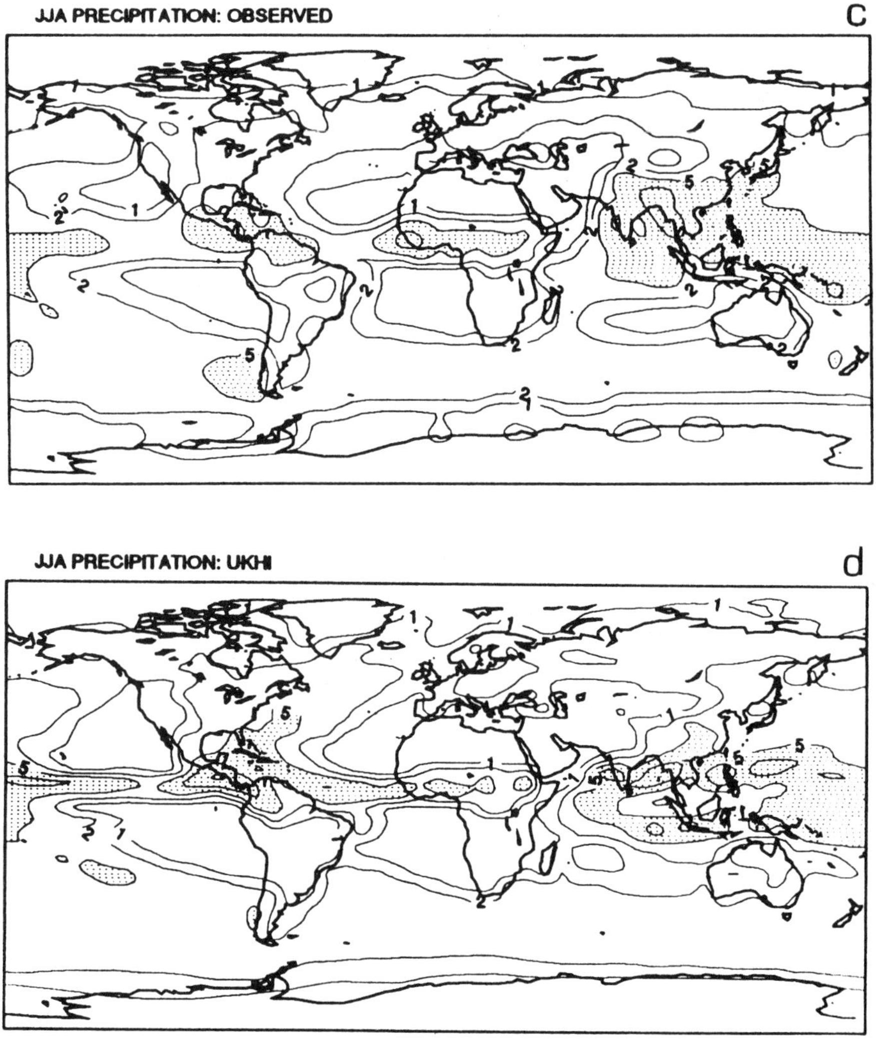

Figure 1-3. *Continued*

In general, physical parameterizations in current GCMs do not consider adequately soil energy exchanges associated with such surface inhomogeneities.

In the next section the principal forcings affecting the climatic regime in the Iberian Peninsula will be described, paying attention only to those related to precipitation, since it is the climatic element of most interest in

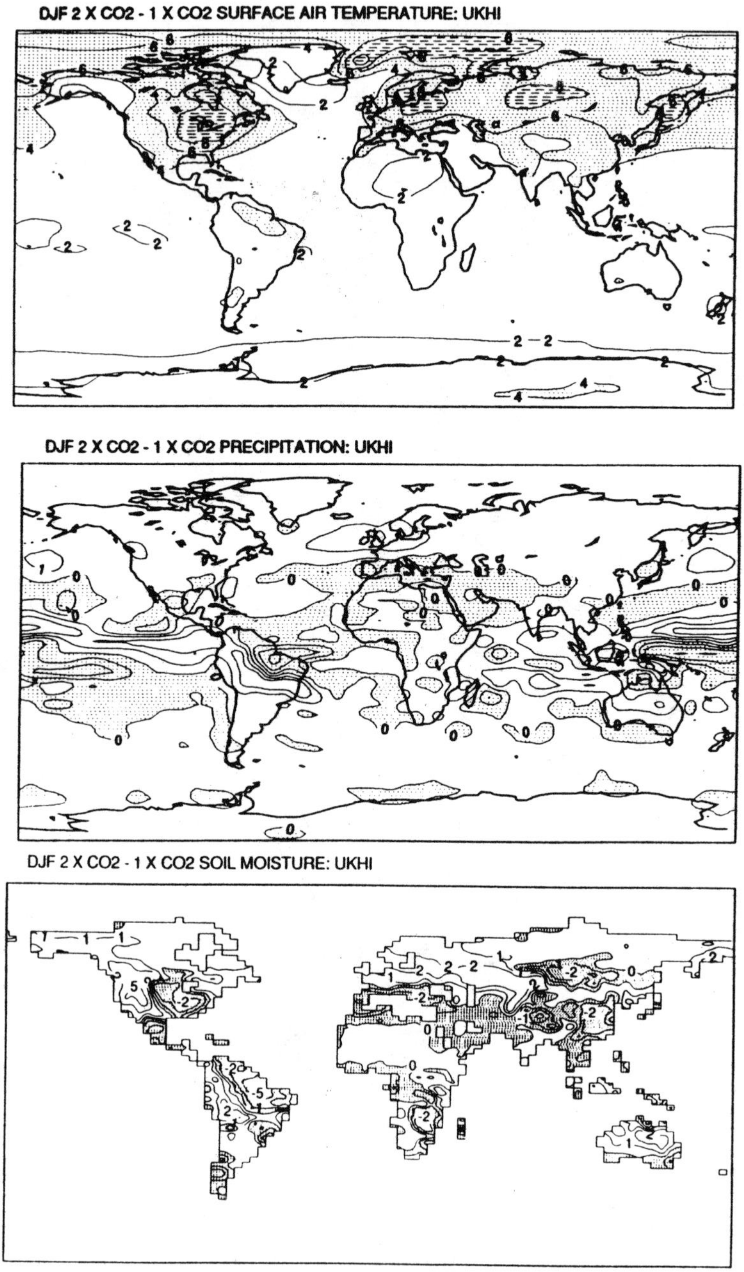

Figure 1-4. Changes in temperature (contours every 2°C), precipitation (contours at +0, 1, 2, 5 mm/day), and soil-moisture (contours at +0, 1, 2, 5 cm) due to doubling CO_2 for months December–January–February and June–July–August, simulated by the UKHI model. (From IPCC, 1990.)

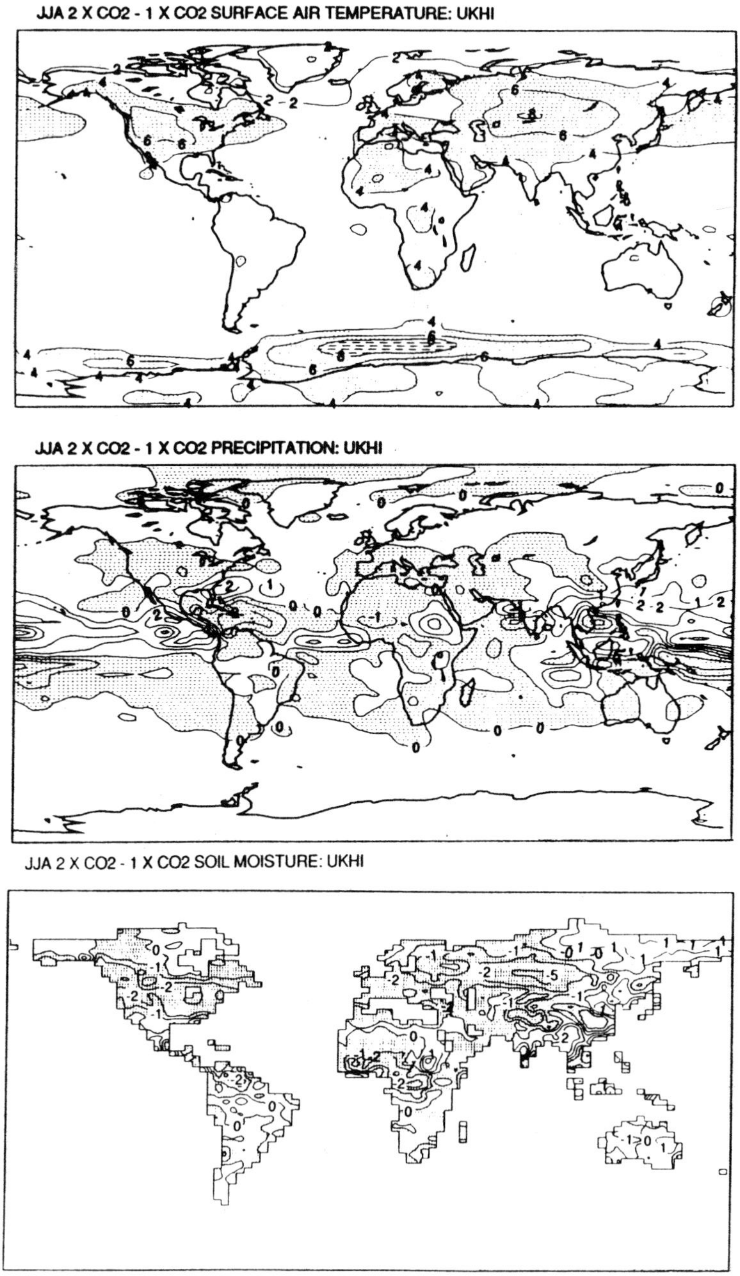

Figure 1-4. *Continued*

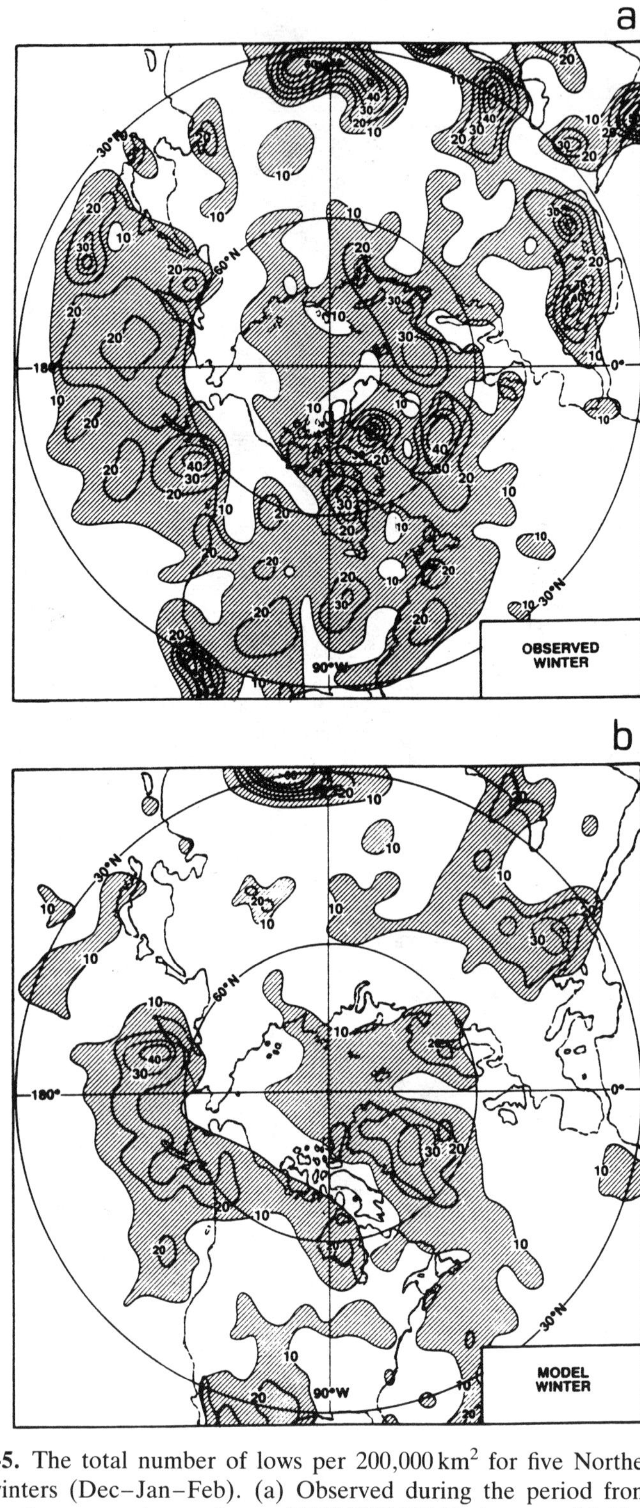

Figure 1-5. The total number of lows per 200,000 km^2 for five Northern Hemisphere winters (Dec–Jan–Feb). (a) Observed during the period from 1980 to 1984 using the once a day analyses of ECMWF-WMO data set. (b) Simulated by the Canadian Climate Center GCM. (From Lambert, 1988.)

this region given the relatively poor vegetation cover in most of its territory and the desertization threat in certain areas.

Therefore, a realistic analysis of regional consequences linked to global change requires the use of more expensive computer-time AOCGCMs with finer resolution than is currently available. For this reason, it does not seem possible to perform such a task within the next 10 to 15 years. Therefore, it is necessary to consider alternative techniques allowing more reliable approaches to model future regional or local climates from current GCM results.

These methods can be grouped into two categories:

1. Semiempirical methods that translate the information given by GCMs into high-resolution statistical descriptions of climatic variables by means of empirical relationships between such large-scale and local surface variables.
2. Modeling techniques that provide a detailed description of regional climate by applying nested limited-area models (LAMs) with finer spatial resolution and more accurate physical parameterizations.

Semiempirical techniques are based on the so-called statistical climate inversion (SCI) method, which accounts for local or regional scale forces in an empirical fashion from large-scale forces provided by GCMs. The simplest of these techniques consists of comparing GCM simulated values of climatic variables for actual and perturbed conditions in a given area (for example a grid box) and appending these differences to data observed in different points within such an area (e.g., Smith and Tirpak, 1989). The most sophisticated are based on regressions between series of climatic values observed in each station within a grid box and corresponding average values of the same variables obtained for such sets of stations throughout this region for an extended period of time. These regression equations are used to infer climatic information at each station location from the output of the GCM corresponding to the grid box considered (e.g., Kim et al., 1984; Wilks, 1989).

However, the results obtained from these techniques have to be treated with care, since in certain cases they may give rise to significant errors. This is essentially due to the fact that such techniques are based on the implicit assumption that the spatial correlation between the climatic variables within a grid box remains unaltered after a climatic change process, even though the local-scale climatic forces have a clearly non-linear character. These errors can become large in case where the surface characteristics fundamentally affecting temperature and precipitation (such as soil moisture, albedo, roughness, etc.) are modified after an alteration of the large-scale climatic regime (Wigley et al., 1990). In other words, semiempirical techniques have the typical limitation of all statistical methods: their predictions are only reliable within the range of the data used to develop them. Outside this range, the statistical relationships lose their significance. In addition to this, data availability as well as the choice of suitable predictors are other weak points of these techniques.

In spite of their relatively recent implementation, regional climate modeling using atmospheric LAMs currently seems to be the most adequate technique for simulating the global change effects at the regional or local scale, at least until the necessary computer power for very high-resolution GCM runs is available. In Regional Climate Modeling the principal characteristics of such models will be briefly described. A discussion of current attempts to apply this technique in the Iberian Peninsula is also included.

Principal Atmospheric Forcings Affecting the Iberian Peninsula Climate

The climate of a subcontinental region is affected by processes of atmospheric circulation with different spatial scales. Although scale separation of the atmospheric motions is not possible, as a consequence of the strong nonlinear character of their mutual interactions, they might be classified by size into two broad groups (Giorgi and Mearns, 1991): the large-scale features determined by the atmospheric general circulation, and the mesoscale processes induced by local characteristics that modify the weather events and initiate regional or local scale circulation. Therefore, the large-scale and the meso-scale motions are due to different forcings. The major large-scale motions are those regulating and controlling energy and momentum transport from the equatorial to the polar regions where solar radiation and rotation speed are smaller. These are the Hadley cell in the tropics and the undulating transient perturbations affecting the middle and high latitudes. Surface–atmosphere exchanges depend on characteristics such as albedo, soil moisture, roughness, etc. Therefore land–ocean distribution and sea surface temperature influence the location of the intertropical convergence zone as well as that of the ridges and troughs that influence westerly winds in the middle and high troposphere at extratropical latitudes, which are in connection with surface meteorological systems that definitively determine the climates of such zones. For more details on this subject, see any text on dynamic meteorology (e.g., Holton, 1989).

The atmosphere–ocean exchanges of sensible heat (thermal convection) and latent heat (condensation-evaporation of water) also give rise to zonal circulations (approximately following geographic parallels), such as the Walker circulation observed in the equatorial Pacific Ocean. The large-scale sea surface temperature anomalies linked to phenomena like El Niño and the quasi-biennial southern oscillation in the eastern Pacific Ocean also affect the general circulation, extending its effect far away from their original zone. Sea ice and snow cover distributions also exert an influence on the large-scale atmospheric features affecting the climate of mid- and high-latitude regions. On the other hand, mountain ranges also modify the upper-westerlies undulation.

Mesoscale atmospheric forcings are induced by each regional surface. They can be mechanical or thermal in origin. The former are due to the topography giving rise to meteorological phenomena such as lee waves, wake circulations, cyclogenetic processes, and gravity waves. The latter are associated either with thermal surface contrasts, such as sea and lake breezes or slope winds, or with vertical air temperature profiles such as local storms or other organized convective systems. The Iberian Peninsula region is characterized by its great variety of climatic subtypes, from wet Atlantic in the north and northwest to the dry Mediterranean type in the east and southeast. This is essentially due to two causes: its location between middle and subtropical latitudes, and its extremely complex topography, with elevated mountain ranges oriented either east-west or north-south, and with 30% of its territory at altitudes greater than 800 m above sea level. Consequently, it can be argued that climate in this southwestern European region is determined by complex interactions between large-scale and mesoscale atmospheric forcings, though the latter seem to exert a decisive influence.

Large-Scale Forcings

The surface large-scale circulation processes that exert an influence on the Iberian Peninsula are closely related to the two centers of action in the eastern Atlantic ocean: the Iceland depression and the anticyclone of the Azores. These mid-latitude systems in the lower troposphere are connected with the upper westerlies, which flow around the pole in a series of waves of variable amplitude and length (Rossby waves). These are essentially due to thermal and topographic influences of the earth's surface. The situation of the planetary wave ridges and troughs determines those of the high- and low-pressure systems at the surface, in such a manner that the upper troughs tend to correspond to cyclones at the surface and the ridges to anticyclones. Nevertheless, this relationship is fairly complex. For example, there are zones within the upper westerlies in which maximum velocities are reached (jet streams). Two major currents may be appreciated in the daily hemispheric maps: the polar front jet and the subtropical jet. The first of these, which is the more powerful, is normally located between 30 and 70 degrees of latitude, while the second varies between 15 and 30 degrees of latitude, flowing at higher altitudes, although the geographic location of both depends on the time of year. The location of these strong upper currents and their wave form heavily influence development of cyclonic and anticyclonic centers at the surface. More details on the formation of these cyclones in the polar front would be beyond the scope of this analysis, and therefore the reader is referred to specialized textbooks (e.g., Palmén and Newton, 1969).

From a climatologic point of view, the main troughs and ridges of the planetary waves are preferentially located in certain regions (Stark, 1965). In the vicinity of the Iberian Peninsula, at the 500 hPa level (approximately 5,500 m altitude) a strong ridge may be observed during the winter over western Europe (30°W–10°E), while a trough is found in eastern Europe (10°–40°E), which is highly sensitive to changes in the intensity of the circulation on a planetary scale. Their relative location is of critical importance with regard to the displacement of Atlantic cyclones and their associated fronts toward the Iberian Peninsula. During the summer this configuration varies. The polar front jet moves toward higher latitudes, weakening its influence over the Iberian Peninsula, so that it is entirely affected by the surface Atlantic anticyclone. In addition to the ridge and trough positions, the wave amplitude plays an important role. The more intense the undulation, the more likely it is that upper cold air from higher latitudes will reach lower latitudes, provoking more instability, or that milder subtropical upper air will flow over higher latitudes, giving rise to more stability.

Few studies on the connection between the climate of the Iberian Peninsula and the characteristics of the general circulation are currently available (e.g., Font, 1983; Zorita et al., 1992). However, the most significant examples of the correspondence between the upper westerlies undulation and the position of surface pressure systems over the Iberian Peninsula and its surroundings are commented on below. In Figure 1-6

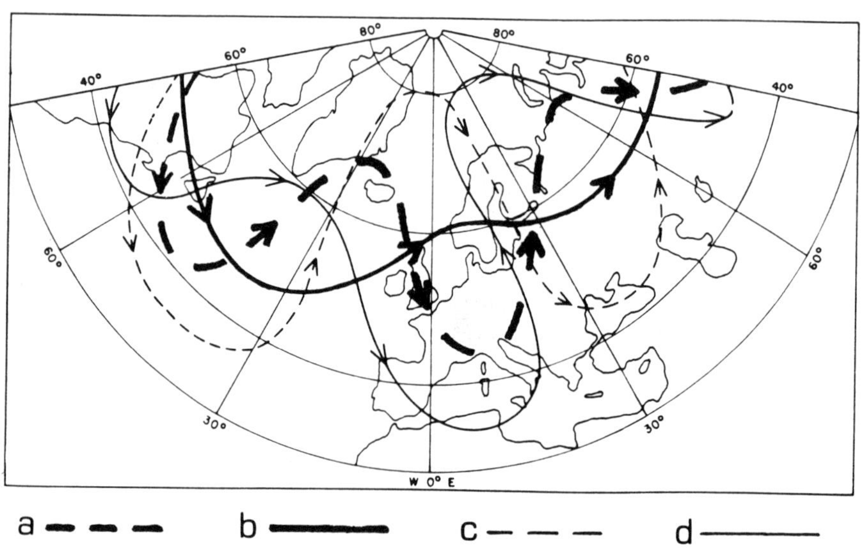

Figure 1-6. Typical cases of undulations of upper-westerlies over western Europe. Lines represent regions of maxima winds. (From Font, 1983.)

the following four more typical cases of upper-wave undulations are shown:

Case a. This upper-wave configuration is frequently related to a succession of fast-moving surface depressions along the southern boundary of the flow axis, whose associated fronts affect the north and northwest of the Iberian Peninsula. The rest of its territory remains under the influence of the Atlantic anticyclone, and clear skies prevail there with no rain. If the upper trough is located over the Mediterranean, the typical cyclogenesis occurs in the Gulf of Genoa and the Tyrrhenian sea. These Mediterranean depressions occasionally affect the Balearic Islands and the northeastern Iberian Peninsula.

Case b. In this case, the undulation is more intense and the ridge axis appears slightly displaced toward the west. With this upper-wave configuration, the western part of the Iberian Peninsula remains under the influence of extratropical depressions coming from the west, which give rise to moderately long periods of rain (typically 2–5 days), mostly in the western Iberian Peninsula.

Case c. This upper configuration usually leads to the establishment of a semistationary, strong surface anticyclone over Scandinavia (blocking anticyclone) while a series of middle-sized depressions are formed west of Portugal and the Gulf of Cadiz that produce abundant precipitations all over the IP except its northern strip.

Case d. This corresponds to the typical summer situation. The Atlantic depression tracks are displaced far to the north, and the Azores anticyclone extends far into the eastern Mediterranean, with practical absence of precipitation in most of the Iberian Peninsula. However, the complex connection between upper waves, the location and displacements of the surface centers of action, and the variability of the general circulation itself make it difficult to establish a classification of weather types over a given region. In an area as complex as the Iberian Peninsula, this task is even more difficult, since apparently similar large-scale situations correspond to rather different regional weather conditions. Probably this behavior is due to alteration of the action of large-scale atmospheric circulation by mesoscale forcings.

Mesoscale Forces

It was mentioned previously that mesoscale forces are mostly induced by inhomogeneities in the distribution of surface characteristics, such as orography, coastlines, and land cover, whose typical variations are observed at scales of 5 to 200 km. These mesoscale forcings of mechanical or thermal origin have a remarkable influence on elements of climate such as wind, temperature, and precipitation. As this article is devoted to the Iberian Peninsula climate, it will only account for the effects of mesoscale forcings on precipitation, since in this semiarid region, soils with sparse vegetatation cover more than 50% of the surface, so that an eventual perturbation in the annual amount of rainfall would have a significant impact on them.

As is well known, the orography of any region decisively affects the precipitation, tending to increase the local amount of rainfall with altitude. This fact can be observed in Figure 1-7 in the case of the Iberian Peninsula. The highest average annual precipitation is recorded along the Pyrenees and Cantabric ranges in the north and northwest and the Central system in the center, while in the central subplateaus, the Ebro Valley in the northeast, and the Mediterranean zones, rainfall is scarce. The proportion between maximum and minimum annual precipitation is approximately 1 to 25, due to the fact that there are often dry areas close to small rainy zones. This is clearly an orographic effect.

The locations of precipitation maxima in mountain areas are essentially due to water vapor condensation in uplifted air inside the traveling storm systems in winter and to enhanced convection in summer. The greater influence of orography on rainfall in the Iberian Peninsula outside the summer is related to the perturbations induced by mountain ranges on frontal systems crossing this region associated with Atlantic storms. Several studies have shown that most of the precipitation on the Atlantic side of the IP is frontal, although it is not always easy clearly to distinguish its origin (Linés, 1970).

However, Hoinka and Seco (1991) pointed out that the orientations of the fronts respect to the Iberian Peninsula mountain ranges influence the distribution of precipitation in this region. Therefore, fronts coming from the north and northwest give rise to significant rain only over the northern third of the Iberian Peninsula, while those from west and southwest produce precipitation on almost all of the peninsula except the northern strip downwind from the Cantabric range. The influence of these fronts on the Mediterranean regions of the Iberian Peninsula after they cross the Iberic and Penibetic ranges appears to be smaller, while convective precipitation predominates there. The annual minimum of rain in the Ebro Valley also seems to be related to a blocking effect produced by the Pyrenees on the Atlantic flows. Summer rainfall is small except in the northern and northwestern Iberian Peninsula where precipitation is essentially not frontal in this season, since the Atlantic storm tracks are displaced to higher latitudes. Such precipitation is thus related to the orographic ascent of northern, maritime air flowing from the Atlantic anticyclone.

Orographic convective precipitation of thermal origin is frequent in certain zones of the Iberian Peninsula during the summer. Such generally stormy rains are more abundant and frequent when very humid low-level air from the Mediterranean flows into these regions in conjunction with adequate tropospheric thermal profiles. This seems to be the reason for the relative abundance of summer rains in the Iberic range, so that this is

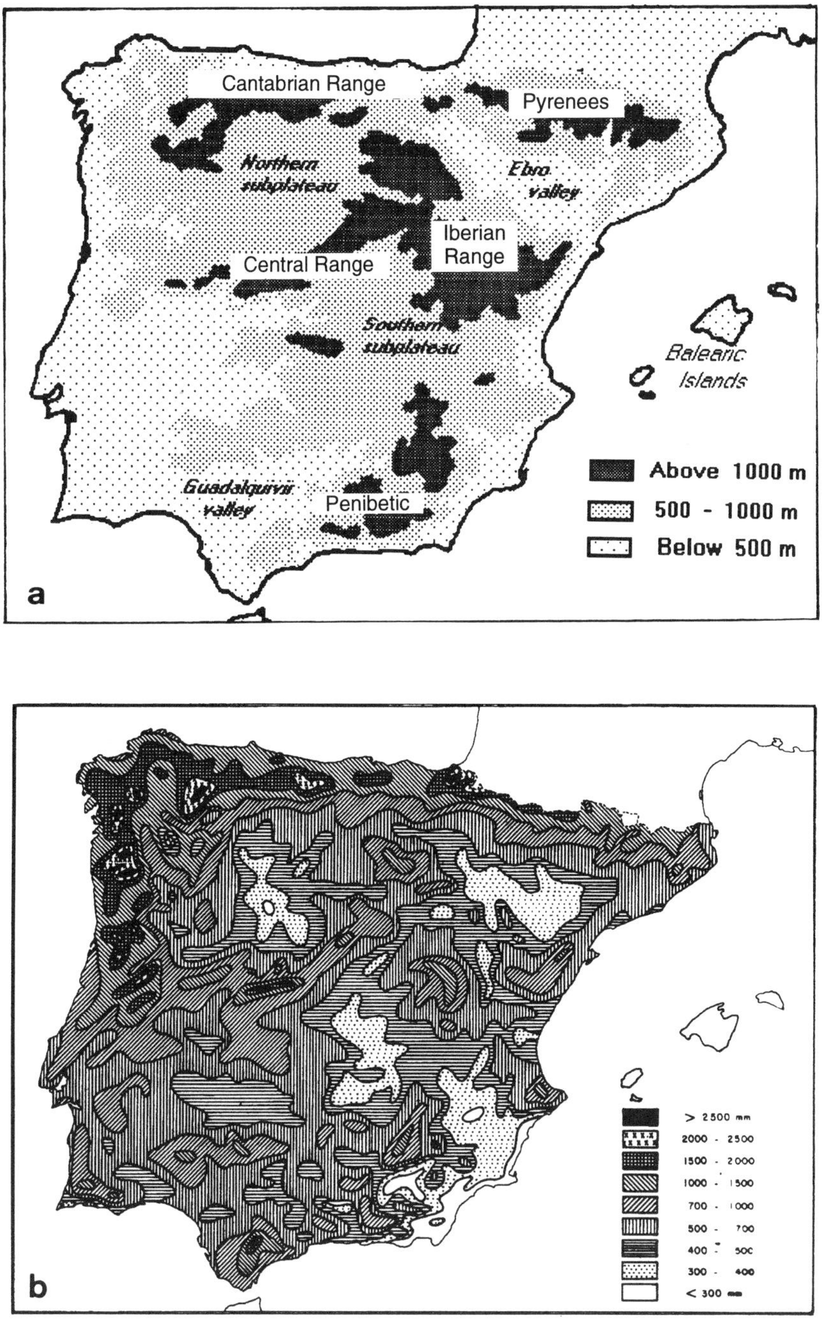

Figure 1-7. (a) Orography of the Iberian Peninsula (contours every 500 m). (b) Distribution of annual average precipitation over the Iberian Peninsula (in mm). (From Font, 1983.)

the unique zone in the Iberian Peninsula where the annual maximum of precipitation is observed during the summer (Font, 1983). Other atmospheric circulations related to orography are registered in the Iberian Peninsula. They do not give rise to precipitation but to strong, dry winds (downslope winds) that might influence the local soil moisture budget. Typical examples are the Cierzo wind in the Ebro Valley originating from the Pyrenees, and the Foehn-like effect observed to the north of the Cantabric range with southern air flows.

Nonorographic mesoscale forcings are generally linked to differential surface warming. The most important are those originating on shorelines from the thermal contrast between land and ocean. This differential warming produces contrasts in the vertical stability characteristics of the atmosphere, with enhanced convection, precipitation, and cloudiness over land in summer and decreased convective activity in winter. Probably such thermal surface contrasts contribute in some way to the typical intense precipitation during autumn registered in the Mediterranean zone of the Iberian Peninsula that start when mid-troposphere cold air "pockets" are formed above this region. When this relatively cold air is located above the surface contrast between the warmer, land air and the humid, cooler sea air, along with a probable contribution of the litoral dry slopes, a well-organized, rapidly developing, vigorous convective activity bursts, giving rise to intense small-size cyclones (100–300 km) that may randomly move up and down the Mediterranean coast for 2 to 4 days. Such torrential rains are responsible for the fact that the maximum annual rainfall amount in these regions is recorded in autumn. Sea breezes also affect local climate by decreasing the surface temperature in coastal zones, although these air circulations do not produce precipitation in the Iberian Peninsula, except in cases when maritime air rises up coastal slopes, as mentioned before.

Finally, numerous theoretical (e.g., Anthes, 1984; Ookouchi et al., 1984; Segal et al., 1988; Castro et al., 1993) and observational (e.g., Barnston and Schickedanz, 1984) studies have been performed concerning the impact of land use on homogeneities in local climate. Simulations with mesoscale meteorologic models demonstrate that local circulations originating from sharp contrasts in vegetation cover characteristics, which are in many aspects similar to sea breezes, may trigger convective precipitation under certain atmospheric conditions. Anthes has shown that certain geometric configurations of cultivated and bare areas may increase local precipitation. Although theoretical or observational results are not currently available, this type of local-scale process must be abundant in the Iberian Peninsula, where great contrasts of land cover characteristics exist. Whereas wet prairies and forests predominate in the north and northwest, in the rest of the Iberian Peninsula soils are mainly of semiarid character, with some forested zones in the mountains and certain watered crop areas.

Regional Climate Modeling

As shown the previous section, simulations with finer resolution GCMs than are currently available are needed to characterize the climates of the Iberian Peninsula, where important mesoscale forces play a decisive role. While atmospheric processes related to orography or sea breezes only can be "captured" with spatial resolutions of 10 to 50 km, or even smaller, current GCM run with grid sizes of 100 to 500 km. Therefore it is clear that the results of GCM climate simulations lack the necessary reliability to analyze the impact of global change on Iberian Peninsula climates. Figure 1-8 shows the topography of western Europe resulting from different grid point spacing. This shows the importance of horizontal resolution in a regional model. As discussed in the Introduction, of the two methods that may be followed to obtain information on regional climates from GCMs, only LAMs offer sufficient reliability. For this reason, and because it is an activity recently initiated, a note on how to apply them to a geographic area as small as the Iberian Peninsula is included below.

In order to describe explicitly the effects of mesoscale forcings on regional climate, LAMs are used embedded or nested in a GCM (Figure 1-9a). The nested LAMs numerically resolve the primitive equations in a fine mesh including the area of interest over the coarser-resolution mesh of a GCM. This nesting procedure may be used in a one-way or a two-way interactive mode. In the former, information from the GCM is used to drive the LAM, but information from the LAM does not feed back into the GCM, while in the latter the information exchange is produced both ways in an interactive way. Two-way nesting (e.g., Zhang et al., 1986) provides a better representation of the regional effects of mesoscale forcings. However, the implementation of such a technique is currently difficult due essentially to aliasing problems when the ratio of the grid size of the coarse and fine resolution is rather large (approximately 3). The two-nesting technique also requires more computer time. Thus, current nested regional climate models use a one-way technique, in which the necessary initial and boundary conditions for predicted variables to run an LAM are provided by the GCM output. This technique has been used for the past decade in operational short-range weather forecasting at several meteorologic offices. But its use for climate studies is rather recent. In Giorgi and Mearns (1991) a review of the few existing regional climate models can be found.

The first results with regional climate models were obtained by Dickinson et al. (1989) who applied their model to the western United States. Later studies have shown that this procedure is able to describe the effects of forcings with a size corresponding to subgrid in a GCM on the circulations and processes determining the climate of a given region. However, with this technique a nested LAM cannot correct large-scale

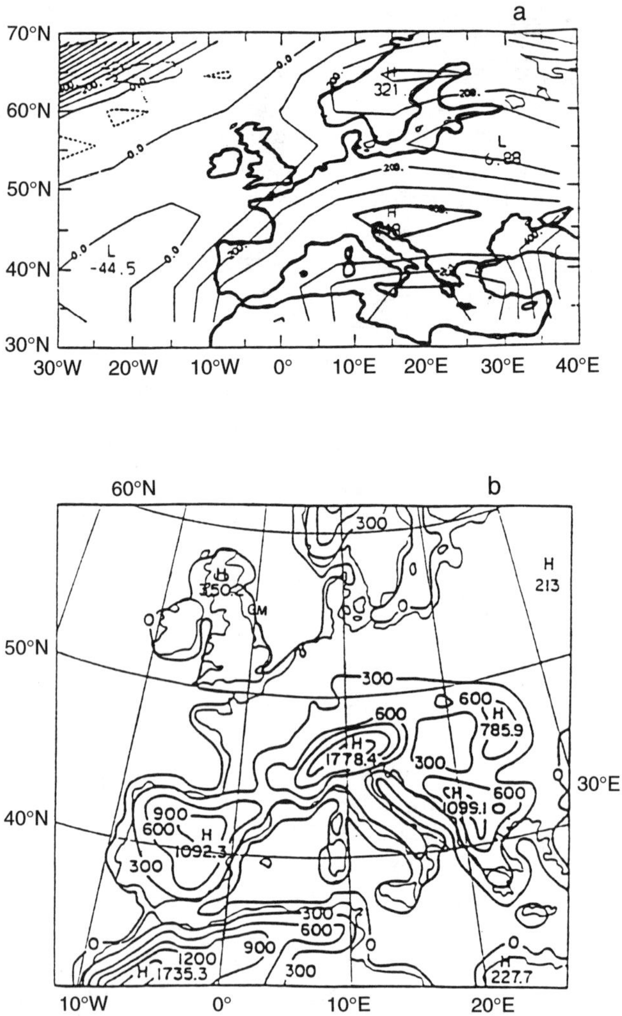

Figure 1-8. Topographic fields of western Europe resulting from various grid-points spacing. (a) 4.5° × 7.5° lat-long (contours every 100 m); (b) 70 × 70 km (contours every 300 m). (From Giorgi and Mearns, 1991.)

errors in GCMs (Giorgi, 1990). For example, if a GCM does not properly simulate the location of polar front jet and storm tracks at mid-latitudes, a nested LAM cannot correct such error, leading to unrealistic regional-scale results. Therefore it is necessary to choose a carefully validated GCM which realistically represents the large-scale circulations affecting the region of interest. It is also important that the LAM includes adequate physical representations to simulate the climate of a given region. There exists a great variety of physical parameterizations to simulate convective precipitation processes, air–soil exchanges, and radiative effects of clouds and greenhouse gases. Parameterizations have to be chosen in accordance with the particular characteristic of the modeled region. Finally, the choice of grid size has to represent a compromise between the size of the most important mesoscale forces influencing regional climate and the computing power available.

Current regional climate models cannot simulate such long time periods as GCMs, because LAMs need more computer time than GCMs. For example an LAM with a 40 × 40 km grid mesh, 80 × 60 grids, and 30 vertical levels needs approximately four times more CPU time than a

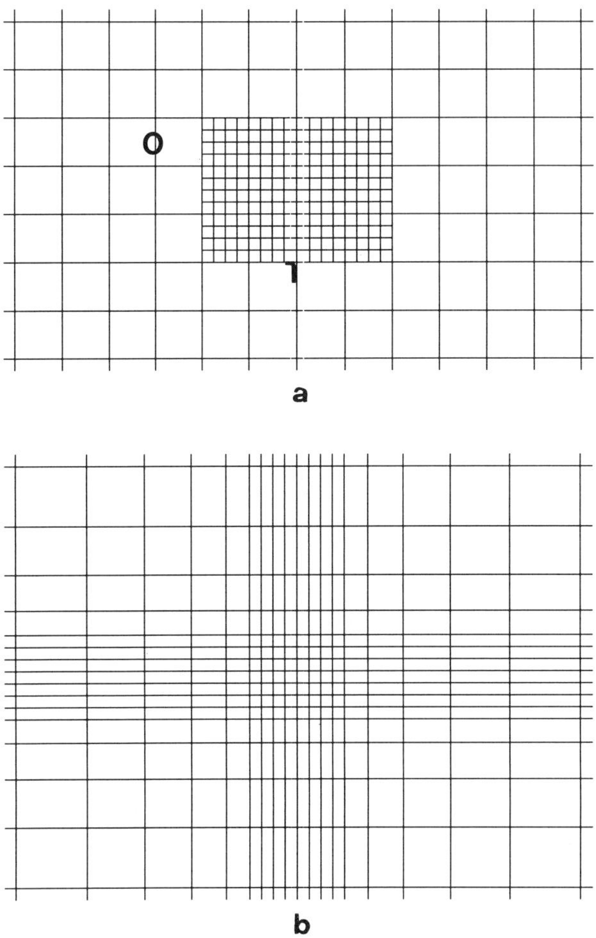

Figure 1-9. Grid-mesh in (a) Limited Area Model nested in a GCM, (b) variable-grid GCM.

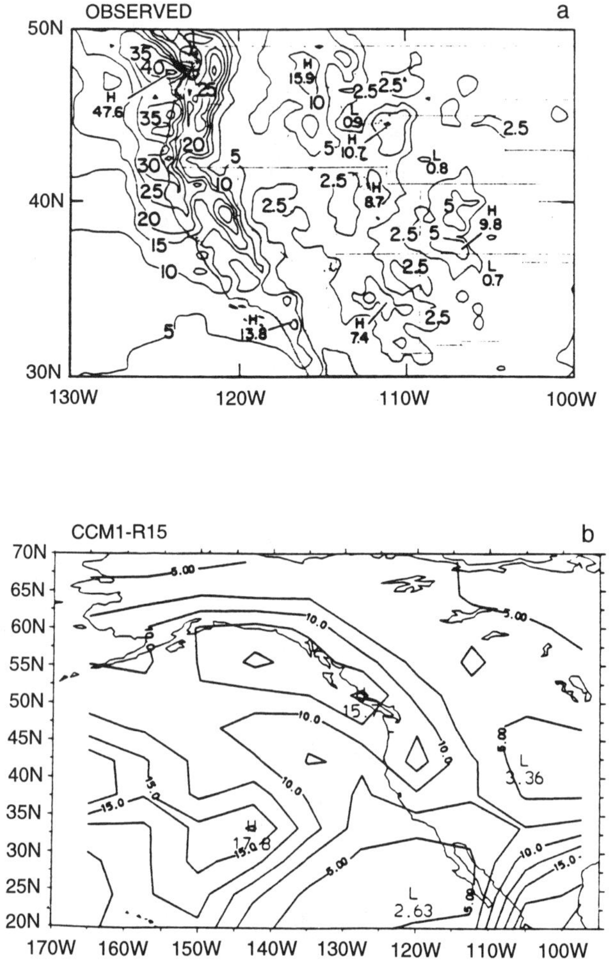

Figure 1-10. Average high-resolution January precipitation over the western US. (a) Observation (0.5° × 0.5° lat-long data). (b) Simulated by the CCM1-R15 global model (4.5° × 7.5° grid).

current high-resolution GCM with a 1.8° × 1.8° horizontal grid and 16 layers. The usual approach consists of simulating monthly or seasonal periods with LAMs. In Figure 1-10 an example of the western US average January precipitation simulated by a nested LAM and a GCM together with observations is presented. This demonstrates that the LAM captures the detail of regional precipitation in a topographically complex area. It should also be pointed out that GCMs with variable resolution grids are

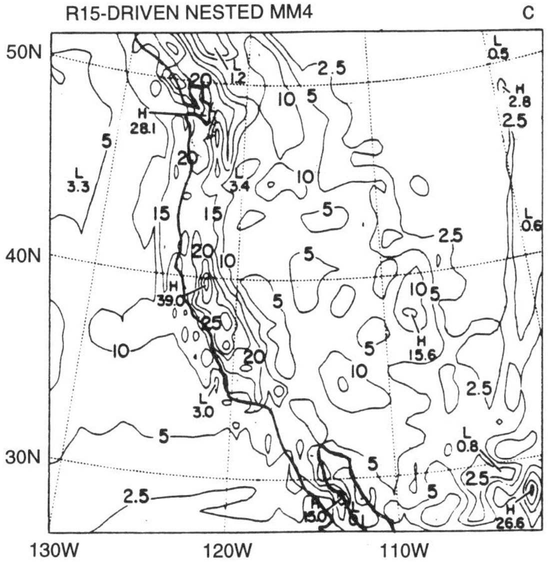

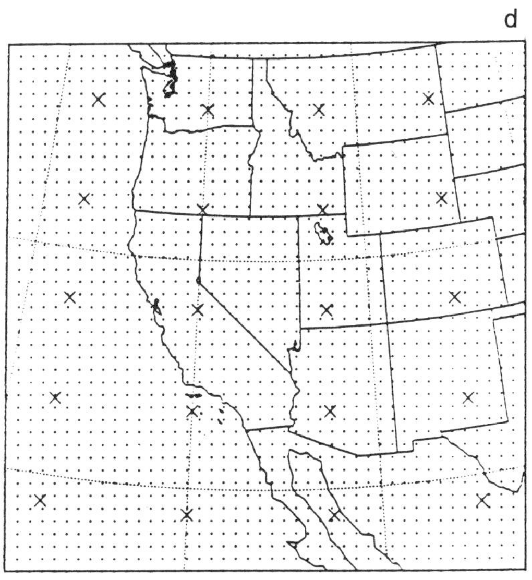

(c) Simulated by nested MM4 regional model (70 × 70 km grid). (d) Grid-points for MM4 (small dots) and CCM1-R15 (crosses). (From Giorgi and Mearns, 1991.)

currently being tested as a promising alternative to nested LAMs for regional climate studies. These models are also capable of explicitly describing the effect of mesoscale forcings by means of an increase in spatial resolution only over the area of interest. Thus, although the number of grid points is similar to that of current GCMs, the spatial resolution is fine enough to simulate local circulations over a given region. Obviously this increase in resolution over any planetary area implies larger grid sizes on the antipodes. The grid-mesh aspect in these models is shown in Figure 1-9b. Currently a good deal of work with such climate models is being performed at MEDIAS Climate Center of the French Meteorologic Institute with a model called ARPEGE (Geleyn et al., 1988). The main advantages of these variable grid GCMs lie in the avoidance of typical nested-LAM boundary and one-way interaction problems as well as the need to run two models per regional simulation (a GCM and a nested LAM). Their principal inconvenience relates to the amount of computer time required, since the integration time step has to be in accordance with the smaller grid size.

Regional Climate Modeling: Preliminary Results

The ECLIPSE project (a Spanish acronym for Climatic Evolution for Southwestern Europe), started in Spain at the end of 1992, essentially deals with the development of a regional climate model in order to simulate the impact in the Iberian Peninsula of the global warming due to the greenhouse gases accumulation. The model consists of a 3-D primitive equations, hydrostatic, high-resolution mesoscale model called PROMES, entirely developed in the Department of Geophysics and Meteorology at Complutense University of Madrid (Castro et al., 1993). To take into account the important feedback-like processes involved in terrain characteristics, which play an essential role in the particular case of the Iberian Peninsula, both an improvement in soil moisture parameterization and a better knowledge of seasonal soil-cover distributions in this semiarid region will be accomplished. The model will be nested in an AOCGCM called ECHAM-MPI developed at the Max-Planck Institute (Hamburg, Germany). The ECHAM-MPI model will be applied with a spectral resolution of T42, equivalent approximately to a grid length of 300 km in a grid model. Since the PROMES regional model will use a grid of 20 × 20 km, it cannot be directly nested into the ECHAM-MPI global model due to the great difference between the respective grid sizes. Instead, a coarser grid meso-alpha model (HIRHAM from Max-Planck Institute) with a grid of approximately 70 × 70 km, covering a quarter hemisphere and centered in western Europe), will be nested in the global model mentioned, and the finer-grid PROMES will be in turn nested in the HIRHAM-MPI model. The research development will be as described in the following.

The first 3-year period (1993–95) of the project concentrates on present-day climate simulations and their evaluation. A planned second part (1995–97) will be devoted to anthropogenic regional future climate change simulations. The first 18 months have been dedicated to carrying out several so-called perfect boundary condition simulations, in order to test the performance of the PROMES model for present-day climate modeling. These "perfect boundary" simulations consist of running the A-LAM with initial fields and time-dependent boundary conditions determined from analyses, i.e., meteorological observations optimally interpolated in a grid mesh, routinely accomplished in the European Centre for Medium-Range Weather Forecast (ECMWF, Reading, UK). The degree of accuracy obtained from these runs will constitute an upper limit of accuracy that can be expected with the PROMES model when applied to regional climate simulations nested in a global model. With these boundary conditions, four monthly periods (July 1990, October 1990, January 1991, and April 1991) will be simulated and the results verified with observations in the Iberian Peninsula. Once this verification has been completed, the PROMES model will be nested in the HIRHAM-MPI model, nesting it in turn in a 100-years-long running of the ECHAM-MPI global model for simulating the present climate. These results will be validated on the basis of current climate in the Iberian Peninsula. Finally, the same procedure will be followed, but with the ECHAM-MPI simulation for double CO_2 future climate, and the results obtained will be compared with those of the present climate simulation.

Following this, some results of the so far unique currently available PROMES model month-long simulation (January 1991) applied in the Iberian Peninsula with "perfect boundary" conditions will be shown. The main characteristics of such simulation are summarized in Table 1-2.

The simulation was completed successfully. No loss of mass, any spurious increase of energy, or any other serious fault has apparently been detected, although this matter requires a more careful analysis. The

Table 1-2. Main Characteristics of the January 1991 Simulation

Geographic area: As presented in Figure 1-13
Coordinate system: Cartesian (x-axis to E and y-axis to W
Grid spacing: 20×20 km
Vertical discretization: 30 sigma-layers
Number of grid points: $88 \times 82 \times 30 = 216{,}480$
Initial conditions: Interpolation from ECMWF analysis mesh ($0.9° \times 0.9°$ lat-lon resolution)
Initialization: Digital filtering (Linch and Huang, 1992)
Boundary values: Updated by a linear interpolation from ECMWF 12-hourly analysis. Boundary conditions: Davies-type ("nudging") for wind, temperature, and pressure. In-outflow dependent scheme (Giorgi et al., 1993b) for specific humidity
Computer used: HP-APOLLO/700 Workstation (40 Mflop)
CPU time required: 300 hours

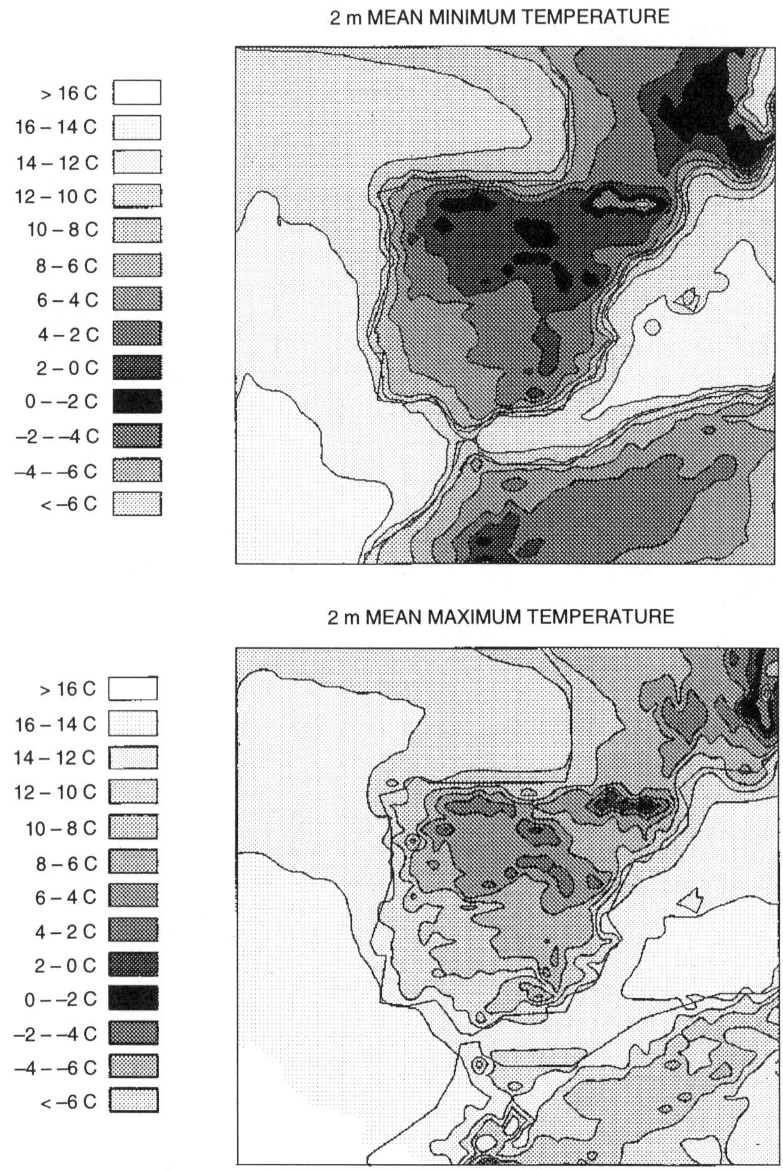

Figure 1-11. Simulated average January 91 minimum and maximum temperatures.

predictability of significant weather events seems to be satisfactorily high, as has been noted by other authors in regional modeling (Giorgi et al., 1990, 1993a,b).

Figure 1-11 presents the distributions of simulated average monthly maximum and minimum temperatures. These distributions seemingly

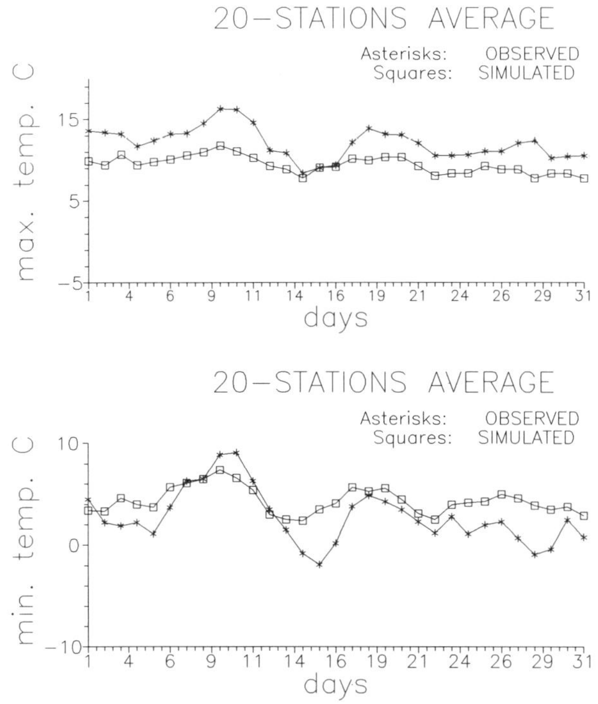

Figure 1-12. January 91 average daily maximum and minimum temperature evolution in 20 selected stations (see Figure 1-13).

meet the climatologic behavior in the Iberian Peninsula. Unfortunately, we cannot compare these maps with observations, due fundamentally to:

1. The temperature network being much less dense than the model resolution, especially in mountainous areas, precisely where more spatial contrasts appear in Figure 1-11.
2. The differences between the actual height of the stations and the corresponding model grid.

Figure 1-12 shows the evolution of observed and simulated average daily maximum and minimum temperatures throughout January 1991 for 20 selected locations (see Figure 1-13). Both curves of simulated temperatures follow more or less closely those corresponding to the observations. This fact seems to reveal a good predictability of significant large-scale events with influence on cold/warm air mass advections over the Iberian Peninsula.

Figure 1-14a and b allow a comparison to be made between the observed and simulated distributions of month-long accumulated precipitation. Figure 1-14a has been developed from data gathered in more than 5,000 rain-gauge stations in Spain and Portugal (see Figure 1-15).

REGIONAL CLIMATE MODEL

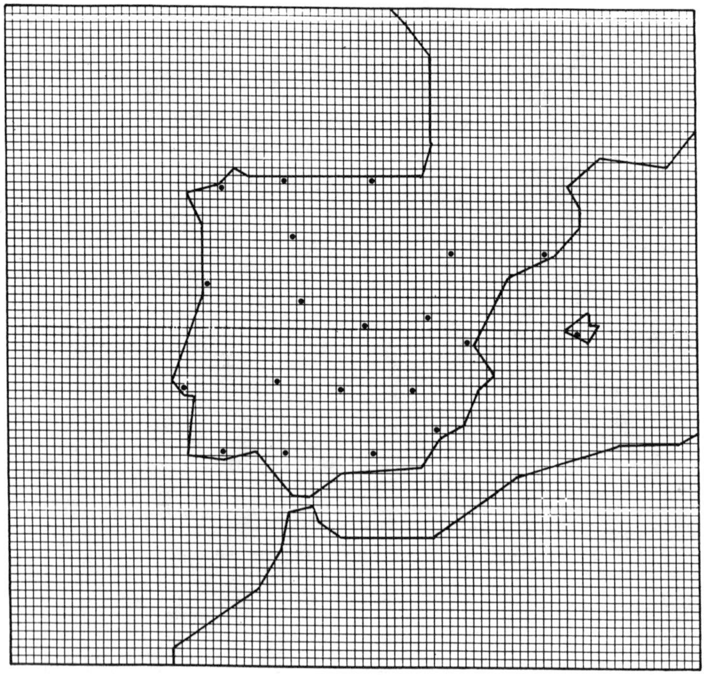

DOMAIN AND GRID-MESH (Lambert projection)

Grid size: 20 x 20 km

Grid number: 88 x 82 x 30

Figure 1-13. Model domain and grid-mesh considered. Points indicate the stations considered in Figure 1-12.

The most important features that this model has successfully reproduced are the maxima observed on the Iberian Peninsula (simulated maximum: 548 mm; observed point maximum: 652 mm) and in the eastern Iberian Peninsula (simulated maximum: 550 mm; observed point maximum: 364 mm). The former is associated with the passage of Atlantic fronts during the first half of the month, while the latter is linked to a 2-day transient easterly flow by the end of the month, which causes moist low-level air to impinge on the mountainous Mediterranean coast, originating abundant convective rains. In both cases, the model maxima are somewhat displaced toward the south, showing a good correlation with the local orography. The model also reproduces successfully a broad band with scarce precipitation crossing the peninsula from NE to SW, including all of the central plateau.

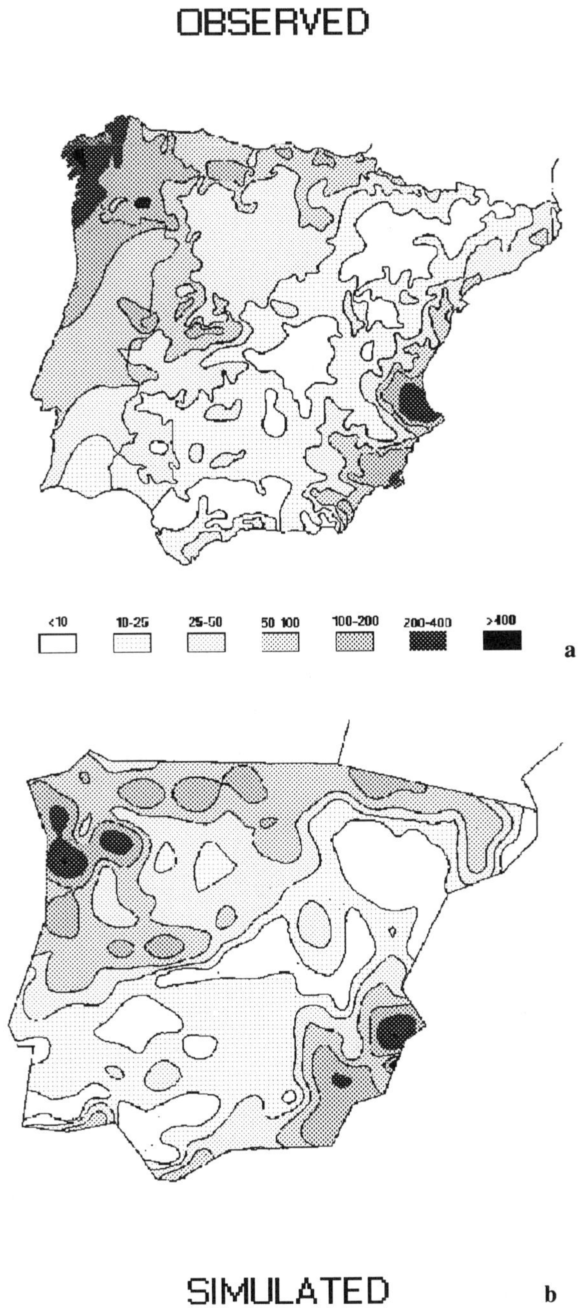

Figure 1-14. Simulated and observed distributions of January 91 accumulated precipitation.

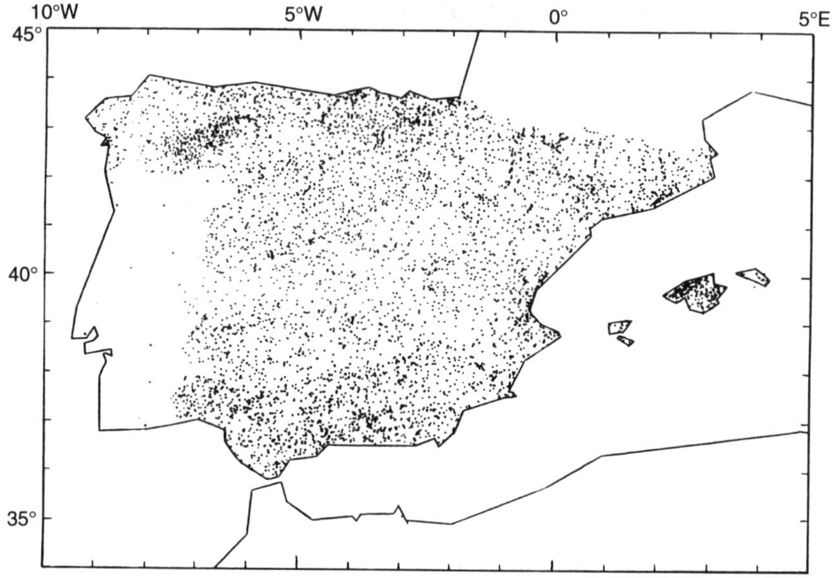

Figure 1-15. Spatial distribution of the rain-gauge stations considered in Figure 1-14b.

These preliminary results show that an LAM nested in a GCM can simulate acceptably well the regional climate detail in a region where high-resolution local climate forces are important, as in the case of the Iberian Peninsula. Therefore the nesting modeling technique seems to be a useful tool for the simulation of future regional climate, with higher degrees of accuracy than a lower-resolution GCM, especially in complex areas.

Acknowledgments. The regional climate modeling research was partially sponsored by the EEC grant EV5V-CT92-0126, the spanish CICYT grant NAT91-02220C04-02 and by the Emp.Nac. Resid. Radiac. S.A. (ENRESA). Comments from the anonymous reviewers have helped to improve this manuscript.

References

Anthes RA. 1984. Enhancement of convective precipitation by mesoscale variations in vegetative covering in semiarid regions. J Clim Appl Meteorol 23: 541–554.

Barnston AG, Schickedanz PT. 1984. The effect of irrigation on warm season precipitation in the Southern Great Plains. J Clim Appl Meteorol 23:865–888.

Boer GJ, Lazare M. 1988. Some results concerning the effect of horizontal resolution and gravity wave drag on simulated climate. J Climate 1:789–806.

Boville BA. 1991. Sensitivity of simulated climate model resolution. J Climate 4:469–485.

Castro M, Fernández C, Gaertner MA. 1993. Description of a meso-scale atmospheric numerical model. Mathematics, Climate and Environment, eds. JI Díaz and JL Lions. Masson, SA, Paris, 230–253.
Crowley TJ. 1990. Are there any satisfactory geologic analogs for a future greenhouse warming? J Climate 3:1282–1292.
Dickinson RE, Errico RM, Giorgi F, Bates GT. 1989. A regional climate model for the western U.S. Clim Change 15:383–422.
Font I. 1983. Climatología de España y Portugal. Inst Nac Meteorología, Madrid, 296 pp.
Geleyn JF, Bougeault P, Rochas M, Cariolle D, Lafore JP, Royer JF, Andre JC. 1988. The evolution of numerical weather prediction and atmospheric modelling at the French Weather Service. J Mec Theor Appl 7:87–110.
Giorgi F. 1990. On the simulation of regional climate using a limited area model nested in a general circulation model. J Clim 3:941–963.
Giorgi F, Mearns LO. 1991. Approaches to the simulation of regional climate change: A review. Rev Geophys 29:191–216.
Giorgi F, Marinucci MR, Bates GT. 1993a. Development of a second-generation regional climate model (RegCM2). Part I: Boundary-layer and radiative transfer processes. Mon Wea Rev 121:2794–2813.
Giorgi F, Marinucci MR, Bates GT. 1993b. Development of a second-generation regional climate model (RegCM2). Part II: Convective processes and assimilation of lateral boundary conditions. Mon Wea Rev 121:2814–2832.
Giorgi F, Marinucci MR, Visconti G. 1990. Use of a limited area model nested in a general circulation model for regional climate simulation over Europe. J Geophys Res 95:18413–18431.
Hoinka KP, Seco J. 1991. Sobre frentes fríos en la Península Ibérica. Rev Geofísica 47:149–168.
Holton JR. 1989. An Introduction to Dynamic Meteorology. Academic Press, New York, 391 pp.
IPCC. 1990. Climate Change: The Intergovernmental Panel on Climate Change, eds. Houghton, Jenkins, and Ephraums. Cambridge University Press, 365 pp.
Kim JW, Chang JT, Baker NL, Wilks DS, Gates WL. 1984. The standard problem of climate inversion: Determination of the relationship between local and large scale climate. Mon Wea Rev 112:2069–2077.
Linch P, Huang X-Y. 1992. Initialization of the HIRLAM model using a digital filter. Mon Wea Rev 120:1019–1034.
Linés A. 1970. The climate of the Iberian peninsula. In: World Survey of Climatology, Vol. 5, ed. CC Wallén. New York, 195–239.
Mitchell JFB. 1990. Greenhouse warming: Is the mid-holocene a good analogue?. J Climate 3:1177–1192.
Ookouchi Y, Segal M, Kessler RC, Pielke RA. 1984. Evaluation of soil moisture effects on the generation and modification of mesoscale circulations. Mon Wea Rev 112:2281–2292.
Palmén E, Newton CW. 1969. Atmospheric Circulation Systems. Academic Press, New York.
Segal M, Avissar R, McCumber MC, Pielke RA. 1988. Evaluation of vegetation effects on the generation and modification of mesoscale circulations. J Atmos Sci 45:2268–2292.
Smith JB, Tirpak DA (eds). 1989. The Potential Effects of Global Climate Change on the United States, Vol. 1. U.S. Environmental Protection Agency, Washington, D.C.
Stark LP. 1965. Positions of monthly mean troughs and ridges in the northern hemisphere, 1949–1963. Mon Wea Rev 93:705–720.

von Storch H, Zorita E, Cubasch U. 1993. Downscaling of global climate change estimates to regional scales: An application to Iberian wintertime rainfall. J Climate 6:1161–1171.

Washington WM, Parkinson CL. 1986. An Introduction to Three-Dimensional Climate Modeling. Oxford University Press, New York, 422 pp.

Wigley TM, Jones PD, Briffa KR, Smith G. 1990. Obtaining sub-grid scale information from coarse-resolution general circulation model output. J Geophys Res 95:1943–1953.

Wilks DS. 1989. Statistical specification of local surface weather elements from large-scale information. Theor Appl Climatol 40:119–134.

Zhang DL, Chang HR, Seaman NL, Warner TT, Fritsch JM. 1986. A two-way interactive nesting procedure with variable terrain resolution. Mon Wea Rev 114:1330–1339.

Zorita E, Kharin V, von Storch H. 1992. The atmospheric circulation and sea surface temperature in the North Atlantic area in winter: Their interaction and relevance for Iberian precipitation. J Climate 5:1097–1108.

2. Modeling Leaf Level Effects of Elevated CO_2 on Mediterranean Sclerophylls

Peter C. Harley

It is now well established that if atmospheric CO_2 continues to increase at the present rate, it will be double its preindustrial concentration by the middle of the next century (Ramanathan, 1988). Although the exact effects of increasing CO_2 on global climate are not known, there is a growing consensus among climate modelers that global average temperatures and precipitation will increase, but that these changes will be nonuniform over the Earth's surface. These change are occurring very rapidly relative to the ability of natural ecosystems to adjust. Because large-scale and long-term effects of CO_2 doubling cannot be studied directly, models will necessarily be required to predict the direction and magnitude of climatic change as well as the effects of these changes on natural systems.

Global climatic change in concert with the direct effects of CO_2 on primary production will exert a significant impact on terrestrial ecosystems, with potentially disruptive societal effects. Ecosystem level changes in primary production and community composition cannot be predicted solely on the basis of individual plant responses, nor can whole plant responses be predicted on the basis of leaf level changes alone. Nevertheless, the primary effects of increasing CO_2 occur at the level of the individual leaf, where physiological processes are well understood at a mechanistic level and effects of increasing CO_2 can be predicted with relative confidence. A leaf level model must be capable of simulating photosynthesis, photorespiration, and stomatal conductance over a wide range of leaf temperatures, light intensities, and relative humidities, and these simulations must then be projected into an elevated CO_2 environment.

Predicting the responses of natural ecosystems to rising CO_2 and changing climate is particularly challenging. The research effort to date has focused primarily on short-lived agronomic species, selected for rapid growth and maximal allocation of carbon to harvestable organs under generally favorable environmental conditions. Naturally occurring species which have evolved to maximize survival in a fluctuating environment in which biotic and abiotic stresses abound are likely to respond quite differently (Oechel and Strain, 1985). It is difficult, therefore, to draw conclusions about likely long-term responses of plants in natural systems from data collected on annual agronomic species.

Mediterranean ecosystems are semiarid and characterized by winter rains and a more or less prolonged summer drought. Plant production in such systems is often limited by cool temperatures in winter and by water stress and high temperatures in summer, and the dominant growth form is the evergreen sclerophyll. Although there exists a substantial literature concerning leaf gas exchange in Mediterranean sclerophylls, the data base describing effects of long-term exposure to elevated CO_2 for Mediterranean species is extremely small. I will describe a range of potential leaf responses and speculate on how Mediterranean sclerophylls may respond, based on newly emerging ideas about those factors which are most likely to control leaf level response to increasing atmospheric CO_2.

Model Theory

Varying the partial pressure of CO_2 (pCO_2) affects photosynthesis in C_3 plants in two fundamental ways, at the level of leaf biochemistry and at the level of CO_2 diffusion. The primary effect is on the carboxylating enzyme, ribulose-1,5-bisphosphate carboxylase-oxygenase (Rubisco), the activity of which is strongly dependent on the pCO_2 at the enzyme active site. The pCO_2 at Rubisco is jointly determined by the rate of carboxylation ("demand" for CO_2) and by the rate at which CO_2 diffuses from the atmosphere to the enzyme site (CO_2 "supply") (Farquhar and Sharkey, 1982). This rate of supply is determined by the CO_2 gradient between the air outside the leaf boundary layer and the site of carboxylation and by the resistance to CO_2 flux occurring along this gradient. This resistance is composed of several resistances (boundary layer, stomatal, mesophyll), of which only stomatal resistance responds directly to varying CO_2. Henceforth, we will utilize conductance to CO_2 flux (g) where conductance is the inverse of resistance.

Figure 2-1A depicts the relationship between leaf biochemistry (the "demand" function) and diffusion processes (the "supply" function). The demand function is curvilinear and depicts net assimilation at a given value of intercellular CO_2 (C_i) based on leaf biochemistry. The supply function is represented by a straight line with x-intercept at C_i equal to ambient pCO_2 (C_a) and slope equal to (−) total leaf conductance to CO_2 flux. The intercept of the supply and demand functions determines the operational C_i and thus the rate of assimilation. (This analysis assumes that C_i is equal to the pCO_2 at the site of carboxylation, i.e., that resistance across the mesophyll is zero. This assumption is commonly

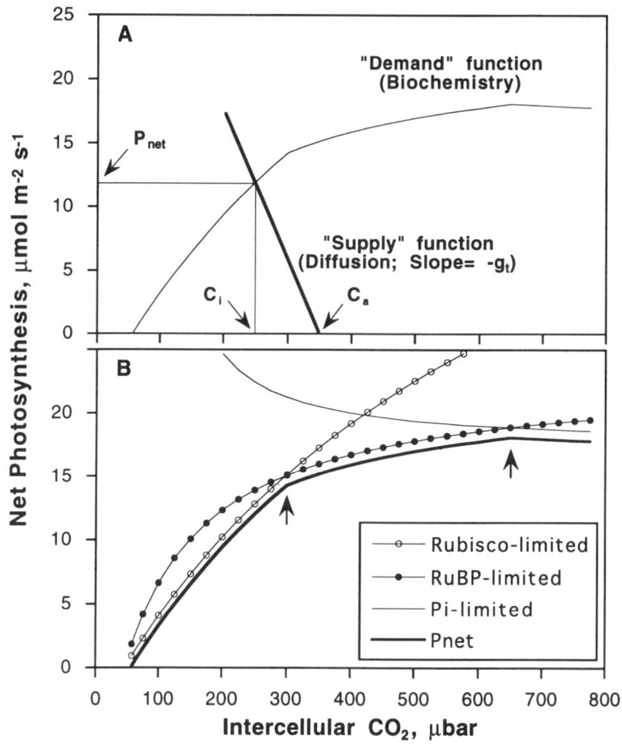

Figure 2-1. (A) Graphic depiction of the interaction between leaf biochemistry ("demand" for CO_2) and CO_2 diffusion ("supply"), which determines the pCO_2 in the intercellular air space (C_i), and the rate of net photosynthesis (P_n). (Adapted from Farquhar and Sharkey, 1982.) (B) Schematic description of the demand function of net photosynthesis. Photosynthesis is the minimum of three functions, representing the situations in which either Rubisco, RuBP regeneration, or P_i is limiting the rate of carboxylation. Arrows denote transitions from one limitation to another.

made, and while not strictly true (Evans et al., 1986) it simplifies the analysis without affecting our conclusions.)

Modeling the "Demand" Function of Photosynthesis

Woodrow and Berry (1988) make a convincing case that the photosynthetic responses of C_3 leaves can be modeled based solely on the kinetics of Rubisco. We use the biochemical model presented by Farquhar et al. (1980), modified to include a potential phosphate limitation (Sharkey, 1985).

Net CO_2 assimilation (A) may be expressed

$$A = V_c - 0.5V_o - R_d = V_c\left(1 - \frac{0.5 \cdot O}{\tau \cdot C_i}\right) - R_d \qquad (2\text{-}1)$$

where V_c and V_o are rates of carboxylation and oxygenation at Rubisco, C_i and O are the concentrations of CO_2 and O_2 in the intercellular air space, R_d is the rate of CO_2 evolution in the light resulting from processes other than photorespiration, and τ is the specificity factor for Rubisco (Jordan and Ogren, 1984).

The model assumes that V_c is limited by one of three processes and that the limiting process changes as light, CO_2, or temperature varies. At low pCO_2 and high irradiance, ribulose-1,5-bisphosphate (RuBP) is generally nonlimiting and V_c is limited only by the amount and kinetic properties of Rubisco and the partial pressures of CO_2 and O_2. The rate of carboxylation when Rubisco is limiting, W_c, obeys Michaelis-Menten kinetics, competitive with respect to CO_2 and O_2:

$$W_c = \frac{Vc_{max} C_i}{C_i + K_c(1 + O/K_o)} \qquad (2\text{-}2)$$

where Vc_{max} is the maximum rate of carboxylation and K_c and K_o are Michaelis constants for carboxylation and oxygenation, respectively. When Rubisco is limiting, increasing CO_2 accelerates the rate of carboxylation, decreases the rate of oxygenation and photorespiration, and results in a substantial stimulation of net photosynthesis.

RuBP is regenerated in the Calvin cycle, which is driven by ATP and NADPH. As pCO_2 increases and/or irradiance decreases, the rate of RuBP regeneration may limit carboxylation and photosynthesis. The rate of carboxylation limited solely by the rate of RuBP regeneration, W_j, is mediated by the rate of electron transport (J); i.e.:

$$W_j = \frac{J \cdot C_i}{4(C_i + O/\tau)} \qquad (2\text{-}3)$$

Implicit in this formulation for W_j is the assumption that four electrons generate sufficient ATP and NADPH for the regeneration of RuBP in the Calvin cycle (Farquhar and von Caemmerer, 1982). Even when the rate of RuBP regeneration is limiting, increasing CO_2 favors carboxylation over oxygenation, and photosynthesis is stimulated, but by substantially less than if Rubisco alone is limiting.

At high pCO_2 and high light, a third limitation may arise as the leaf becomes incapable of fully utilizing the products of the Calvin cycle. The immediate products of photosynthesis are triose phosphates, which are either exported from the chloroplast in equimolar exchange for P_i and used in sucrose production, or used in starch synthesis inside the chloroplast with the release of P_i. If rates of starch and sucrose formation are inadequate to keep up with the rate of triose phosphate production, phosphate is sequestered and P_i availability limits photosynthesis (Sharkey, 1985; Sage and Sharkey, 1987). W_p, the rate of carboxylation limited solely by P_i, is expressed

$$W_p = 3 \cdot \mathrm{TPU} + \frac{V_o}{2} = 3 \cdot \mathrm{TPU} + \frac{V_c \cdot 0.5 \cdot O}{C_i \cdot \tau} \tag{2-4}$$

where TPU is the rate of phosphate release in triose phosphate utilization. When P_i is limiting, increasing CO_2 either has no effect on net photosynthesis or leads to a slight inhibition (Sharkey, 1985; Harley and Sharkey, 1991).

The rate of carboxylation, V_c, is described as the minimum of these three potentially rate-limiting processes, i.e.:

$$V_c = \min\{W_c, W_j, W_p\} \tag{2-5}$$

Combining Equations 1 and 5:

$$A = \left(1 - \frac{0.5 \cdot O}{\tau \cdot C_i}\right) \cdot \min\{W_c, W_j, W_p\} - R_d \tag{2-6}$$

The interaction of these three processes with C_i is depicted in Figure 2-1b. Clearly, the extent to which a short-term doubling of CO_2 increases photosynthesis depends on which of these three processes is limiting for a given set of conditions (Stitt, 1991).

The only light dependency in the model is on electron transport, based on Smith (1937):

$$J = \frac{\alpha \cdot I}{\left(1 + \frac{\alpha^2 I^2}{J_{max}^2}\right)^{1/2}} \tag{2-7}$$

where I is incident PPFD ($\mu mol\,m^{-2}\,s^{-1}$), α is the efficiency of light energy conversion on an incident light basis (mol electrons/mol photons), and J_{max} is the light-saturated rate of electron transport.

The temperature dependencies of K_c, K_o, R_d, and τ are described by an exponential function:

$$\text{Parameter}\ (K_c, K_o, R_d, \tau) = \exp(c\text{-}\Delta H_a/(R \cdot T_k)) \tag{2-8}$$

where c is a scaling constant, ΔH_a is an activation energy, R is the gas constant ($0.00831\,kJ\,K^{-1}\,mol^{-1}$), and T_k is leaf temperature (K). The temperature dependencies of J_{max}, Vc_{max}, and TPU are described as follows, after Johnson, Eyring, and Williams (1942):

$$\text{Parameter}\ (J_{max}, Vc_{max}, \text{TPU}) = \frac{\exp(c - \Delta H_a/(R \cdot T_k))}{1 + \exp((\Delta S T_k - \Delta H_d)/(R \cdot T_k))} \tag{2-9}$$

where ΔH_d is the energy of deactivation and ΔS is an entropy term.

Modeling the "Supply" Function

The "demand" function uses C_i as a driving variable. However, C_i results from the interaction of A and leaf conductance to H_2O flux, g_s, according to the familiar Fick's law relationship:

$$C_i = C_a - \frac{A \cdot 1.6}{g_s} \tag{2-10}$$

where C_a is the concentration of CO_2 in the air outside the leaf boundary layer, and the factor 1.6 corrects for the difference in diffusivity between CO_2 and H_2O.

A mechanistic understanding of the regulation of stomatal behavior remains elusive. We describe stomatal conductance as a function of relative humidity (RH) and C_a (Ball et al., 1987):

$$g_s = g_0 + g_1 \cdot A \cdot \frac{\text{RH}}{C_a} \tag{2-11}$$

The parameter g_0 is the minimum stomatal conductance to H_2O when $A = 0$ at the light compensation point, and g_1 is an empirical coefficient which represents the composite sensitivity of conductance to assimilation, pCO_2, humidity, and temperature (Ball et al., 1987). Because A and g_s are not independent, the value of C_i must be determined in an iterative fashion.

In order to predict changes in photosynthesis in response to increased CO_2, the effect of CO_2 on both the "demand" and "supply" functions must be addressed. Furthermore, we need to distinguish between the immediate or short-term effects of increasing CO_2, which are easily determined experimentally, and the more relevant, but much harder to establish, longer-term effects resulting from growth in an elevated CO_2 atmosphere.

A parameterization scheme for the combined biochemical and stomatal models has been described (Harley and Tenhunen, 1991), and the model successfully simulates observed leaf behavior under field conditions in two Mediterranean sclerophylls, *Arbutus unedo* (Harley et al., 1986) and *Quercus coccifera* (Harley and Tenhunen, 1991). The model has also been used successfully to describe the photosynthetic characteristics of leaves of cotton plants grown both at current ambient CO_2 and at 650 μbar CO_2 (Harley et al., 1992).

Modeling the Short-Term Effects of Increased CO_2

Both Rubisco and stomatal aperture are responsive to CO_2. Thus, the photosynthetic response to an increase in ambient CO_2 depends on the shape of the demand function (fixed in the short term) and the response of the supply function to altered CO_2. One hypothetical result of a short-term doubling of CO_2 is illustrated in Figure 2-2. At C_a equal to 340 μbar CO_2, using supply function (g_1), photosynthesis is approximately 9.5 μmol m^{-2} s^{-1} with a C_i of ≈240 μbar. If we double CO_2, but maintain the same rate of supply ($g_2 = g_1$), the leaf attains a photosynthetic rate of approximately 17.0. Typically, stomata close somewhat when exposed to

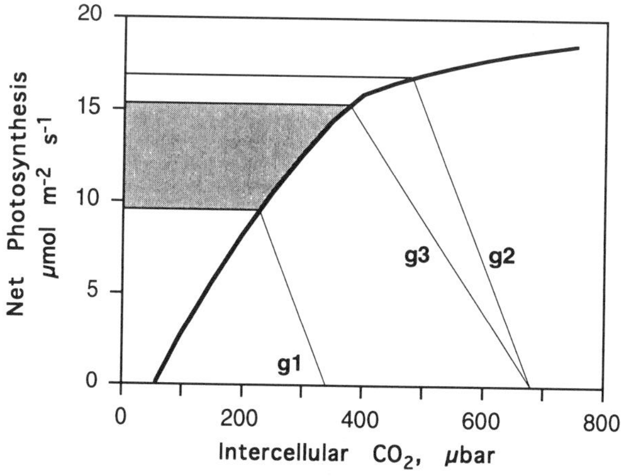

Figure 2-2. Supply and demand functions of photosynthesis illustrating the hypothetical response to short-term doubling of CO_2. Supply functions g_1 and g_2 represent identical values of leaf conductance. Supply function g_3 represents a 40% decrease in leaf conductance.

elevated CO_2 (by an average 40% if CO_2 is doubled; Morison, 1985) as depicted by supply function (g_3), which lowers C_i and leads to a slight decline in photosynthesis to approximately 15.5, which nevertheless represents an increase of over 50% from the original rate (stippled area).

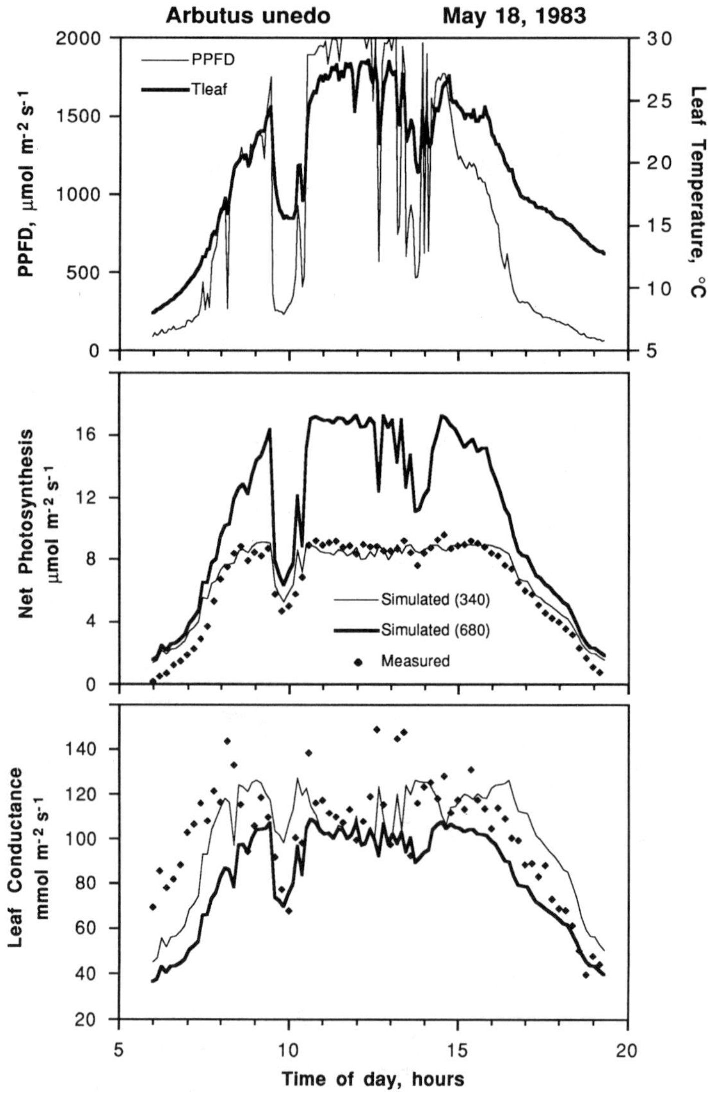

Figure 2-3. Daily time courses of incident PPFD, leaf temperature, net photosynthesis, and leaf conductance measured on a shoot of *Arbutus unedo* under natural conditions. Also shown are model simulations assuming that the ambient pCO_2 is either 340 or 680 µbar.

In this example, the leaf is operating largely on the Rubisco-limited portion of the demand curve; had the leaf been operating on the RuBP regeneration-limited portion of the demand function, the result of a doubling of CO_2 would have been much less. This simple example illustrates the complexity of predicting leaf response to changing CO_2 and the necessity of modeling the effects of CO_2 on both leaf biochemistry and stomatal behavior.

The integrated models were parameterized previously for *Arbutus unedo*, growing in a natural macchia in Portugal, and used to simulate plant response under both ambient CO_2 (Harley et al., 1986; Harley and Tenhunen, 1991) and saturating CO_2 (Harley et al., 1986; Figure 2-6). We simulate here the effects of a short-term doubling of CO_2 on photosynthesis and conductance in *Arbutus* (Figure 2-3) during spring when water availability is adequate and temperatures are moderate. Instantaneous photosynthesis is enhanced by as much as 107%, whereas conductance is reduced an average 18%. This predicted decline in conductance is substantially less than the average 40% decline found by Morison (1985), but is not inconsistent with changes in conductance measured in *Arbutus* (Figure 2-4). These results indicate that *Arbutus* is capable of taking significant advantage of elevated CO_2, at least in the short term, due to the shape of the demand function. At high light and 30°C, photosynthesis is Rubisco limited, increasing almost linearly to C_i values above 400 μbar. However, the effect of increased pCO_2 on photosynthesis is strongly dependent on both light and temperature. The light response of photosynthesis at both C_a = 340 and 680 μbar is plotted in Figure 2-5 along with the percent increase at 680 μbar. In the initial quantum yield portion of the light response, electron transport limits

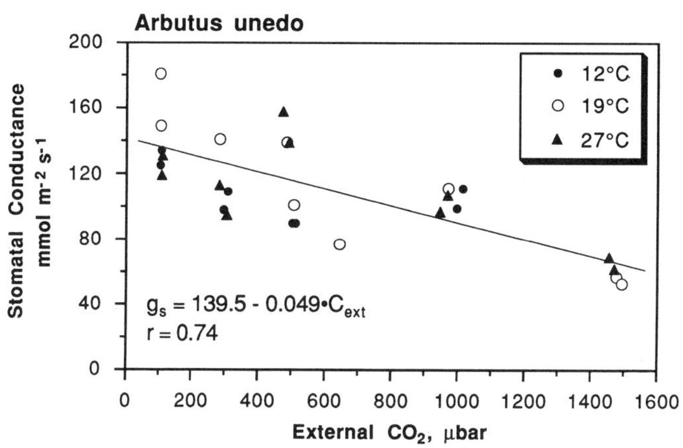

Figure 2-4. Measured stomatal conductance of *Arbutus unedo* as a function of external pCO_2 at three different leaf temperatures.

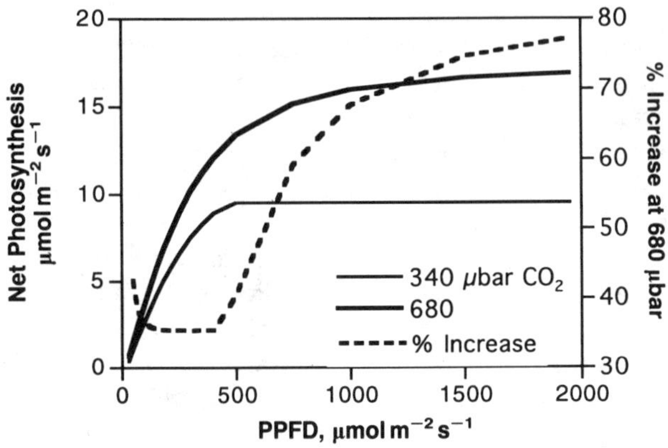

Figure 2-5. Model simulations of net photosynthesis as a function of incident PPFD at an ambient pCO_2 of either 340 or 680 μbar, based on parameter values used for simulations in Figure 2-3. Also shown is the percent stimulation of photosynthesis at 680 μbar CO_2. Leaf temperature is 25°C.

photosynthesis, and doubling CO_2 results in a 35% stimulation of CO_2 assimilaton; above PPFD of 500 $\mu mol\,m^{-2}\,s^{-1}$, Rubisco is limiting (at $C_a = 340$) and the stimulation increases up to a maximum of 77% at full sunlight.

Leaf temperature also controls the magnitude of the photosynthetic response to a doubling of CO_2 (Figure 2-6), due primarily to Rubisco kinetics. Oxygenation increases more rapidly with temperature than does

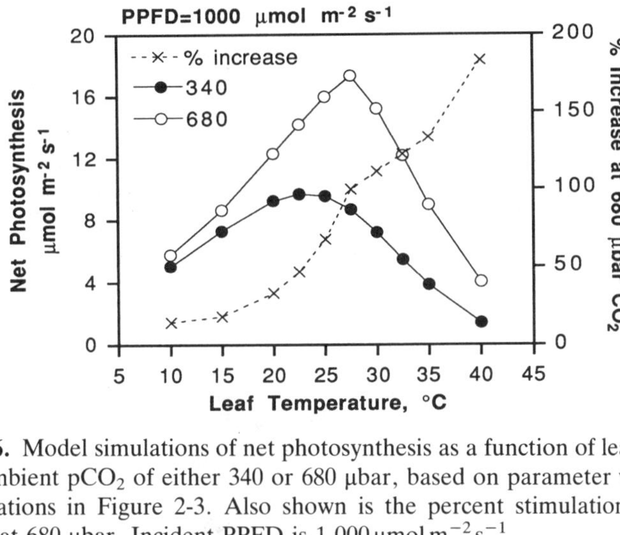

Figure 2-6. Model simulations of net photosynthesis as a function of leaf temperature at ambient pCO_2 of either 340 or 680 μbar, based on parameter values used for simulations in Figure 2-3. Also shown is the percent stimulation of photosynthesis at 680 μbar. Incident PPFD is 1,000 $\mu mol\,m^{-2}\,s^{-1}$.

carboxylation. The effect of increasing CO_2 and reducing the rate of oxygenation relative to carboxylation thus becomes increasingly significant as temperature increases.

Integrated over the entire daylight period, the simulation in Figure 2-3 predicts a 59% increase in carbon gain and a reduction in evapotranspiration of 13% (Table 2-1). The combined effects of increased assimilation and decreased water loss result in an 83% increase in water use efficiency.

Although instructive, these results should not be taken as representative, either for *Arbutus* or for Mediterranean sclerophylls in general. They are based on measurements of a single branch under optimal conditions of water availability and temperature; effects of increased CO_2 during the summer drought may be very different.

Model Shortcomings

Although the model presented above has been used successfully, it has not been validated for a wide variety of species and environmental situations, and is not without shortcomings. Although the biochemical model of Farquhar et al. (1980) is widely accepted and has been used extensively to interpret photosynthetic data, the Ball model of conductance remains empirical and has been verified for only a few experimental situations (Leuning, 1990; Harley and Tenhunen, 1991; Harley et al., 1992). The use of relative humidity rather than explicit dependencies on temperature and leaf to air vapor pressure deficit has been criticized (Aphalo and Jarvis, 1993). In addition, the response of the model to elevated CO_2 has received little attention. If relative humidity is held

Table 2-1. Integrated Daytime Totals of CO_2 Assimilation, Evapotranspiration, and Water Use Efficiency of *Arbutus unedo*

Growth Status and Measurement CO_2	Net Daily CO_2 Assimilation ($mmol\,m^{-2}d^{-1}$)	Daily Transpiration ($mol^{-2}m^{-2}d^{-1}$)	Water Use Efficiency (mmol CO_2/mol H_2O)
Grown at 340 Measured at 340	653	130.3	5.01
Grown at 340 Measured at 680	1039 (+59%)	113.5 (−13%)	9.16 (+83%)
Acclimated Measured at 680	866 (+33)	97.8 (−25%)	8.85 (+77%)
Homeostasis Measured at 680	631 (−3%)	77.7 (−40%)	8.12 (+62%)

Based on model simulations using input weather data data for May 18, 1983 and assuming either 340 or 680 μbar CO_2. Figures in parentheses are the percent change from the results of the simulation based on nonacclimated plants at 340 μbar CO_2.

constant, conductance is proportional to the ratio A/C_a. In Figure 2-3, a doubling of C_a leads to nearly a doubling of A, with the result that g_s is virtually unchanged (e.g., Figure 2-3, midday). This is unlikely to be realistic (Morison, 1985), although the data in Figure 2-4 suggest that conductance in *Arbutus* may be less sensitive to high CO_2 than in many species. The validity of this stomatal model in elevated CO_2 scenarios needs further investigation.

Although the combined model has been used successfully to simulate leaf responses in nonstress situations, Mediterranean species under stress exhibit behavior which is not presently captured in a realistic fashion. Harley and Tenhunen (1991) were able to simulate the diurnal photosynthetic behavior of another Mediterranean sclerophyll, *Quercus coccifera*, during both nonstress periods and times of severe water stress, characterized by midday depression of conductance and photosynthesis (Tenhunen et al., 1985). In order to do so, however, it was necessary to reparameterize the model for each simulation in order to capture stress effects; i.e., there was no single parameter set that could successfully capture leaf behavior over an annual cycle, and the parameterization at each period was an ad hoc attempt to simulate measured data. Tenhunen et al. (1990) were able to successfully capture the essential changes in photosynthetic behavior of *Q. coccifera* over an annual cycle by choosing a single set of biochemical parameters and changing only the proportionality constant used in the conductance model, which in effect alters the relationship between photosynthesis, C_i, and conductance. Although successful in describing photosynthetic behavior, this empirical approach fails to capture the true nature of physiological changes during periods of stress.

For example, midday depression of photosynthesis in *Arbutus* and in *Quercus suber* has been shown by Beyschlag et al. (1992) to be largely due to nonhomogeneous stomatal closure, which, in effect, changes the active photosynthetic leaf area over the course of the day. Nonhomogeneous stomatal closure has been identified in several other Mediterranean sclerophylls, including *Pistacia vera*, *Olea europaea*, *Quercus ilex*, *Eucalyptus ficifolia*, and *Nerium oleander* (cited with references in Terashima, 1992). Presumably (though unproven), the active portion of the leaf continues to behave as described by the model, but in order to model whole leaf behavior, we need to understand the controls over "stomatal patchiness" and model these effects. These controls are likely to be extrinsic to the leaf itself, involving whole-plant water balance and hormonal signals, perhaps arising in the roots (Blackman and Davies, 1985). Incorporating these whole-plant effects is a challenge for the future. For the present, it remains extremely difficult to predict how a doubling of CO_2 is likely to affect stomatal closure and leaf physiology under stress conditions.

Modeling Long-Term Effects of Growth in Elevated CO_2

Although the simulated effects of a short-term exposure to doubled CO_2 are often large, it is well established that when plants are grown continuously under high CO_2, changes in leaf biochemistry often occur which result in lowered photosynthetic capacity (Cure and Acock, 1986; Stitt, 1991); i.e., the demand function is lower over a portion of or the entire range of C_i values. This phenomenon has been referred to as "acclimation" to elevated CO_2.

A large number of studies have examined the question of leaf level response to relatively long-term exposure (weeks to months) to elevated CO_2 and the results, summarized elsewhere (e.g., Cure and Acock, 1986; Stitt, 1991), are by no means consistent. In general, large initial increases in photosynthesis when plants were transferred into an elevated CO_2 environment were not sustained over periods of weeks to months. Based on a literature survey of crop plants, Cure and Acock (1986) report an average 52% initial stimulation of photosynthesis upon transfer to doubled CO_2, which fell over time to an average 29% stimulation. In some studies, however, large initial increases in photosynthesis were sustained or increased over time, as in potato (Sage et al., 1989), soybean (Clough et al., 1981; Campbell et al., 1988), cotton (Radin et al., 1987), and a sedge, *Scirpus olneyi* (Arp and Drake, 1991). On the other hand, in a few cases, the increase in C_i which occurs when plants are grown in 680 μbar CO_2 is insufficient to counteract the effects of a greatly decreased demand function, and net photosynthesis of plants grown at 680 μbar is less than that of 340-grown plants (Wong, 1979; Larigauderie et al., 1988; Sage et al., 1989).

Despite inconsistent experimental results, Stitt (1991) makes the following generalizations: (1) the initial slope of the photosynthesis vs. C_i response curve tends to decline in plants grown in high CO_2, and (2) there is no indication that photosynthesis of plants grown in high CO_2 is more likely to be limited by P_i than under ambient CO_2. The reduction in the initial slope indicates reduced Rubisco activity, which may result from partial deactivation of the enzyme (Sage et al., 1989), reduced amounts of enzyme (Wong, 1979; Peet et al., 1985) or both (Rowland-Bamford et al., 1991). Typically, leaves that develop in high CO_2 contain less nitrogen than leaves grown in ambient CO_2 and devote a smaller percentage of that nitrogen to Rubisco (Sage et al., 1989; Rowland-Bamford et al., 1991). In addition, amounts of other key photosynthetic enzymes are frequently lower (Porter and Grodzinski, 1960; Besford, 1990).

Two hypothetical acclimation responses for *Arbutus* are shown in Figure 2-7, one labeled "acclimated" which results in a substantial increase in photosynthesis, despite a reduction in the demand function, and a second labelled "homeostasis," which results in a rate of photo-

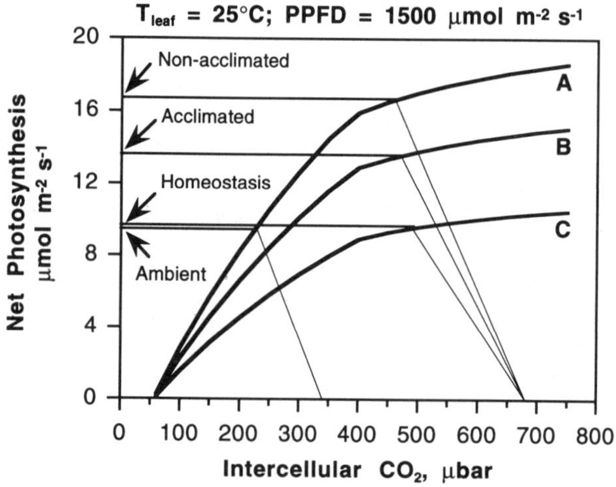

Figure 2-7. Three demand functions of photosynthesis. Curve A is that used in the simulations in Figure 2-3 and yields the photosynthetic rate labeled Ambient at C_a = 340 and that labeled Non-acclimated at C_a = 680 μbar. Curves B and C represent hypothetical acclimation scenarios. For B, the rate of photosynthesis at 680 μbar CO_2 (Acclimated) is substantially greater than the original ambient rate. For C, the rate of photosynthesis at 680 μbar (Homeostasis) is virtually the same as the original ambient rate at 340 μbar.

synthesis virtually the same as the ambient scenario. (To achieve the acclimation and homeostasis scenarios, model parameters describing the rate of Rubisco activity, electron transport, and dark respiration were reduced by 20% and 45%, respectively. P_i has never been found to play a role in limiting photosynthesis in plants grown in elevated CO_2, and P_i limitations are not considered here.)

Figure 2-8 illustrates the potential effects of these different levels of acclimation on the diurnal course of photosynthesis and leaf conductance; the integrated daily effects on carbon gain, transpiration, and water use efficiency are summarized in Table 2-1. Even in the homeostasis scenario, where daily carbon gain is slightly reduced, there are sustained increases in water use efficiency, due to reduction in stomatal conductance and transpiration. Figure 2-9 illustrates how incident PPFD affects the photosynthetic response at 680 relative to 340 μbar for all three scenarios. Even in the case of near homeostasis, where there is virtually no effect of doubling CO_2 on photosynthesis at light saturation, there remains a distinct advantage to elevated CO_2 when light is strongly limiting. This is an inevitable consequence of the positive effect of increased CO_2 on the quantum yield of photosynthesis in C_3 plants (Ehleringer and Björkman, 1977) and has important implications for leaves in low light environments under elevated CO_2. This stimulation of quantum yield by high CO_2 increases with increasing temperature (Long, 1991).

Figures 2-8 and 2-9 illustrate two possible acclimation scenarios for a single Mediterranean sclerophyll species, *Arbutus unedo*, but we have little experimental data on which to base a decision as to which of the infinite number of possible acclimation responses is likely to occur as plant species adapt to gradually increasing CO_2. Furthermore, most experiments have been relatively short-term (weeks to months), and performed on well-watered, well-fertilized annual agronomic species. How then can one extrapolate these results to perennial plants of natural ecosystems, where nutrients are frequently limiting, and levels of biotic and abiotic stress may be severe?

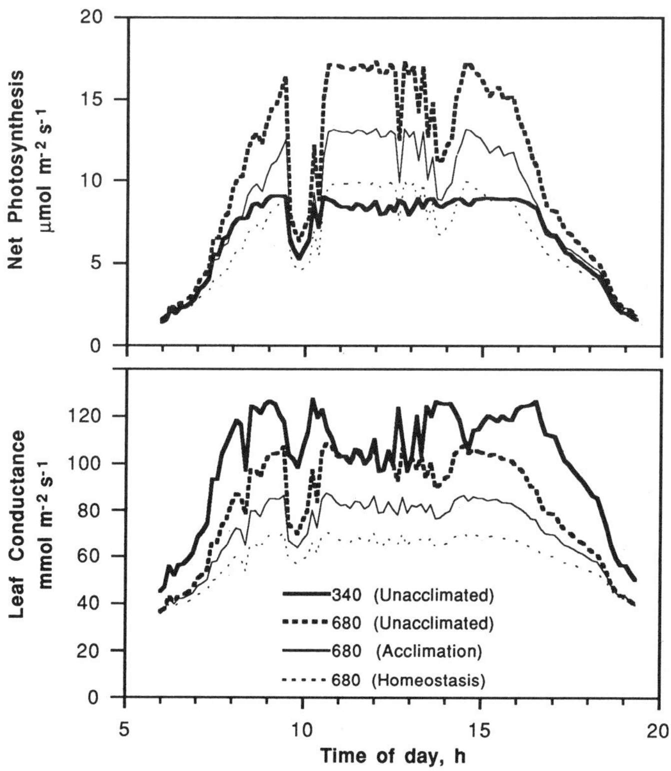

Figure 2-8. Simulations of the diurnal time course of net photosynthesis and leaf conductance of *Arbutus unedo* using the values of PPFD and leaf temperature shown in Figure 2-3. The original simulations at C_a = 340 and 680 μbar for the unacclimated plant are compared to simulations at 680 μbar based on parameter sets used to generate the hypothetical acclimation scenarios depicted in Figure 2-7.

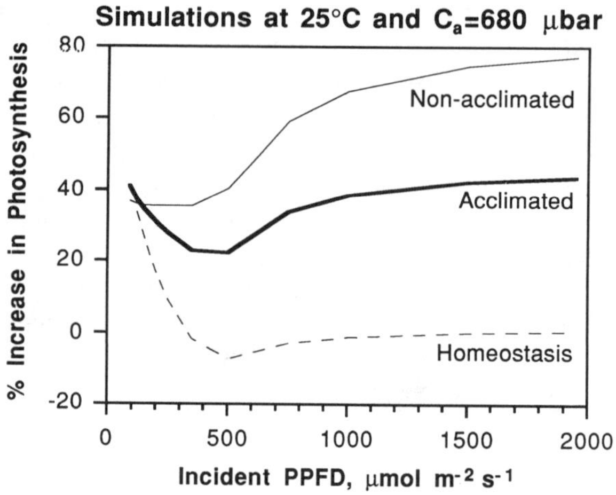

Figure 2-9. Graph depicting the percent stimulation of net photosynthesis achieved by doubling the pCO_2 from 340 to 680 μbar as a function of incident PPFD. Simulations at 680 μbar are based on parameter sets used to generate the hypothetical acclimation scenarios depicted in Figure 2-7. Leaf temperature is 25°C.

There is growing evidence that the extent to which a plant can take advantage of increased CO_2 is influenced by the relative magnitude of carbohydrate sources and sinks (Arp, 1991; Stitt, 1991). Doubling CO_2 results in enhanced photosynthesis and accumulation of carbohydrates in leaves, indicative of at least a short-term imbalance between the increased carbon source and preexisting sinks. If the whole plant can generate sufficient carbon sinks to utilize the increased carbon supply over a sustained period, leaf photosynthetic performance is expected to remain high and contribute to enhanced whole plant production. If, on the other hand, the plant is incapable of fully utilizing the increased supply of carbon, sources and sinks will remain in imbalance unless source strength declines, whereby resources (especially Nitrogen) are reallocated, resulting in lowered leaf photosynthetic capacity. Sink strength may take the form of metabolically active meristematic regions or storage sinks, such as tubers, seeds, roots, or rhizomes. Sink strength changes in the course of plant phenology, and thus the response to increasing CO_2 will vary over the life of a plant. These ideas are summarized in Figure 2-10. Stitt (1991) reviews experiments in support of this hypothesis and suggests mechanisms by which leaf level processes might be altered.

Although these ideas cannot explain all experimental results involving long-term effects of growth in elevated CO_2, they do provide an effective context in which to interpret a number of observed phenomena. Stimulation of photosynthesis by elevated CO_2 is often greater in source-limited seedlings than in older plants (Ackersen et al., 1984). Experiments with soybeans in which sink strength (Clough et al., 1981) or source strength (Peet, 1984) is manipulated give results consistent with this hypothesis. Photosynthetic rates in leaves of Valencia orange trees grown in an enriched CO_2 atmosphere were 23% higher than those in ambient CO_2 at the end of flowering, 77% higher when fruits were young, and 18% higher as fruits neared maturity (Downton et al., 1987), emphasizing the importance of phenological changes in source-sink balance.

The possibility that restricted root growth in pots could lead to a decrease in sink strength in cotton was demonstrated conclusively by Thomas and Strain (1991), and Arp (1991) suggests that unintended sink strength manipulations arising from growth in pots of inadequate size have biased a large number of studies towards declining photosynthetic capacity with elevated CO_2.

Photosynthesis was reduced much more in elevated CO_2 relative to ambient CO_2 controls in cotton plants grown in limiting nitrogen than in plants grown with high nitrogen (Wong, 1979), presumably because growth (i.e., sink strength) was reduced in low nitrogen. Larigauderie et al. (1988) report similar interactions of nitrogen nutrition with elevated CO_2 in the annual grass *Bromus mollis*. Only if growth is solely carbon limited would plants be expected to respond maximally to increasing CO_2.

Finally, there is necessarily a strong genetic component to whole-plant sink strength. Determinate plants are intrinsically less able to increase sink strength in response to increased carbohydrate source than are indeterminate plants that can initiate new meristematic activity. Some

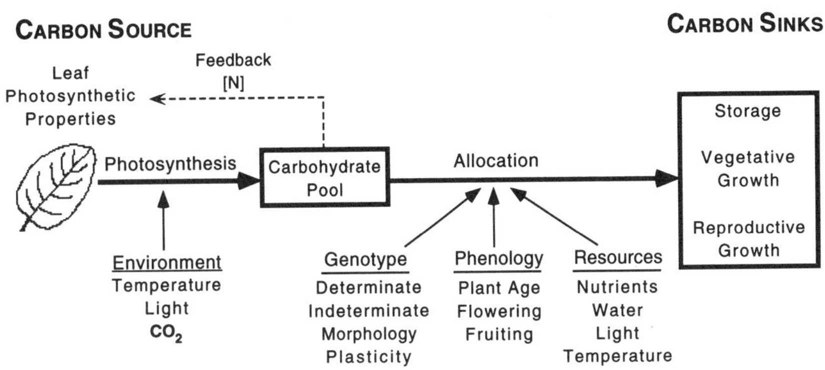

Figure 2-10. Diagram depicting the relationship between carbon sources (photosynthetic tissue) and carbon sinks (growth and storage) and factors which may affect the relative size of each. Source-sink imbalance may feed back on leaf level photosynthetic properties.

plants possess large storage organs for carbohydrate, and different species invest differing amounts of carbon in reproductive effort.

Based largely on the results of a 4-year field study of the effects of CO_2 enrichment on a salt marsh community (Arp and Drake, 1991) in which photosynthetic capacity of *Scirpus olneyi* increased relative to ambient CO_2 controls, Arp (1991) suggests that high rates of photosynthesis are likely to be sustained in natural and agro-ecosystems. It may well be that salt marsh species, which receive daily influxes of nutrients and which have large sinks for carbon in rapidly spreading rhizome systems, suffer no water stress, and agricultural crops, fertilized, watered, and selected for rapid growth, are poised to take maximal advantage of increasing atmospheric CO_2. But what about plants of Mediterranean ecosystems?

Mediterranean Ecosystems

Mediterranean systems are semiarid and characterized by winter rains and a more or less prolonged summer drought. Plant production is limited by cool temperatures in winter and by water stress and high temperatures in summer. Soil nutrients are frequently limiting for growth, except immediately following fire. Mediterranean sclerophyll vegetation tends to be slow growing, with growth concentrated in the short period between winter rains and summer drought. Whether or not they respond strongly to increasing CO_2 may depend on whether slow growth is carbon (source) limited or whether it is a genetic trait which has evolved to maximize survival in a resource-limited, periodically stressful environment. Whether this life-form can develop new sinks to take advantage of increased carbohydrate is a question that needs to be addressed experimentally.

The experimental data base on plant response to long-term exposure to elevated CO_2 remains quite small, and I am aware of only two relatively short-term studies on Mediterranean species, two California chaparral sclerophylls and an annual grass. Jenkins (1992) conducted a 30-week experiment on *Adenostema fasciculatum* and *Ceanothus greggii* (see also Chapter 3) in which seedlings were grown in pots with soil N levels comparable to that found under natural stands following fire. A 15-week drought treatment was imposed to assess the interactive effects of drought and elevated CO_2. After 7 weeks, photosynthesis of plants of both species grown and measured at 700 μbar CO_2 was greater than that of plants grown and measured at ambient CO_2; however, some downward adjustment of photosynthetic capacity had occurred, evidenced by the fact that photosynthetic rates measured at 700 μbar CO_2 were less in plants grown at 700 μbar than in those grown at 350 μbar, particularly for *A. fasciculatum*. After 30 weeks of treatment, both species demonstrated almost perfect homeostatic adjustment, and there were no significant differences between photosynthetic rates of plants when measured at the CO_2 concentration under which they were grown.

After 16 weeks of drought, photosynthesis in both species was drastically reduced and highly variable. Although rates were reduced in both ambient and high CO_2-grown plants, plants grown at 700 μbar maintained a slight absolute advantage.

Larigauderie et al. (1988) studied the interaction between increased CO_2 and soil Nitrogen on *Bromus mollis*, an annual grass of the Mediterranean basin, naturalized in California. When nitrogen availability was high, photosynthetic rates were higher in plants grown and measured at 650 μbar CO_2 than in those grown and measured at ambient CO_2, although considerable downward adjustment of photosynthetic capacity occurred, accompanied by loss of leaf Nitrogen. When Nitrogen was strongly limiting, downward adjustment was extreme such that photosynthetic rates measured at growth CO_2 levels were lower in plants grown at elevated CO_2 than in those grown at ambient.

Thus, the limited amount of data suggests that leaf photosynthetic response to elevated CO_2 in Mediterranean species may involve considerable downward acclimation of photosynthetic capacity, particularly in nutrient-limited sites. Both these experiments were carried out during the seedling stage, a source-limited period when the response to CO_2 fertilization would be expected to be relatively strong. It is important to point out that growth was enhanced in both of these studies. The acclimation of photosynthesis merely suggests that these plants are less capable of taking full advantage of CO_2 enrichment than plants from systems where other resources are less limiting. It would be foolish to generalize based on these studies, but it appears unlikely that periodically stressful, low-resource environments will respond as strongly to increasing atmospheric CO_2 as studies from agronomic or salt marsh ecosystems might suggest.

Conclusions

Although an understanding of how increased CO_2 affects leaf gas exchange is critical to understanding whole plant responses, it is clear that whole plant responses cannot be predicted from leaf-level changes alone and that prediction of leaf level changes cannot be made except in the context of the whole plant. Effects of water stress at the leaf level, such as midday depression of conductance and photosynthesis, may only be interpretable in the context of whole plant water balance, root activity, and hormonal control. Feedback effects on leaf processes seem to be related to source-sink balance, which can only be understood as a whole plant phenomenon.

The impact of doubled CO_2 is strongly light and temperature dependent; in order to integrate leaf level effects over the entire plant, the light and temperature environment within the plant canopy must be described in some detail. Although leaves frequently adjust their photosynthetic capacity downward in response to increased CO_2, it is important to note that whole plant carbon fixation usually increases substantially as reductions in carbon uptake per unit leaf area are more than compensated for by increased leaf area, which is a typical response to growth in elevated CO_2. Indeed, given a limited supply of nitrogen and certain canopy structures, whole plant carbon uptake may perhaps be maximized by reducing nitrogen (and photosynthesis) on a per leaf basis, but using the nitrogen saved to produce additional leaves. Similarly, although water use by individual leaves is reduced in high CO_2, some or all of these savings may be lost on a whole plant or stand basis if leaf area increases proportionally.

In summation, changes in leaf level processes—increased photosynthesis and decreased stomatal conductance—are the primary initial effects of increasing CO_2. Increases in the amount of fixed carbon available to the plant then set in motion a cascade of poorly understood events—carbon allocation and growth of different organs—which exert an impact throughout the plant and may ultimately feed back on leaf processes. It is crucial to elucidate the physiological and biochemical nature of these feedback effects—to couple leaf level and whole plant effects—if we are to understand and ultimately predict long-term plant response.

If the goal is to predict ecosystem level effects of increasing atmospheric CO_2 and associated climate change, however, it will be necessary to conduct long-term and necessarily expensive CO_2-enrichment studies on entire plant communities. Study of leaf level processes, as a component of multiyear field CO_2 enrichment studies under natural conditions, should certainly continue. However, it seems unlikely that leaf level phenomena alone, no matter how well understood, will allow prediction of ecosystem level change, independent of whole plant and ecosystem level research.

References

Ackersen RC, Havelka VD, Boyle MG. 1984. CO_2 enrichment effects on soybean physiology. II. Effects of stage-specific CO_2 exposure. Crop Sci 24: 1150–1154.

Aphalo PJ, Jarvis PG. 1993. An analysis of Ball's empirical model of stomatal conductance. Ann Bot 72:321–327.

Arp WJ. 1991. Effects of source-sink relations on photosynthetic acclimation to elevated CO_2. Plant Cell Env 14:869–875.

Arp WJ, Drake BG. 1991. Increased photosynthetic capacity of *Scirpus olneyi* after four years of exposure to elevated CO_2. Plant Cell Env 14:1004–1008.

Ball JT, Woodrow IE, Berry JA. 1987. A model predicting stomatal conductance and its contribution to the control of photosynthesis under different environ-

mental conditions. In: Progress in Photosynthesis Research, (I Biggins, ed), Vol IV. Proceedings of the VII International Congress on Photosynthesis, Martinus Nijhoff, Dordrecht, The Netherlands, pp. 221–224.
Besford RT. 1990. The greenhouse effect: acclimation of tomato plants growing in high CO_2: relative changes in Calvin cycle enzymes. J Plant Physiol 136: 458–463.
Beyschlag W, Pfanz H, Ryel RJ. 1992. Stomatal patchiness in Mediterranean sclerophylls. Phenomenology and consequences for the interpretation of the midday depression in photosynthesis and transpiration. Planta 187:546–553.
Blackman PG, Davies WJ. 1985. Root to shoot communication in maize plants of the effects of soil drying. J Exp Bot 35:174–179.
Campbell WJ, Allen LH Jr, Bowes G. 1988. Effects of CO_2 concentration on Rubisco activity, amount, and photosynthesis in soybean leaves. Plant Physiol 88:1310–1316.
Clough JM, Peet MM, Kramer PJ. 1981. Effects of high atmospheric CO_2 and sink size on rates of photosynthesis of a soybean cultivar. Plant Physiol 67:1007–1010.
Cure JD, Acock B. 1986. Crop responses to carbon dioxide doubling: a literature survey. Agric Forest Meteor 38:127–145.
Downton WJS, Grant WJR, Loveys BR. 1987. Carbon dioxide enrichment increases yield of Valencia orange. Aust J Plant Physiol 14:493–501.
Ehleringer J, Björkman O. 1977. Quantum yields for CO_2 uptake in C_3 and C4 plants. Dependence on temperature, CO_2 and O2 concentration. Plant Physiol 59:86–90.
Evans JR, Sharkey TD, Berry JA, Farquhar GD. 1986. Carbon isotope discrimination measured concurrently with gas exchange to investigate CO_2 diffusion in leaves of higher plants. Aust J Plant Physiol 13:281–292.
Farquhar GD, Sharkey TD. 1982. Stomatal conductance and photosynthesis. Ann Rev Plant Physiol 33:317–345.
Farquhar GD, von Caemmerer S. 1982. Modelling of photosynthetic response to environment. In: *Encyclopedia of Plant Physiology, NS vol. 12B: Physiological Plant Ecology II* (OL Lange, PS Nobel, CB Osmond, and H Ziegler, eds.), pp. 549–587. Springer-Verlag, Berlin.
Farquhar GD, von Caemmerer S, Berry JA. 1980. A biochemical model of photosynthetic CO_2 assimilation in leaves of C_3 species. Planta 149:78–90.
Harley PC, Sharkey TD. 1991. An improved model of C_3 photosynthesis at high CO_2: Reversed O2 sensitivity explained by lack of glycerate reentry into the chloroplast. Photosyn Res 27:169–178.
Harley PC, Tenhunen JT. 1991. Modeling the photosynthetic response of C_3 leaves to environmental factors. In: Modeling Crop Photosynthesis (KJ Boote and RS Loomis, eds.), pp. 17–39. Special Publication of the American Society of Agronomy, Madison, WI.
Harley PC, Tenhunen JT, Lange OL. 1986. Use of an analytical model to study limitations to net photosynthesis in *Arbutus unedo* under field conditions. Oecologia 70:393–401.
Harley PC, Thomas RB, Reynolds JF, Strain BR. 1992. Modelling photosynthesis of cotton grown in elevated CO_2. Plant Cell Env 15:271–282.
Jenkins M. 1992. Effects of atmospheric carbon dioxide level on photosynthesis, water-use efficiency, and allocation patterns of seedlings of two chaparral shrubs. M. Sc. Thesis, San Diego State University, San Diego, CA.
Johnson F, Eyring H, Williams R. 1942. The nature of enzyme inhibitions in bacterial luminescence: Sulfanilamide, urethane, temperature, and pressure. J Cell Comp Physiol 20:247–268.

Jordan DB, Ogren WL. 1984. The CO_2/O_2 specificity of ribulose 1,5-bisphosphate carboxylase/oxygenase. Dependence on ribulose-bisphosphate concentration, pH and temperature. Planta 161:308–313.

Larigauderie A, Hilbert DW, Oechel WC. 1988. Effect of CO_2 enrichment and nitrogen availability on resource acquisition and resource allocation in a grass, *Bromus mollis*. Oecologia 77:544–549.

Leuning R. 1990. Modelling stomatal behavior and photosynthesis of *Eucalyptus grandis*. Aust J Plant Physiol 17:159–175.

Long SP. 1991. Modification of the response of photosynthetic productivity to rising temperature by atmospheric CO_2 concentrations: has its importance been underestimated? Plant Cell Env 14:729–739.

Morison JIL. 1985. Sensitivity of stomata and water use efficiency to high CO_2. Plant Cell Env 8:467–474.

Oechel W, Strain BR. 1985. Native species responses to increased carbon dioxide concentration. In: Direct effects of increasing carbon dioxide on vegetation, pp. 117–154 (BR Strain and JD Cure, eds), U.S. Dept. of Energy. DOE/ER-0238.

Peet MM. 1984. CO_2 enrichment of soybeans. Effects on Leaf/Pod ratio. Physiol Plantarum 60:38–42.

Peet MM, Huber SC, Patterson DT. 1985. Acclimation to high CO_2 in monoecious cucumbers. II. Alterations in gas exchange rates, enzyme activities and starch and nutrient concentrations. Plant Physiol 80:63–67.

Porter MA, Grodzinski B. 1960. Acclimation to high CO_2 in bean. Plant Physiol 74:413–416.

Radin JW, Kimball RA, Hendrix DL, Mauney JR. 1987. Photosynthesis of cotton plants exposed to elevated levels of carbon dioxide in the field. Photosyn Res 12:191–203.

Ramanathan V. 1988. The greenhouse theory of climate change: A test by inadvertent experiment. Science 240:293–299.

Rowland-Bamford AJ, Baker JT, Allen LH Jr, Bowes G. 1991. Acclimation of rice to changing atmospheric carbon dioxide concentration. Plant Cell Env 14:577–583.

Sage RF, Sharkey TD. 1987. The effect of temperature on the occurrence of O_2 and CO_2 insensitive photosynthesis in field grown plants. Plant Physiol 84: 658–664.

Sage RF, Seemann JR, Sharkey TD. 1989. Acclimation of photosynthesis to elevated CO_2 in five C_3 species. Plant Physiol 89:590–596.

Sharkey, TD. 1985. Photosynthesis in intact leaves of C_3 plants: physics, physiology and rate limitations. Bot Rev 51:53–106.

Smith E. 1937. The influence of light and carbon dioxide on photosynthesis. Gen Physiol 20:807–830.

Stitt M. 1991. Rising CO_2 levels and their potential significance for carbon flow in photosynthetic cells. Plant Cell Env 14:741–762.

Tenhunen JD, Lange OL, Harley PC, Beyschlag W, Meyer A. 1985. Limitations due to water stress on leaf net photosynthesis of *Quercus coccifera* in the Portuguese evergreen scrub. Oecologia 67:23–30.

Tenhunen JD, Sala Serra A, Harley PC, Dougherty RL, Reynolds JF. 1990. Factors influencing carbon fixation and water use by Mediterranean sclerophyll shrubs during summer drought. Oecologia 82:381–393.

Terashima I. 1992. Anatomy of non-uniform leaf photosynthesis. Photosyn Res 31:195–212.

Thomas RB, Strain BR. 1991. Root restriction as a factor in photosynthetic acclimation of cotton seedlings grown in elevated CO_2. Plant Physiol 96: 627–634.

Wong SC. 1979. Elevated atmospheric partial pressure of CO_2 and plant growth. I. Interactions of nitrogen nutrition and photosynthetic capacity in C_3 and C4 plants. Oecologia 44:68–74.

Woodrow IE, Berry JA. 1988. Enzymatic regulation of photosynthetic CO_2 fixation in C_3 plants. Ann Rev Plant Physiol Plant Mol Biol 39:533–594.

3. Direct Effects of Elevated CO_2 in Chaparral and Mediterranean-Type Ecosystems

Walter C. Oechel, Steven J. Hastings, George L. Vourlitis, Mitchell A. Jenkins, and Cara L. Hinkson

Atmospheric CO_2 has increased by about 25% since the beginning of the industrial revolution, and is currently rising by about 1.8 ppm per year (Waston et al., 1990). The concentration at the end of the next century will depend both on human use of fossil fuels and the response of natural ecosystems, and could range from 450 ppm CO_2 to more than 800 ppm CO_2 (Waston et al., 1990). Although the specific nature and regional pattern are uncertain, climate change associated with increases in atmospheric CO_2 and other greenhouse-active gases is generally expected (Folland et al., 1992). For much of the contiguous U.S., this may translate to higher temperatures and decreasing soil moisture (Mitchell et al., 1990). In the chaparral of southern California, higher temperatures are likely to result in increased evapotranspiration, and if precipitation does not increase markedly, available soil moisture should decrease (Rind et al., 1990).

In general, plants are more responsive to elevated CO_2 at higher temperatures (Idso et al., 1987), under water stress (Gifford, 1979; Tolley and Strain, 1985; Hogan et al., 1991), in the presence of air-borne pollutants (Coyne and Bingham, 1977; Carlson and Bazzaz, 1982), and under nutrient adequacy (Oberbauer et al., 1986; Larigauderie et al., 1988; Strain, 1992). This indicates that plants in ecosystems which are warm, drought stressed, and moderately fertile should be more responsive to elevated CO_2 than are plants in ecosystems which are cold, wet, and strongly nutrient deficient. Photosynthesis and plant growth in chaparral and related Mediterranean-type ecosystems should therefore be fairly responsive to elevated CO_2.

Significance of Chaparral Ecosystems

Chaparral is important for its wildlife, aesthetic, and recreational values, and it borders on other important ecosystems including desert, montane forest, oak woodlands, and grasslands. Chaparral is a major vegetation type in California, in the southwestern U.S., and in northern Mexico (Keeley and Keeley, 1988). Chaparral and its associated woodland, grassland, and forest ecosystems comprise part of a vegetation form, the Mediterranean-type ecosystems (MTEs), which have worldwide importance and distribution (di Castri, 1981). Mediterranean-type ecosystems are found in North America, the Mediterranean Basin, Chile, Australia, and South Africa. They are often associated with current or historic centers of population and cultural development, and are frequently severely impacted by man (le Houérou, 1981). For these reasons and others, the European Economic Community considers Mediterranean-type ecosystems as the most threatened of the major European ecosystems, and the most in need of further research (LICC Conference, 1989).

Hypothetical Response of Chaparral Ecosystems to Elevated CO_2

The buffering capacity and potentially large carbon sink in woody perennial vegetation suggest that woody vegetation may respond more strongly to elevated CO_2 than herbaceous vegetation (Bazzaz, 1990; Mooney et al., 1991). Potential storage of carbon in the below-ground lignotuber characteristic of chaparral vegetation may provide the sink necessary to allow for long-term carbon storage. We expect chaparral to be responsive to elevated CO_2, since it is water limited (Strain and Bazzaz, 1983; Oechel and Strain, 1985; Mooney et al., 1991). Elevated CO_2 should allow chaparral to stretch its water resource further, with a net positive effect on canopy production (Gifford, 1979; Tolley and Strain, 1985; Hogan et al., 1991). However, some of the reduction in transpiration could be offset by increased leaf area, so that canopy transpiration is affected minimally by elevated CO_2. Symbiotic and asymbiotic nitrogen fixation are expected to increase due to greater carbohydrate supply to N-fixing bacteria (Bentley, 1992). Some homeostatic adjustment of leaf-level photosynthesis is expected; however, it will probably be much less compared to ecosystems which are wetter, colder, and/or more nutrient limited. Although the effects of elevated CO_2 on population and community level dynamics are largely speculative, there may be a shift in relative dominance to species best able to respond to the elevated CO_2 under water stress (Bazzaz, 1990). This includes species with strong sinks for photosynthate, those that strongly limit water use, and those that increase nutrient status through either symbiotic or asymbiotic fixation or increased nutrient use efficiency (Luxmoore, 1981; Norby et al., 1986a).

Photosynthetic Response of Chaparral Plants to Elevated CO_2

Elevated CO_2 appears to initially stimulate photosynthesis in some chaparral plants. In controlled-environment studies conducted over 7.5 months, there was partial homeostatic adjustment of photosynthesis to elevated CO_2 (Figure 3-1). Photosynthetic capacity decreased by about 40–65% for *Adenostoma fasciculatum* and *Ceanothus greggii* seedlings, respectively, after 7 weeks of exposure to double ambient CO_2 (Jenkins, 1993). This indicates that after 7 weeks of treatment, the seedlings were still able to respond to elevated CO_2. Presumably, the rapid growth rate and strong sinks for photosynthate in seedlings allowed use of the higher photosynthate supply. However, after 30 weeks, and in the absence of drought, there was about 80% homeostatic adjustment (Figure 3-1). That is, only 20% of the instantaneous stimulation remained in plants exposed to elevated CO_2 for 7.5 months. We interpret this result to indicate that photosynthate supplies are becoming saturated under conditions of double ambient CO_2 and adequate moisture within 7.5 months of exposure.

In contrast, droughted *C. greggii* plants experienced "up-regulation" under elevated CO_2, while droughted *A. fasciculatum* plants experienced reduced "down-regulation" (Figure 3-1). The relative stimulation in photosynthesis of droughted plants exposed to elevated CO_2 was probably due to decreased leaf transpiration, which stretched the water reserve and allowed higher photosynthetic rates for a longer period of time. In addition, photosynthetic rates were lower at all CO_2 concentrations under water stress, which decreased the photosynthate supply to the plant, and reduced the likelihood that there would be some form of end-product inhibition or feedback at elevated CO_2.

Coast live oak, *Quercus agrifolia*, exhibited a different response to elevated CO_2 in similar controlled-environment studies. In this experiment, drought conditions were established 14 weeks into the experiment by allowing the soil to dry, so at 22 weeks of the exposure, drought conditions had been established for 8 weeks. After 22 weeks of exposure to elevated CO_2, well-watered plants showed increased stimulation to elevated CO_2 over time, while droughted plants experienced lower photosynthetic capacity in the high-CO_2 treatment (Figure 3-2). Interestingly, high CO_2-grown droughted plants experienced greater mortality after 7.5 months of exposure compared to ambient-grown droughted plants. It is conceivable that the high CO_2-grown plants initially accumulated greater

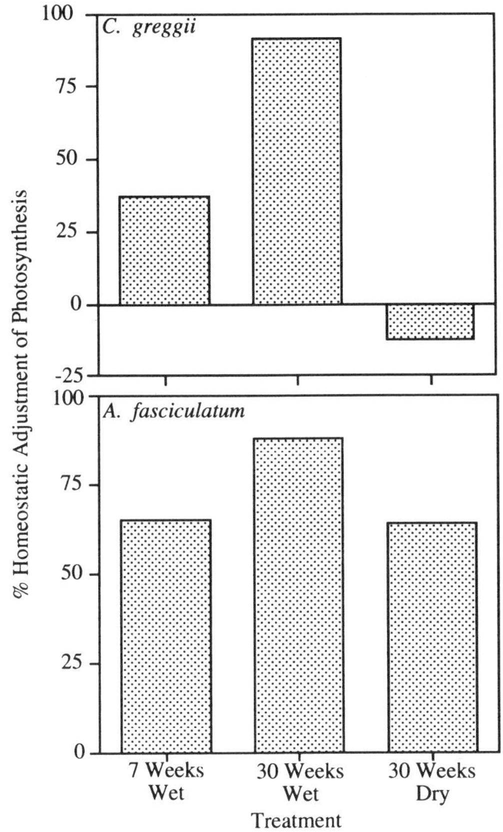

Figure 3-1. Percent homeostatic adjustment of photosynthesis per unit leaf area of high-CO_2 grown *Adenostoma fasciculatum* and *Ceanothus greggii* plants relative to ambient-grown plants after 7 weeks of treatment, and under drought and well-watered conditions 30 weeks after treatment. Percent homeostatic adjustment is calculated as: $[((Ps_i - Ps_a) - (Ps_e - Ps_a))/(Ps_i - Ps_a)]*100$; where Ps_i = initial photosynthesis rate of ambient-grown plants exposed to elevated CO_2, Ps_a = photosynthesis rate of plants grown and measured in ambient CO_2, and Ps_e = photosynthesis rate grown and measured in elevated CO_2.

leaf area and biomass that could not be supported as drought conditions increased. The positive response of watered *Q. agrifolia* may be due to an increase in tissue nitrogen status from asymbiotic nitrogen-fixation (Gluckman and Oechel, 1992; Hinkson and Oechel, in ms.). A number of chaparral shrub and post-fire herbaceous species are symbiotic nitrogen fixers, including *C. greggii* (Zinke, 1969; Kummerow et al., 1978; Poth, 1982), and recent data suggest that *Q. agrifolia* is associated with asymbiotic nitrogen-fixing organisms (Gluckman and Oechel, 1992; Hinkson and Oechel, in ms.).

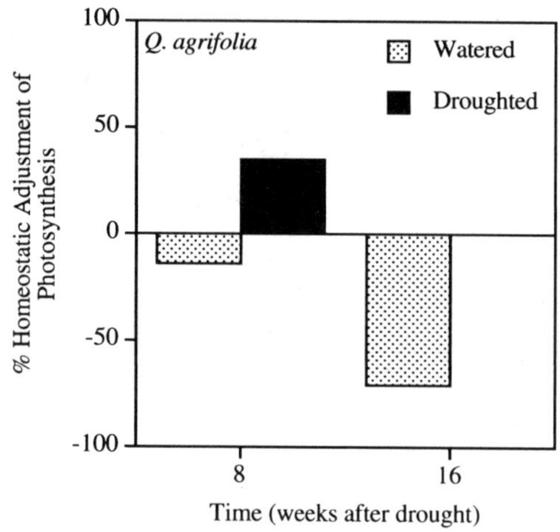

Figure 3-2. Percent homeostatic adjustment of photosynthesis per unit leaf area of high-CO_2 grown *Quercus agrifolia* plants relative to ambient-grown plants under drought (closed bars) and well watered (stippled bars) conditions 8 and 16 weeks after treatment. Droughted plants exposed to elevated CO_2 after 16 weeks of treatment experienced heavy mortality. Percent homeostatic adjustment is calculated as in Figure 3-1.

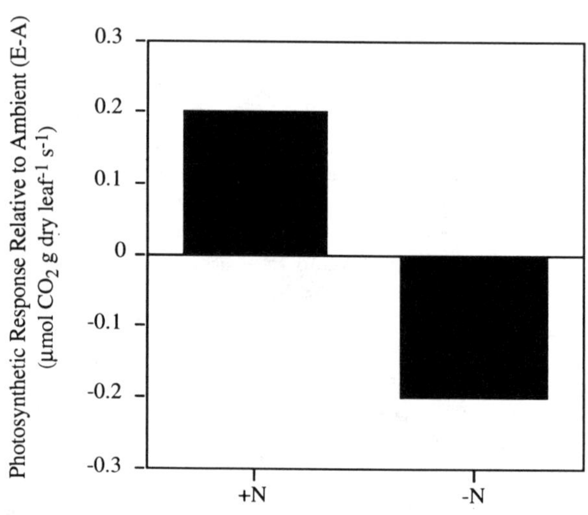

Figure 3-3. Effect of elevated CO_2 and nitrogen availability on the relative photosynthetic response (elevated photosynthesis-ambient photosynthesis) of the C_3 perennial grass *Bromus mollis*. Data are from Larigauderie et al. (1988).

Changes in nitrogen availability will have profound effects on the physiology of Mediterranean-type vegetation. Added nitrogen was found to prolong the enhancement of elevated CO_2 in the C_3 perennial grass *Bromus mollis* (Larigauderie et al., 1988). Under added nitrogen high CO_2 grown plants had higher rates of photosynthesis than ambient-grown plants (Figure 3-3). Under nitrogen-limited conditions, ambient-grown plants had higher rates of photosynthesis than high CO_2-grown plants (Figure 3-3). Because photosynthesis is linearly related to leaf nitrogen concentration (Field and Mooney, 1986), the loss of photosynthetic capacity of high CO_2-grown plants in nitrogen-limited conditions was due primarily to the dilution of leaf tissue nitrogen reserves (Larigauderie et al., 1988). In chaparral, where periodic fire acts to enhance nutrient availability, early successional shrubs and forbs may experience a large stimulation in growth and photosynthesis. This stimulation, however, may decline over time as nutrients become incorporated into biomass, and available nitrogen pools decline.

Effects of Elevated CO_2 on Plant Growth and Allocation

In general, lower transpiration rates and greater water-use efficiency at elevated CO_2 should act to maintain greater leaf area in water-limited areas such as the chaparral (Youngner et al., 1972; Jenkins, 1993). Chaparral shrubs *A. fasciculatum*, *C. greggii*, and *Q. agrifolia* grown at elevated CO_2 show an increase in leaf area under both droughted and watered conditions (Figure 3-4; Gluckman and Oechel, 1992; Jenkins, 1993). The stimulation of leaf area at elevated CO_2 is sufficient to at least partially offset the effects of drought on leaf area development. In fact, after extended drought, the leaf area of plants grown at 750 ppm CO_2 meets (*Q. agrifolia*, *C. greggii*) or exceeds (*A. fasciculatum*) the leaf area of watered plants grown at preindustrial CO_2 levels (Gluckman and Oechel, 1992; Jenkins, 1993).

In chaparral, elevated CO_2, with and without drought, results in greater shoot biomass (Figure 3-4). In the well-watered case, the increased biomass may be the result of early stimulation of growth and photosynthesis. Root growth significantly increased in high-CO_2 chaparral plants (Jenkins, 1993), and in *Q. agrifolia* as well (Gluckman and Oechel, 1992). Because the relative increase was similar for root and shoot growth, root:shoot ratio was not affected by elevated CO_2 as suggested (Bazzaz, 1990). Although allocation patterns remain relatively unchanged in elevated CO_2, larger root systems may allow plants to acquire water from deeper reserves and increase nutrient acquisition by allowing plants to explore a greater volume of soil (Rogers et al., 1994).

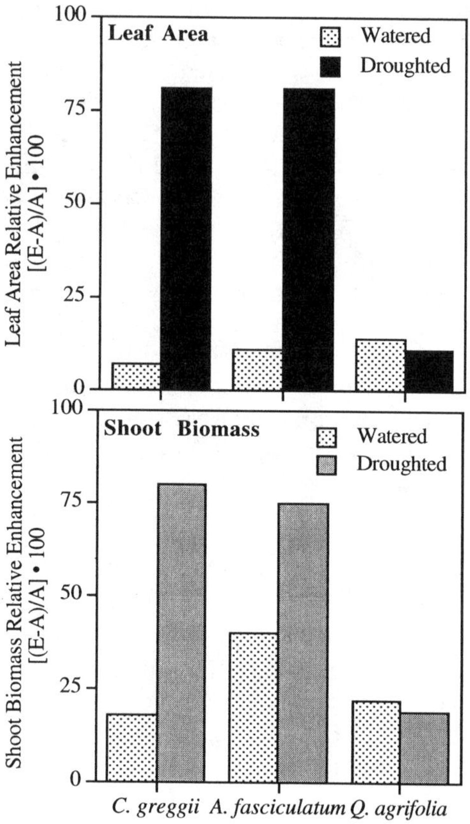

Figure 3-4. Percent increase in leaf area (top panel) and shoot biomass (bottom panel) in high-CO_2 grown *Adenostoma fasciculatum*, *Ceanothus greggii*, and *Quercus agrifolia* plants relative to ambient-grown plants under drought (closed bars) and well-watered (stippled bars) conditions. Relative increase is calculated as: $[(E - A)/A]*100$; where E = leaf area or biomass of elevated CO_2 grown plants and A = leaf area or biomass of elevated CO_2-grown plants.

Effects of Elevated CO_2 on Water Flux

There is little field information on the effects of elevated CO_2 and associated climate change on the water balance of woody perennial ecosystems. In chaparral, atmospheric CO_2 generally increases water-use efficiency (WUE, expressed as mass of C fixed per mass of H_2O transpired) but may either increase or decrease rates of transpiration per unit leaf area. For example, leaf level transpiration of well-watered *A. fasciculatum* and *Q. agrifolia* increased by approximately 16% and 40% and declined by 8% in *C. greggii* plants growing in elevated CO_2 (Figure 3-5;

Jenkins and Oechel, 1991; Gluckman and Oechel, 1992; Jenkins, 1993). WUE, however, increased in all three species, with the greatest increase (ca. 55%) observed in *C. greggii* (Figure 3-5). Under drought conditions, however, transpiration per unit leaf area in *Q. agrifolia* decreased by

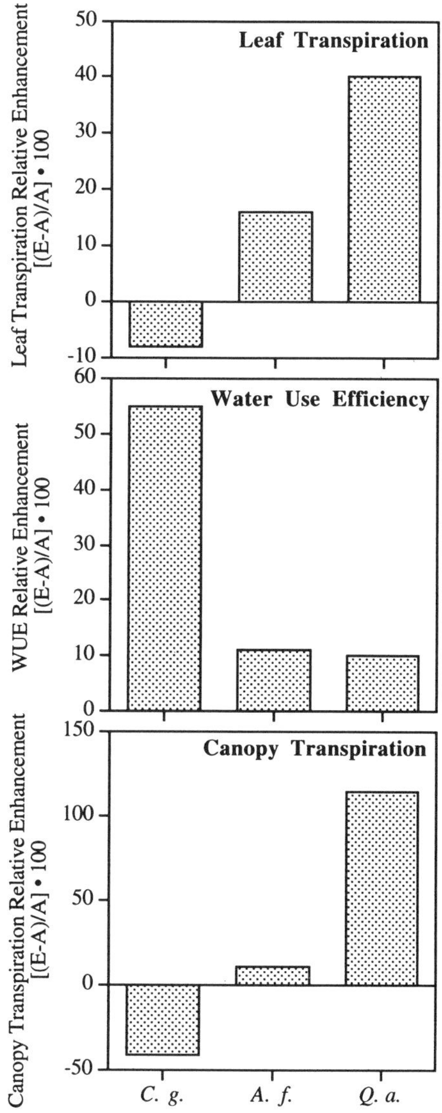

Figure 3-5. Percent change in transpiration per unit leaf area (top panel), water use efficiency (middle panel), and canopy transpiration (bottom panel) in high-CO_2 grown *Adenostoma fasciculatum* (*A.f.*), *Ceanothus greggii* (*C.g.*), and *Quercus agrifolia* (*Q.a.*) plants relative to ambient grown plants under well-watered conditions. Relative increase is calculated as in Figure 3-4.

84%, supporting the hypothesis that plants under significant water stress may be benefited by elevated CO_2 (Bazzaz, 1990; Mooney et al., 1991).

Reduction in transpiration per unit leaf area have the potential to significantly affect leaf energy balance. Decreased stomatal conductance results in elevated leaf temperature and an increase in water vapor pressure inside the leaf (Larcher, 1980; Ehleringer, 1989). As a result, transpiration rate may not be reduced in direct proportion to the decrease in stomatal conductance. In hot environments, which frequently occur during the summer in MTEs, this temperature increase could result in heat stress to leaves (Larcher, 1980; Ehleringer, 1989).

Other factors being equal, greater WUE would decrease stand transpiration in water-limited systems and cause an increase in water yield to streams and reservoirs which would tend to offset the negative effects of predicted decreasing soil moisture availability (Rind et al., 1990). However, the effects of elevated CO_2 on whole plant water loss are not entirely straightforward. The increase in leaf area could offset the savings in water use from greater WUE (Oechel and Strain, 1985; Norby et al., 1986a). For example, canopy transpiration was slightly to markedly increased for *A. fasciculatum* and *Q. agrifolia*, respectively, under elevated CO_2, whereas canopy transpiration in *C. greggii* declined (Figure 3-5). The same pattern was found in *Quercus alba* seedlings exposed to double ambient CO_2 levels (Norby et al., 1986a). This could have a large impact on community composition by increasing the demand for available soil moisture, and by competitive exclusion of plants unable to survive prolonged periods of drought (Bazzaz, 1990). Increased canopy leaf area and water use coupled with decreased precipitation and/or increased temperatures could result in decreased water yields to streams and reservoirs, thereby reducing the total amount of water available for aquatic plant and animal species and for human populations.

Effects of Elevated CO_2 on Population and Community Level Dynamics

Little is known regarding the effects of elevated CO_2 on population and community dynamics in MTEs. Because chaparral ecosystems are largely water limited, the effects of elevated CO_2 on canopy water use and soil moisture levels should be reflected in community level dynamics. Experiments with successional species indicate that community composition is greatly influenced by CO_2 concentration and soil moisture levels (Bazzaz and Carlson, 1984; Tolley and Strain, 1984, 1985; Wray and Strain, 1986). In mixed species studies with the old field successional perennials *Aster pilosus* (C_3) and *Andropogon virginicus* (C_4), Wray and Strain (1986) found that under drought stress and elevated CO_2, the C_3 *Aster* was able to outcompete the C_4 perennial *Andropogon*. The majority of the chaparral species are C_3 plants, which are expected to respond to a combination of drought stress and CO_2 (Oechel and Strain, 1985). On the basis of their individual responses in WUE and whole canopy water loss to elevated CO_2 alone, one would expect *C. greggii* to increase in abundance and *A. fasciculatum* to decline in the community (see Figure 3-5). However, individual species response to elevated CO_2 cannot necessarily be extrapolated to infer community level responses (Williams et al., 1986), because of competition from interacting species in the community, and because the potential for species to respond to elevated CO_2 may be constrained by other traits that enable them to survive in stressful environments (Bazzaz, 1990).

Differential growth responses to elevated CO_2 may interact with physiological responses and either mask or accentuate plant competitive abilities and interactions (Bazzaz, 1990). For example, initial differences in growth responses of ambient- and high CO_2 grown plants may be observed long after homeostatic adjustment or "down-regulation" is complete. Differential above-ground growth may lead to shading, which could differentially increase the competitive ability of shade-tolerant plants (Bazzaz et al., 1990). Greater root biomass will also affect competitive dynamics below ground, leading to differential capability of both nutrient and water acquisition, and possible effects on growth and community structure (Norby et al., 1986a,b).

Many chaparral plants reproduce from resprouts initiated from a subsurface lignotuber (Hanes, 1971). The below-ground root system and lignotuber represent a significant sink for fixed C (Sparks et al., 1993) and may allow for the long-term utilization and storage of additional carbohydrate produced under elevated CO_2. This large C sink may benefit resprouting species relatively more than obligate seeders, especially if climate change associated with increased atmospheric CO_2 results in soil drying from increased evapotranspiration. Although seed size is expected to increase in an elevated CO_2 environment due to an increase in C allocation to reproduction (Bazzaz, 1990), soil surface drying may then inhibit seed germination and establishment, and thus feed back on the proportion of resprouting and obligate seeding species in the community.

Ecosystem Level Effects of Elevated CO_2

The many feedback between elevated CO_2, ecosystem physiology, and nutrient availability are complex, however, changes in nutrient availability will undoubtedly be one of the most important variables in determining the response of natural ecosystems to elevated CO_2 (Bazzaz, 1990; Mooney et al., 1991; Oechel and Billings, 1992; Shaver et al., 1992). Changes in plant and litter C:N ratios are expected in an elevated CO_2 environment (Norby et al., 1986a), and this change is expected to significantly affect nutrient cycling (Figure 3-6A; Melillo et al., 1990; Mooney et al., 1991). Increased CO_2 is expected to result in litter that is Carbon and Nitrogen poor (Melillo et al., 1990), and as a result, soil decomposition and nutrient mineralization are expected to occur at slower rates (Strain and Bazzaz, 1983; Williams et al., 1986). For chaparral, lower mineralization rates should reduce the initial photosynthetic and growth stimulation to elevated CO_2 (Jenkins and Oechel, 1991; Gluckman and Oechel, 1992; Jenkins, 1993).

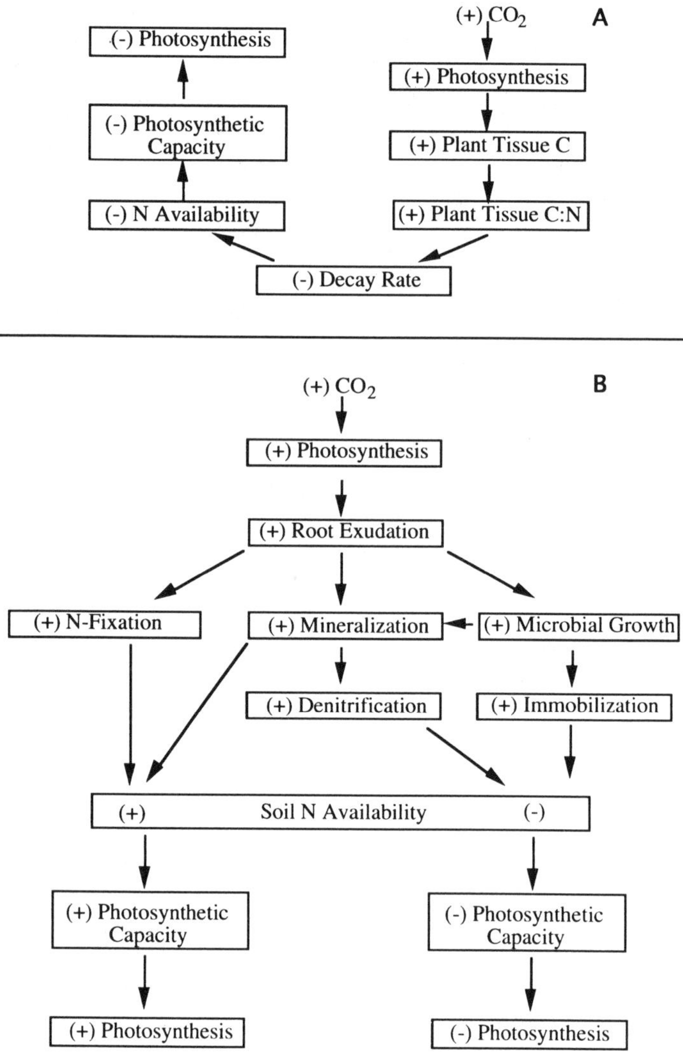

Figure 3-6. Hypothetical effects of elevated CO_2 on litter quality, nutrient cycling, and feedbacks on ecosystem productivity.

In contrast, work by Norby et al. (1986a) suggests that litter decomposition rates may not be significantly altered at elevated CO_2, as the forms of carbon in litter from high CO_2 grown plants may be richer in soluble sugars and less in lignin content, and therefore more labile and more easily metabolized by soil microorganisms. In addition, increased photosynthate produced under elevated CO_2 should increase root exudation (Luxmoore, 1981), resulting in greater microbial growth, mineralization, and nitrogen fixation (Figure 3-6B; Bentley, 1992). This could lead to increased nitrogen availability through nitrogen fixation, or may act to increase internal cycling of nitrogen through mineralization (Figure 3-6B). However, increases in microbial growth could lead to increased immobilization (Figure 3-6B), especially if litter quality decreases (Mellilo et al., 1990), and could lead to increased denitrification during the rainy winter and spring seasons (Sprent, 1987). This would result in lower nitrogen availability, which would lead to a concomitant decrease in photosynthetic rates (Figure 3-6B). Plants may be able to compensate for decreased nitrogen availability with increased nutrient-capture efficiency and nutrient-use efficiency at elevated CO_2 (Luxmoore, 1981; Norby et al., 1986a) which may increase plant productivity and plant competitive ability.

Both symbiotic and asymbiotic nitrogen fixation has been demonstrated in a number of chaparral shrubs and post-fire herbaceous species (Zinke, 1969; Kummerow et al., 1978; Poth, 1982; Ellis and Kummerow, 1988), and in *Q. agrifolia* (Gluckman and Oechel, 1992; Hinkson and Oechel, in ms.). Increases in photosynthate supply to the roots could result in greater exudation into the rhizosphere, higher nitrogen fixation (Luxmoore, 1981), and increased mycorrhizal infection (Strain and Bazzaz, 1983), which could increase both nutrient availability and capture. In *Q. agrifolia*, nitrogen fixation (determined from $\delta^{15}N$ natural abundance techniques; Marion, 1987) increased with increasing CO_2 growth concentration (Gluckman and Oechel, 1992; Hinkson and Oechel, in ms.). This would tend to offset processes reducing nitrogen availability with increased CO_2 supply, and might result in substantial growth and photosynthetic response in unmanaged MTEs, including those of southern California.

The effects of elevated CO_2 on net ecosystem CO_2 flux are largely unknown in MTEs. Experiments using controlled environment chambers (Oechel et al., 1992) in southern California chaparral are currently under way. On one hand, whole ecosystem CO_2 flux may be relatively unaffected if nutrient availability or nutrient use efficiency does not increase. Under drought conditions, which occur frequently during the summer months, elevated CO_2 may allow for greater carbon acquisition relative to previous (e.g., preindustrial) rates for any given soil moisture availability. These conclusions are based on individual species response to elevated CO_2 in short-term growth-chamber studies (Jenkins and Oechel,

1991; Jenkins, 1993), and it is at best problematic to extrapolate these results to whole ecosystems and over longer time scales.

Effects of Climatic Change Associated with Elevated CO_2

Changes in climate will obviously tend to alter competitive interaction, stand composition, and the position of vegetation boundaries between various ecosystems (ecotones). However, it is not a simple task to predict the rate of change or even the full nature of the change in these attributes. Tolerance to drought, frost, and air pollution may all change at elevated CO_2 (Oechel and Strain, 1985). Fire frequency and intensity are likely to be significantly affected by the prevalence of climatic conditions that lead to fire as well as elevated CO_2. Stand flammability is closely tied to fuel accumulation (Riggan et al., 1988). Fuel accumulation is, in turn, affected by rates of primary productivity (Riggan and Dunn, 1982; Tyrrel, 1982). Other factors being equal, greater productivity resulting from higher atmospheric CO_2 would tend to increase stand flammability and fire frequency. Stand flammability and fire frequency would be further exaggerated if, in addition to higher productivity, climatic conditions changed, it increasing the number of days conducive to fire (i.e., days with low fuel moisture, high winds, and/or low relative humidity; Tyrrel, 1982).

Warmer temperatures will undoubtedly interact with elevated CO_2 to affect the water and energy balance of chaparral plants. As mentioned earlier, lower rates of transpiration per unit leaf area in an elevated CO_2 environment may act to increase the heat load of leaves, leading to heat stress (Larcher, 1980; Ehleringer, 1989), and elevated temperatures could act to accentuate this stress (Mooney et al., 1991). The direct effects of elevated temperatures on photosynthesis per unit leaf area are not expected to be large, however, as the temperature optimum for photosynthesis in many chaparral plants is quite broad (Oechel et al., 1981), unless increased heat load raises leaf temperature well above the optimum range.The most significant effects of climate change will likely be through changes in the patterns, amount, and seasonality of precipitation (Folland et al., 1992), and through the effects of temperature on evapotranspiration and soil moisture, decomposition of organic matter, and nutrient mineralization (Marion et al., 1981).

Conclusions

The effects of elevated CO_2 on plants from MTEs are variable and are strongly dependent upon the levels of other resources, such as nutrients and water. Photosynthetic acclimation to elevated CO_2 in southern California chaparral species resulted in a down-regulation of photosynthetic capacity within 8 months, possibly due to the effects of low nutrient availability on growth and the lack of available sinks for the excess carbohydrate fixed under elevated CO_2. For the coast live oak, associations with asymbiotic nitrogen-fixing bacteria may have alleviated nitrogen stress, allowing the initial stimulatory effects of elevated CO_2 on photosynthesis to be prolonged. This initial stimulation of photosynthesis resulted in relatively greater biomass production; however, proportional allocation of fixed carbon to root tissues was not affected.

WUE increased due to elevated CO_2; however, transpiration per unit leaf area either decreased slightly or increased in high CO_2 grown plants. In addition, increases in leaf area resulted in greater whole plant water use in two of the three species studied, indicating that the hypothesized increase in water yield from chaparral watersheds under elevated CO_2 may be limited.

The effects of elevated CO_2 were relatively greater for droughted plants than for well-watered plants. In the case of *Q. agrifolia*, the interactive effects of elevated CO_2 and drought resulted in increased mortality, while droughted *C. greggii* and *A. fasciculatum* benefited from higher CO_2 concentrations. However, the effects of elevated CO_2 coupled with temperature rise are largely unknown for plants adapted to these ecosystems. Higher WUE and leaf temperature will likely affect leaf energy balance, increasing the heat load of leaves. One scenario may be that the increased heat load will feed back to increase transpiration and subsequent water loss. During the summer months, when soil moisture availability is low, the change in energy balance will make plants more susceptible to heat stress.

Overall, the effects of elevated CO_2 on energy balance at the macroscale are largely speculative. Over the landscape and regional scales, changes in latent and sensible heat fluxes (*Le* and *H*, respectively) may act to feed back on regional water yield, and possibly even climate. If canopy level *Le* decreases due to the direct effects of elevated CO_2, water yield should increase. However, this scenario is unlikely, as whole canopy water loss is relatively unaffected by elevated CO_2, and depending on the direction of climate change, ecosystem evapotranspiration may actually increase (Rind et al., 1990). Plant species which are susceptible to desiccation may decline in abundance due to elevated *H* and the resulting heat load at the leaf level. Depending on the large-scale climatic changes, local climate patterns could be significantly affected by the energy balance of MTEs. Changes in *Le* can affect local weather patterns, which can feed back on radiation and energy fluxes (Groisman et al., 1994). With the use of fast-response sensors and eddy correlation methodology, changes in large-scale CO_2, H_2O, and energy fluxes can be monitored. These methods will be undoubtedly provide insight into the questions raised above, and should be implemented in future flux sampling programs in MTEs.

Community and ecosystem level effects have not been studied. At present, we can only infer from individual species responses for predictions regarding the effects of elevated CO_2 on population dynamics and community structure. In general, plants that are more responsive to elevated CO_2 under drought conditions, and plants that have associations with N-fixing bacteria, are likely to gain more prominence in the chaparral community. Ecosystem level attributes such as the rate of nutrient cycling are likely to be altered through changes in litter quality. Greater rates of denitrification and nitrogen immobilization may further limit nitrogen availability and the positive effects of nitrogen fixation. Plants that are more efficient at acquiring nutrients or more efficient in nutrient use are likely to be more dominant in the community. At present, predictions regarding the effect of elevated CO_2 on net ecosystem carbon balance are largely speculative. Whole ecosystem flux may be relatively unaffected if nutrient availability or nutrient use efficiency does not increase significantly, especially during the wet winter and spring seasons. During the drier summer period, elevated CO_2 may allow for greater carbon acquisition relative to previous (e.g., preindustrial) rates.

The need for long-term, in situ CO_2 manipulations has never been more acute. Presently, controlled environment greenhouses (Oechel et al., 1992) are being used to manipulate CO_2 concentration from preindustrial to double ambient (750 ppm) levels in 100-ppm increments. This experiment was recently initiated on a chaparral stand regenerating following a recent prescribed burn, and is expected to run for 10 years. In addition, a FACE (Free Air CO_2 Enrichment) facility is being initiated in the southern California chaparral and adjacent oak woodland community. Elevated CO_2 springs in Italy are providing information on the long-term effects of elevated CO_2 on intact MTEs. With these whole ecosystem experiments, the community and ecosystem level effects of elevated CO_2 will be addressed directly.

References

Bazzaz FA. 1990. The response of natural ecosystems to the rising global CO_2 levels. Ann Rev Ecol Syste 21:167–196.

Bazzaz FA, Carlson RW. 1984. The response of plants to elevated CO_2. I. Competition among an assemblage of annuals at two levels of soil moisture. Oecologia 62:196–198.

Bazzaz FA, Colman JS, Morse SR. 1990. Growth responses of seven major co-occurring tree species of the northeastern United States to elevated CO_2. Can J For Res 20:1479–1484.

Bentley B. 1992. The effects of increased atmospheric CO_2 on *Lupinus* and its herbivores. Am J Bot Supple 79:59–60.

Carlson RW, Bazzaz FA. 1982. Photosynthetic and growth response to fumigation with SO_2 at elevated CO_2 for C_3 and C_4 plants. Oecologia 54:188–193.

di Castri F. 1981. Mediterranean shrublands of the world. In: di Castri F, DW Goodall, and RL Specht (eds). Ecosystems of the World 11: Mediterranean-type Shrublands. Elsevier Scientific Publishing Co., Amsterdam, The Netherlands.

Coyne PI, Bingham GE. 1977. Carbon dioxide correlation with oxidant air pollution in the San Bernardino mountains of California. J Air Pollution Control Assoc 27:782–784.

Ehleringer J. 1989. Temperature and energy budgets. Pages 117–136 in Pearcy RW, J Ehleringer, HA Mooney, and PW Rundel (eds). Plant Physiological Ecology: Field Methods and Instrumentation. Chapman and Hall, New York.

Ellis BA, Kummerow J. 1988. $N_2(C_2H_2\text{-}C_2H_4)$ fixation in two species of *Ceanothus* seedlings in second year postfire chaparral. Plant Soil 109:207–213.

Field CB, Mooney HA. 1986. The photosynthesis-nitrogen relationship in wild plants. In: Givnish TJ (ed). On The Economy of Plant Form and Function. Cambridge University Press, Cambridge, UK.

Folland CK, Karl TR, Nicholls N, Nyenzi BS, Parker DE, Vinnikov K Ya. 1992. Observed climate variability and change. Pages 135–170 in Houghton JT, BA Callander, and SK Varney (eds). Climate Change 1992: The Supplementary Report to the IPCC Scientific Assessment. WMO/UNEP, Cambridge University Press, New York.

Gifford RM. 1979. Growth and yield of CO_2-enriched wheat under water-limited conditions. Aust J Plant Physiol. 6:367–369.

Gluckman CH, Oechel WC. 1992. The effects of atmospheric CO_2 and drought on the nutrient relations of *Quercus agrifolia*. 43rd AIBS Annual Meeting, Honolulu, Hawaii, 9–13 August, 1992.

Groisman PY, Karl TR, Knight RW. 1994. Observed impact of snow cover on the heat balance and the rise of continental spring temperatures. Science 263: 198–200.

Hanes TL. 1971. Succession after fire in the chaparral of southern California. Ecol Monogr 41:27–52.

Hinkson CL, Oechel WC. Asymbiotic nitrogen fixation in association with *Quercus agrifolia* and responses to elevated atmospheric CO_2 (in manuscript, n.d.).

Hogan KP, Smith AP, Zizka LH. 1991. Potential effects of elevated CO_2 and changes in temperature on tropical plants. Plant Cell Environ 14:763–778.

Le Houérou HN. 1981. Man and ecosystems. In: di Castri F, DW Goodall, and RL Specht (eds). Ecosystems of the World 11: Mediterranean-type Shrublands. Elsevier Scientific Publishing Co., Amsterdam, The Netherlands.

Idso SB, Kimball BA, Anderson MG, Mauney JR. 1987. Effects of atmospheric CO_2 enrichment on plant growth: the interactive role of air temperature. Agric Ecosystems Environ 20:1–10.

Jenkins MA. 1993. Effect of atmospheric CO_2 level and water stress on growth and physiology of two species of chaparral shrubs. MSc. Thesis. San Diego State University, San Diego, CA.

Jenkins MA, Oechel WC. 1991. Effect of atmospheric CO_2 level and water stress on growth and physiology of two species of chaparral shrubs. 42nd Annual Meeting of the Ecological Society of America, San Antonio, Texas. 4–8 August, 1991. Abstract Bull Ecol Soc Amer 71:154.

Keeley J, Keeley S. 1988. Chaparral. In: Barbour MG and WC Billings (eds). North American Terrestrial Vegetation. Cambridge University Press.

Kummerow J, Krause D, Jow W. 1978. Seasonal changes of fine root density in the southern California chaparral. Oecologia 37:201–212.

Larcher W. 1980. Plant Physiological Ecology, 2nd edition. Springer-Verlag, Berlin, Germany. 303 pages.

Larigauderie A, Hilbert DW, Oechel WC. 1988. Effect of CO_2 enrichment and nitrogen availability on resource acquisition and resource allocation processes in a grass, *Bromus mollis*. Oecologia 77:544–549.

Luxmoore RJ. 1981. CO_2 and phytomass. BioScience 31:626.

Marion GM. 1987. Use of nitrogen-15 to assess terrestrial nitrogen cycling processes. Pages 129–155 in Tenhunen JD (ed). Plant Response to Stress. Springer-Verlag, Berlin.

Marion GM, Kummerow J, Miller PC. 1981. Predicting nitrogen mineralization in chaparral soils. Soil Sci Soc Am J 45:956–961.

Melillo JM, Callaghan TV, Woodward FI, Salati E, Sinha SK. 1990. Effects on Ecosystems. In: Houghton JT, GJ Jenkins, and JJ Ephraums (eds). Climate Change, The IPCC Scientific Assessment. Cambridge University Press, Cambridge, UK.

Mitchell JFB, Manabe S, Meleshko V, Tokioka T. 1990. Equilibrium climate change and its implications for the future. In: Houghton JT, GJ Jenkins, and JJ Ephraums (eds). Climate Change, The IPCC Scientific Assessment. Cambridge University Press, Cambridge, UK.

Mooney HA, Drake BG, Luxmoore RJ, Oechel WC, Pitelka LF. 1991. Predicting ecosystem responses to elevated CO_2 concentrations. BioScience 41:96–104.

Norby RJ, O'Neill EG, Luxmoore RJ. 1986a. Effects of atmospheric CO_2 enrichment on the growth and mineral nutrition of *Quercus alba* seedlings in nutrient-poor soil. Plant Physiol 82:83–89.

Norby RJ, Pastor J, Melillo JM. 1986b. Carbon-nitrogen interactions in CO_2-enriched white oak: physiological and long-term perspectives. Tree Physiol 2:233–241.

Oberbauer SF, Sionit N, Hastings SJ, Oechel WC. 1986. Effects of CO_2 enrichment and nutrition on growth, photosynthesis and nutrient concentration of Alaskan tundra species. Can J Bot 64:2993–2998.

Oechel WC, Billings WD. 1992. Anticipated effects of global change on carbon balance of arctic plants and ecosystems. Pages 139–168 in Chapin FS III, R Jefferies, J Reynolds, G Shaver, and J Svoboda (eds). Arctic Physiological Processes in a Changing Climate. Academic Press, New York.

Oechel WC, Strain B. 1985. Response of native species and ecosystems. In: Direct Effects of Increasing Carbon Dioxide on Vegetation. Department of Energy DOE/ER-0238.

Oechel WC, Lawrence W, Mustafa J, Martínez J. 1981. Energy and carbon acquisition. Pages 151–183 in Miller PC (ed). Resource Use by Chaparral and Matorral. Springer-Verlag, New York.

Oechel WC, Riechers G, Lawrence WT, Prudhomme TI, Grulke N, Hastings SJ. 1992. "CO_2LT", A closed, null-balance system for long-term *in situe* ecosystem manipulation and measurement of CO_2 level, CO_2 flux, and temperature. Functional Ecol. 6:86–100.

Poth M. 1982. Biological dinitrogen fixation in chaparral. Pages 285–290 in Conrad CE and WC Oechel (eds). Proceedings of the Symposium on Dynamics and Management of Mediterranean-type Ecosystems (June 22–26, 1981). Pacific Southwest Forest and Range Experiment Station, Berkeley, CA.

Riggan PJ, Dunn PH. 1982. Harvesting chaparral biomass for energy—an environmental assessment. Pages 149–157 in Conrad CE and WC Oechel (eds). Proceedings of the Symposium on Dynamics and Management of Mediterranean-type Ecosystems, San Diego State University, San Diego, CA.

Riggan PJ, Goode S, Jacks PM, Lockwood RN, 1988. Interaction of fire and community development in chaparral of southern California. Ecol Monog 58(3):155–176.

Rind D, Goldberg R, Hansen J, Rosenzweig C, Ruedy R. 1990. Potential evapotranspiration and the likelihood of future drought. J Geophysical Res 95:9983–10004.
Rogers HH, Runion GB, Krupa SV. 1994. Plant responses to atmospheric CO_2 enrichment with emphasis on roots and the rhizosphere. Environ Pollut 83: 155–189.
Shaver GR, Billings WD, Chapin FS III, Giblin AE, Nadelhoffer KJ, Oechel WC, Rastetter EB. 1992. Global change and the carbon balance of arctic ecosystems. BioScience 42:433–441.
Sparks SR, Oechel WC, Mauffete Y. 1993. Photosynthate allocation patterns along a fire-induced age sequence in two shrub species from the California chaparral. Int J Wildland Fire 3:21–30.
Sprent JI. 1987. The Ecology of the Nitrogen Cycle. Cambridge University Press, Cambridge, UK. 151 pages.
Strain BR. 1992. Field measurements of CO_2 enhancement and change in natural vegetation. Water Air Soil Pollut 64:45–60.
Strain BR, Bazzaz F. 1983. Terrestrial plant communities. Pages 177–222 in Lemon ER (ed). CO_2 and Plants: The Response of Plants to Rising Levels of Atmospheric Carbon Dioxide. AAAS Selected Symposium 84. Westview Press, Inc., Boulder, CO.
Tolley LC, Strain BR. 1984. Effects of CO_2 enrichment on growth of *Liquidambar styraciflua* and *Pinus taeda* seedlings under different irradiance levels. Can J For Res 14:343–350.
Tolley LC, Strain BR. 1985. Effects of CO_2 enrichment and water stress on gas exchange of *Liquidambar styraciflua* and *Pinus taeda* seedlings grown under different irradiance levels. Oecologia 65:166–172.
Tyrrel RK. 1982. Chaparral in southern California. Pages 56–59 in Conrad CE and WC Oechel (eds). Proceedings of the Symposium on Dynamics and Management of Mediterranean-type Ecosystems, San Diego State University, San Diego, CA.
Watson RT, Rodhe H, Oeschger H, Siegenthaler U. 1990. Greenhouse gases and aerosols. In: Houghton JT, GJ Jenkins, and JJ Ephraums (eds). Climate Change, The IPCC Scientific Assessment. Cambridge University Press, Cambridge, UK.
Williams WE, Garbutt K, Bazzaz FA, Vitousek PM. 1986. The response of plants to elevated CO_2: IV. Two deciduous-forest tree communities. Oecologia 69: 454–459.
Wray SM, Strain BR. 1986. Response of two old field perennials to interactions of CO_2 enrichment and water stress. Am J Bot 73:1486–1491.
Youngner VB, Kesner WD, Berg AR, Green LR. 1972. Progress report of the Maloney Canyon greenbelt project. Contract No. 140. University of California Water Resources Center.
Zinke PJ. 1969. Nitrogen storage of several California forest soil-vegetation systems. Pages 40–53 in Biology and Ecology of Nitrogen. National Academy of Sciences, Washington, DC.

4. Biomass Partitioning and Resource Allocation of Plants from Mediterranean-Type Ecosystems: Possible Responses to Elevated Atmospheric CO_2

David W. Hilbert and Josep Canadell

Predicting the effects of global change on Mediterranean-type ecosystems requires an understanding of the direct effects of elevated atmospheric CO_2 on plant growth and development. While the immediate effects of CO_2 on leaf photosynthesis are well known, these effects can lead to changes in allocation patterns and other properties at the whole-plant level (Strain, 1985) that are not well known or understood. For example, elevated CO_2 can modify leaf photosynthetic capacity and alter gross morphology (Larigauderie et al., 1988), influence interactions among plants (Bazzaz and Carlson, 1984), and alter net ecosystem carbon exchange with the atmosphere (Hilbert et al., 1987). This chapter discusses the importance of allocation in determining whole-plant responses to CO_2, reviews some of the general characteristics of woody plants from Mediterranean-type ecosystems that may influence their sensitivity to increased CO_2, and applies some modeling approaches to consider how biomass partitioning and resource allocation in Mediterranean plants may influence their responses to elevated CO_2.

At first sight, the question of plant response to elevated CO_2 is quite simple. Since carbon is acquired through photosynthesis and photosynthetic rate is known to respond to increasing CO_2 concentration at the site of fixation, one would expect that increasing atmospheric CO_2 concentrations should elevate the photosynthetic rate and consequently stimulate plant growth. Indeed, many studies have demonstrated increased growth rates of several plant species when grown under higher than ambient CO_2 concentrations (see reviews by Kimball, 1983; Oechel and Strain, 1985). On the other hand, growth is not always stimulated by CO_2 (Oberbauer et al., 1986; Williams et al., 1986), the degree of stimulation varying depending on plant species and growth conditions (Cure, 1985), and many plant properties appear to change when plants are grown for long periods at elevated CO_2. These changes in plant characteristics are observed at many levels of organization, from the biochemical to the whole-plant level, and greatly affect the integrated growth response that is observed in any particular experimental condition. Understanding these responses is a major challenge that must be met in order to arrive at a general answer to the question of how plant growth will response to increasing CO_2 concentrations. The study of resource allocation, both within and among organs, is particularly important if we are to understand and predict these responses.

Importance of Allocation

During the past two decades, plant ecophysiology has been very successful in describing the controls on leaf-level processes such as photosynthesis and transpiration (Mooney and Chiariello, 1984; Mooney, 1987; Pearcy et al., 1987), and progress has been made on the physiological behavior of roots (e.g., Schrader, 1978; Weatherley, 1982; Clarkson, 1985). Thus, we now have a reasonably good understanding of how plants acquire carbon, nitrogen, and water and the environmental controls on rates of assimilation of these basic resources. While many practical difficulties still limit our ability to measure and predict root activity, many ecophysiologists now stress the importance of whole-plant studies that integrate physiological knowledge at the organ level with the controls on resource allocation (Mooney and Chiariello, 1984; Ehleringer et al., 1986). One difficulty is that when we are able to predict the rates of resource acquisition, our ability to predict the allocation of these resources to plant parts or processes is still limited. Consequently, plant growth in response to patterns of resource availability (elevated CO_2 is one example) is poorly understood and difficult to predict. While an understanding of organ-level processes is necessary for prediction of whole-plant behavior, it is not sufficient, and this is why understanding allocation is critical. Put simply, photosynthetic rates alone often are not correlated with growth rates because of the importance of carbon allocation to subsequent carbon gain (see discussions by Monsi, 1968; Ledig, 1969; Mooney, 1972; Wilson, 1975; Charles-Edwards, 1982; Loomis, 1983; Mooney, 1987). Furthermore, long-term adjustments in physiological processes, also mediated by allocation patterns, need to be predicted.

The uptake and allocation of nitrogen and other nutrients is just as essential as that of carbon. Nitrogen and carbon, in particular, are linked through the strong relationship between leaf nitrogen content and light-saturated photosynthetic rate (Field, 1983; Field and Mooney, 1983, 1986). It is known that plants vary their photosynthetic capacity and, consequently, their shoot specific activity (σ_s, $g\,C\,g^{-1}\,shoot\,d^{-1}$), in response to resource availability in the environment. At the chemical level, this seems to be due to changes in the amount of RuBPcarboxylase and electron transport components in the leaves which are affected by growth light level (Björkman, 1981) and nitrogen nutrition (Medina, 1971; von Caemmerer and Farquhar, 1981). Similarly, rates of nutrient assimilation can be affected by carbon and nitrogen availability in the root. Thus, specific activity of roots (σ_r, $g\,N\,g^{-1}\,root\,d^{-1}$) may be modifiable by adjustments in resource allocation. Plant water relations also influence greatly the physiological activity of organs and the partitioning of biomass between roots and shoots. Givnish (1986) presents a detailed analysis of the interrelationships among photosynthesis, water use, and allocation to absorbing roots.

Patterns of biomass partitioning and resource allocation to roots and shoots, therefore, are critical in determining the physiological efficiency, relative growth rate, ultimate size, and reproductive effort of plants. In fact, changing allocation patterns may be one of the most effective means by which plants deal with environmental stresses (Chapin et al., 1987) and one of the main challenges in understanding growth of plants is the quantitative description of this partitioning of growth between roots and shoots (Johnson, 1985). A comprehensive understanding of allocation and its controls would significantly increase our basic understanding of the mechanisms by which plants maximize their resource acquisition and greatly improve our ability to predict responses to global change.

The Balanced Activity Hypothesis

A useful theoretical approach to allocation is the balanced activity hypothesis (Brouwer, 1962), which states that the assimilation of carbon by the shoot is kept in balance with the uptake of nitrogen by the roots. It was quantified by Davidson (1969) as

$$\sigma_s W_s = \lambda \sigma_r W_r \tag{4-1}$$

where σ_s is the shoot specific activity (net carbon uptake per unit shoot weight), σ_r is the root specific activity (usually defined as nitrogen uptake per unit root weight), W_s and W_r are the weight of shoots and roots, respectively, and λ is a proportionality constant (the C:N ratio in dry matter). Balanced activity is assumed to be maintained through changes in allocation to shoot and root biomass. Changing specific activities can also balance root and shoot activity, but, unfortunately, this hypothesis cannot predict precisely the degree to which specific activity changes simultaneously with altered biomass allocation. The balanced activity hypothesis has been used as the basis of a variety of plant growth models (Thornley, 1972, 1976; Reynolds and Thornley, 1982; Johnson, 1985; Mäkelä and Sievänen, 1987; Hilbert and Reynolds, 1991). The hypothesis and models have been applied to observations made with several herbaceous plant species with reasonable success (Charles-Edwards, 1976; Cooper and Thornley, 1976; Edwards and Barber, 1976; Raper et al., 1977; Richards et al., 1979), and a recent review of these approaches (Wilson, 1988) concludes that the general approach, epitomized by Thornley's work, is consistent with the majority of observed root:shoot allocation patterns, while more tests of the hypothesis are needed.

Whether the balanced activity theory may be applied to growth forms other than herbs has received little attention, although general validity has often been assumed (Borchert, 1973; Reynolds, 1986). Some recent work with tree seedlings has built on the balanced activity concept (Ågren and Ingestad, 1987). They suggest that partitioning follows directly from the requirement for a balance between carbon assimilating structures and carbon utilization, which is determined by nitrogen nutrition. In herbaceous plants, 90% or more of the root system is made up of absorbing fine roots, subject to more or less intensive turnover. In contrast, more than 90% of the root biomass of woody perennials is in the form of perennial roots, having substantial secondary growth and limited nutrient and water uptake capacities. Only the remaining 10% or so of root biomass consists of fine roots which are active in uptake and subject to seasonal turnover. Similar differences between herbaceous species and woody perennials occur in above-ground structure and function, where woody stems are analogous to structural woody roots and leaves to fine roots.

Establishing the validity of the balanced activity principle for woody chaparral shrubs (or trees and shrubs of any other ecosystem) would greatly increase our understanding of growth dynamics in ecosystems dominated by woody plants. Furthermore, since fine-root biomass and fine-root surface area are closely correlated (Kummerow et al., 1988) it might be feasible to predict the effect of environmental factors on fine root surface area based on knowledge of leaf area changes. Temporal deviations from balanced function may of course occur due to adjustments in shoot or root specific activities and in functional shoot or root biomass, or both simultaneously. Thus, a balance in function may only be apparent by integration of activity over a relatively long period, especially in Mediterranean-climate regions where there are large annual environmental fluctuations. Finally, large structural and storage organs such as lignotubers complicate the application of the balanced activity hypothesis to woody plants. For example, storage of nutrients and water may buffer the internal resource pool of the plant against temporal changes in organ specific activities. Another problem is that conducting tissues in woody structure may directly limit leaf water use and consequently carbon gain.

Allocation Patterns in Mediterranean-Type Ecosystems

Mediterranean-type ecosystems have been characterized by a period of plant water stress, soil mineral deficiencies, and—especially in the Californian, south African, and south Australian ecosystems—by a high fire frequency (Mooney et al., 1981; Kruger et al., 1983; Booysen and Tainton, 1984). These stresses have resulted in a series of adaptive morphological traits and, therefore, biomass partitioning patterns which are thought to be common, although not exclusive, to plants in all Mediterranean-type ecosystems. For example, thick cuticles, sclerophylly, shrubby forms, and an extensive root system are often considered convergent adaptive traits for plants in the various Mediterranean climate ecosystems. In other cases, investigators have identified morphological traits (e.g., lignotubers) whose adaptive significance is not completely known (Canadell and Zedler, in press).

Some of the general patterns of carbon allocation in various Mediterranean-type communities are known, especially the accumulation and partitioning of above-ground biomass following a disturbance (Rundel and Parsons, 1979; Schlesinger and Gill, 1980; Groves et al., 1986; Black, 1987; Mesleard, 1987). Very few studies, however, have considered the root system, so our knowledge of allocation below ground is rather limited. In addition, some of the available studies of roots only provide a partial picture of the below-ground component, either because the root excavation was limited to the upper soil horizons or because lignotubers and burls were not included. Finally, some of these studies have completely ignored the fine-roots biomass, which can account for 10% of the total root biomass (Kummerow, 1981). We also know relatively little about how the biomass allocation patterns differ for plants with different life history strategies, such us resprouters and seeders (Pate et al., 1990) which are very important in determining community dynamics following a disturbance.

Mediterranean-type ecosystems are dominated by evergreen trees and shrubs, although other forms are also present. For instance, drought-deciduous shrubs frequently occur in the most arid or highly disturbed areas in California and Chile (Mooney et al., 1977). Annual and perennial herbs are also present, e.g., in heavily grazed areas or shallow soils. In this section we summarize some of the available data on biomass allocation to the four major plant components (leaves, stems, roots, and reproductive structures) for forests, heathlands, and shrublands in the five Mediterranean-climate regions.

Root : Shoot Biomass Ratio

Because of the common occurrence of large underground woody structures and the possession of an extensive root system exploiting large soil volumes, plants in Mediterranean-type ecosystems are expected to show higher root:shoot biomass ratios than plants in more temperate ecosystems. Table 4-1 shows the R:S ratio for 13 evergreen Mediterranean communities ranging from 0.1 to 4.8. The mean R:S ratio for the forest communities is 0.64 ± 0.38, whereas for the heathlands and shrublands (numbers 4 through 13 in Table 4-1) it is 2.0 ± 0.5. From data compiled by Cannell (1982), the mean R:S ratio for coniferous forests (n = 70) is 0.26 ± 0.03, for deciduous forests (n = 31) it is 0.25 ± 0.02, and for tropical forests (n = 34) it is 0.31 ± 0.05. These figures are much lower than the R:S ratio of 1.4 for the holm oak forest dominated by mutistemmed trees with large root crowns (Canadell and Rodà, 1991) which is a fairly common stand structure for oak forests of the Mediterranean region and California, as well as for eucalypt forests of Australia. Whittaker and Marks (1975) gathered biomass data for seven shrubs from the Smoky Mountains (Tennessee) and the Brookhaven Forest (New York) with the mean R:S ratio of 1.18 (ranging from 0.54 to 2.69). So, although the R:S range for these temperate shrubs mostly overlaps with that for the Mediterranean shrublands, the latter have a much higher mean R:S ratio.

It is well known that high R:S ratios are associated with environments where water and nutrients are limited (Chapin, 1980). Although both of these factors occur to different extents in Mediterranean-type communities, the ecological interpretation of the R:S ratio of 2.0 for shrubland communities must be questioned to some extent. There are two major aspects for this concern. Firstly, because of the high fire frequency of some Mediterranean regions, communities have been studied before above-ground biomass has reached a steady state with the roots. Secondly, lignotubers and root crowns enlarge with each disturbance event, so that comparisons among ecosystems with different disturbance regimes have to be made carefully.

Allocation to Leaves and Stems

The proportion of the above-ground biomass allocated to leaves is greater for shrubs than for trees, primarily because shrubs lack a central trunk. Taking an average of the three forest communities in Table 4-1, leaf biomass accounts for approximately 6% of the above-ground biomass. For the 10 heath and shrubland communities dominated by evergreen species, the mean proportion of leaf biomass is 20% of the above-ground biomass. For drought-deciduous dominated shrublands (not shown in Table 4-1), the biomass allocated to leaves is clearly lower than for evergreen shrubs. Gray and Schlesinger (1981) reported a figure of 12% leaf biomass for a Californian coastal sage scrub dominated by drought-

Table 4-1. Biomass Characteristics of 13 Evergreen Mediterranean Communities from the Various Mediterranean-type Regions of the World

No.	Type of Community	Dominant Species	Locality	Stand Age (yr)	Height (m)	Biomass (gm^{-2}) Shoot	Biomass (gm^{-2}) Root	Allocation (%) Shoot Leaves	Allocation (%) Total Shoot	Allocation (%) Total Root	Root : shoot Ratio	References
1	Pinus forest	*Pinus pinea*	France	35	10–11	15,680	2,200	8.1	87.7	12.3	0.14	Rapp and Cabanettes, 1981
2	Oak forest	*Quercus ilex*	Spain	60–90	9–12	16,090	6,300	3.8	71.9	28.1	0.39	Escarré et al., 1987 Canadell and Rodà; 1991
3	Oak forest	*Quercus ilex*	Spain	25	6–8	7,910	10,800	6.3	42.2	57.8	1.4	Canadell et al., 1988 Canadell and Rodà; 1991
4	Coastal fynbos	*Phylica cephalantha*	S Africa	11	—	1,458	1,016	—	58.9	41.1	0.7	Low; 1983
5	Mountain fynbos	*Protea merisfolia* *Protea repens*	S Africa	26	4	3,650	1,223	—	74.9	25.1	0.34	Higgins et al., 1987
6	Heath	*Banksia ornata* *Casuarina sps*	SE Australia	25	2.4	3,683	13,355	—	21.6	78.4	3.6	Specht and Rayson, 1957; Specht et al., 1958

7	Scrub-heath	*Banksia hookeriana* *Banksia attenuata*	SW Australia	12	2.5	1,076	2,529	21.5	29.8	70.2	2.35	Low and Lamont; 1990
8	Chaparral	*Adenostoma fasciculatum* *Arctostaphylos pungens*	California	—	1.7	1,865	1,126	18	62.4	37.6	0.6	Kummerow et al., 1977, 1978; Mooney et al., 1977
9	Chaparral	*Adenostoma fasciculatum* *Quercus dumosa*	California	50	2–3	6,381	4,144	12.7	60.6	39.4	0.65	Kummerow and Wright, 1988
10	Matorral	*Lithraea caustica* *Cryptocarya alba*	Chile	—	1.2	2,988	11,314	28	20.9	79.1	3.7	Hoffman and Kummerow, 1978; Mooney et al., 1977
11	Garrigue	*Quercus coccifera*	France	17	1	2,355	4,600	17	33.9	66.1	1.9	Rapp and Lossaint, 1981
12	Garrigue	*Quercus coccifera*	France	36	1.2	1,496	7,200	26	17	83	4.81	Malanson and Trabaud, 1987; Kummerow et al., 1990
13	Phrygana	*Phlomis fruticosa* *Euphorbia acanthothamnos*	Greece	55	1	1,095	1,618	19.1	40.4	59.6	1.48	Margaris, 1976

deciduous species. Gill and Mahall (1986) reported a figure of 14% for the drought-deciduous *Salvia mellifera*.

Rodin and Bazilevich (1967) found, in a survey of 15 deciduous forests, that over 2% of the above-ground biomass consisted of leaves, whereas the leaf biomass proportion was about 10% for 12 coniferous forests. Whittaker et al. (1974) found for several deciduous trees that about 1% of the aboveground biomass was photosynthetic tissue. Thus, leaf biomass proportion of the broadleaf evergreen forest seems to be intermediate between deciduous and coniferous forests.

Interpretation of these data is complicated, since the proportion of biomass in leaves or other organs does not necessarily reflect the proportion of resources allocated in a given time period. The latter will be greatly affected by the life span of the organ of interest. Unfortunately, more detailed studies that directly measure resource allocation are very rare.

Another important factor that should be taken into account when considering allocation of carbon to leaves is the frequent occurrence of sclerophylly in plants of the Mediterranean-type ecosystems. The amount of carbon used per unit of leaf area is as significant as the total carbon allocation to leaves, and the ecological significance of sclerophyllous evergreen leaves has been discussed with relation to nutrient limitation, water stress, and herbivory (Rundel, 1988a; Specht and Rundel, 1990). Two major characteristics of the sclerophyllous leaf are higher specific weight (SLW) and the lower nitrogen concentration (N_w, weight basis) compared with mesophyllous leaves, especially deciduous ones. The first trait results in the high carbon cost (per unit area) of producing sclerophyllous leaves (Miller and Stoner, 1979; Merino et al., 1982) while the second characteristic influences the photosynthetic capacity (Field and Mooney, 1986). A mean SLW of $22.8 \pm 1.2\,mg/cm^2$ (mean $\pm$ standard deviation) was calculated from a sample of 28 sclerophyllous Mediterranean species (from Rundel, 1988b), while the mean SLW for 14 evergreen shrubs of the Great Smoky Mountains was $8.5 \pm 0.9\,mg/cm^2$ (Whittaker, 1962). From data collected by Rundel (1988b), the mean N_w for 17 Mediterranean-type communities was $1.0\% \pm 0.1$, while nitrogen concentrations for Mediterranean deciduous leaves commonly range from 1.5% to 3% (Margaris et al., 1984; Escudero et al., 1987). Rundel (1988a) also pointed out that even among sclerophylls, those in Mediterranean-type ecosystems tend to have lower concentrations of nitrogen.

Allocation to the Root System

The possession of an extensive root system has been considered an adaptive trait of plants in Mediterranean-type ecosystems. Although root systems have been described with a great diversity of morphology and rooting pattern, the possession of a tap root and well-developed lateral roots is very common in the various Mediterranean communities (Hellmers et al., 1955; Specht and Rayson, 1957; Dodd et al., 1984); thus, a high proportion of the total biomass is allocated to the root system. Oechel and Lawrence (1981) found that 30% to 40% of the carbon gained by *Arctostaphylos glauca* was used to build and maintain its root system. The proportion of the total biomass allocated to the root system is between 12% and 83% for the 13 Mediterranean communities presented in Table 4-1.

In addition, a fairly common trait of root systems in Mediterranean trees and shrubs, although not exclusive, is the possession of lignotubers or burls (James, 1984). These have been found in woody species in South Africa (Carlquist, 1977), South Australia (Lacey, 1983), California (Kummerow, 1981), Chile (Montenegro et al., 1983), and the Mediterranean Basin (Canadell and Rodà, 1991). This structure plays a decisive role in the resprouting response after above-ground destruction (e.g., fire, cutting). Functions of bud protection (Zammit, 1988; Canadell et al., 1991) as well as storage of nutrients and carbohydrates (Mullette and Bamber, 1978; Miyanishi and Kellman, 1986) have also been ascribed to this structure.

The possession of a lignotuber can greatly affect the biomass allocation pattern in the plant, since the lignotuber sometimes can account for more than 50% of the total below-ground biomass. A 67-kg burl of *Lithraea caustica*, a Chilean evergreen shrub, accounted for four times the total root biomass (Hoffmann and Kummerow, 1978), and a lignotuber as heavy as 317 kg made up 67% of the total below-ground biomass for a multistemmed *Quercus ilex* in Spain (Canadell and Rodà, 1991). Kummerow and Mangan (1981) reported a 15-kg burl of *Quercus dumosa* in California which made up 81% of the total below-ground biomass.

Allocation to Reproductive Tissues

The biomass allocation to the reproductive tissues is highly variable through the year and with the age of the plant, but overall the annual carbon investment is not generally higher than 10% (Gray, 1982; Ferrés, 1984). However, a remarkably high investment of carbon is made by plants with woody fruits, as are commonly found in Western Australia (Cupressaceae, Casuarinacea, Proteacea, and Myrtiaceae) and South Africa (Mooney, 1983). Some proteaceous shrubs in South Africa were found to have almost 25% of their above-ground biomass in reproductive organs (Kruger, 1977). A proportion as high as 30% to 40% was found both in a 25-year-old heath (Specht et al., 1958) and in a 50-year-old *Banksia* scrub-heath (Groves et al., 1986) in South Australia. Low and Lamont (1990) found that 21% of the above-ground biomass was inflorescences and cones in a 12-year-old heathland in Western Australia.

Woody fruits are also found on conifers, which are remarkably important in the Mediterranean Basin but only of local importance in the other Mediterranean regions. For a 35-year-old *Pinus pinea* stand in France, reproductive tissues accounted for 23% of the above-ground annual production (Rapp and Cabanettes, 1981).

Modeling Approaches to Allocation in Response to Elevated CO_2

A Simulation Model of Allocation

Hilbert and Reynolds (1991) developed a model (ALLOCATE) that considers the allocation of carbon and nitrogen substrates to a protein compartment in the shoots, shoot structural components, and root biomass. The inclusion of a shoot-protein compartment allows variation in shoot specific activity as a function of leaf nitrogen concentration. Since the actual mechanisms that control allocation are poorly understood, the model is teleonomic, i.e., the allocation parameters are defined so as to produce a specific goal—that balanced activity, as defined by Equation 4-1, will be achieved. Thus, the model produces balanced activity where the shoot specific activity, as well as root and shoot biomass, vary in response to the environment, both above ground (light and CO_2) and below ground (nitrogen).

Applying the model to the question of the effects of elevated CO_2, it was shown (Hilbert and Reynolds, 1991) that increasing internal CO_2 concentrations (C_i) increases the root:shoot ratio and decreases the leaf nitrogen concentration on an area basis (N_a). Increasing C_i elevates σ_s, so in order to achieve balanced growth, a plant has various options: decrease shoot fraction, increase root fraction, and/or decrease N_a to reduce the photosynthetic stimulation. The model generally responds to elevated C_i in all of these ways—as do real plants—but the specific responses depend on the levels of above- and below-ground resource availability.

An important consideration in Mediterranean-type ecosystems is their low soil nitrogen availability relative to the majority of conditions under which elevated CO_2 experiments have been carried out. We have run ALLOCATE under a variety of soil nitrogen availabilities (represented by root specific activity) in order to get a general impression of the interactions among allocation patterns, CO_2, and soil nitrogen levels. The results are presented in Figure 4-1. When nitrogen uptake is high, allocation to roots, in response to increasing CO_2, increases only slightly and leaf nitrogen concentration remains high. Consequently, the photosynthetic response to elevated CO_2 is relatively high. On the other hand, when the nitrogen uptake rate is low, root allocation is more stimulated, leaf nitrogen concentration decreases significantly, and photosynthesis is somewhat less stimulated by elevated CO_2. In general, then, we might

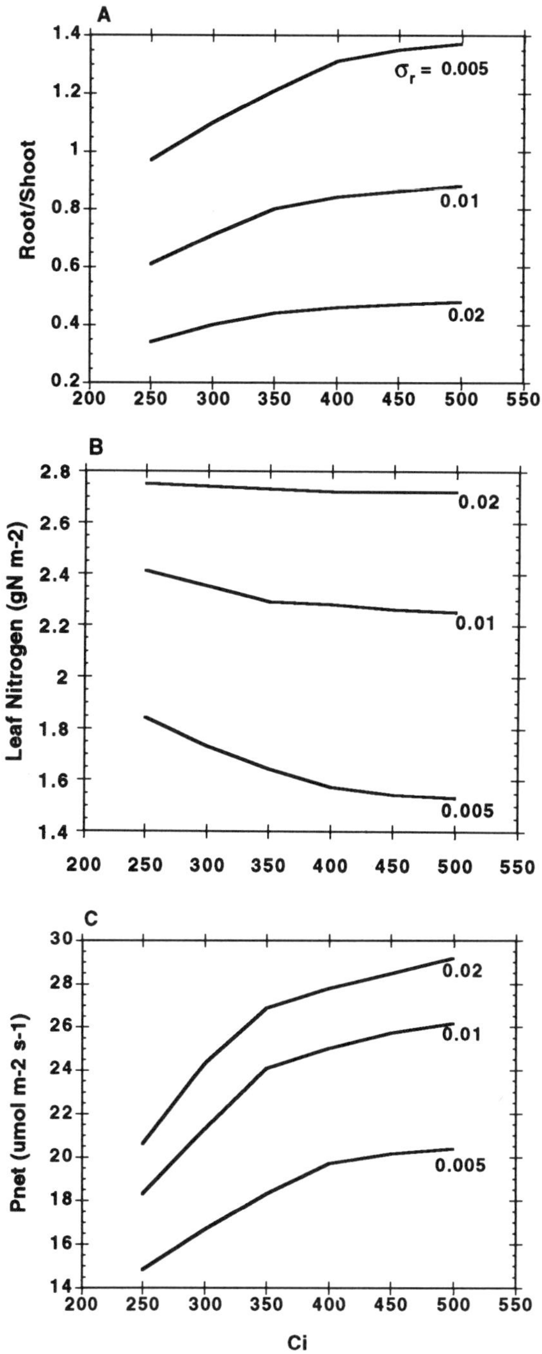

Figure 4-1. Runs of ALLOCATE with the standard parameters but varying σ_r from 0.005 to 0.01 to 0.02 (g N g root^{-1} d^{-1}) and C_i from 250 to 500 (μL L^{-1}). Parameters were the same as in Hilbert and Reynolds (1991). (A) Predicted root : shoot ratio; (B) leaf nitrogen (g N m^{-2}); (C) net photosynthesis (μmol m^{-2} s^{-1}).

reasonably expect smaller stimulation of growth at high CO_2 in nutrient-limited ecosystems (due to the lower photosynthetic stimulation) and lower investment of resources to produce more leaf area.

Optimal Leaf Nitrogen Concentration

Generally, analyses of optimal allocation patterns in plants have emphasized single traits in isolation—optimal root:shoot partitioning or optimal leaf nitrogen concentration, for example—but there has been little consideration of the interactions inherent in allocation. Hilbert (1990) and Hilbert et al. (1991), however, stress the tradeoff between allocation to shoot mass (thus, leaf area) and the allocation of nitrogen to proteins in the leaf (thus, photosynthetic capacity). Photosynthetic capacity is limited in part by the amount of nitrogen that can be supplied to the leaves. Since nitrogen is generally acquired by the roots and the resources allocated to roots cannot be used to grow leaf area, a plant can produce a large leaf area with relatively low photosynthetic capacity or a smaller leaf area with high photosynthetic capacity. In the first case, the root:shoot ratio will be lower, whereas in the latter case, the root:shoot ratio will be higher. Determining the optimal strategy in a given environment can be accomplished by considering the costs and revenue (benefits) associated with allocation of nitrogen to leaves.

Assuming that the whole-plant growth rate is proportional to the rate of carbon gain, relative growth rate (RGR, d^{-1}) can be written as

$$(1/W)dW/dt = kh\sigma_s f_s \tag{4-2}$$

where W is the total plant dry weight (g), k is the proportionality constant between carbon and biomass (g biomass g^{-1} C fixed), h is specific leaf area (m^2 leaf g^{-1} shoot), σ_s is the shoot specific activity on an area basis (g C $m^{-2} d^{-1}$), and f_s is the fraction of biomass in the shoot. Thus, RGR is directly proportional to both the fractional allocation of biomass to the shoots and the photosynthetic rate of those shoots. The costs and revenue of leaf nitrogen can be expressed in terms of the plant's relative growth rate by assuming that σ_s and f_s are, in part, functions of leaf nitrogen concentration. Since $f_s = 1 - f_r$ where f_r is the fraction of biomass in the roots, Equation 4-2 can be rewritten as

$$(1/W)\ dW/dt = kh\sigma_s - kh\sigma_s f_r. \tag{4-3}$$

Revenue ($kh\sigma_s$) is the RGR (d^{-1}) that would be possible if there was no need for allocation to roots, while cost ($kh\sigma_s f_r$) is the decrease in RGR due to biomass partitioning to the roots.

The leaf nitrogen concentration that optimizes the tradeoff between leaf area and photosynthetic capacity occurs when the marginal revenue

(MR) and marginal cost (MC) (units: g biomass g^{-1} N d^{-1}) are equal. These are obtained from the derivative of the cost and revenue functions with respect to leaf nitrogen concentration (N_a) (g N m^{-2} leaf). Assuming that both σ_s and f_r are functions of leaf nitrogen, implicit differentiation of Equation 4-3 yields

$$MR = d(\text{revenue})/dN_a = kh\ d\sigma_s/dN_a \tag{4-4}$$

and

$$MC = d(\text{cost})/dN_a = kh(f_r d\sigma_s/dN_a + \sigma_s df_r/dN_a). \tag{4-5}$$

Setting Equation 4-4 equal to Equation 4-5 and simplifying, gives the optimization criterion

$$f_s d\sigma_s/dN_a = \sigma_s df_r/dN_a. \tag{4-6}$$

Using this analysis [see Hilbert (1990) and Hilbert et al. (1991) for details], some of the changes in allocation patterns that might be expected in response to elevated CO_2 were predicted. Figure 4-2 illustrates the response of N_{opt} to internal carbon dioxide concentration. N_{opt} should decrease at elevated CO_2 levels. This response is the direct result of the strong dependency of the rate of RuBP saturated carboxylation on CO_2.

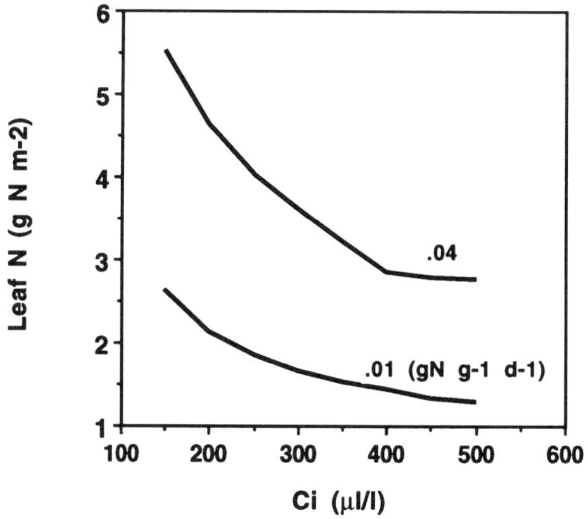

Figure 4-2. Predictions of optimal leaf nitrogen (g N m^{-2}) as a function of internal CO_2 concentration (C_i) and two values of root specific activity (g N g $root^{-1}$ d^{-1}) using daily net photosynthesis calculated from the Farquhar model (parameters as in Hilbert, 1990).

The marginal revenue and marginal cost curves intersect at lower N_a when plants are grown at elevated CO_2 levels. Although photosynthetic capacity is reduced, this is more efficient than paying the higher cost of increased root:shoot ratios that would be necessary to maintain a high N_a. Several studies on a variety of plants grown at high levels of CO_2 show decreased photosynthetic capacity compared with plants grown at ambient levels (Madsen, 1975; Wong, 1979; Peet and Willits, 1984; Hocking and Meyers, 1985; Norby et al., 1986; Williams et al., 1986; Larigauderie et al., 1988), and this response is correlated with decreased N_a.

Possible Effects of Sclerophylly

The sclerophyllous nature of evergreen Mediterranean plants can be expected to greatly influence their response to elevated atmospheric CO_2. Using the optimization analysis described above, we can investigate the optimal allocation of nitrogen to leaves in plants with high vs. low specific leaf weight (g leaf m^{-2}) or its inverse, specific leaf area (h in $m^2 g^{-1}$ leaf). An analytic solution to Equation 4-6 can be found by using an equation for df_r/dN_a derived in Hilbert et al. (1991) and the assumption that shoot specific activity is a linear function of leaf nitrogen on an area basis given by

$$\sigma_s = a(N_a - n_0) \tag{4-7}$$

where a is the slope of the response and n_0 is the nitrogen concentration where there is zero net photosynthesis. The solution for optimal leaf nitrogen concentration (N_{opt^a}) is then

$$N_{opt^a} = \mathrm{SLW}(\sigma_r k^{-1} a^{-1})^{1/2} + n_0 \tag{4-8}$$

or, for nitrogen concentration on a weight basis (N_{opt^w}),

$$N_{opt^w} = (\sigma_r k^{-1} a^{-1})^{1/2} + n_0/\mathrm{SLW} \tag{4-9}$$

where SLW is the specific leaf weight, which equals h^{-1}.

This suggests that, all else being equal, plants with high SLW will tend to have higher nitrogen concentrations on an area basis and lower nitrogen concentrations on a weight basis, since N_{opt^w} is inversely proportional to SLW. The optimal responses of N_a and N_w to increasing SLW are plotted in Figure 4-3 where $\sigma_r = 0.001$, $k = 1.54$, $a = 4$, and $n_0 = 0.5$. These results are consistent with the general observation that evergreen leaves, compared to deciduous mesophylls, have similar or even higher nitrogen concentrations on an area basis but, usually, much lower concentrations on a weight basis.

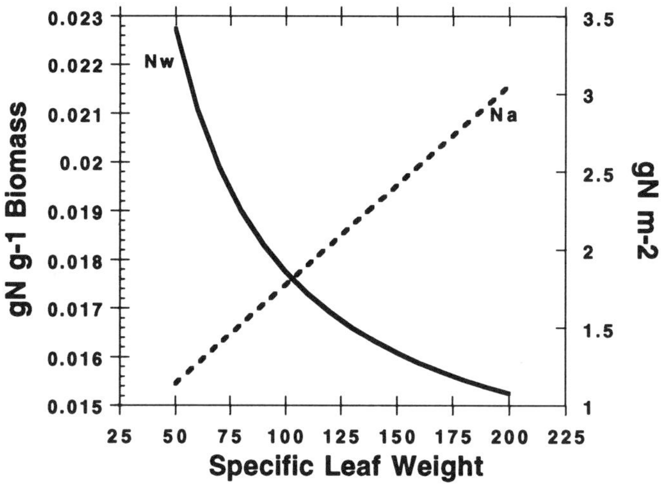

Figure 4-3. Optimal leaf nitrogen on an area basis (dashed line) and on a weight basis (solid line) as a function of specific leaf weight. The curves were calculated from Equations 4-8 and 4-9 with $\sigma_r = 0.001$, $k = 1.54$, $a = 4$, and $n_0 = 0.5$.

As a result of these nitrogen-related characteristics, we can reasonably expect that photosynthetic responses to elevated CO_2 will be less in sclerophyllous plants than in nonsclerophyllous ones. This can be seen by examining the Farquhar photosynthesis model (Farquhar et al., 1980). In this model, photosynthesis is assumed to be RuBPcarboxylase saturated at high light, with the rate of assimilation (A) given by

$$A = V_{cmax}(C_i - k_1)/(C_i + k_2) - R_d \tag{4-10}$$

where R_d is the "dark" respiration rate, C_i is the CO_2 concentration at the site of fixation, k_1 and k_2 can be considered constant for our purposes, and V_{cmax} is the maximum rate of carboxylation. V_{cmax} is itself a linear function of leaf nitrogen concentration, $V_{cmax} = aN + b$ (Field, 1983). We can see the effect of leaf nitrogen concentration on the response to CO_2 by taking the partial derivative of A with respect to C_i:

$$\partial A/\partial C_i = (aN + b)(k_1 + k_2)/(C_i + k_2)^2. \tag{4-11}$$

Since $V_{cmax}(aN + b)$ is low in sclerophyllous species, we would expect from Equation 4-11 that the increase in photosynthesis for an increase in CO_2 would be lower in these species than in plants with high intrinsic photosynthetic capacities. Note that this effect could be caused by either lower leaf nitrogen concentration or a lower slope (a) of the response to leaf nitrogen. To further illustrate this point, we plot in Figure 4-4 the photosynthetic responses to CO_2 of a plant with a relatively high capacity

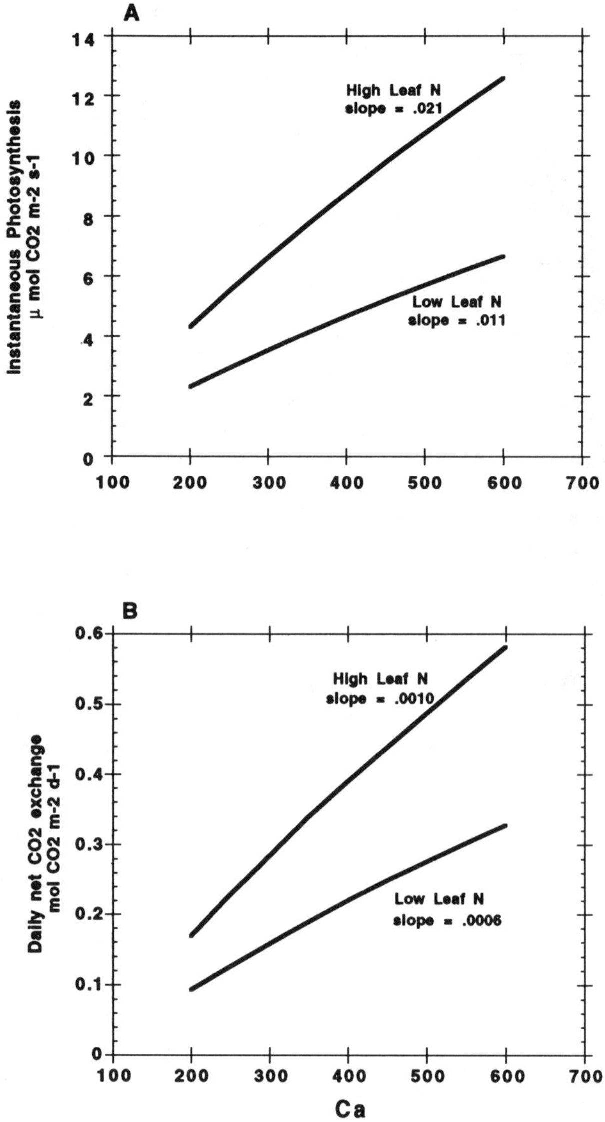

Figure 4-4. Photosynthetic responses to atmospheric CO_2 of a plant with a relatively high capacity and one with a low capacity for photosynthesis (v_{cmax}, j_{max}, and r_d reduced by 1/2 from the standard run), parameters from Harley and Tenhunen (1991): (A) Predicted rates of instantaneous photosynthesis are plotted where PPFD = 1,800 μmol m^{-2} s^{-1}, air temperature = 25°C, and relative humidity = 45%. (B) Daily rates of carbon uptake, with temperature and PPFD varying diurnally.

and one with a low capacity for photosynthesis. Here we used a model coupling the Farquhar model with an empirical model of stomatal conductance proposed by Ball (1988) (see Harley and Tenhunen, 1991 for a description of the model). In Figure 4-4A, the predicted rates of instantaneous photosynthesis are plotted where PPFD = 1,800 μmol m^{-2} s^{-1}, air temperature = 25°C, and relative humidity is 45%. The daily rates of carbon uptake, with temperature and PPFD varying diurnally, are plotted in Figure 4-4B. In both cases the responses are almost linear and the slope of the response is roughly double for the plant with higher leaf nitrogen and, consequently, a photosynthetic capacity approximately double that of the plant with low leaf nitrogen. Based on this very simple analysis, we would expect that photosynthesis would be less stimulated by high CO_2 in evergreen plants than in drought deciduous plants or grasses.

We have also run ALLOCATE to compare the responses of a nonsclerophyllous plant [relatively high root specific activity, high nitrogen content in the tissues, and low specific leaf weight (50 g m^{-2})] with a more sclerophyllous plant growing in a soil with low nitrogen availability [low root specific activity, low nitrogen content, and high specific leaf weight (210 g m^{-2})]. The results are presented in Figure 4-5, where the percent changes in several plant characteristics following an increase in C_i from 240 to 500 μL/L are plotted. As previously predicted, the leaf nitrogen concentration (here on a weight basis) declines much more for the sclerophyll, and the photosynthetic stimulation is about 15% less than for the nonsclerophyll. The root : shoot ratio of the plant growing in more

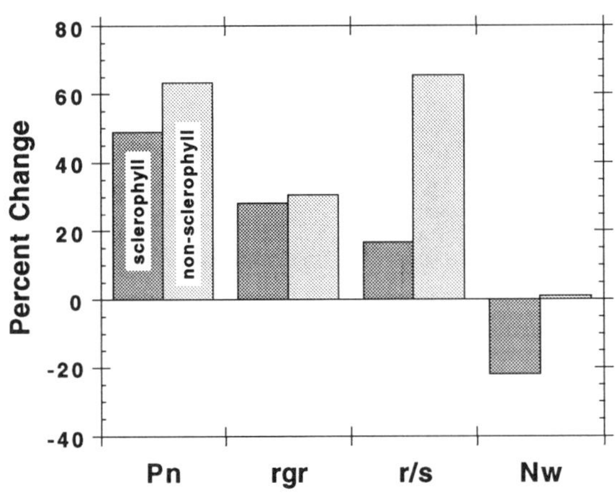

Figure 4-5. Percent changes, predicted by ALLOCATE, in net photosynthesis (Pn), relative growth rate (RGR), root : shoot ratio (R/S), and leaf nitrogen concentration on an area basis (Nw) after a doubling of atmospheric CO_2 for sclerophyllous Mediterranean plants in soils of low nitrogen availability with mesophylls in more fertile soils.

fertile soil has a far greater increase in allocation to root biomass after an increase in CO_2, which is consistent with some experimental results at high and low nitrogen nutrition (Larigauderie et al., 1988).

Allocation to Storage and Water Availability

Another potentially important characteristic of sclerophyllous shrubs and trees in Mediterranean-type ecosystems is their high allocation of carbon to storage in stems and burls. It is thought that the response of many plants to elevated CO_2 may be decreased when sink strength is limiting (Thomas and Strain, 1991), so the existence of large storage sinks in resprouting perennials may predispose them to respond more positively to higher CO_2 levels. Woody fruits similarly represent a large C sink. Currently our models are not well designed to address this area of research. What effects, if any, increased carbon storage would have on growth, survival, and reproduction is unknown.

Finally, plant water relations have not been considered in any of our models. Clearly, water stress is a significant factor that influences both physiology and morphology in all Mediterranean-type ecosystems. Furthermore, one would expect important interactions with CO_2, since stomatal conductances are invariably reduced by high atmospheric CO_2 concentrations. To the extent that this conserves water, root:shoot ratios may decrease, but complex interactions with nutrient uptake may also occur. It is clearly important to further understand the role of water availability in biomass partitioning and resource allocation to photosynthetic enzymes.

Conclusions

In this chapter we have stressed the importance of resource allocation, both among and within organs, in determining the response of plants to elevated atmospheric CO_2, reviewed some of the allocation patterns of plants from Mediterranean ecosystems, and used models to assess the degree to which these plants might be expected to respond to high CO_2 levels. Although we do not yet have a comprehensive understanding of how plants allocate resources and integrate their growth responses to resource availability, the principle of balanced activity has been helpful in understanding these responses in herbaceous plants. Further work is necessary, however, to determine the applicability of this principle (or develop new paradigms) for woody plants. Relatively little is known about the allocation patterns that are prevalent in Mediterranean-climate shrubs and trees, and few species have been studied in detail. This is due, in part, to the extreme difficulty of harvesting roots. New experimental approaches, especially for fine roots, are needed to overcome these difficulties (Kummerow and Ellis, 1989). Several generalizations, however, are possible for evergreen sclerophyllous shrubs. These plants tend to have a relatively high proportion of above-ground biomass allocated to leaves, and these leaves have a high LSW and low leaf nitrogen concentration on a weight basis (indicating a high ratio of C/N allocation to leaves). Furthermore, allocation below ground to structural, storage, and absorbing tissues is very large, but the data are coarse and should be interpreted with caution. In some cases, carbon allocation to reproduction is very high. Almost nothing is known about the responses of allocation patterns to resource availability in these plants, and although this will be challenging, we believe that studies of this kind are essential if we are ever to understand the coordination of structure and function in these systems.

We have applied several modeling approaches to make an initial assessment of how some of the characteristics of Mediterranean—climate plants may influence their responses to elevated CO_2. In general, we predict decreased photosynthetic capacities in plants grown at high CO_2 and increased allocation to roots. Sclerophylls, which already have low photosynthetic capacities, may show an even larger decline in leaf nitrogen concentration and, thus, be less stimulated by elevated CO_2 than the herbaceous mesophylls that have previously been studied. On the other hand, the high demand for carbon in roots and reproductive organs, plus the high carbon cost of leaves, may increase the sensitivity of these plants to increased atmospheric CO_2. Certainly, other characteristics, such as water use efficiency (the topics of other chapters in this volume), will be very important as well in determining overall responses to CO_2. Although these results are little more than educated guesses—in the near total absence of experimental results—we hope they will be of some use in the design of future experiments and, at the very least, emphasize the importance of considering resource allocation in studying the responses of Mediterranean-climate plants to global change.

Acknowledgments. We thank two anonymous reviewers for their helpful suggestions. Funding was provided by the United States Department of Energy, Carbon Dioxide Research Division, grant DE-FG60386ER60490, and by Subprogram of P.D.T.E., Ministerio de Educación y Ciencia, Spain.

References

Ågren GI, Ingestad T. 1987. Root: shoot ratio as a balance between nitrogen productivity and photosynthesis. Plant Cell Environ 10:579–586.

Ball JT. 1988. An analysis of stomatal conductance. PhD dissertation, Stanford University, Stanford, CA, 88 pp.

Bazzaz FA, Carlson RW. 1984. The response of plants to elevated CO_2. I. Competition among an assemblage of annuals at two levels of soil moisture. Oecologia 62:196–198.

Björkman O. 1981. Responses to different quantum flux densities, pp. 57–107. In: Encyclopedia of Plant Physiology Vol. 12A, Physiological Plant Ecology I, Responses to the physical environment. Lange OL, PS Nobel, CB Osmond, and H Ziegler (eds). Springer-Verlag, Berlin, Heidelberg, New York.

Black CH. 1987. Biomass, nitrogen, and phosphorus accumulation over a southern California fire cycle chronosequence. In: Plant Response to Stress, Tenhunen JD, FM Catarino, OL Lange, and WC Oechel (eds). Springer-Verlag, Berlin, pp. 445–458.

Booysen PV, Tainton NM. 1984. Ecological Effects of Fire in South African Ecosystems. Springer-Verlag, Berlin.

Borchert R. 1973. Simulation of rhythmic tree growth under constant conditions. Physiol Plant 29:173–180.

Brouwer R. 1962. Nutritive influences on the distribution of dry matter in the plant. Neth J Agric Sci 10:399–408.

von Caemmerer S, Farquhar GD. 1981. Some relationships between the biochemistry of photosynthesis and the gas exchange of leaves. Planta 153:376–387.

Canadell J, Rodà F. 1991. Root biomass of *Quercus ilex* in a montane Mediterranean forest. Can J For Res 21:1771–1778.

Canadell J, Zedler PH. 1995. Underground structures of woody plants in the Mediterranean ecosystems of Australia, California, and Chile. In: Ecology and Biogeography of Mediterranean Ecosystems in Chile, California and Australia, Arroyo MTK, PH Zedler, and MD Fox (eds). Springer-Verlag, New York, pp. 177–210.

Canadell J, Lloret F, López-Soria L. 1991. Resprouting vigor of two Mediterranean shrub species after experimental fire treatments. Vegetatio 95:119–126.

Canadell J, Riba M, Andrés P. 1988. Biomass equations for *Quercus ilex* L. in the Montseny massif, northeastern Spain. Forestry 61:137–147.

Cannell, MGR. 1982. World Forest Biomass and Primary Production Data. Academic Press, London.

Carlquist S. 1977. Wood anatomy of Penaeaceae (Myrtales): comparative, phylogenetic, and ecological implications. Bot J Linn Soc 75:211–227.

Chapin FS III. 1980. The mineral nutrition of wild plants. Ann Rev Ecol Syst 11:233–260.

Chapin FS III, Bloom AJ, Field CB, Waring RH. 1987. Plant responses to multiple environmental factors. BioScience 37:49–57.

Charles-Edwards DA. 1976. Shoot and root activities during steady state plant growth. Ann Bot 40:767–772.

Charles-Edwards DA. 1982. Physiological Determinants of Crop Growth. Academic Press, London, New York, Toronto, Sydney, San Francisco, 161 pp.

Clarkson DT. 1985. Factors affecting mineral nutrient acquisition by plants. Ann Rev Plant Physiol 36:77–115.

Cooper AJ, Thornley JHM. 1976. Responses of dry matter partitioning, growth, and carbon and nitrogen levels in the tomato plant to changes in root temperature: experiment and theory. Ann Bot 40:1139–1152.

Cure JD. 1985. Carbon dioxide doubling responses: A crop survey. In: Direct effects of increasing carbon dioxide on vegetation, Strain BR and JD Cure (eds). U.S. DOE, Nat. Tech. Info. Service, Springfield, VA, pp. 99–116.

Davidson RL. 1969. Effects of soil nutrients and moisture on root/shoot ratios in *Lolium perenne* L. and *Trifolium repens* L. Ann Bot 33:571–577.

Dodd J, Heddle EM, Pate JS, Dixon KW. 1984. Rooting patterns of sand plain plants and their functional significance. In: Kwongan—Plant Life of the Sand plain, Pate JS and JS Beard (eds). University of Western Australia Press, Nedlands, pp. 146–177.

Edwards JH, Barber SA. 1976. Nitrogen flux into corn roots as influenced by shoot requirements. Agron J 689:471–473.
Ehleringer JR, Pearcy RW, Mooney HA. 1986. Recommendations of the workshop on the future development of plant physiological ecology. Bull Ecol Soc Am 67:48–58.
Escarré A, Ferrés Ll, López R, Martin J, Rodà F, Terradas J. 1987. Nutrient use strategy by evergreen-oak (*Quercus ilex* ssp. *ilex*) in NE Spain. In: Plant Response to Stress, Tenhunen JD, FM Catarino, OL Lange, and WC Oechel (eds). Springer-Verlag, Berlin, pp. 429–435.
Escudero A, Manzano JJ, del Arco JM. 1987. Nitrogen concentrations in the leaves of different Mediterranean woody species. Ecol Mediterr 13:11–17.
Farquhar GD, von Cammerer S, Berry JA. 1980. A biochemical model for photosynthetic CO_2 assimilation in leaves of C_3 species. Planta 149:78–90.
Ferrés L. 1984. Biomassa, producción y mineralomasa del encinar montano de La Castanya (Montseny). PhD Dissertation. Universitat Autònoma de Barcelona, Spain.
Field C. 1983. Allocating leaf nitrogen for the maximization of carbon gain: leaf age as a control on the allocation program. Oecologia 56:341–347.
Field C, Mooney HA. 1983. Leaf age and seasonal effects on light, water, and nitrogen use efficiency in a California shrub. Oecologia 56:348–355.
Field C, Mooney HA. 1986. The photosynthesis-nitrogen relationship in wild plants. In: On the Economy of Plant Form and Function, Givnish TJ (ed). Cambridge University Press, Cambridge, UK, pp. 25–55.
Gill DS, Mahall BE. 1986. Quantitative phenology and water relations of an evergreen and a deciduous chaparral shrub. Ecol Monog 56:127–143.
Givnish TJ. 1986. On the Economy of Plant Form and Function. Cambridge University Press, Cambridge, UK, 717 pp.
Gray JT. 1982. Community structure and productivity in *Ceanothus* chaparral and coastal sage scrub of southern California. Ecol Monog 52:415–435.
Gray JT, Schlesinger WH. 1981. Biomass, production, and litterfall in the coastal sage scrub of southern California. Am J Bot 68:24–33.
Groves RH, Hocking PJ, McMahon J. 1986. Distribution of biomass, nitrogen, phosphorus and other nutrients in *Banksia marginata* and *B. ornata* shoots of different ages after fire. Aust J Bot 34:709–725.
Harley PC, Tenhunen JD. 1991. Modeling the photosynthetic response of C_3 leaves to environmental factors. In: Modeling Crop Photosynthesis—From Biochemistry to Canopy, Boote KJ (ed). Proceedings of American Society of Agronomy Symposium.
Hellmers H, Horton JS, Juhren G, O'Keefe J. 1955. Root systems of some chaparral plants in southern California. Ecology 36:667–678.
Higgins KB, Lamb AJ, van Wilgen BW. 1987. Root systems of selected plant species in mesic mountain fynbos in the Jonkershoek Valley, South-western Cape Province. S Afr J Bot 53:249–257.
Hilbert DW. 1990. Optimization of plant root:shoot ratios and internal nitrogen concentration. Ann Bot 66:91–99.
Hilbert DW, Reynolds JF. 1991. A model allocating growth among leaf proteins, shoot structure, and root biomass to produce balanced activity. Ann Bot 68:417–425.
Hilbert DW, Larigauderie A, Reynolds JF. 1991. Influence of carbon dioxide and daily photon-flux density on optimal leaf nitrogen concentration and root:shoot ratio. Ann Bot 68:365–376.
Hilbert DW, Prudhomme TI, Oechel WC. 1987. Response of tussock tundra to elevated carbon dioxide regimes: analysis of ecosystem CO_2 flux through nonlinear modeling. Oecologia 72:466–472.

Hocking PJ, Meyers CP. 1985. Responses of Noogoora Burr (*Xanthium occidentale* Bertol.) to nitrogen supply and carbon dioxide enrichment. Ann Bot 55:835–844.

Hoffmann A, Kummerow J. 1978. Roots studies in the Chilean matorral. Oecologia 32:57–69.

James S. 1984. Lignotubers and burls—their structure, function and ecological significance in Mediterranean ecosystems. Bot Rev 50:225–266.

Johnson IR. 1985. A model of the partitioning of growth between the shoots and roots of vegetative plants. Ann Bot 55:421–431.

Kimball BA. 1983. Carbon dioxide and agricultural yield: an assemblage and analysis of 430 prior observations. Agron J 75:779–788.

Kruger FJ. 1977. Ecology of Cape fynbos in relation to fire. In: Symposium on the Environmental Consequences of Fire and Fuel Management in Mediterranean-Climate Ecosystems, Mooney HA and CE Conrad (eds). USDA Forest Service General Technical Report WO-3, pp. 230–244.

Kruger FJ, Mitchell DT, Jarvis JUM (eds). 1983. Mediterranean-Type Ecosystems. The Role of Nutrients. Springer-Verlag, Berlin.

Kummerow J. 1981. Structure of roots and root systems. In: Mediterranean-type Shrublands, di Castri F, DW Goodall, and RL Specht (eds). Elsevier Scientific Publications Company, Amsterdam, pp. 269–288.

Kummerow J, Ellis BA. 1989. Structure and function in chaparral shrubs. In: The California Chaparral—Paradigms Reexamined, Keeley SC (ed). Natural History Museum of Los Angeles County, Los Angeles, CA, pp. 140–150.

Kummerow J, Mangan R. 1981. Root systems in *Quercus dumosa* Nutt. dominated chaparral in southern California. Ac Oecol/Oecol Plant 2:177–188.

Kummerow J, Wright CD. 1988. Root distribution and resource availability in mixed chaparral of southern California. Proceedings 5th International Conference on Mediterranean Ecosystems (MEDECOS). International Union of Biological Sciences, Paris, pp. 225–259.

Kummerow J, Krause D, Jow W. 1977. Root systems of chaparral shrubs. Oecologia 29:163–177.

Kummerow J, Krause D, Jow W. 1978. Seasonal changes of fine root density in the southern California chaparral. Oecologia 37:201–212.

Kummerow J, Kummerow M, Trabaud L. 1990. Root biomass, root distribution and the fine-root growth dynamics of *Quercus coccifera* L. in the garrigue of southern France. Vegetatio 87:37–44.

Kummerow J, Mills JN, Ellis BA, Kummerow A. 1988. Growth dynamics of cotton-grass (*Eriophorum vaginatum*). Can J Bot 66:253–256.

Lacey CJ. 1983. Development of large plate-like lignotubers in *Eucalyptus batryoides* Sm. in relation to environmental factors. Aust J Bot 31:105–118.

Larigauderie A, Hilbert DW, Oechel WC. 1988. Effects of CO_2 enrichment and nitrogen availability on resource acquisition and resource allocation in a grass, *Bromus mollis*. Oecologia 77:544–549.

Ledig FT. 1969. A growth model based on the rate of photosynthesis and distribution of assimilates. Photosynthetica 3:263–275.

Loomis RS. 1983. Productivity of agricultural systems. In: Encyclopedia of Plant Physiology, NS vol. 12D: Physiological Plant Ecology IV: Ecosystems Processes: Mineral Cycling, Productivity, and Man's Influence, Lange OL, PS Nobel, CB Osmond, and H Ziegler (eds). Springer, Berlin-Heidelberg, New York, Tokyo, pp. 151–172.

Low AB. 1983. Phytomass and major nutrient pools in an 11-year post-fire coastal fynbos community. S Afr J Bot 2:98–104.

Low AB, Lamont BB. 1990. Aerial and below-ground phytomass of banksia scrub-heath at Eneabba, South-western Australia. Aust J Bot 38:351–359.

Madsen E. 1975. Effect of CO_2 enrichment on growth, development, fruit production and fruit quality of tomato from a physiological viewpoint. Phytotronics III:318–330.

Mäkelä AA, Sievänen RP. 1987. Comparison of two shoot-root partitioning models with respect to substrate utilization and functional balance. Ann Bot 59:129–140.

Malanson GP, Trabaud L. 1987. Post-fire development of canopy structure in a Mediterranean shrub, *Quercus coccifera* L. Phys Geogr 8:266–274.

Margaris NS. 1976. Structure and dynamics in a phryganic (east Mediterranean) ecosystem. J Biog 3:249–259.

Margaris NS, Adamandiadou S, Siafacaf L, Diamantopoulos J. 1984. Nitrogen and phosphorus content in plants species of Mediterranean ecosystems in Greece. Vegetatio 55:29–35.

Medina E. 1971. Effect of nitrogen supply and light intensity during growth on the photosynthetic capacity and carboxydismutase activity of leaves of *Atriplex hastata* spp. *hastata*. Carnegie Institute Washington Yearbook 70: 551–559.

Merino J, Field C, Mooney HA. 1982. Construction and maintenance costs of Mediterranean-climate evergreen and deciduous leaves. I. Growth and CO_2 exchange analysis. Oecologia 53:208–213.

Mesleard F. 1987. Dynamique, après pertubations, de peuplements de deux éricacées (*Arbutus unedo* L. et *Erica arborea* L.) en Corse. PhD Dissertation. Université des Sciences et Techniques du Languedoc, France, 146 pp.

Miller PC, Stoner WA. 1979. Canopy structural and environmental interactions. In: Topics in Plant Population Biology, Solbrig OT, S Jain, GB Johnson, and PH Raven (eds). Columbia University Press, New York, pp. 428–458.

Miyanishi K, Kellman M. 1986. The role of root nitrient reserves in regrowth of two savanna shrubs. Can J Bot 64:1244–1248.

Monsi M. 1968. Mathematical models of plant communities. In: Functioning of Terrestrial Ecosystems at the Primary Production Level, Eckardt FE (ed). UNESCO, Paris, pp. 131–149.

Montenegro G, Avila G, Schatte P. 1983. Presence and development of lignotubers in shrubs of the Chilean matorral. Can J Bot 61:1804–1808.

Mooney, HA. 1972. The carbon balance of plants. Ann Rev Ecol Syst 3:315–346.

Mooney HA. 1983. Carbon-gaining capacity and allocation patterns of Mediterranean-climate plants. In: Mediterranean-Type Ecosystems. The Role of nutrients, Kurger FJ, DT Mitchell, and JUM Jarvis (eds). Springer-Verlag, Berlin, pp. 103–119.

Mooney HA. 1987. The impact of environmental stress on plant performance in Mediterranean-climate: Differing levels of analysis. In: Plant Response to Stress—Functional Analysis in Mediterranean Ecosystems, Tehnunen JD, FM Catarino, OL Lange, and WC Oechel (eds). Springer-Verlag, Berlin, Heidelberg, New York, Tokyo, pp. 661–668.

Mooney HA, Chiariello NR. 1984. The study of plant function—the plant as a balanced system. In: Perspectives On Plant Population Ecology, Dirzo R and J Sarukhán (eds). Sinauer Associates Inc., Sunderland, MA, pp. 305–323.

Mooney HA, Bonicken TM, Christensen NL, Lotan JE, Reiners WA (eds). 1981. Fire Regimes and Ecosystems Properties. General Technical Report WO-26.

Mooney HA, Kummerow J, Johnson AW, Parsons DJ, Keeley S, Hoffmann A, Hays RI, Giliberto J, Chu C. 1977. The producers—Their resources and adaptive responses. In: Convergent Evolution in Chile and California, Mooney HA (ed). Dowden, Hutchinson & Ross, Inc., Stroudsburg, pp. 85–143.

Mullette KJ, Bamber RK. 1978. Studies of the lignotubers of *Eucalyptus gummifera* (Gaert. & Horchr.). III Inheritance and chemical composition. Aust J Bot 26:23–28.

Norby RJ, O'Neill EG, Luxmoore RJ. 1986. Effects of atmospheric CO_2 enrichment on the growth and mineral nutrition of *Quercus alba* seedlings in nutrient-poor soil. Plant Physiology 82:83–89.

Oberbauer SF, Sionit N, Hastings SJ, Oechel WC. 1986. Effects of CO_2 enrichment and nutrition on growth, photosynthesis, and nutrient concentration of Alaskan tundra plant species. Can J Bot 64:2993–2998.

Oechel WC, Lawrence L. 1981. Carbon allocation and utilization. In: Resource Use by Chaparral and Matorral, Miller PC (ed). Springer-Verlag, Berlin, pp. 185–235.

Oechel WC, Strain BR. 1985. Native species responses to increased atmospheric CO_2 concentration. In: Direct Effects of Increasing Carbon Dioxide on Vegetation, Strain BR and JD Cure (eds). U.S. DOE, Nat. Tech. Info. Service, Springfield, VA, pp. 117–154.

Pate JS, Froend RH, Bowen BJ, Hansen A, Kuo J. 1990. Seedling growth and storage characteristics of seeder and resprouter species of Mediterranean-type ecosystems of S.W. Australia. Ann Bot 65:585–601.

Pearcy RW, Björkman O, Caldwell MM, Keeley JE, Monson RK, Strain BR. 1987. Carbon gain by plants in natural environments. BioScience 37:21–29.

Peet MM, Willits DH. 1984. CO_2 enrichment of greenhouse tomatoes using a closed-loop heat storage: effects of cultivar and nitrogen. Scientia Horticulturae 24:21–32.

Raper CD, Parson LR, Patterson DT, Kramer P. 1977. Relationship between growth and nitrogen accumulation for vegetative cotton and soybean plants. Bot Gaz 138:129–137.

Rapp M, Cabanettes A. 1981. Biomass and productivity of a *Pinus pinea* L. stand. In: Components of Productivity of Mediterranean-Climate Regions—Basic and Applied Aspects, Margaris NS and HA Mooney (eds). Dr. W. Junk Publishers, The Hague, pp. 131–134.

Rapp M, Lossaint P. 1981. Some aspects of mineral cycling in the garrigue of southern France. In: Mediterranean-Type Shrublands, di Castri F, DW Goodall, and RL Specht (eds). Elsevier Scientific Publishing Company, Amsterdam, pp. 289–301.

Reynolds JF. 1986. Adaptive strategies of desert shrubs with special reference to the creosotebush. In: Pattern and Process in Desert Ecosystems, Whitford WG (ed). University of New Mexico Press, Albuquerque, NM, pp. 19–49.

Reynolds JF, Thornley JHM. 1982. A shoot:root partitioning model. Ann Bot 49:585–597.

Richards D, Goubran FH, Collins KE. 1979. Root-shoot equilibria in fruiting tomato plants. Ann Bot 43:401–404.

Rodin LE, Bazilevich NI. 1967. Production and Mineral Cycling in Terrestrial Vegetation. Oliver and Boyd, Edinburgh.

Rundel PW. 1988a. Leaf structure and nutrition in Mediterranean-climate sclerophylls. In: Mediterranean-Type ecosystems. A Data Source Book, Speht RL (ed). Kluwer Academic Publishers, Dordrecht, pp. 157–167.

Rundel PW. 1988b. Vegetation, nutrition and climate-data tables. (2) Foliar analyses. In: Mediterranean-Type Ecosystems. A Data Source Book, Speht RL (ed). Kluwer Academic Publishers, Dordrecht, pp. 63–80.

Rundel PW, Parsons DJ. 1979. Structural changes in chamise (*Adenostoma fasciculatum*) along a fire-induced age gradient. J Range Manage 32:462–466.

Schlesinger DH, Gill DS. 1980. Biomass, production and changes in the availability of light, water and nutrient during the development of pure stands of the chaparral shrub, *Ceanothus megacarpus*, after fire. Ecology 61:781–789.

Schrader LE. 1978. Uptake, accumulation, assimilation and transport of nitrogen in higher plants. In: Nitrogen in the Environment, Vol 2, Nielson DR and JG MacDonald (eds). Academic Press, New York, pp. 101–141.

Specht RL, Rayson P. 1957. Dark Island Heath (Ninety-Mile Plain, south Australia). III. The root systems. Aust J Bot 5:103–114.

Specht RL, Rundel PW. 1990. Sclerophylly and foliar nutrient status of Mediterranean-climate plant communities in southern Australia. Aust J Bot 38:459–474.

Specht RL, Rayson P, Jackman ME. 1958. Dark Island Heath (Ninety-Mile Plain, south Australia). VI. Pyric succession: changes in composition, coverage, dry weight and mineral nutrient status. Aust J Bot 6:59–88.

Strain BR. 1985. Physiological and ecological controls on carbon sequestering in ecosystems. Biogeochemistry 1:219–232.

Thomas RB, Strain BR. 1991. Root restriction as a factor in photosynthetic acclimation of cotton seedlings grown in elevated carbon dioxide. Plant Physiol 96:627–634.

Thornley JHM. 1972. A balanced quantitative model for root:shoot ratios in vegetative plants. Ann Bot 36:431–441.

Thornley JHM. 1976. Mathematical Models in Plant Physiology. 318 pp. Academic Press, London.

Weatherley PE. 1982. Water uptake and flow in roots. In: Physiological Plant Ecology II: Water Relations and Carbon Assimilation, Lange OL, PS Nobel, CB Osmond, and H Ziegler (eds). Springer-Verlag, Berlin, Heidelberg, New York, pp. 79–109.

Whittaker RH. 1962. Net production relations of shrubs in the Great Smoky mountains. Ecology 43:357–377.

Whittaker RH, Marks PL. 1975. Methods of assessing terrestrial productivity. In: Primary Productivity of the Biosphere, Lieth H and RH Whittaker (eds). Springer-Verlag, Berlin, pp. 55–118.

Whittaker RH, Bormann FH, Likens GE, Siccama TG. 1974. The Hubbard Brook ecosystem study: forest biomass and production. Ecol Monog 44:233–252.

Williams WE, Garbutt K, Bazzaz FA. Vitousek PM. 1986. The response of plants to elevated CO_2. IV. Two deciduous-forest tree communities. Oecologia 69:454–459.

Wilson JB. 1988. A review of evidence on the control on shoot:root ratio, in relation to models. Ann Bot 61:433–449.

Wilson JR. 1975. Comparative response to nitrogen deficiency of a tropical and temperate grass in the interrelation between photosynthesis, growth and the accumulation of non-structural carbohydrate. Neth J Agric Sci 23:104–112.

Wong SC. 1979. Elevated atmospheric partial pressure of CO_2 and plant growth. I. Interactions of nitrogen nutrition and photosynthetic capacity in C_3 and C_4 plants. Oecologia 44:68–74.

Zammit C. 1988. Dynamics of resprouting in the lignotuberous shrub *Banksia ablongifolia*. Aust J Ecol 13:311–320.

5. Preliminary Studies of the Long-Term CO_2 Response of Mediterranean Vegetation Around Natural CO_2 Vents

F. Miglietta, M. Badiani, I. Bettarini, P. van Gardingen, F. Selvi, and A. Raschi

One of the major concerns about the effect of increasing carbon dioxide concentration in the atmosphere is for the long-term response of photosynthesis and biomass accumulation in natural vegetation. Knowledge about this response is critical while attempting to predict the effect of global change on primary production, on vegetation spatial distribution, and on potential sinks for carbon which may be provided by terrestrial ecosystems. Some recent experiments suggested that carbon sequestering, water use efficiency, and nutrient capture of terrestrial ecosystems will be enhanced by increasing atmospheric CO_2 concentrations (Long and Drake, 1991), but some other studies showed that growth stimulation did not occur in some natural (Grulke et al., 1990) and artificial (Körner, 1992) ecosystems under elevated CO_2. Whatever the case, it has been recently suggested that small spatial-scale community experiments which simulate a future greenhouse climate will be hardly capable of predicting ecosystem functioning and nutrient and water cycling under global change because the necessary range of natural disturbances and relevant information about the long-term response of the system are lacking (Woodward, 1992).

The occurrence of long-existing natural carbon dioxide vents, termed CO_2 springs, was considered to have the potential to overcome some of these experimental limitations, in the sense that these may provide a unique opportunity to study and understand adjustments of existing natural ecosystems to long-term CO_2 enrichment (Miglietta and Raschi, 1993; Miglietta et al., 1993c), where long-term indicates a time period of decades or more. In CO_2 springs, carbon dioxide of deep origin is naturally released to the atmosphere by one or more vents, and this often produces significant anomalies in the atmospheric CO_2 concentration of the surrounding area where natural vegetation occurs. This natural vegetation may therefore be assumed to have been exposed to elevated CO_2 since its primitive establishment and to have been subjected during this time to the largest range of natural disturbances of both physical and biological origin. These key factors are likely to enhance the predictive value of observations that can be made about the response of plants to elevated CO_2 (Woodward, 1992). In this paper, a brief explanation of the origin of deep carbon dioxide is given. Subsequently, conditions of natural atmospheric CO_2 enrichment at the Italian CO_2 springs are illustrated and, finally, some examples of the potential use of these springs for research on the impact of increasing atmospheric CO_2 concentrations of Mediterranean ecosystems are shown and discussed.

The CO_2 Spring Resource

The Origin of Deep Carbon Dioxide

There are more than 100 sites in central Italy where CO_2 freely vents at the surface. Theories on the origin of carbon dioxide in natural manifestations may most conveniently be divided into two general groups: organic and inorganic. The CO_2 of organic origin can be produced both from the decay or combustion of organic matter and from the oxidation of hydrocarbons by mineralized waters (Panichi and Tongiorgi, 1975). In this last case, carbon is lighter than the carbonate rocks, and its stable isotope ratio $\delta^{13}C$ falls in the range of isotope ratios of organic material. On the contrary, carbon dioxide of inorganic origin is at most related to the occurrence of volcanic activity in the sedimentary sequences of carbonate rocks, and the CO_2 produced is normally heavier. Other theories, such as igneous emanations and dissolution of limestone by ground waters, represent the main mechanisms of CO_2 production. In the case of the manifestations occurring in the central part of Italy, it may be reasonably assumed that the CO_2 originates from the dissolution of limestones (Panichi and Tongiorgi, 1975). Nevertheless, there is some variation in the isotopic composition of the CO_2 venting at the surface, and this probably depends on the fractionation occurring at different temperatures as well as the presence of organic sediments or underground reservoirs of methane contaminating the CO_2 released at the surface. In a study of the whole geothermal region of central Italy, Panichi and Tongiorgi (1975) found that the $\delta^{13}C$ composition of CO_2 ranged from $+1.8$ to -21.3‰ with an average of -4‰.

The existence of gas emissions in central Italy was already reported in publications of the Eighteenth and Nineteenth Centuries (Targioni-Tozzetti, 1835). Although it may be assumed that these CO_2 emissions have existed for a longer time, these reports provide sufficient evidence that for the large majority of springs, there has not been significant change in CO_2 emission rate for more than a century.

Natural Atmospheric CO_2 Enrichment

More than 20 CO_2 springs were visited in central Italy between October 1990 and December 1991 while searching for conditions of natural atmospheric CO_2 enrichment. Ten of these were considered to be relevant for the study of long-term adjustment of natural vegetation to CO_2 enrichment (Figure 5-1). The occurrence of natural CO_2 enrichment was

Figure 5-1. Map of Italy showing the location of the principal CO_2 springs. (1) Solfatara di Pozzuoli (Naples); (2) Acque Albule (Rome); (3) Caldara di Manziana (Viterbo); (4) Biscione, Monterano (Rome); (5) Solfatara (Viterbo); (6) Bossoleto, Rapolano Terme (Siena); (7) Castiglioni (Siena); (8) Vagliagli (Siena); (9) Carpineto, Vagliagli (Siena); (10) Gambassi (Florence).

carefully determined in each site by means of direct CO_2 concentration measurements as well as by indirect isotopic methods. Direct measurements were based on the combined use of infrared gas analyzers (EGM-1, PP-System) and Dräeger diffusion tubes. These tubes allow a rapid and accurate determination of the mean atmospheric CO_2 concentration over a given time period on the basis of a visual estimation of the progressive color change of an alkaline reactant along a graduated scale (Miglietta and Raschi, 1993). These direct measurements revealed, at first, that anomalies in atmospheric CO_2 were occurring on rather large areas around each spring, but that the size of this enriched area was different depending on the strength of the CO_2 vents, the topography of the site, and the weather (Miglietta and Raschi, 1993). The presence of steep gradients of CO_2 concentrations with distance from the vents was also observed, indicating that plants occurring at different distances from the vents were likely exposed to different average CO_2 concentrations. The reconstruction of these gradients on the basis of direct CO_2 concentration measurements was, however, very difficult. Short-term fluctuations in CO_2 concentration and their variability associated with conditions of atmospheric turbulence or stability prevented, in fact, a correct determination of average concentrations, since the number of sampling points needed for that purpose was too high.

The gradients could instead be better defined using indirect methods based on isotopic signals provided by the occurrence of significant anomalies in both the radiocarbon and stable carbon isotopic ratio of plant tissues due to the mixing of ambient CO_2 and the carbon dioxide released by the vents.

The radiocarbon method is based on the fact that carbon dioxide emitted from the vents, which originates from the dissolution of limestone, is completely depleted in the radioactive carbon isotope ^{14}C. When this CO_2 mixes with air, the atmospheric CO_2 concentration of the resulting mixture increases while its ^{14}C content decreases. This signal is detectable in plant tissues, since their radiocarbon content depends on the discrimination against ^{14}C occurring during photosynthesis and on ^{14}C abundance in source air. Discrimination against ^{14}C may be simply calculated as is currently done in radiocarbon dating methods (Geyh and Scheicher, 1990) so that ^{14}C abundance can provide an estimation of the average effective CO_2 exposure according to the following equation:

$$[CO_2] = 350 * (M_a/M_p) \qquad (5\text{-}1)$$

where

M_a = abundance of ^{14}C in the atmosphere,
M_p = measured ^{14}C abundance in plant material corrected for discrimination.

This method has recently been successfully used in the determination of the average exposure CO_2 concentration (Table 5-1) along a CO_2 gradient on plants of *Erica arborea* growing in a CO_2 spring in southern Italy (Figure 5-1, site 1).

The stable isotope method is based on the fact that carbon released from vents is often heavier than atmospheric carbon (Panichi and Tongiorgi, 1975). Again, the mixing of CO_2 from the vents and from the air determines an increase in atmospheric CO_2 concentration as well as a change in the stable carbon isotope ratio of the mixture. In this case, plants with C_4 metabolism, which are known to have a conservative discrimination versus $\delta^{13}C$ across a range of environmental conditions (Farquhar, 1983; Polley et al., 1993), may be used as biosensors to measure the average effective isotope ratio of the air near vents and from this the average CO_2 concentration using the following equation (Farquhar, 1983):

$$\Delta C_4 = a + (b_4 + b\Phi - a)C_i/Ca \qquad (5\text{-}2)$$

Table 5-1. CO_2 Exposure Levels Calculated from ^{14}C Abundance Measurements (Mp) of Plant Material of *Erica arborea* Growing at Solfatara di Pozzuoli (Figure 5-1) (Bettarini et al., 1995)

Location	MP	$[CO_2]$ ($\mu mol\ mol^{-1}$)
18	1.048	369
16	0.989	392
17	0.976	397
19	0.973	398
15	0.960	403
14	0.955	406
9	0.909	426
12	0.870	446
13	0.853	454
8	0.808	479
1	0.805	481
3	0.775	500
2	0.717	540
11	0.697	556
4	0.690	561
5	0.675	574
6	0.672	577
20	0.657	589
7	0.654	592
10	0.521	744

The numbers of location refer to selected shrubs occurring all over the area, at different distances from vents.

where

ΔC_4 = carbon isotope discrimination in plants with C_4 metabolism,
a = −4.4‰, discrimination associated with CO_2 diffusion,
b_4 = −0.7‰, discrimination of PEP carboxylase against ^{13}C,
b = −27‰, discrimination associated with Rubisco carboxylation,
Φ = proportion of carbon fixed by PEP which leaks out of the bundle sheath,
C_i/C_a = ratio of intercellular to ambient CO_2 concentrations.

Since the second term of Equation 5-2 equals zero in the majority of cases (Farquhar, 1983) ΔC_4 can be assumed to be equal to a, and the isotope ratio of source air (δ_{sa}) is given by

$$\delta_{sa} = \Delta C_4 - a \tag{5-3}$$

where

δC_4 = isotope ratio measured on C_4 plants.

Finally, CO_2 concentration of source air ($[CO_2]$) can be calculated by mass balance according to the following equation:

$$[CO_2] = 350 * (\delta_{air} - 0.1/(\delta_{sa} - 0.1) \tag{5-4}$$

where

δ_{air} = current $\delta^{13}C$ value of the atmospheric air (−8‰).

This stable isotope method was successfully tested in a CO_2 spring (Figure 5-1) where the CO_2 released by the vent had an isotopic ratio close to zero ($\delta^{13}C = -0.2‰$). In this case, plants of *Pennisetum americanum* (pearl millet, a C_4 species) were sown along a CO_2 gradient in small groups 5 m apart. The carbon isotope ratio of leaf tissues was then measured 3 months after sowing. Isotopic ratio (δC_4) of millet plants decreased with increasing distance from the vents, together with atmospheric CO_2 concentrations (Figure 5-2). These ranged from 360 $\mu mol\,mol^{-1}$ to 742 $\mu mol\,mol^{-1}$ (Miglietta et al., 1993a).

The reconstruction of CO_2 gradients by means of both radiocarbon and stable isotope methods provides an invaluable tool to understand the effective level of CO_2 exposure for the different individual plants or populations studied. These plants can be then used to assess the long-term response to the different levels of CO_2 enrichment and to search for acclimatory or even adaptative traits of their metabolism.

Phytotoxic Gases

Hydrogen sulfide is the only phytotoxic component of gases emitted by a number of CO_2 springs in central Italy (Duchi et al., 1985). The presence

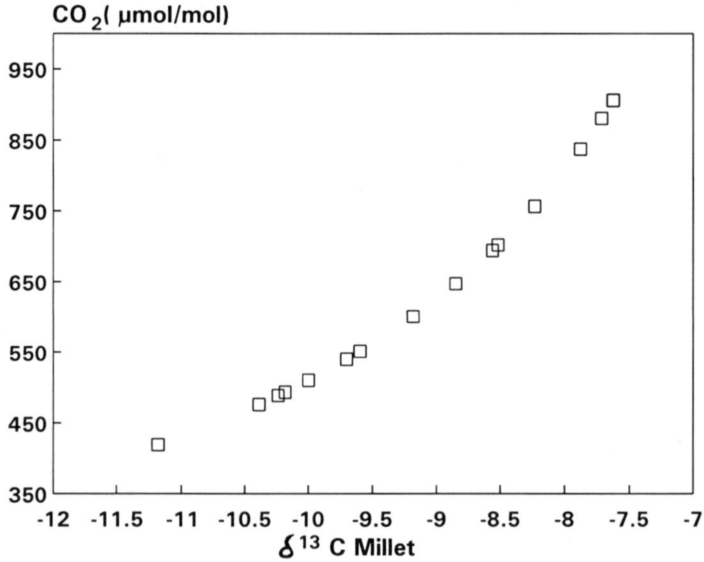

Figure 5-2. Isotopic ratio values ($\delta^{13}C$) measured on leaves of pearl millet are plotted versus the calculated average CO_2 concentrations along a transect at the CO_2 spring Solfatara.

of H_2S in the gas emissions led researchers to wonder about the occurrence of toxic effects on vegetation. It is known, in fact, that even low concentrations (0.03 μmol mol^{-1}) of H_2S in the air may have a potential toxic effect on plants (De Kok, personal communication). Difficulties in the measurement of average H_2S concentrations in the air surrounding the plants were encountered (Miglietta and Raschi, 1993) because very low concentrations together with rapid fluctuations made it difficult to obtain reliable information. Although H_2S is a known inhibitor of plant respiration (Maas and De Kok, 1988), it has been shown that an important toxic effect of H_2S exposure is the increase at the cellular level of nonprotein sulfhydryl compounds (TF-SH), of cysteine/sulfhydryl ratio, and in the size of the glutathione pool (GSH + GSSG), and a reduction in the capacity to oxidize exogenous NADH (Rennenberg, 1984; Maas et al., 1987, 1988; De Kok et al., 1988). Hence the presence of H_2S toxicity was evaluated by biochemical analysis of plant tissues. In a recent experiment (Miglietta et al., 1993b; Badiani et al., 1993), plants of soybean, which is known to be a H_2S-sensitive species (De Kok, 1990), were exposed to increasing CO_2 concentrations and, presumably, also to increased H_2S concentrations in a spring (Solfatara, Figure 5-1), where the H_2S concentration in the emitted gas is known to be the highest (1,200 μmol mol^{-1}) and then compared with control plants grown at ambient CO_2 concentration (350 μmol mol^{-1}) where H_2S may be assumed to be absent. During growth, uniform and apparently healthy leaves were collected. Total water-soluble nonprotein sulfhydryl compounds and cysteine contents were determined on leaf extracts using the methods of Maas et al. (1987) and De Kok et al. (1986); glutathione (GSH) and glutathione disulfide levels (GSSG) were determined according to the method of Griffith (1989). NADH oxidation by leaf extracts was measured according to the method of Maas and De Kok (1988). Compared with control, neither accumulation of total free thiols, GSH pool, and cysteine, nor increase in the cysteine /sulfhydryl ratio were apparent in leaves grown at elevated CO_2 (Miglietta et al., 1993b). The titers of total free thiols were very similar to those observed in soybean leaves grown in H_2S-free environment (Buwalda, De Kok and Stulen, personal communication) and no foliar accumulation of sulfur was apparent. Also, the GSH content was not altered by the presence of H_2S and, together with the substantial constancy of NADH oxidation capacity, this suggests the absence of direct toxic effects due to H_2S in the gas emissions.

These results suggested that due to the rapid dilution and oxidation of H_2S in the air, its atmospheric concentrations at the CO_2 springs may be considered subtoxic even in worst cases where H_2S contamination is the highest. This conclusion was also confirmed by similar measurements made on a natural population of *Erica arborea* growing around another CO_2 spring (Figure 5-1, site 1).

Vegetation Around CO_2 Springs

The spatial arrangement, structure, and diversity of vegetation around CO_2 springs are largely determined by local ecological factors. A distributional pattern in concentric rings formed by four community types (Figure 5-3) is repeated in most of the sites, despite their localization in different phytoclimatic areas. The overwhelming ecological gradient occurring in CO_2 vents involves three major factors decreasing in intensity with the distance from vents: CO_2 atmospheric concentration, soil moisture, and soil acidity (pH values from 2.3 to 3.7). The combined effects of these factors exert a strong selective pressure which leads to low diversity, repetitivity, and directional variation in the qualitative and quantitative floristic composition of the vegetation (Selvi et al., 1993).

At Bossoleto (Figure 5-1) steep banks of limestone outcrops determine a rather different ecological situation, although the arrangement of the community is consistent with that of other springs. Small fragments of species-rich xerocalcicolous Mediterranean grassland occur behind a narrow ring of herbaceous vegetation dominated by *Agrostis stolonifera* closely surrounding the vents. A list of species occurring around each CO_2 spring is reported in Table 5-2.

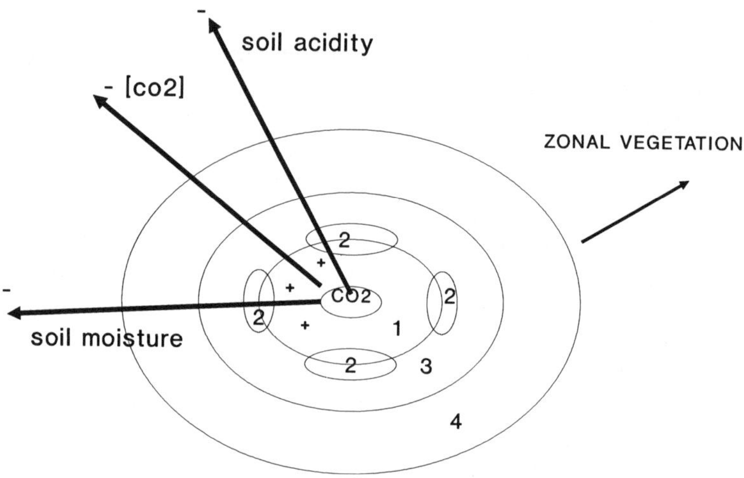

Figure 5-3. Schematic representation of the spatial arrangement of vegetation around CO_2 vents. (1) Aphytoic zone: the innermost ring with no vascular plants. (2) *Scirpus-Juncus* community: a patch-forming wetland community around vents with permanent water bodies formed by almost pure and dense stands of *Scirpus lacustris* subsp. *tabernaemontani.* (3) *Agrostis* community type: a ring of dense herbaceous vegetation formed by pure stands of *Agrostis canina* subsp. *monteluccii* Selvi (Selvi, 1994). (4) *Cytisus-Adenocarpus* community type: an open shrubby zone with acid-loving species (*Adenocarpus complicatus* ssp. *complicatus, Cytisus scoparius, Calluna vulgaris*).

The Long-Term CO_2 Response of Mediterranean Species

Experimental work at the CO_2 springs started in 1991. In the following a brief overview of results obtained over the last 2 years and their interpretation is given. This includes data from a study of stomata density and dimensions in *Quercus pubescens*, on photosynthetic capacity of *Phragmites australis*, and on leaf nitrogen content in *Agrostis canina*, *Scirpus lacustris*, and *Erica arborea*. Although their preliminary nature will be evident, these examples will provide a wider justification of the use of CO_2 vents in research.

Leaf Morphoanatomy in *Quercus pubescens*

The observations reported under this heading were made in winter 1990–1991 at Solfatara (Figure 5-1, site 5). Fourteen evenly aged oak trees (*Quercus pubescens*) were selected near and far away from the vents. Samples for stomata observation and measurement were taken from seven sunny leaves located in the upper (3 to 4 m above ground) and lower (1 to 2 m above ground) part of the crown (Abrams and Kubiske, 1990). Microphotographs of Rodhoit impressions (Vazzana et al., 1988)

Table 5-2. List of Vascular Species Occurring at the Different CO_2 Springs in Western Italy

Species
Adenocarpus complicatus (L.) Gay ssp. *complicatus* (5)
Agrostis canina L. ssp. *monteluccii* Selvi (1,2,3,4,5,7,8,9)
Agrostis stolonifera L. (6,10)
Allium sphaerocephalon L. (6)
Arbutus unedo L. (7)
Asparagus officinalis L. (1)
Buxus sempervirens L. (6)
Calluna vulgaris (L.) Hull. (7,8,9)
Calycotome spinosa (L.) Link (1)
Carex hirta L. (3,5)
Castanea sativa Miller (8)
Centaurea alba L. ssp. *deusta* (Ten.) Nyman (6)
Cerastium semidecandrum L. (6)
Cistus incanus L. (6)
Cyperus esculentus L. (1)
Cytisus scoparius (L.) Link (3,4,5,8)
Deschampsia flexuosa (L.) Trin (4,9)
Erica arborea L. (1,7)
Fraxinus ornus L. (6,7)
Galium corrudifolium Vill. (6)
Globularia punctata Lapeyr (6)
Hedera helix L. (6)
Helianthemum apenninum (L.) Miller (6)
Holcus lanatus L. (3,4,5,8)
Imperata cylindirica L. (1)
Juncus articulatus L. (2,3,4,9,10)
Juncus effusus L. (2,3)
Molinia caerulea L. ssp. *arundinacea* (Schrank) Paul (7)
Myrtus communis L. (1,7)
Ostrya carpinifolia Scop. (6)
Phragmites australis (Cav.) Trin ex Steud. (2,5,6)
Plantago lanceolata L. (6)
Poa annua L. (4,8,9)
Poa trivialis L. (5,8)
Populus tremula L. (5,6)
Quercus cerris L. (4,5,9)
Quercus ilex L. (6,7,8)
Quercus pubescens Willd. (5,6,7)
Quercus robur L. (5)
Rubia peregrina L. (6,7)
Ruscus aculeatus L. (6)
Salix alba L. (5)
Sanguisorba minor L. (6)
Scabiosa columbaria L. (6)
Scirpus holoschoenus L. (2,6,7,10)
Scirpus lacustris L. ssp. *tabernaemontani* (Gmelin) Syme (2,3,5,10)
Sherardia arvensis L. (6)
Silene vulgaris (Moench) Garke ssp. *angustifolia* (Miller) Hayek (6)
Typha latifolia L. (2,3,5)

Numbers in parentheses, see Figure 5-1. Nomenclature follows Flora Europea (Tutin et al., 1964–1980).

Table 5-3. Results of Two-Sample Statistical Analysis (Mean ± Standard Deviation and Random Samples *t*-Test) of Stomatal Characteristics of Oak Trees Grown at Elevanted (N) and at Ambient (F) CO_2

	N	*F*	*t*-Test
Guard cells size (μm^{-2})	355.40 ± 50.6	263.40 ± 59.1	+8.08084**
Stomata (mm^{-2})	514.65 ± 99.5	536.40 ± 99.5	−0.76665
Epidermal cells (mm^{-2})	2605.09 ± 356.4	2665.75 ± 628.8	−0.55715
Stomatal index	0.16 ± 0.02	0.16 ± 0.03	−0.43121

** Significance level < 0.001.

obtained from leaf samples were then observed under the microscope. Stomatal and epidermal cell density and the mean areal size of guard cells were measured. Stomatal and epidermal cell numbers were the same in the two groups of plants, but the areal size of the guard cells appeared to be significantly reduced in the leaves of plants grown in enriched atmosphere (Table 5-3).

When ambient CO_2 is raised, stomatal resistance increases, and net CO_2 assimilation may be increased and water loss reduced (Morison, 1987). Stomatal closure or a reduction in the stomatal pore area per unit surface increases the exchange resistance between leaf and ambient air (Goudriaan and Unsworth, 1990). Similarly to what has been reported to have occurred in plants from herbaria collected in preindustrial times (Woodward, 1992; Peñuelas and Matamala, 1990; Paoletti and Gellini, 1993), acclimation or even genetic adaptation to elevated CO_2 may have been the cause of these permanent changes in the anatomy and distribution of stomata, which likely limited the maximal conductance of leaves (Laisk, 1983; Körner et al., 1979).

Photosynthesis and Stomatal Conductance in *Phragmites australis*

Plants of *Phragmites australis* were selected for a study of photosynthetic acclimation to increased ambient CO_2 concentration. Plants growing at a CO_2 spring were compared in July 1992 with a population growing at a reference site approximately 8 km distant. Average atmospheric CO_2 concentrations were measured during the day using infrared gas analysis and found to be approximately 350 $\mu mol\,mol^{-1}$ at the control site and to range between 800 and 1,500 $\mu mol\,mol^{-1}$ at the spring. Measurements of leaf gas exchange were obtained using portable photosynthesis system (LI-6200, Li-COR, Lincoln, Nebraska, USA) to investigate long-term photosynthetic response of leaves to elevated CO_2 in the field. Assimilation rate (A) was compared at identical intercellular CO_2 concentrations (C_i) by generating A/C_i curves for the leaves. The approach was modified as described by McDermitt et al. (1989). Leaves were placed in a small leaf chamber ($0.3\,dm^3$) and allowed to acclimate for 15 minutes with the gas exchange system in open mode in an air stream of $1{,}800\,\mu mol\,mol^{-1}$ and water vapor pressure deficit of less than 1.0 KPa. The gas exchange system was then changed to closed mode and the CO_2 concentration in the chamber fell as a result of photosynthesis. The proportion of the air passing through a magnesium perchlorate desiccant was varied to control the water vapor pressure deficit in the chamber at a relatively constant level. Chamber heating was minimized by blowing air over the chamber using an electric fan mounted on a tripod. Data were recorded commencing at a chamber CO_2 concentration of $1{,}500\,\mu mol\,mol^{-1}$ and then at regular intervals. Estimates of the photosynthetic rate, stomatal conductance, and leaf internal CO_2 concentrations were calculated by the system console using standard formulas (von Caemmerer and Farquhar, 1981) as modified by McDermitt et al. (1989) to correct for the effects of air leaks in a closed system. Examples of the results from this approach are shown in Figure 5-4 for typical leaves from the enriched and control sites. The maximum rate of photosynthesis at saturating CO_2 concentrations is clearly significantly lower in plants from the spring, as is the initial slope of the A/C_i relationship, indicating that photosynthetic acclimation occurred. It has not yet been tested if such decreased capacity in plants

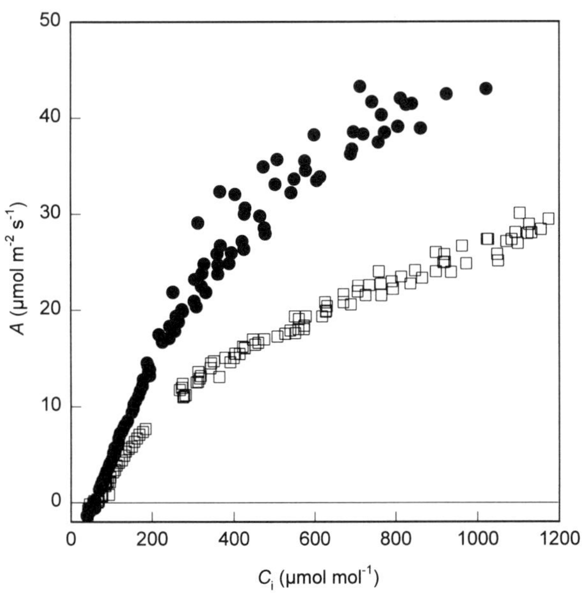

Figure 5-4. A/C_i relationship for typical *Phragmites australis* leaves growing at a CO_2 spring (□) and in a control site (●). Data were obtained from plants in situ using a Li-COR LI-6200 Portable Photosynthesis System.

Table 5-4. Stomatal Density for *Phragmites australis* Measured Using Silicon Rubber Impression

$[CO_2]$	Abaxial Surface	Adaxial Surface
1500 μmol mol^{-1}	738 ± 36	537 ± 40
600 μmol mol^{-1}	921 ± 52	713 ± 50
350 μmol mol^{-1}	774 ± 55	978 ± 42

Data are the mean of 5 leaves at Bossoleto (Figure 5-1) and of 6 leaves at the control site ± se. The values of CO_2 levels are average daytime concentrations measured both by infra-red gas analysis and the radiocarbon method described in the text.

exposed to elevated CO_2 was an acclimatory or a genetic adaptative response. This problem is currently under investigation by transplanting experiments. Plants of *P. australis* growing at the CO_2 spring and in the control site were also compared in terms of stomatal conductance and stomatal density. Stomatal conductance was estimated from the surface temperature difference between a normal transpiring leaf and a non-transpiring leaf that had been coated with petroleum jelly. Rates of transpiration were proportional to the temperature difference, and estimates of stomatal conductance were calculated using data for air temperature and humidity (Leuning and Foster, 1990). Preliminary analysis of these data showed clearly that stomatal conductances of plants from the CO_2 site were almost one-half of the values observed at the control site. Part of the reduction in stomatal conductance was explained by lower stomatal density, particularly on the adaxial leaf surface (Table 5-4).

Photosynthesis and Nitrogen Relationship

Photosynthesis measurements made on *P. australis* revealed that downward regulation may occur in response to long-term CO_2 exposure. This suggested a broader investigation of different species occurring at the different CO_2 springs.

All of the biochemical and photobiological processes of photosynthesis require nitrogen, and it may be estimated that more than 70% of leaf nitrogen is invested in the chloroplast and in photosynthetic processes (Field and Mooney, 1986). Not surprisingly, a robust relationship between nitrogen content of the leaves and their photosynthetic capacities has often been found in natural vegetation (Field and Mooney, 1986) and, in particular, in Mediterranean vegetation (De Lillis and Federici, 1991). It has been argued that this relationship reflects a fundamental relationship that is independent of species and growth conditions (Field, 1991) although a different interpretation of the data has been made (Schulze and Chapin, 1987) and a few exceptions have been found (Evans, 1989). In C_3 plants, the photosynthetic enzyme Rubisco, which is responsible for carboxylation, also has a high affinity for oxygen; when the oxygenation reaction occurs, this leads to photorespiration and a net loss of CO_2 (Stitt, 1991). When plants are exposed to elevated CO_2 concentrations, photorespiration decreases and carbon assimilation increases. The reduction in photorespiration suggests that reduced nitrogen investment in Rubisco at elevated CO_2 does not necessarily lead to a decrease in current photosynthetic assimilation, especially if intercellular CO_2 concentrations rise and photosynthesis is limited more by RuBP regeneration capacity than by Rubisco activity (van Caemmerer and Farquhar, 1981), but it may be speculated that even a small decline in nitrogen investment in the leaves, expressed on a leaf area basis, does necessarily reflect the occurrence of a downward regulation of photosynthetic capacity (Sage et al., 1989).

A study of the variation in leaf nitrogen content on native species occurring at the various CO_2 sites has been recently undertaken. Leaf nitrogen content has been monitored throughout the growing season in one Mediterranean species (*Erica arborea* L.), in a grass species (*Agrostis canina* L.), and in an emergent wetland species (*Scirpus lacustris tabernaemontani*) growing in the various CO_2 springs as well as in control sites. Specific leaf area (SLA $m^2 g^{-1}$) and percent of leaf nitrogen were determined. The results showed that leaf nitrogen did not significantly decrease with increasing CO_2 in the species studied (Figure 5-5).

These observations contrasted with results of other, shorter-term experiments in elevated CO_2, in which leaf nitrogen content and thereby the photosynthetic capacities of leaves were decreased (Sage et al., 1989; Woodin et al., 1992).

Such a striking difference suggested that photosynthetic acclimation, as it was observed in laboratory or other enclosured experiments, could just be the result of a transient phenomenon. In natural ecosystems, after long-term exposure to elevated CO_2 concentrations, an adjustment of the source /sink ratio of the plants might instead have occurred, thus preventing the accumulation of nonstructural carbohydrates in the leaves and the feedback mechanisms leading to a downward regulation of photosynthesis and of nitrogen investment in the photosynthetic processes. Increased root growth, which has often been observed to be associated with a rise in CO_2 concentration (Norby et al., 1986), is likely to be the major factor responsible for this adjustment, especially when nutrient availability is limiting and the mycorrhizal symbiosis becomes critical for nutrient acquisition, as occurs in Mediterranean ecosystems. A larger root system and increased carbohydrate availability sustaining the mycorrhizal symbiosis may in fact increase the volume of soil explored by the roots and the nutrient acquisition by the plant (Mooney and Winner, 1991).

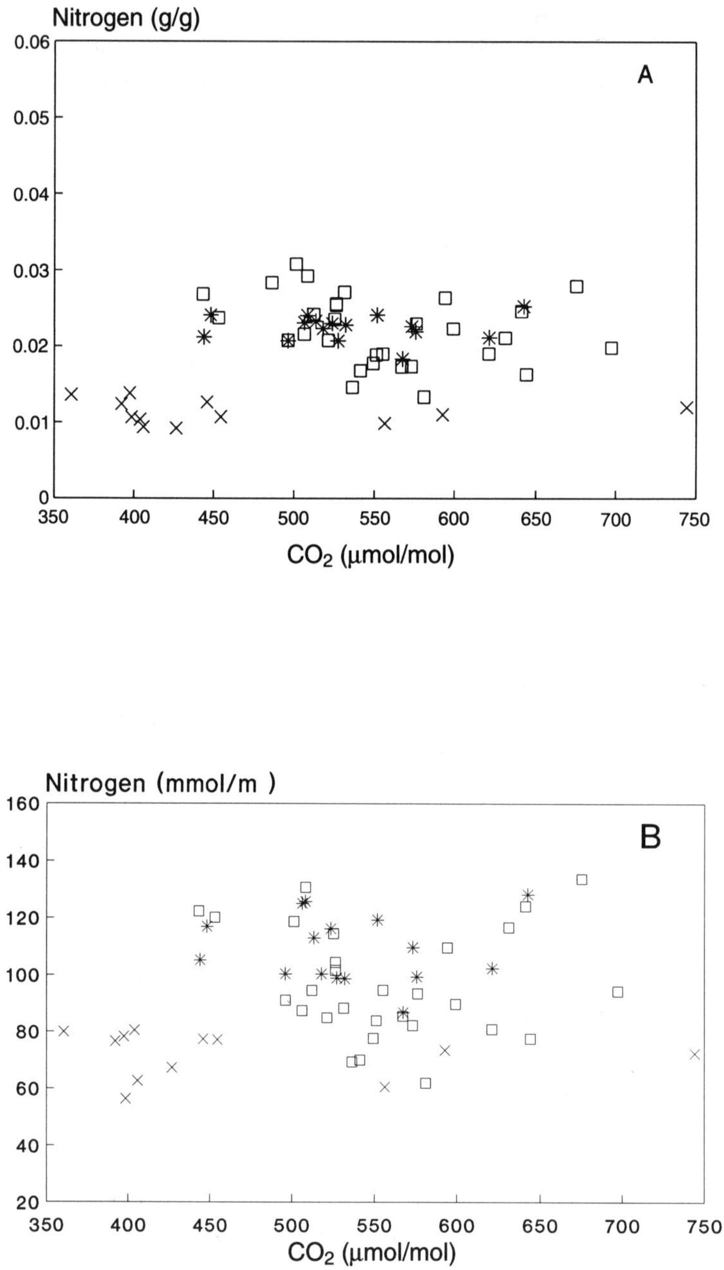

Figure 5-5. Leaf nitrogen concentration expressed on a weight basis (A) and on a leaf area basis (B) is plotted versus mean atmospheric CO_2 concentrations for three native species growing at two CO_2 springs (n. 5, 6, 7 Figure 1) (□) *Agrostis canina*, (*) *Scirpus lacustris*, (×) *Erica arborea*.

Conclusions

The preliminary results from research conducted on vegetation growing around CO_2 springs that have been briefly summarized in this chapter may give an idea of the potentials offered by CO_2 springs for studying the long-term response of species to chronic CO_2 enrichment. Naturally, CO_2-enriched environments have the potential to address some important questions about acclimation versus adaptation to elevated CO_2 in Mediterranean ecosystems. A better knowledge of differences and similarities in the response of different species to CO_2 or of the same species under different climatic conditions is a priority of research, in view of predicted global atmospheric changes. The integration of the various methods of environmental physiology and of relevant theories concerning ecosystem functioning is likely to enhance the scientific value of the resource provided by the CO_2 springs.

Some of the observations made on natural vegetation at the sites, which are briefly reported in this chapter, suggested that homeostatic adjustment of natural vegetation to elevated CO_2 may play a fundamental role in the long-term response of ecosystems. However, the photosynthetic capacity of *P. australis* suggested that downward regulation of photosynthesis may occur in the long term in some species. This apparent contradiction between experimental results may depend on differences existing between species or on the fact that disturbances other than elevated CO_2 may sometimes occur in these natural ecosystems. Provided that the effect of elevated CO_2 and of these disturbances may be separated, the occurrence of natural vegetation around CO_2 springs has the potential to provide an unrivaled resource to study adjustment mechanisms and, potentially, the occurrence of genetic adaptation in the flora in response to elevated CO_2 concentrations.

Greater international cooperation in the search for other CO_2 springs in other areas of the Mediterranean and of the world may provide a good basis for improving our understanding of the impact of global change on terrestrial vegetation.

References

Abrams MD, Kubiske ME. 1990. Leaf structural characteristics of 31 hardwood and conifer tree species in central Wisconsin: influence of light regime and shade-tolerance rank. For Ecol Manage 31:245–253.

Badiani M, D'Annibale A, Paolacci AR, Miglietta F, Raschi A. 1993. The antioxidant status of soybean (*Glycine max*) leaves grown under natural CO_2 enrichment in the field. Aust J Plant Physiol 20:275–284.

Bettarini I, Calderoni G, Miglietta F, Raschi A, Ehleringer J. 1995. Isotopic discrimination and leaf nitrogen content of *Erica arborea* L. along a CO_2 concentration gradient in a carbon dioxide spring in Italy. Tree Physiol 15: 327–332.

von Caemmerer S, Farquhar GD. 1981. Some relationships between the biochemistry of photosynthesis and the gas exchange of leaves. Planta 153:376–387.

De Kok LJ. 1990. Sulfur metabolism in plants exposed to atmospheric sulfur. In: Sulfur Nutrition and Sulfur Assimilation in Higher Plants, Rennenberg H, (ed), pp. 111–130. SPB Academic Publishing, The Hague.

De Kok LJ, Buwalda F, Bosma W. 1988. Determination of cysteine and its accumulation in spinach leaf tissue upon exposure to excess sulfur. J Plant Physiol 133:502–505.

De Kok LJ, Maas FM, Godeke J, Haaksma AB, Kuiper PJL. 1986. Glutathione, a tripeptide which may function as a temporary storage compound of excessive sulphur in H_2S fumigated spinach plants. Plant Soil 91:349–352.

De Lillis M, Federici F. 1991. Vegetation dynamics and ecophysiological trends in the Mediterranean succession. In: Symposium on mechanisms in vegetation dynamics. In 34th I.A.V.S. Pro. Conf., 26–30 August 1991, Eger, Hungary.

Duchi V, Minissale A, Romani L. 1985. Studio geochimico su acque e gas dell'area geotermica lago di Vico-M. Cimini (Viterbo). Atti Società Toscana Scienze Naturali, Memorie, Serie A 92:237–254.

Evans JR. 1989. Photosynthesis and nitrogen relationship in leaves of C3 plants. Oecologia 78:9–19.

Farquhar GD. 1983. On the nature of carbon isotope discrimination in C_4 species. Aust J Plant Physiol 10:205–226.

Field C. 1991. Ecological scaling of carbon gain to stress and resource availability. In: Response of Plants to Multiple Stresses, Mooney HA, WE Winner, and J Pell (eds). Academic Press.

Field C, Mooney H. 1986. The photosynthesis-nitrogen relationship in wild plants. In: On the Economy of Plant Form and Function, Givnish TJ (ed), pp. 25–55. Cambridge University Press, Cambridge, UK.

Geyh MA, Scheicher H. 1990. Absolute age determination: physical and chemical dating methods and their application. Springer Verlag, Heidelberg, p. 503.

Goudriaan J, Unsworth MH. 1990. Implications of increasing carbon dioxide and climate change for agricultural productivity and water resources. In: Impact of Carbon Dioxide, Trace Gases and Climate Change on Global Agriculture, Kimball BA, NJ Rosenberg, and LH Allen (eds), pp. 111–130. ASA Special Publication n. 53.

Griffith OW. 1989. Determination of glutathione and glutathione disulphide using glutathione reductase. Ann Biochem 106: 207–212.

Grulke NE, Riechers GH, Oechel WC, Hjelm U, Jaeger C. 1990. Carbon balance in tussock tundra under ambient and elevated atmospheric CO_2. Oecologia 83:485–494.

Körner C. 1992. Responses to elevated carbon dioxide in artificial tropical ecosystems. Science 257:1672–1675.

Körner C, Scheel A, Bauer H. 1979. Maximum leaf diffusive conductance in vascular plants. Photosynthetica 13:45–82.

Laisk A. 1983. Calculation of photosynthetic parameters considering the statistical distribution of stomatal apertures. J Exp Bot 149:1627–1635.

Leuning R, Foster IJ. 1990. Estimation of transpiration by single trees: comparison of a ventilated chamber, leaf energy budgets and a combination equation. Agric For Meteorol 51:63–86.

Long SP, Drake BG. 1991. Effects of the long-term elevation of CO_2 concentration in the field on the quantum yield of photosynthesis of the C3 sedge *Scirpus olney*. Plant Physiol 96:221–226.

Maas FM, De Kok LK. 1988. In vitro NADH oxidation capacity as an early indicator for growth reduction in spinach exposed to H_2S in the ambient air. Plant Cell Physiol 29:523–526.

Maas FM, van Loo EN, van Hasselt PR. 1988. Effect of long-term H_2S fumigation on photosynthesis in spinach. Correlation between CO_2 fixation and chlorophyll fluorescence. Physiol Plantarum 72:77–83.

Maas FM, De Kok LJ, Peters JL, Kuiper PJC. 1987. A comparative study on the effects of H_2S and SO_2 fumigation on the growth and accumulation of sulphate and sulphydryl compounds in *Trifolium pratense* L., *Glycine max* Merr. and *Phaseolus vulgaris* L. J Exp Bot 38:1459–1469.

McDermitt DK, Norman JM, Davis JT, Ball TM, Arkebauer TJ, Welles JM, Roemer SR. 1989. CO_2 response curves can be measured with a field portable closed loop photosynthesis system. Ann Science Forestieres 46:416–420.

Miglietta F, Raschi A. 1993. Studying the effect of elevated CO_2 in the open in a naturally enriched environment. Vegetatio 104:290–301.

Miglietta F, Bettarini I, Raschi A. 1993a. Differences in isotopic discrimination between native *Scirpus lacustris* and *Agrostis canina* growing in a naturally elevated CO_2 site. J Exp Bot 44(supplement):14.

Miglietta F, Raschi A, Resti R, Badiani M. 1993b. Growth and onto-morphogenesis of soybean (*Glycine max* Merril) grown in an open naturally CO_2 enriched environment. Plant Cell Environ 16:909–918.

Miglietta F, Raschi A, Bettarini I, Resti R, Selvi F. 1993c. Natural CO_2 springs in Italy: a resouce for examining the long-term response of vegetation to rising atmospheric CO_2 concentrations. Plant Cell Environ 16:873–878.

Mooney HA, Winner WE. 1991. Partitioning response of plants to stress. In: Response of Plants to Multiple Stresses, Mooney HA, WE Winner, and J Pell (eds). Academic Press.

Morison JIL. 1987. Intercellular CO_2 concentration and stomatal response to CO_2. In: Stomatal function, Zeiger E, GD Farquahar, and IR Cowan (eds), pp. 129–251. Stanford University Press, Stanford, CA.

Norby RJ, O'Neil EG, Luxmoore RJ. 1986. Effects of atmospheric CO_2 enrichment on the growth and mineral nutrition of *Quercus alba* seedlings in nutrient poor soil. Plant Physiol 82:83–89.

Panichi C, Tongiorgi E. 1975. Carbon isotopic composition of CO_2 from springs, fumaroles, mofettes, and travertines of Central and Southern Italy: a preliminary prospection method of geothermal area. In: Second United Nations Symposium on the Development and Use of Geothermal Resources, pp. 815–825, Vol. 1. San Francisco, CA.

Paoletti E, Gellini R. 1993. Stomatal density vatiation in beech and holm oak leaves collected over the past 200 years. Acta Oecol 14:173–178.

Peñuelas J, Matamala R. 1990. Changes in N and S leaf content, stomatal density and specific leaf area of 14 plant species during the last three centuries of CO_2 increase. J Exp Bot 41:1119–1124.

Polley HW, Johnson HW, Marino BD, Mayeux HS. 1993. Increase in C3 plant water-use efficiency and biomass over glacial to present CO_2 concentrations. Nature 361:61–64.

Rennenberg H. 1984. The fate of excess sulfur in higher plants. Ann Rev Plant Physiol 35:121–153.

Sage RF, Sharkey TD, Seemann JR. 1989. Acclimation of photosynthesis to elevated CO_2 in five C3 species. Plant Physiol 89:590–596.

Schulze ED, Chapin FS. 1987. Plant specialization to environments of different resource availability. In: Potentials and Limitations in Ecosystem Analysis, Schulze ED and H Zwolfer (eds). Springer-Verlag, Berlin.

Selvi F. 1994. *Agrostis Canina* L. Subsp. *Monteluccii* Selvi, subsp. nova (Poaceae). Webbia 49:51–59.

Selvi F, Bettarini I, Miglietta F, Raschi A. 1993. Vegetation survey in Italian CO_2 springs. In: International Workshop on Carbon Dioxide Springs and Their Use in Biological Research, 14–17 October 1993, S.Miniato, Italy.

Stitt M. 1991. Rising CO_2 levels and their potential significance for carbon flow in photosynthetic cells. Plant Cell Environ 14:741–762.

Targioni-Tozzetti A. 1835. Analisi chimica delle acque solfureo-termali di Rapolano nella Provincia Superiore Senese. Biblioteca Botanica, Università di Firenze, pp. 80.

Tutin TG, et al (eds). 1964–1980. Flora Europea, Vol 1–5. Cambridge University Press.

Vazzana C, Puliga S, Bochicchio A, Raschi A. 1988. Studio della densità stomatica in un ecotipo di erba medica (*Medicago sativa* L.). Riv. Agron 22(2):127–132.

Woodin S, Graham B, Killick A, Skiba U, Cresser M. 1992. Nutrient limitations of the long-term response of heather (*Calluna vulgaris* (L.) Hull) to CO_2 enrichment. New Phytol 122:635–642.

Woodward FI. 1987. Stomatal numbers are sensitive to increases in CO_2 concentration from pre-industrial levels. Nature 327:617–618.

Woodward FI. 1992. Predicting plant response to environmental change. New Phytol 122:239–251.

6. Anticipated Effects of Elevated CO_2 and Climate Change on Plants from Mediterranean-Type Ecosystems Utilizing Results of Studies in Other Ecosystems

Boyd R. Strain and Richard B. Thomas

The purpose of this paper is to review our knowledge of plant responses to atmospheric carbon dioxide partial pressure. This review follows earlier reviews on the direct effects of atmospheric CO_2 on plants (Strain, 1985, 1987) in which it was pointed out that plants and ecosystems respond in a hierarchical fashion. Increasing CO_2 causes primary physiological responses which generate secondary metabolic responses which lead to tertiary responses of growth and development. Changes in plant biomass and structure induce changes in plant–animal and plant–microbial interactions and eventually changes will appear at the ecosystem level (Table 6-1).

We have previously reported that global atmospheric CO_2 partial pressure is currently limiting biological productivity and is increasing dramatically following the historically low levels of the Pleistocene (Strain, 1991). In the current review, interactions with other components of global change will also be considered. Knowledge will be reviewed without regard to species, phenological stage, or ecosystem. Following the general review, we will compare generalities from research on plants from various ecosystems, and finally we will speculate on anticipated responses of Mediterranean ecosystems to CO_2 increase and other components of global change.

Table 6-1. Hierarchy of Responses of Plants and Ecosystems to Atmospheric CO_2 Enhancement

Physiological and growth responses of plants
Primary physiological responses
Secondary metabolic responses
Tertiary growth responses
Ecological and evolutionary responses of ecosystems
Competition among plants
Plant–animal and plant–microbial responses
Ecological succession through time

A CO_2-induced change in the primary physiological responses will unavoidably cause changes throughout the ecosystem eventually leading to successional and evolutionary changes.

Generalized Plant Responses

In most cases, CO_2 has the properties of a plant fertilizer (Strain, 1992). In an environment that is optimal for plant growth, an increase in CO_2 partial pressure will increase photosynthesis, carbohydrate allocation to growing and storage points, growth respiration, and plant growth (Baker and Enoch, 1983; Pearcy and Bjorkman, 1983). Recent studies in our laboratory and others suggest that there is a relationship between atmospheric CO_2, plant growth stage, and carbon acquisition and allocation (Clough et al., 1981; Peet, 1984; Sasek et al., 1985; Thomas and Strain, 1991). Although regulatory mechanisms are not clearly understood, it is clear that plant source-sink balance is an important determinant on plant response to atmospheric CO_2 increase (Thomas and Strain, 1991; Stitt, 1991). Figure 6-1 is a schematic model of the principal processes and major control points involved in photosynthetic carbon assimilation and carbon allocation of C_3 plants (Sharkey, 1985; Stitt et al., 1987; Woodrow and Berry, 1988).

The effect of CO_2 is similar to the effect of solar irradiance. Increases in both factors will generate increases in photosynthesis, growth, and reproductive output (Figure 6-2). Thus, growth is positively correlated with increases in light and CO_2 partial pressure. Dependent upon the species and the habitat of origin, continual increase in either factor may eventually cause a decline in response. Among growth types, however, there is considerable variation in all characters. In fact, there is considerable variation within species. This inter- and intraspecific variation in response to environmental change is the source of variation for the change we expect to be observed as global change occurs in the future (Strain, 1991).

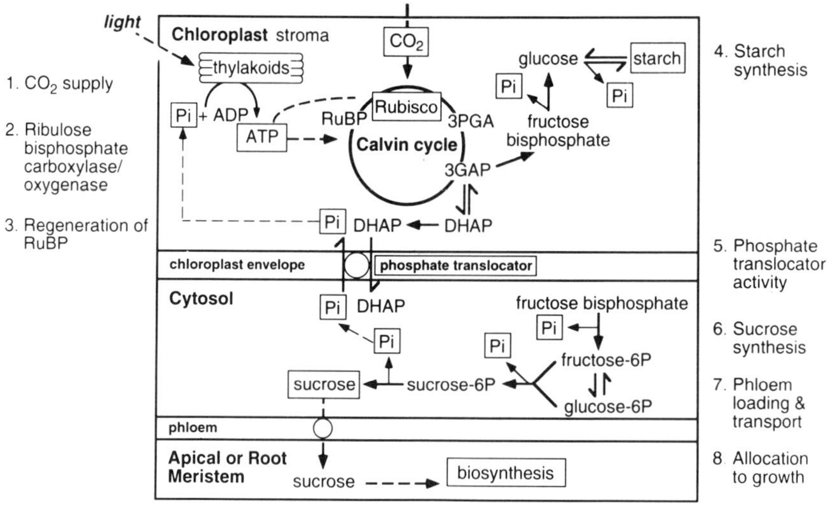

Figure 6-1. Schematic diagram of the principal processes and the major control points involved in photosynthetic CO_2 assimilation and carbon allocation of C_3 plants. Key enzymes are Rubisco (A), sucrose phosphate synthase (B), and cytosolic and chloroplastic fructose bisphosphatases (C). Major control points include CO_2 supply (1), Rubisco content and activity (2), regeneration of RuBP (3), and the interaction between the use of inorganic phosphate in production of triose phosphate and the release of inorganic phosphate in production of sucrose and starch (4, 5, 6, 7, 8). Abbreviations: Pi = inorganic phosphate, ADP = adenosine diphosphate, ATP = adenosine triphosphate, NADP = nicotinamide adenine dinucleotide phosphate, 3PGA = 3-phosphoglyceric acid, 3PGAl = 3-glyceraldehyde phosphate, DHAP = dihydroxyacetone phosphate, RuBP = ribulose-1,5-bisphosphate, UDP-glucose = uridine diphosphate-glucose, sucrose-6P = sucrose 6-phosphate, Rubisco = ribulose-1,5-bisphosphate carboxylase oxygenase.

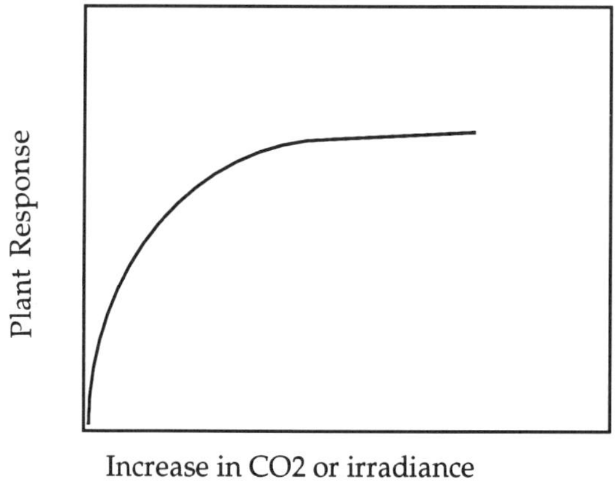

Figure 6-2. Generalized plant response to increasing solar irradiance and to increasing atmospheric CO_2 concentration. The curve applies to photosynthesis, plant growth, and reproductive output.

Summary of Hierarchical Responses in Plants

Primary Physiological Responses

Photosynthesis and Photorespiration

Perhaps the best-known physiological response of plants to increased atmospheric CO_2 is net carbon dioxide exchange. There are two major reasons why CO_2 partial pressure affects photosynthesis: (1) at present atmospheric levels, CO_2 is a limiting substrate for Rubisco activity; and (2) atmospheric O_2 competes for active sites because of Rubisco's oxygenase activity (Sage and Sharkey, 1987). Even though Rubisco has a higher affinity for CO_2 than for O_2, the high ratio of O_2 to CO_2 results in a significant limitation of photosynthesis. As CO_2 increases in partial pressure within the leaves, O_2 is displaced and photorespiration declines (Figure 6-3). A reduction in carbon loss via photorespiration results in an increase in apparent photosynthesis.

In the range of preindustrial CO_2, partial pressure to 2× that level (27 to 54 MPa), the average response for C_3 and C_4 plants is shown in Figure 6-4 (Pearcy and Bjorkman, 1983). The long-term photosynthetic response of many species to CO_2 is often lower than that after short-term exposure (Wong, 1979; DeLucia et al., 1985; Thomas and Strain, 1991). Acclimation to CO_2 may occur in response to changes in source-sink balance (Thomas and Strain, 1991) and may involve changes in nitrogen allocation. In a recent study in our laboratory with cotton (Harley et al., 1992), we found a linear correlation between the modeled photosynthetic parameters, such as the maximum rate of caboxylation and the rate of electron

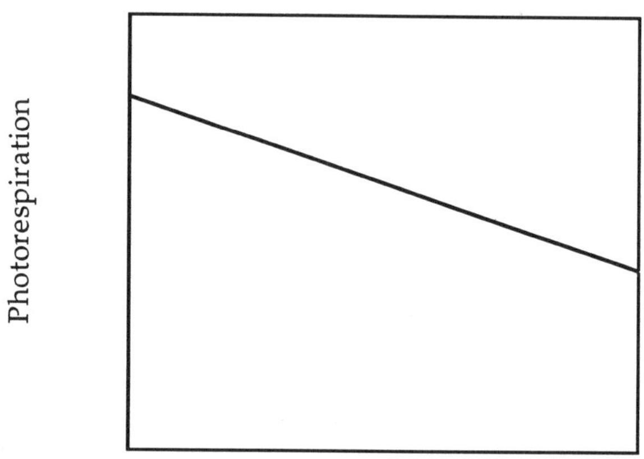

Figure 6-3. Photorespiration declines slightly with a doubling of atmospheric CO_2. (Concept from Musgrave and Strain, 1988.)

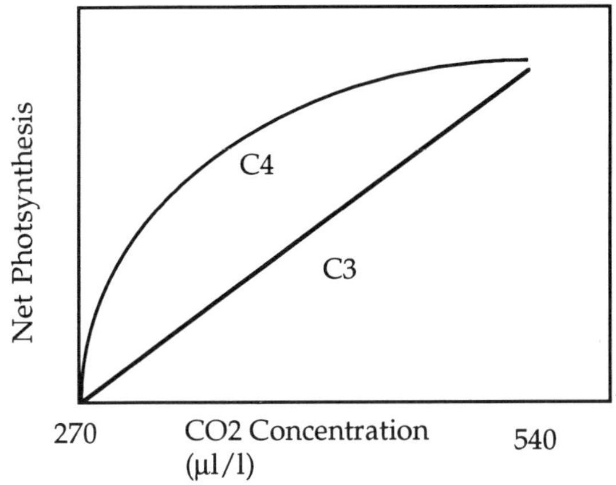

Figure 6-4. Generalized response of C_3 and C_4 plants to increasing atmospheric CO_2 concentration. (From Pearcy and Bjorkman, 1983.)

transport, and leaf nitrogen content. In addition to the Rubisco and RuBP limitations normally required to model photosynthetic capacity (von Caemmerer and Farquhar, 1981), the modeled response to internal leaf CO_2 required a phosphate limitation (Sharkey, 1985) arising from the failure of triose phosphate utilization (starch and sucrose production) to keep up with triose phosphate production in the Calvin cycle (Harley et al., 1992). This photosynthetic model will be discussed in depth by Harley (this volume).

It is also known that carbon dioxide and oxygen compete for the active sites on the carboxylating enzyme of C_3 plants (Sage and Sharkey, 1987). As CO_2 increases in partial pressure within the leaves, O_2 is displaced and photorespiration will decline (Figure 6-3). A reduction in carbon loss via photorespiration results in an increase in apparent photosynthesis.

Stomatal Aperture

Stomatal guard cells are sensitive to CO_2 changes in the atmosphere. As CO_2 partial pressure increases, stomatal conductance decreases (Pallas, 1965) with average reductions being 40% for a doubling of atmospheric CO_2 (Morison and Gifford, 1983). Instantaneous water-use efficiency (WUE), calculated as moles CO_2 fixed per mole of water transpired, may double with a 2× increase in atmospheric CO_2 (Oberbauer et al., 1985; Smith et al., 1987) as shown in Figure 6-5.

Respiration

An important question remains on the possibility that a relevant increase in global CO_2 partial pressure will have a significant effect on dark

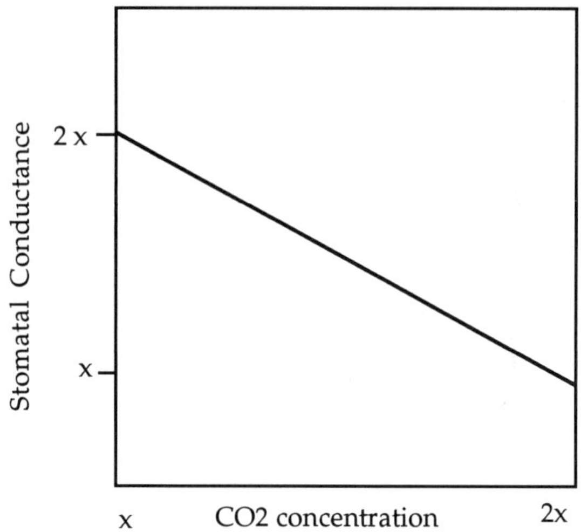

Figure 6-5. Stomatal conductance declines by half with a doubling of atmospheric CO_2. (Concept from Pallas, 1965.)

respiration. Dark respiration uses a significant fraction of the carbon fixed daily via photosynthesis and is, therefore, an important component of plant productivity and carbon balance (Lambers, 1985). Data are emerging from CO_2 enrichment studies which show that increases in atmospheric CO_2 may have a substantial effect on plant respiration. Unfortunately, the direction of the effect of CO_2 is not certain. Depending on the particular study, doubling the current ambient CO_2 partial pressure has been shown either to increase plant respiration rates (Hrubec et al., 1985; Du Cloux et al., 1987; Poorter et al., 1988; Nijs et al., 1989; Baker et al., 1992) or to decrease respiration rates (Silsbury and Stevens, 1984; Gifford et al., 1985; Reuveni and Gale, 1985; Bunce, 1990; Bunce and Caulfield, 1991).

Carbon dioxide may affect plant respiration either directly or indirectly (Amthor, 1991; Ryan, 1991). Direct effects are those by which a short-term change in CO_2 partial pressure affects respiratory metabolism, including changes in enzyme activities or diffusion rates of CO_2. It has been clearly shown that extremely high CO_2 partial pressures will inhibit plant respiration (Kerbel et al., 1988; Palet et al., 1991). Recently, direct inhibitory effects of CO_2 on leaf respiration have been found with CO_2 partial pressures between 5 and 95 Pa (Amthor et al., 1992). This response has been confirmed in our laboratory and others, but it remains unclear what biochemical mechanisms are involved and if the overall carbon balance of plants grown in long-term CO_2 enrichment affects the net respiration rate. It may be that processes that produce or take up CO_2

other than respiration are involved, such as PEP carboxylase. As far as root respiration is concerned, it is hard to conceive that a doubling of atmospheric CO_2 will directly affect below-ground CO_2 partial pressures sufficiently to affect any metabolic reactions.

Indirect effects of CO_2 are those by which respiration is affected by growth in elevated CO_2 partial pressures. Long-term CO_2 enrichment may differentially affect the respiratory component associated with new tissue synthesis (growth respiration) and the component associated with maintaining existing tissues (maintenance respiration). Whereas the respiratory cost per unit of tissue is unlikely to be altered by CO_2, changes in nonstructural carbohydrate content or nitrogen allocation, both typically associated with growth in elevated CO_2, suggest potential modifications in maintenance respiration (Amthor, 1991; Ryan, 1991). We have recently found an increase in maintenance respiration of cotton leaves grown in elevated CO_2, associated with increased starch accumulation, while growth respiration is unchanged by growth CO_2 (Thomas et al., 1992).

Secondary Metabolic Responses

Photosynthate Concentration and Composition

The above primary physiological responses of plants to atmospheric CO_2 partial pressure generate many secondary metabolic responses (Strain, 1985, 1987). Starch accumulation indicates that the plant cannot use photosynthate at the rate at which it is being produced and, when correlated with decreased net photosynthesis, reflects possible feedback effects on photosynthesis (Delucia et al., 1985; Sasek et al., 1985; Thomas and Strain, 1991). This reduction in photosynthesis may be reversible if sink strength increases (Thomas and Strain, 1991) or if CO_2 is reduced (Sasek et al., 1985). On the other hand, there is some evidence that, in extreme conditions, excessive starch grain production may ultimately disrupt chloroplast structure (Cave et al., 1981; Wulff and Strain, 1982).

Plant Water Status

Soluble sugars increase sufficiently to raise osmotic concentrations, increase turgor potentials, and increase whole-plant water potential (Sionit et al., 1980). Idso has shown that declining stomatal conductance, as discussed above, will decrease transpiration and the subsequent decrease in evaporative cooling will allow increases in leaf temperature (Idso et al., 1986). The net effect on ultimate evapotranspiration, however, is a very complex matter of temperature, vapor pressure, and leaf area interactions, and one can only speculate on the eventual plant and soil water status that will develop.

Plant Nutrient Status

The increase in carbohydrate compounds relative to nitrogen compounds (amino acids and proteins) is sufficient to generate increasing C/N ratios and affect secondary metabolism. In general, plant C/N ratios increase in response to CO_2 enrichment. This increase in the ratio of carbon to nitrogen can be explained either by an increase in nonstructural carbohydrate, a decrease in nitrogen, or a combination of these two. The extent to which nitrogen is reallocated to nitrogen-limiting processes is not clear. However, the increase in nonstructural carbohydrate is very common (Sasek et al., 1985; Thomas and Strain, 1991), which dilutes the nitrogen concentration of tissues.

Tertiary Plant Growth Responses

Growth

If other environmental conditions are not seriously limiting, increasing CO_2 will accelerate plant growth in mass and height (Sionit et al., 1980). Extensive reviews (Kimball, 1983; Cure and Acock 1985) have revealed that total biomass increases an average of 40% with a doubling of atmospheric CO_2 for C_3 crops in agricultural systems. A recent review by Poorter (1993) found ecological species, including C_4 grasses and C_3 trees and shrubs, are less stimulated, with average increases being in the range of 15–20% at $2\times$ CO_2. Other growth studies have shown that leaf area will increase (Sionit et al., 1981) and increasing leaf-area duration (LAD) has been observed (Brun and Cooper, 1967). On the other hand, determinate chaparral annuals have been shown to reach maturity earlier and, consequently, senesce earlier, usually with no net increase in final plant biomass realized under CO_2 enhancement (St. Omer and Horvath, 1983).

Growth Form

Plants with indeterminate growth form increase in node number and branches and thus experience changes in growth form and structure (e.g., Thomas and Strain, 1991). More branched plants have larger leaf area, and cotton has been shown to gain greater leaf area indexes under high CO_2 (Idso et al., 1986). Determinate annual plants accelerate in phenological development but senesce earlier without an ultimate increase in size, change in form, or increase in reproductive output (Paez et al., 1980; St. Omer and Horvath, 1983). Indeterminate plants accelerate in growth, become more branched, and increase in fruiting and reproductive output (Sionit et al., 1980; Paez et al., 1984; Peet, 1984).

Reproduction

Increasing size and node number generate more reproductive meristems, and increases in number, size, and seed content of fruits are the result

(Wulff and Alexander, 1987). Phenology in general is accelerated by CO_2 fertilization, with earlier dates of flowering (Garbutt and Bazzaz, 1984), anthesis (Krenzer and Moss, 1975), and seed maturation (Wulff and Alexander, 1987) having been observed.

Plant–Plant Interactions

Competition

Differential responses of species provide the mechanism for changing competitive interactions among plant species. When competing annual plants were challenged with different CO_2, soil nutrient, and water stress conditions, differential improvements in water status and soil fertility under high CO_2 changed the outcome of the competition (Bazzaz and Carlson, 1984). Among crops and their weeds, C_3 species outperform their C4 competitors in all cases analyzed (Patterson and Flint, 1990). In old-field weed populations, C_3 species become more competitive as CO_2 increases, causing a displacement of C_4 species by C_3 species (Carter and Peterson, 1983; Zangerl and Bazzaz, 1984; Wray and Strain, 1986, 1987; Marks and Strain, 1989).

Interference

The shading of slower-responding and smaller-stature species by rapidly expanding leaves of species which quickly respond to CO_2 enrichment (Sasek and Strain, 1989) increases light interference and may lead to species displacement. A comparative study with the invasive vines, Japanese honeysuckle and kudzu, led us to predict that these vines will increase in aggressiveness and displace more native species as global change occurs (Sasek and Strain, 1990).

Symbiosis

Although it was predicted several years ago (Strain, 1985) that symbiotic relationships among vascular plants would be affected by CO_2 enrichment, we still do not know of any study that has directly examined the issue. It seems logical to assume, however, that all types of symbioses (commensalism, mutualism, and parasitism) would be affected by differential environmental response of the symbionts (see Pate in this volume).

Plant–Animal Interactions

Leaf Herbivory

David Lincoln of the University of South Carolina and his colleagues have clearly demonstrated that nutritional changes occurring in the leaves of plants growing under long-term CO_2 enrichment decrease the feeding efficiency of insect herbivores (Lincoln et al., 1984, 1986; Lincoln and Couvet, 1989). Fajer and colleagues have confirmed Lincoln's studies

(Fajer, 1989; Fajer et al., 1989). These studies have shown that insect behavior can be influenced by the effect of CO_2 on their food source and suggest that the removal of increasing amounts of leaf tissue may decrease plant yields.

Pollination

Beverly Ratchke of the University of Michigan (unpublished) is examining the effect of CO_2 enrichment on flowering, pollen and nectar production, and seed production in *Convolvulus* sp. Her preliminary results show accelerated phenological development, more and larger flowers, more nectar of higher osmotic potential, and greater seed production. The observations on pollen and nectar are particularly important to associated pollinators (see Bond in this volume).

Shelter

It seems probable that changing branching habit and the production of more leaves, as cited above, will produce more bushy canopies and, thereby, affect animal habitat in a future high-CO_2 world. We are, however, unaware of any data to support the hypothesis or of any study under way to examine it.

Plant–Microbe Interactions

Symbiosis

Nitrogen-fixing bacteria have been shown to be stimulated by increased carbohydrate flux to host roots (Hardy and Havelka, 1977; Norby, 1987; Norby and Sigal, 1989; Thomas et al., 1991). Barbara Bentley (1992) is conducting experiments in the Duke Phytotron to determine if CO_2 stimulation will occur in low-fertility soils and if symbiotic nitrogen fixation will raise leaf nitrogen levels to overcome the nitrogen limitation problems explained by Lincoln and colleagues (Lincoln et al., 1986; Lincoln and Couvet, 1989).

Mycorrhizal stimulation by carbon dioxide enhancement has been shown by Norby and colleagues at Oak Ridge National Laboratory (Norby et al., 1986a,b, 1987; O'Neill et al., 1987). Lewis et al. (1994) reported mycorrhizal stimulation by high CO_2 with *Pisolithus tinctorius* on two ecotypes of *Pinus taeda*, but a complex interaction between soil phosphorus and mycorrhizal infection which leaves in doubt the magnitude of mycorrhizal stimulation of growth in phosphorus-limited native soils.

Decomposition

Bacterial and fungal decomposers are known to be inhibited by C/N ratios in the ranges observed by Lincoln et al. (1984) for nonnodulated soybeans. We have speculated that poor nutritional quality might also affect the quality of leaf litter, slowing decomposition and mineral cycling in ecosystems (Strain and Bazzaz, 1983). Several studies are currently under way but this issue is still not resolved. If carbohydrates are translocated from senescing tissue prior to abscission, the resulting leaf litter may not retain the high C/N ratios reached in fresh, green leaves. On the other hand, Couteaux and coworkers (Couteaux et al., 1991, 1992) have shown that when the entire complement of soil microbes, including Protozoa, Collembola, Isopoda, and Nematoda are present, the decomposition rate may not be impeded. These larger soil microbes macerate the recalcitrant tissues, apparently making the material more accessible to fungal and bacterial decomposers.

Disease

Coakley (1988) has pointed out that many phytopathogens are stimulated by increased humidity, temperature, and carbon status of the plant tissues. She predicts changing phytopathogenic virulence with climate change. The only specific study we know of CO_2 level and disease is one by Marks and Clay (1990) in which they compared disease susceptibility in C_3 and C_4 grasses to a pathogenic fungus. They found no statistically significant relationship between CO_2 partial pressure and disease.

Ecological Succession and Evolutionary Effects Through Time

Ecological Succession

Differential species responses which lead to changes in plant competition and changes in other organism interactions will change known successional patterns. If the changes persist through decades and centuries, we can predict ecosystem changes relative to the patterns that we have studied to date. Thus, empirical models of species and stand response derived under contemporary environmental conditions probably will not be successful in predicting ecosystem response under future CO_2 atmospheres. Idso and Quinn (1983) have predicted that future vegetation boundaries in New Mexico and Arizona will respond directly to atmospheric CO_2, but their model was empirically primed to differentially increase height growth and competitive potential of the tallest species.

Evolutionary Response

Theoretically, the occurrence of intra- and interspecific variation in the response of individual plants to environmental change, including direct responses to CO_2, and the probability that global environmental change is likely to proceed for a century or more, make it likely that gene frequencies will begin to shift and that survival and sexual reproduction of the best-adapted individuals will lead to genetic changes (Strain, 1991). If the base of the food chain, the green plants, changes, we predict subsequent changes in coevolved animals and microbes and accelerating genetic change in ecosystems.

Summary of Responses in "Natural" Ecosystems

Few system-level studies have been attempted (Strain and Thomas, 1992). Overdieck (1984; Overdieck and Reining, 1986a,b) conducted a study of a simulated meadow system by growing clover and rye grass in small greenhouses under two levels of atmospheric CO_2. The two species responded differentially, with the nitrogen-fixing clover quickly gaining dominance over the grass. In 1982, Walter Oechel initiated the first long-term study of CO_2 enrichment and thermal warming in a "natural" ecosystem with a study in the Alaskan Tundra (Oechel and Strain, 1985; Oechel and Riechers, 1986; Oechel et al., 1992). At first, a doubling of atmospheric CO_2 stimulated carbon uptake by the system (Tissue and Oechel, 1987), but within 3 years physiological homeostasis reduced the response to no stimulation (Oechel and Billings, 1992). Apparently, nutrient limitations in a saturated and cold soil profile prevented further carbon sequestering. The only other system-level study initiated to date is in a coastal salt marsh (Drake et al., 1989). After 4 years of CO_2 enrichment, the salt marsh species still responded to the CO_2 with increased annual productivity (Arp and Drake, 1991). Salt marshes are inundated frequently with nutrient-rich sea water which apparently keeps the system fertile. There were differences among the salt marsh species, with C_3 species outperforming C_4. There are some other efforts under way to understand ecosystem response to CO_2 enhancement and global change (e.g., Clinton Owensby in tall-grass prairie, William Parton and colleagues in short-grass prairie, Harold Mooney and colleagues in Californian annual grassland), but no published information is yet available from those studies. Additional studies, using autecological techniques, are under way in many laboratories around the world, but they will not be reviewed here because they are not of a system nature.

Predicted Responses in Mediterranean Ecosystems

Mediterranean ecosystems can be expected to exhibit large and significant responses to CO_2 increase and other factors of global change (Figure 6-6). The system is extremely sensitive to changes in summer drought. Any climate change which increases or decreases available system moisture will affect plant physiology and growth and, therefore, the entire ecosystem as argued above. In addition, Mediterranean ecosystems can be predicted

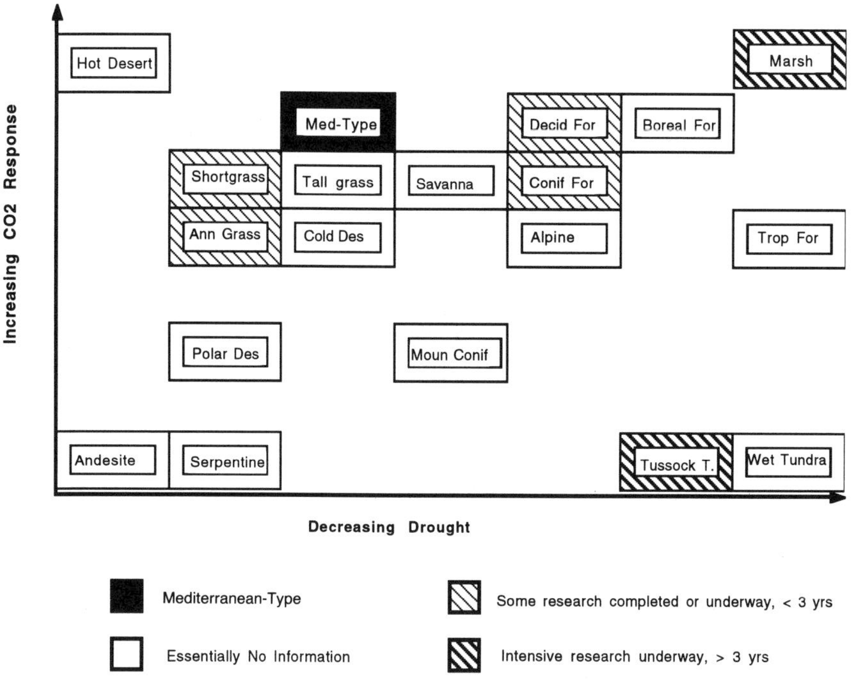

Figure 6-6. Ecosystem types representing the major biomes of the earth arranged on a matrix of decreasing drought prevailing in the ecosystem and our anticipated degree of CO_2 response. The largest CO_2 responses are expected to occur in the ecosystems to the top of the figure for various reasons as discussed in the text. (Modified from Oechel and Strain, 1985.)

to have a large response to atmospheric CO_2 enhancement, because stomatal closure to high CO_2 will greatly increase WUE and improve plant and ecosystem water status, as argued above. As cited above, Idso and Quinn (1983) showed the possible magnitude of this effect by simulation of desert species, which are similar to Mediterranean species in their sensitivity to summer moisture.

In addition, we predict that the fire-dominated Mediterranean ecosystems will experience more frequent fires. If increasing WUE leads to increasing annual productivity, fuel will accumulate more rapidly, leading to more frequent fires. It is difficult to speculate on the eventual outcome of this effect. It could initiate an overall decline in the systems by accelerating erosion. On the other hand, increased rates of carbon fixation in the systems might gradually increase biogeochemical cycling by accelerating mineralization, and net annual ecosystem productivity could increase.

Research Recommended

1. Field studies should be conducted in representative natural Mediterranean ecosystems with normal soil and rooting environments.
2. Long-term studies should be conducted using several CO_2 concentrations to obtain a dose-response-type relationship.
3. Factor interaction studies should be conducted to obtain the data base required to understand mechanisms of overall response to all of the factors of global environmental change.
4. Results should be used in organism response and in ecosystem response models to develop a time-based understanding of the global change responses.

References

Amthor JS. 1991. Respiration in a future, higher-CO_2 world. Pl Cell Environ 14:13–20.

Amthor JS, Koch GW, Bloom AJ. 1992. CO_2 inhibits respiration in leaves of *Rumex crispus* L. Pl Physiol 98:757–760.

Arp WJ, Drake BG. 1991. Increased photosynthetic capacity of *Scirpus olneyi* after 4 years of exposure to elevated CO_2. Pl Cell Environ 14:1004–1008.

Baker DN, Enoch HZ. 1983. Plant growth and development. In: Lemon ER (ed). CO_2 and Plants. AAAS Select Symposium 84. Westview Press, Boulder, CO, 280 pp.

Baker JT, Laugel F, Boote KJ, Allen LH Jr. 1992. Effects of daytime carbon dioxide concentration on dark respiration in rice. Pl Cell Environ 15:231–239.

Bazzaz FA, Carlson RW. 1984. The response of plants to elevated CO_2. I. Competition among an assemblage of annuals at different levels of soil moisture. Oecologia 62:196–198.

Bentley B. 1992. The effects of increased atmospheric CO_2 on *Lupinus* and its herbivores. Am J Bot Supple 79(6):59–60.

Brun WA, Cooper RL. 1967. Effects of light intensity and carbon dioxide concentration on photosynthetic rate of soybean. Crop Sci 7:451–454.

Bunce JA. 1990. Short- and long-term inhibition of respiratory carbon dioxide efflux by elevated carbon dioxide. Ann Bot 65:637–642.

Bunce JA, Caulfield F. 1991. Reduced respiratory carbon dioxide efflux during growth at elevated carbon dioxide in three herbaceous perennial species. Ann Bot 67:325–330.

von Caemmerer S, Farquhar GD. 1981. Some relationships between the biochemistry of photosynthesis and the gas exchange of leaves. Planta 153: 376–387.

Carter DR, Peterson KM. 1983. Effects of a CO_2 enriched atmosphere on the growth and competitive interaction of a C_3 and a C_4 grass. Oecologia 58: 188–193.

Cave G, Tolley LC, Strain BR. 1981. Effect of carbon dioxide enrichment on chlorophyll content, starch content and starch grain structure in *Trifolium subterraneum*. Physiol Planta 51:171–174.

Clough JM, Peet MM, Kramer PJ. 1981. Effects of high atmospheric carbon dioxide and sink size on rates of photosynthesis of a soybean cultivar. Pl Physiol 67:1007–1010.

Coakley SM, McDaniel LR, Line RF. 1988. Quantifying how climatic factors affect variation in plant disease severity: A general method using a new way to analyze meteorological data. Climatic Change 12:57–75.

Couteaux MM, Bottner P, Rouhier H, Billes G. 1992. Atmospheric CO_2 increase and plant material quality: production, nitrogen allocation and litter decomposition of sweet chestnut. In: Teller A, P Mathy, and JMR Jeffers (eds). Responses of Forest Ecosystems to Environmental Changes. Elsevier.

Couteaux MM, Mousseau M, Celerier M-L, Bottner P. 1991. Increased CO_2 and litter quality: decomposition of sweet chestnut leaf litter with animal food webs of different complexities. Oikos 61:54–64.

Cure JD, Acock B. 1985. Crop responses to carbon dioxide doubling: A literature survey. Agric and For Meteorol 38:127–145.

DeLucia EH, Sasek TW, Strain BR. 1985. Photosynthetic inhibition after long-term exposure to elevated levels of atmospheric carbon dioxide. Photosyn Res 7:175–184.

Drake BG, Leadley PW, Arp WJ, Nassiry D, Curtis PS. 1989. An open top chamber for field studies of elevated atmospheric CO_2 concentration on saltmarsh vegetation. Funct Ecol 3:363–371.

Du Cloux HC, Andre M, Daguenet A, Massimino J. 1987. Wheat response to CO_2 enrichment: growth and CO_2 exchanges at two densities. J Exp Bot 38:1421–1431.

Fajer ED. 1989. The effects of enriched CO_2 atmospheres on plant-insect herbivore interactions: growth responses of larvae of the specialist butterfly, *Junonia coenia* (Lepidoptera: Nymphalidae). Oecologia 81:514–520.

Fajer ED, Bowers MD, Bazzaz FA. 1989. The effects of enriched carbon dioxide atmospheres on plant-insect herbivore interactions. Science 243:1198–1200.

Garbutt K, Bazzaz FA. 1984. The effects of elevated CO_2 on plants. III. Flowers, fruits and seed production and abortion. New Phytol 98:433–446.

Gifford RM, Lambers H, Morison JIL. 1985. Respiration of crop species under CO_2 enrichment. Physiol Planta 63:351–356.

Hardy RWF, Havelka UD. 1977. Possible routes to increase the conversion of solar energy to food and feed by grain legumes and cereal grains (crop production): CO_2 and N_2 fixation, foliar fertilization and assimilate partitioning. In: Mitsui A, S Miyachi, A San Peitro, and S Tamura (eds). Biological Solar Energy Conversion, pp. 299–322. Academic Press, New York.

Harley PC, Thomas RB, Reynolds JF, Strain BR. 1992. Modelling photosynthesis of cotton grown in elevated CO_2. Pl Cell Environ 15:271–282.

Hrubec TC, Robinson JM, Donaldson RP. 1985. Effects of CO_2 enrichment and carbohydrate content on the dark respiration of soybeans. Pl Physiol 79:684–689.

Idso SB, Quinn JA. 1983. Vegetational redistribution in Arizona and New Mexico in response to a doubling of the atmospheric CO_2 concentration. Climatological Publ Sci Paper 16.7. Laboratory of Climatology, Arizona State University, Tempe, AZ 85287. 52 pp.

Idso SB, Clawson KL, Anderson MG. 1986. Foliage temperature: effects of environmental factors with implications for plant water stress assessment and the CO_2/climate connection. Water Resource Res 22:1702–1706.

Kerbel EL, Kader AA, Romani RJ. 1988. Effects of elevated CO_2 concentrations on glycolysis in intact "Bartlett" pear fruit. Pl Physiol 86:1205–1209.

Kimball BA. 1983. Carbon dioxide and agricultural yield: an assemblage and analysis of 430 prior observations. Agron J 75:779–788.

Krenzer EG, Moss DN. 1975. Carbon dioxide enrichment effects upon yield and yield components in wheat. Crop Sci 15:71–74.

Lambers H. 1985. Respiration in intact plants and tissues: its regulation and dependence on environmental factors, metabolism and invaded organisms. In: Douce R and DA Day (eds). Encyclo Pl Physiol, pp. 418–473. Springer-Verlag, Heidelberg.

Lewis JD, Thomas RB, Strain BR. 1994. Effect of elevated CO_2 on mycorrhizal colonization of loblolly pine (*Pinus taeda* L.) seedlings. Plant and Soil 165:81–88.

Lincoln DE, Couvet D. 1989. The effect of carbon supply on allocation to allelochemicals and caterpillar consumption of peppermint. Oecologia 78:112–114.

Lincoln DE, Couvet D, Sionit N. 1986. Response of an insect herbivore to host plants grown in carbon dioxide enriched atmospheres. Oecologia 69:556–560.

Lincoln DE, Sionit N, Strain BR. 1984. Growth and feeding response of *Pseudoplusia includens* (Lepidoptera: Noctuidae) to host plants grown in controlled carbon dioxide atmospheres. Environ Entomol 13:1527–1530.

Marks S, Clay K. 1990. Effects of CO_2 enrichment, nutrient addition, and fungal endophyte-infection on the growth of two grasses. Oecologia 84:207–214.

Morison JIL, Gifford RM. 1983. Stomatal sensitivity to carbon dioxide and humidity. A comparison of two C_3 and two C_4 species. Pl Physiol 71:789–796.

Musgrave ME, Strain BR. 1988. Response of two wheat cultivars to CO_2 enrichment under subambient oxygen conditions. Pl Physiol 87:346–350.

Nijs I, Impens I, Behaeghe T. 1989. Leaf and canopy responses of *Lolium perenne* to long-term elevated carbon-dioxide concentration. Planta 177: 312–320.

Norby RJ. 1987. Nodulation and nitrogenase activity in nitrogen-fixing woody plants stimulated by CO_2 enrichment of the atmosphere. Physiol Planta 71:77–82.

Norby RJ, Sigal LL. 1989. Nitrogen fixation in the lichen in elevated atmospheric carbon dioxide. Oecologia 79:566–568.

Norby RJ, O'Neill EG, Luxmoore RJ. 1986a. Effects of atmospheric CO_2 enrichment on the growth and mineral nutrition of *Quercus alba* seedlings in nutrient poor soil. Pl Physiol 82:83–89.

Norby RJ, Pastor J, Melillo J. 1986b. Carbon-nitrogen interactions in CO_2-enriched white oak: physiological and long-term perspectives. Tree Physiol 2:233–241.

Norby RJ, O'Neill EG, Hood WG, Luxmoore RJ. 1987. Carbon allocation, root exudation and mycorrhizal colonization of *Pinus echinata* seedlings grown under CO_2 enrichment. Tree Physiol 3:203–210.

Oechel WC, Billings WD. 1992. Effects of global change on the carbon balance of arctic plants and ecosystems. In: Chapin FS III, et al. (eds). Physiological Ecology of Arctic Plants: Implications for Climate Change. Academic Press, New York, pp. 139–168.

Oechel WC, Riechers GH. 1986. Impacts of increasing CO_2 on natural vegetation, particularly the tundra. Proceedings of Climate-Vegetation Workshop, NASA/Goddard Space Flight Center, Greenbelt, MD, pp. 36–42.

Oechel WC, Strain BR. 1985. Native species responses to increased carbon dioxide. In: Strain BR and JD Cure (eds). Direct Effects of Increasing Carbon Dioxide on Vegetation. Nat Tech Infor Ser US Dept Comm Springfield, VA, pp. 117–154.

Oechel WC, Riechers G, Lawrence WT, Prudhomme TI, and Grulke N. 1992. CO_2LT, A closed, null-balance system for long-term *in situ* ecosystem manipulation and measurement of CO_2 level, CO_2 flux, and temperature. Functional Ecology 6:86–100.

O'Neill EG, Luxmoore RJ, Norby RJ. 1987. Increases in mycorrhizal colonization and seedling growth in *Pinus echinata* and *Quercus alba* in an enriched CO_2 atmosphere. Can J For Res 17:878–883.
Overdieck D. 1984. Longterm effects of an increased CO_2 concentration level on terrestrial plants in model ecosystems. I. Phytomass production and competition of *Trifolium repens* L. and *Lolium perenne* L. Progr in Biometerol 3:344–352.
Overdieck D, Reining E. 1986a. Effect of atmospheric CO_2 enrichment on perennial ryegrass (*Lolium perenne*) and white clover (*Trifolium repens* L.) competing in managed model-ecosystems. I. Phytomass production. Acta Oecologica 7:26–35.
Overdieck D, Reining E. 1986b. Effect of atmospheric CO_2 enrichment on perennial ryegrass (*Lolium perenne* L.) and white clover (*Trifolium repens* L.) competing in managed model-ecosystems. II. Nutrient uptake. Acta Oecologica 7:35–46.
Paez A, Hellmers H, Strain BR. 1980. Carbon dioxide effects on apical dominance in *Pisum sativum*. Physiol Planta 50:43–46.
Paez A, Hellmers H, Strain BR. 1984. CO_2 enrichment and water stress interaction on growth of two tomato cultivars. J Agric Sci 102:687–693.
Palet A, Ribas-Carbo M, Argiles JM, Azcon-Bieto J. 1991. Short-term effects of carbon dioxide on carnation callus cell respiration. Pl Physiol 96: 467–472.
Pallas JE. 1965. Transpiration and stomatal opening with changes in carbon dioxide content of the air. Science 147:171–173.
Patterson DT, Flint EP. 1990. Implications of increasing carbon dioxide and climate change for plant communities and competition in natural and managed ecosystems. In: Kimball B, NJ Rosenberg, and LH Allen (eds). Impact of Carbon Dioxide Trace Gases and Climate Change on Global Agriculture. ASA Publ. No. 53 Amer Soc Agron, Madison, WI, pp. 83–110.
Pearcy RW, Bjorkman O. 1983. Physiological effects. In: Lemon ER (ed). CO_2 and Plants. AAAS Select Symposium 84. Westview Press, Boulder, CO, 280 pp.
Peet MM. 1984. CO_2 enrichment of soybeans. Effects of leaf/pod ratio. Physiol Planta 60:38–42.
Poorter H. 1993. Interspecific variation in the growth response of plants to an elevated ambient CO_2 concentration. Vegetatio 104/105:77–97.
Poorter H, Pot S, Lambers H. 1988. The effect of an elevated atmospheric CO_2 concentration on the growth, photosynthesis and respiration of *Plantago major*. Physiol Planta 73:553–559.
Reuveni J, Gale J. 1985. The effect of high levels of carbon dioxide on dark respiration and growth of plants. Pl Cell Environ 8:623–628.
Ryan MG. 1991. Effects of climate change on plant respiration. Ecol Applic 1:157–167.
Sage RF, Sharkey TD. 1987. The effect of temperature on the occurrence of O_2 and CO_2-insensitive photosynthesis in field grown plants. Pl Physiol 84:658–664.
Sasek TW, Strain BR. 1989. Effects of carbon dioxide enrichment on the expansion and size of Kudzu (*Pueraria lobata*) leaves. Weed Sci 37:23–28.
Sasek TW, Strain BR. 1990. Implications of atmospheric CO_2 enrichment and climatic change for the geographical distribution of two introduced vines in the U.S.A. Climatic Change 16:31–51.
Sasek TW, DeLucia EH, Strain BR. 1985. Reversibility of photosynthetic inhibition in cotton after long-term exposure to elevated CO_2 concentrations. Pl Physiol 78:619–622.

Sharkey TD. 1985. Photosynthesis in intact leaves of C_3 plants: physics, physiology and rate limitations. Bot Rev 51:53–105.

Silsbury JH, Stevens R. 1984. Growth efficiency of *Trifolium subterraneum* at high [CO_2]. In: Sybesma C (ed). Adv Photosyn Res pp. 134–136. Dr W Junk, The Hague, Netherlands.

Sionit N, Hellmers H, Strain BR. 1980. Growth and yield of wheat under CO_2 enrichment and water stress. Crop Sci 20:687–690.

Sionit N, Strain BR, Hellmers H. 1981. Effects of different concentrations of atmospheric CO_2 on growth and yield components of wheat. J Agric Sci 79:335–339.

Smith SD, Strain BR, Sharkey TD. 1987. Effects of CO_2 enrichment on four Great Basin grasses. Funct Ecol 1:139–143.

St. Omer L, Horvath SM. 1983. Elevated carbon dioxide concentration and whole plant senescence. Ecology 64:1311–1314.

Stitt M. 1991. Rising CO_2 levels and their potential significance for carbon flow in photosynthetic cells. Pl Cell Environ 14:741–762.

Stitt M, Huber S, Kerr P. 1987. Control of photosynthetic sucrose formation. In: The Biochemistry of Plants, Vol. 10, pp. 327–409. Academic Press, New York.

Strain BR. 1985. Background statement. In: Strain BR and JD Cure (eds). Direct effects of increasing carbon dioxide on vegetation. US DOE/ER-0238. Springfield, VA, Natl Tech Info Ser, US Dept Comm, p. xvii.

Strain BR. 1987. Direct effects of increasing atmospheric CO_2 on plants and ecosystems. Trends Ecol Evol 2:18–21.

Strain BR. 1991. Possible genetic effects of continually increasing atmospheric CO_2. In: Taylor GE, LF Pitelka, and MT Clegg (eds). Ecological Genetics and Air Pollution. Springer-Verlag, New York, pp. 237–244.

Strain BR. 1992. Atmospheric carbon dioxide: a plant fertilizer? New Biologist 4:87–89.

Strain BR, Bazzaz FA. 1983. CO_2 and plants, Chapter 7. Terrestrial plant communities. In: Lemon ER (ed). CO_2 and Plants: The Response of Plants to Rising Levels of Atmospheric Carbon Dioxide, pp. 177–222. Westview Press, Boulder, CO.

Strain BR, Thomas RB. 1992. Field measurements of CO_2 enhancement and climate change in natural vegetation. Water Air Soil Pollut 64:45–60.

Thomas RB, Strain BR. 1991. Root restriction as a factor in photosynthetic acclimation of cotton seedlings grown in elevated carbon dioxide. Pl Physiol 96:627–634.

Thomas RB, Reid CD, Ybema R, Strain BR. 1992. Growth and maintenance components of leaf respiration of cotton grown in elevated carbon dioxide. Pl Cell Environ 16:539–546.

Thomas RB, Richter DD, Ye H, Heine PR, Strain BR. 1991. Nitrogen dynamics and growth of seedlings of an N-fixing tree *Gliricidia semium* (Jacq. Walp.) exposed to elevated atmospheric carbon dioxide. Oecologia 88:415–421.

Tissue DT, Oechel WC. 1987. Physiological response of *Eriophorum vaginatum* to elevated CO_2 and temperature in the Alaskan tussock tundra. Ecology 68:401–410.

Wong SC. 1979. Elevated atmospheric partial pressure of CO^2 and plant growth. I. Interactions of nitrogen nutrition and photosynthetic capacity in C_3 and C_4 plants. Oecologia 44:68–74.

Woodrow IE, Berry JA. 1988. Enzymatic regulation of photosynthetic CO_2 fixation in C_3 plants. Ann Rev Pl Physiol Pl Molec Biol 39:533–594.

Wray SM, Strain BR. 1986. Response of two old field perennials to interactions of CO_2 enrichment and drought stress. Am J Bot 73:1486–1491.

Wray SM, Strain BR. 1987. Interaction of age and competition under CO_2 enrichment. Funct Ecol 1:145–149.
Wulff RD, Alexander HM. 1987. Intraspecific variation in the response to CO_2 enrichment in seeds and seedlings of *Plantago lanceolata*. Oecologia 66:458–460.
Wulff RD, Strain BR. 1982. Effects of CO_2-enrichment on growth and photosynthesis in *Desmodium paniculatum*. Can J Bot 60:1084–1091.
Zangerl AR, Bazzaz FA. 1984. The response of plants to elevated CO_2. II. Competitive interactions among annual plants under varying light and nutrients. Oecologia 62:412–417.

7. Plant Responses to Drought Under Climate Change in Mediterranean-Type Ecosystems

J.S. Pereira and M.M. Chaves

The vegetation of Mediterranean-type ecosystems has adapted to a marked seasonality characterized by a mildly cold and rainy winter and a hot dry summer. However, it is drought in the hot summer that has more dramatically influenced evolution and plant life. Because of the peculiar climatic conditions, Mediterranean-type ecosystems have a great diversity of plant life forms dominated by C_3 drought avoiders, ranging from winter annuals to deep-rooted perennials or summer-deciduous shrubs. The lack of water in the warmer part of the year does not allow C_4 plants to compete successfully under natural conditions, even though they may be highly productive in irrigated agriculture.

In summer, the depletion in soil moisture and the increase in water vapor deficit of the air not only affect plant physiological processes but also increase the incidence of catastrophic fires, leading to the loss of nutrients and a decrease in soil depth (see, e.g., di Castri and Mooney, 1973). These tendencies have been exacerbated by the intense human occupation of the land, leading to deforestation for agriculture and urban uses with consequent soil erosion. If we disregard riparian vegetation, most plants in Mediterranean-type ecosystems have evolved to face environmental stress caused by summer drought and a moderate frequency of disturbance (Grime, 1979) linked to fire and herbivory.

Climate change may modify the intensity and the interactions of environmental stresses on plants and lead eventually to changes in vegetation. Even though there is some uncertainty about exactly how global changes will affect the environment in areas with a Mediterranean type of climate, some atmospheric warming is likely to occur, and elevated CO_2 will certainly have an effect on plant life. It is very unlikely that rainfall will remain unchanged if temperature changes but it is uncertain whether there will be a decrease in summer rainfall as predicted by some models (Tegart et al., 1990) or a shift in seasonality towards more rain in summer. If the last hypothesis is verified, it is also uncertain whether this change in the seasonality of rain will be enough to offset the increase in potential evapotranspiration resulting from a warmer climate. In any case, drought will very probably remain a major factor in Mediterranean regions.

The effects of water deficits on the physiology and ecology of plants in Mediterranean-type environments have been reviewed on repeated occasions (e.g., di Castri and Mooney, 1973; Tenhunen et al., 1987). In the past, most attention has been given to plant response to tissue dehydration during summer. However, the depletion of soil moisture occurs slowly, often intermittently, from the spring months up to the summer, in this type of climate. In terms of plant physiology, this suggests that acclimation to increasing water deficits may occur before summer drought brings about plant dehydration. In this chapter we discuss the nature and the adaptive value of acclimation to decreasing soil water availability, as well as the survival of plants during periods of dehydration. The possible impact of changes in the amount or seasonality of rainfall will be discussed in the context of warming of a CO_2-enriched atmosphere.

Adjusting Plant Size and Metabolism to Decreasing Water Availability

The vegetation of Mediterranean-type ecosystems includes a large variety of plant life forms. Not all plants withstand the same types or the same intensities of environmental stress. Woody plants normally survive summer drought in the vegetative state, whereas shallow-rooted herbaceous annuals die after producing seed by early summer and quite often do not possess any tolerance to dehydration. In any case, a limitation in soil water availability will result in a sequence of plant responses that usually involve reductions in growth, changes in root/shoot ratio, and stomatal closure before the onset of tissue dehydration (Sinclair and Ludlow, 1985; Pereira and Chaves, 1993). Woody perennials acclimate to the slowly developing soil water deficits before they face the dehydration stress once the rooting zone has been depleted of moisture. In a given location, the rate of plant dehydration varies with depth and extension of the root systems, transpiration rate and soil properties.

The two processes that are important in the early phase of acclimation to soil water deficits, i.e., reduction in growth and increased stomatal closure, tend to reduce the ratio of transpiration to water uptake from the soil, especially because root growth is less sensitive than shoot growth to water deficits (Sharp, 1990). Therefore, plant water balance may be temporarily improved during acclimation, delaying the onset of the dehydration phase when plant survival is endangered. However, we may hypothesize that to have adaptive value, acclimation has to be triggered by small stepwise changes in soil moisture before any important tissue dehydration occurs. This means that the plant should be able to "sense" minor soil water deficits and possess a feed-forward type of response, such as root to shoot signaling (Davies and Pereira, 1992; Pereira and Chaves, 1993).

Growth Responses

Growth is generally more sensitive to water deficits than gas exchange, and reductions in growth (mainly foliage) may occur at moderate soil water deficits even if plant water relations do not change, as observed in a number of species and experimental systems (Passioura, 1988; Gowing et al., 1990; Kuang et al., 1990; Pereira and Chaves, 1993). In general, mild water deficits result in less foliage, greater apical dominance, and frequently in smaller leaves with lower specific leaf area (SLA). Root growth is often less affected by water shortage than shoots (Sharp, 1990) and this results in the increase in plant root/shoot ratio under water deficits. Although this leads to a decrease in the ratio between the capacities for transpiration and water uptake, it may contribute to further decrease plant growth as a result of the decrease in the ratio of leaf area to total plant biomass (Pereira, 1993). For example, there was a reduction in biomass production of 30% to 60% in ramets of three *E. globulus* clones grown in a soil kept at one-fourth of field capacity, as compared to well-watered controls (Osório and Pereira, 1994). During most of the experiment, there were no differences in predawn leaf water potential (Ψ_b between -0.17 and -0.27 MPa) or in midday leaf water potential (Ψ) of all clones irrespective of the treatments. In addition, all plants had similar photosynthetic capacity (measured at saturating light and CO_2) and the same leaf nitrogen concentration. The differences in the growth response to water deficit of the clones were strongly dependent on the capacity for each clone to produce new leaves and expand its foliage area under limited water availability (see Pereira and Chaves, 1993).

Evidence has been accumulating that foliage growth responds to hormonal rather than to hydraulic signals from the roots. Abscisic acid (ABA) is a likely candidate for this role (Gowing et al., 1990; Davies and Zhang, 1991). The increase in root/shoot ratio with moderate water deficits has been explained by the difference in sensitivity of roots and shoots to the increase in endogenous ABA and the osmotic adjustment of root systems (Sharp, 1990). Recent data on wheat genotypes containing different levels of endogenous ABA indicate that high-ABA genotypes produce relatively more roots than low-ABA genotypes under drought (W.J. Davies, personal communication).

The regulation of growth under moderate soil water deficits is one area that needs further research. It is known that low nutrient availability, especially low nitrogen, may have an effect on growth and plant structure similar to drought. Water deficits reduce nutrient uptake and transport to the leaves. For example, Gollan et al. (1992) showed that the concentration of nitrate (and phosphate) in the xylem sap decreased with decreasing soil water content in plants kept fully turgid. We may speculate that the lower supply of nitrate to the whole root system under moderate water deficits could result in increased apical dominance possibly mediated by reduced cytokinin levels (Samuelson et al., 1992). On the other hand, it was suggested that in some plants at least, xylem cavitation and the reduction in whole-plant hydraulic conductance may result in growth inhibition even without major soil water deficits by affecting the water balance of growing meristems (Schultz and Matthews, 1988a,b).

Another process that is likely to be affected by water deficits and that may lead to a decrease in productivity is leaf senescence and abscission. However, the importance of drought-induced leaf shedding may be quite variable with species and with the timing and intensity of water stress (Pereira and Pallardy, 1989; Davies and Pereira, 1992; Thomas, 1992). Although severe plant dehydration almost always triggers early leaf senescence, it is likely that moderate water deficits that do not cause severe dehydration may delay leaf senescence in evergreen plants, as happens with nutrient stress (Thomas, 1992). The same was observed, experimentally when irrigated and rain-fed *Eucalyptus globulus* stands were compared in Portugal (see Pereira et al., 1989). In irrigated stands, where trees were kept well watered throughout the year, most leaves were shed in summer. In rain-fed stands, where trees suffered water deficits in summer, the timing of maximum monthly litterfall occurred earlier than in the irrigated plots, that is, in late spring–early summer at the time of maximum shoot expansion (Madeira et al., 1995).

In a large number of Mediterranean tree species, most leaf shedding occurs at the time of intense spring shoot growth rather than in the summer (Pereira et al., 1987). As in other cases in the Mediterranean flora, substantial differences should be expected between evergreen trees and winter annuals or summer deciduous shrubs. For example, it is likely that leaf senescence and shedding in shallow-rooted shrubs which grow mostly in winter and shed their leaves in summer (e.g., some *Cistus* species; see Pereira et al., 1987) are accelerated by water deficits. This requires further experimental confirmation, however, especially because in many winter annuals and semideciduous species the timing of leaf senescence seems to be mostly temperature dependent (see below).

Gas Exchange Responses

The seasonality of leaf gas exchange of Mediterranean vegetation has been described in many species and locations (e.g., Tenhunen et al., 1986). Characteristically, gas exchange rates reach absolute minima in the summer. During the acclimation process to decreasing water availability, stomata close for increasingly longer periods during the day (see, e.g., Pereira et al., 1986; Tenhunen et al., 1987) as a result of complex and sometimes subtle interactions of water depletion in the rooting zone and increased vapor pressure deficits of the air and temperature (see, e.g., Schulze et al., 1986). In the early morning, stomatal conductance (g_s) of plants under moderate soil water deficits may be indistinguishable from g_s of well-watered plants. This is the period when most of the daily carbon assimilation may take place (Schulze and Hall, 1982).

During the early phase of soil moisture depletion, stomatal closure may be enough to prevent tissue dehydration, and there are situations in which midday leaf water potential in dry soils may be equal to or higher than that in well-watered plants (Pereira and Chaves, 1993). Other advantageous consequences of midday stomatal closure would be to prevent xylem cavitation during hours of high evaporative demand (Winkel and Rambal, 1993) and to reduce water loss when photosynthetic carbon assimilation may be depressed as a result of high temperatures and high light in the Mediterranean summer, as discussed below.

Water use efficiency (WUE) may be increased as a consequence. For example, in the *E. globulus* clones mentioned above (Osório and Pereira, 1994), long-term WUE (i.e., biomass per unit of water transpired) was significantly increased by decreasing the water supply to one-fourth without significantly changing leaf water status. The long-term WUE was well correlated with the quotient of the daily integrals of carbon assimilation rate (A) and g_s but not with maximum g_s measured in the morning, which was not different among clones. WUE also correlated positively with the $\delta^{13}C$ (i.e., $^{13}C/^{12}C$ ratio measured against the Pee Dee Belemnite, PDB, standard) and was negatively correlated with growth rate.

The causes for the decline in stomatal aperture during the day are complex and seem to be exacerbated by water deficits. In field-grown grapevine plants, gas exchange measured during the day under constant near optimal conditions of temperature, irradiance, and leaf to air water vapor pressure deficit showed a continuous decline in g_s and photosynthesis in the afternoon in water-stressed plants, but this did not occur in well-watered plants (Correia et al., 1995).

As in the case of growth, the effects of soil water deficits on the stomata are very likely mediated by a "positive root signal," possibly ABA (Davies and Zhang, 1991), especially during the phase of acclimation to soil water deficits (Wartinger et al., 1990). For example, the differences in daily maximum leaf conductance between grapevines grown with different soil water availability, were closely related to changes in ABA concentration in the xylem sap and could be reproduced by applying ABA to the leaves of well-watered plants, even though leaf water potentials were not significantly different during the day (Correia et al., 1995). However, the diurnal variation in g_s is not a simple function of the ABA concentration in the xylem sap, with g_s being normally lower in the afternoon, for the same ABA concentration, than in the morning. Often this cannot be fully explained by the diurnal variation in temperature or air humidity. Although a number of reasons have been invoked as an explanation, recent research suggests that, for a given rate of ABA delivery through the transpiration stream, the rate of ABA catabolism in leaf tissues may change the amount reaching the guard cells (Trejo et al., 1993). It seems possible that diurnal changes in the temperature or epidermis water status may determine the rate of ABA metabolism and therefore the amount of ABA effectively controlling stomatal aperture (Gowing et al., 1993; Trejo et al., 1993). The interaction of ABA with other components of the xylem sap that change diurnally and may also act as root signals (cytokinins, ion concentrations) cannot be disregarded.

In conclusion, reducing the supply of water to the roots leads to plants with slow growth rates as a result of a decreased foliage growth, high root/shoot ratios, low specific leaf areas and reduced daily carbon balance due to stomatal closure. Acclimation to decreasing water availability may occur without major tissue dehydration and contributes to plant drought avoidance (Sinclair and Ludlow, 1985; Pereira and Chaves, 1993). The relative importance of stomatal closure or growth reduction during this process may vary with genotype (Winkel and Rambal, 1993; Osório and Pereira, 1994). In the ecological context, this acclimation process would not have any value in the competitive environment if herbaceous annuals and woody summer-green plants were competing for the same water resources. It may be advantageous, however, if perennial plants are able to extract water from deeper soil layers, out of reach of the roots of herbaceous annuals or summer-deciduous shrubs. The survival of summer-green plants requires acclimation to decreasing water availability coupled with tolerance to dehydration, because complete drought avoidance is virtually impossible in the present Mediterranean type of climate.

Carbon Assimilation and Survival in the Mediterranean Summer

Effects of Dehydration at the Leaf Level

Plants facing a prolonged water deficit will drastically reduce carbon assimilation. Even though the major limitation to carbon assimilation for plants in a drought environment is stomatal closure, since the photosynthetic machinery is very resistant to dehydration (Kaiser, 1987; Cornic et al., 1989; Chaves, 1991), there is some evidence that mesophyll resistance may also increase under water stress (Tourneux and Peltier, 1995). In fact, stomata may respond to soil drying even in the absence of leaf dehydration (Davies and Zhang, 1991) whereas mesophyll photosynthesis begins to drop only after relative water content (RWC) in the leaf has declined below 70% (Kaiser, 1987).

In spite of the high resistance of the photosynthetic apparatus to dehydration, it was recognized that under some circumstances, such as during prolonged moderate water deficit or among the older leaves of the canopy, a decrease in photosynthetic capacity may occur at higher RWC than indicated above, in association with a decrease in ribulose bisphosphate carboxylase (Rubisco) (Chaves, 1991). Some variability among species has also been observed, with grapevines exhibiting a much higher susceptibility to dehydration of the mesophyll than lupin, sunflower, or eucalyptus (Quick et al., 1992). It is likely that some effects attributed to dehydration may result from enhancement of leaf senescence in droughted plants or from leaves sampled in comparable positions in the shoots of droughted plants being older than in well-watered plants, due to reduced growth in the former.

When stomata close in water-stressed leaves, CO_2 concentration in the intercellular spaces tends to decline, since photosynthesis at the chloroplast level is maintained. As a consequence, photorespiration is increased. This effect, together with an enhancement in the thermal dissipation of excitation energy, leads to a decrease in the quantum yield of Photosystem II (PSII) in dehydrated leaves (Cornic and Briantais, 1991).

Another major change induced by dehydration at the leaf level is the altered carbohydrate metabolism. This often comprises both an increase in the rate of newly synthesized sucrose at the expense of starch synthesis and a stimulation of starch degradation (Zrenner and Stitt, 1991; Quick et al., 1992). Both alterations lead to a relative increase in soluble sugars and therefore may contribute to the amelioration of the leaf water status through osmotic adjustment. The regulatory processes responsible for this response are not fully understood but seem to be related to changes in the activation state of sucrose phosphate synthase on one hand and in the timing of starch degradation on the other. In particular, water deficits allow hydrolysis of starch to begin at higher endogenous sucrose levels than in well-watered leaf tissues (Zrenner and Stitt, 1991). Additionally, growth inhibition in water-stressed plants reduces export of sucrose out of the leaf, thus enabling the maintenance of soluble sugar concentrations at levels similar to those in well-watered leaves, in spite of the very low rates of carbon uptake (Quick et al., 1992). One question that remains to be answered is whether the maintenance or the increase in soluble sugars in dehydrated leaves leads to a "down-regulation" in photosynthesis, as was observed in several plants under sink-limited conditions (Krapp et al., 1991).

Recovery of photosynthesis after rehydration will depend on the intensity and duration of drought, genotype, plant and leaf age, and the superimposition of other environmental stresses. Most of the irreversible effects will take place in the older leaves of the canopy. Stomata may take some days to recover to unstressed patterns after drought (presumably as a result of chemical signals coming from roots still partially dehydrated), but photosynthesis at the chloroplast level in younger leaves either has not been affected or may recover in a shorter time.

Interaction of Water Deficits with High Light and High Temperature

In the Mediterranean summer, dehydration stress usually coincides with high light and high temperatures. Under these conditions, leaves are the most exposed and vulnerable plant organs. These environmental conditions accentuate the negative effects of drought on the photosynthetic machinery, usually as a result of the inability of leaves to utilize and dissipate excess intercepted energy when stomata are closed. This is well illustrated by the increased susceptibility of water-stressed lupin plants to photoinhibition when they are maintained at high light and supraoptimal temperatures (Chaves et al., 1992a). In fact, it was shown that at optimal and suboptimal leaf temperatures for carbon assimilation in lupins (respectively 25° and 15°C), photosynthetic capacity decreases only at RWCs about 60%, whereas this inhibition occurred at RWCs around 80% when leaf temperatures were supraoptimal (35°C). Cornic and Ghashghaie (1991) also showed that low temperature can relieve the effects of water deficits on leaf net carbon uptake.

Under field conditions at midday, when temperature and light reach their maxima, photoinhibitory processes may occur in conjunction with stomatal closure. The sclerophyllous shrub *Arbutus unedo* growing under natural conditions in Portugal showed changes in photochemical activity during midday depression of stomatal conductance and photosynthesis (Demmig-Adams et al., 1989). Several lines of evidence indicate that photoinhibition under field conditions can be reversed overnight (Chaves et al., 1992b; Epron et al., 1992). Therefore, it may be considered more as a protective mechanism adjusting the activity of PSII to the electron requirement for photosynthesis, rather than a damaging effect. Plants have, however, a range of mechanisms that may protect them from excessive excitation energy and are particularly important under drought conditions when energy utilization through carbon reduction is impaired by stomatal closure. These include short-term events such as state transition and nonphotochemical quenchings (occurring within seconds and minutes), intermediate-term events such as protection against oxygen free radicals, and long-term events such as photorespiration and photoinhibition (Masojidek et al., 1991).

Many plants have the potential to acclimate to higher temperatures by increasing both the optimal and the critical temperatures for photosynthesis (Berry and Björkman, 1980). This happens mainly in species native to habitats where large temperature differences occur, but it has also been detected in species growing during the winter and spring months in the Mediterranean, such as *Lupinus albus* (Osório and Chaves, 1992). An increase of about 2.5°C in the leaf temperature, causing irreversible damage to the photosynthetic apparatus, was observed in *L. albus* plants growing at a temperature regime of 25°C (day)/20°C (night) as compared to plants growing at 15°/10°C. This acclimation response increases the ability of plants to endure extreme temperature events, which will tend to increase in frequency in the Mediterranean areas under predicted climate change. However, a dramatic increase in the rate of leaf senescence as well as a shorter growth cycle may occur at higher temperatures, as observed in *Lupinus albus* growing at 25°/20°C as compared to those growing at 15°/10°C (Figure 7-1). In many species, these effects may be more pronounced in plants subjected to water deficits.

In addition to the protection mechanisms at the cellular level, plants adapted to arid environments may also undergo leaf movements which protect photosynthetic components in midsummer (Gamon and Pearcy, 1989). Leaf orientation and position in the canopy are important determinants of intercepted light and leaf temperature. As a consequence, leaf movements may allow a compromise at the canopy level between intercepted light for maximal carbon gain and stress damage. The "buffering

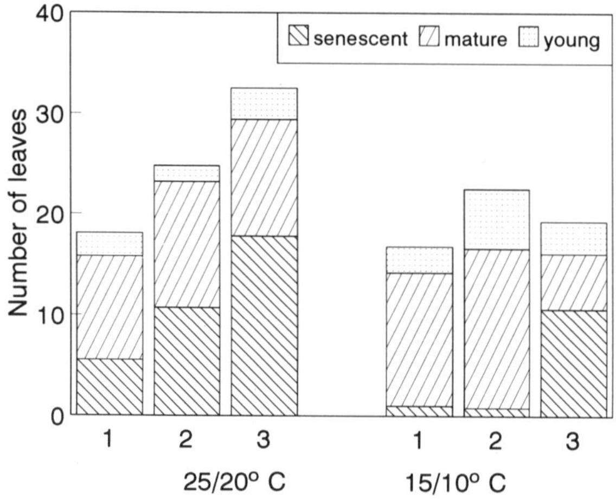

Figure 7-1. Number of young, mature, and senescent leaves of *Lupinus albus* plants grown at 25°C (day)/20°C (night) and 15°C/10°C at three different times during the life cycle (Passarinho, Osório, Faria and Chaves, unpublished).

effect" induced by the canopy in the microclimate is also illustrated when comparing, for example, water stress effects on photosynthesis of grapevines growing in pots in the open air and under field conditions (Figure 7-2). In the first case the effects are much more pronounced, partly because leaf temperatures in potted plants are usually higher than in the field, where most leaves are in a denser canopy undergoing mutual shading.

Whole-Plant Response

Survival of perennial plants under a Mediterranean-type climate requires a certain degree of drought tolerance. In addition to metabolic tolerance

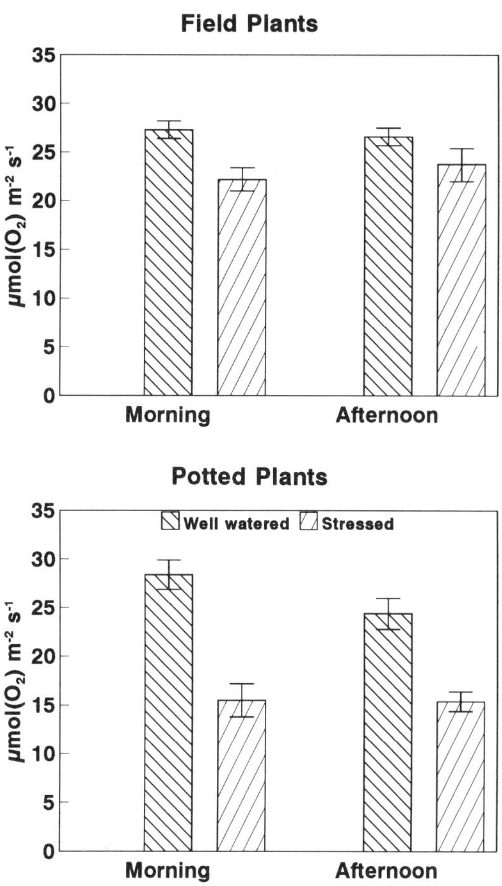

Figure 7-2. Leaf photosynthetic capacity, measured by O_2 evolution at saturating CO_2 and light, measured in the morning (10 h) and in the afternoon (14 h) in well-watered and water-stressed grapevine plants grown in the field and in pots (Chaves et al, 1992b).

and capacity of leaves to recover from the summer stress, plants may tolerate water deficits up to the limits imposed by the capacity for osmoregulation and by the hydraulic sufficiency of the xylem.

Osmoregulation has been described as a process enabling plants to adjust their tissue water relations and eventually maintain plant metabolism in an increasingly dry soil (Turner and Jones, 1980; Morgan, 1984). This has been found to occur during the summer in Mediterranean plants. For example, in droughted grapevines in Portugal, a decrease in osmotic potential at full turgor was found to correspond to an osmotic adjustment of ca. 0.45 MPa (Rodrigues et al., 1993). This was the result of an increase in soluble sugars relative to a decrease in starch in combination with the increase in concentration of other soluble compounds in droughted plants. Nevertheless, osmotic adjustment in the conditions of Mediterranean summer is unlikely to restore carbon assimilation rates in water-stressed plants (Rodrigues et al., 1993).

Water deficits also lead to the increase of soluble sugars in heterotrophic plant cells (Dancer et al., 1990), which may be considered similar to meristems. Since carbon assimilation is reduced in water-stressed plants, osmotic adjustment in the meristems, which represent a very small proportion of plant biomass, may be easier to achieve and more important to long-term survival than the loss of turgor in leaves.

Although water transport in the xylem of well-watered plants may not be a limiting factor for water uptake, evidence has accumulated showing that cavitation is a common event in the xylem (Milburn, 1979). With increasing water deficit, runaway cavitation and xylem embolism may result in dehydration and death of tree branches and whole plants (Pallardy, 1989; Tyree and Sperry, 1989). There is, however, a large variation among plants in the vulnerability of xylem to embolism. Even though only a few studies dealt with species living in Mediterranean-type environments, (Schultz and Matthews, 1988a,b; Salleo and Lo Gullo, 1989a,b; Lo Gullo and Salleo, 1993), it is possible that climate change may result in the increase in the ratio of potential evapotranspiration to soil water availability and enhance the risk of runaway xylem cavitation during the summer.

In *Quercus ilex*, Lo Gullo and Salleo (1993) showed that cavitation occurred both in winter with freezing temperatures and in summer due to drought. Xylem embolism and repair could explain the distribution of this species in the Mediterranean. Differences among species are likely to occur because of differences in xylem hydraulic vulnerability. Branching patterns and "hydraulic architecture" may also contribute to species differences, because internodes are more vulnerable and branching points may limit the propagation of embolism in the stem. Figure 7-3 shows the increase in nonconducting xylem in the cross section of nodes and internodes of *Vitis vinifera* plants with increasing water deficits (Ψ reached ca.

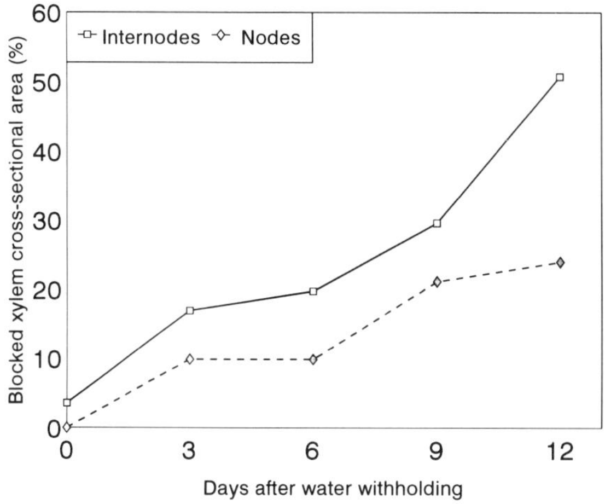

Figure 7-3. Increase in xylem blockage in the cross section of nodes and internodes of *Vitis vinifera* plants measured near midday (at 11 h) with increasing water deficits (Ψ reached ca. -2 MPa at 12 days after water withholding). (Modified from Salleo and Lo Gullo, 1989a.)

-2 MPa at 12 days after withholding of water) with nodes functioning as hydraulic constriction and protective zones in the stem (Salleo and Lo Gullo, 1989a). Optimal stomatal functioning may have been dictated during evolution by the need to prevent runaway xylem embolism that is likely to occur with severe water stress (Jones and Sutherland, 1991). This is an area requiring more research to evaluate the impact of climate changes in Mediterranean-type vegetation.

Interaction Between Water Deficits and Plant Nutrition

Drought and low nutrient availability are common to Mediterranean-type ecosystems, and the two interact strongly. Water deficits will inhibit, to a certain extent, uptake and transport in the transpiratory stream of several ions (Gollan et al., 1992). However, this may not result in nutrient deficiency because growth is concomitantly inhibited. On the other hand, the efficiency of use of absorbed nutrients may be modified by water deficits. Photosynthetic "nitrogen use efficiency" (i.e., the rate of photosynthesis per unit of leaf nitrogen) decreases with water deficits (Fredeen et al., 1991; Pereira et al., 1994). The same tendency is often encountered when nutrient use efficiencies are calculated over long periods.

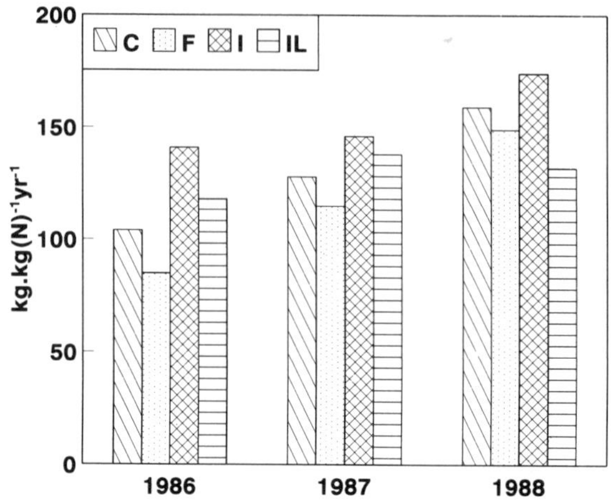

Figure 7-4. Nitrogen use efficiency of *Eucalyptus globulus* trees subjected to different treatments of water and nutrient supply (see text for details) calculated as the above-ground biomass produced per unit of nitrogen taken up during three consecutive growing seasons in Portugal (unpublished).

One clear example of the interaction of water availability and nutrient supply is given in Figure 7-4. Four treatments were applied to *Eucalyptus globulus* experimental plantations in Portugal: (1) irrigation plus addition of fertilizers with a near optimal nutrition protocol (IL), (2) irrigation (I), (3) fertilizers added without irrigation (F), and (4) a control (C), rain-fed without fertilizers added according to Pereira et al. (1989) and Madeira et al. (1995). In the IL treatment, net annual productivity was 50% greater than in the control at 3 years of age. Productivities were 24% and 14% higher than in the control in treatments I and F, respectively (Tomé and Pereira, 1991). In every case, high water supply (I) resulted in higher nitrogen use efficiency (NUE) calculated as biomass produced per unit of mass of nitrogen absorbed, than both nonirrigated treatments C and F. The same happened with phosphorus use efficiency (calculated in a manner similar to NUE). As discussed in Pereira et al. (1994), this may be explained by reduced carbon assimilation in the summer due to water deficits in these treatments (C and F). Because nutrient concentrations in the foliage did not decline in summer in the two rain-fed treatments, an obvious lower productivity per unit of mass of absorbed nutrients was observed (Madeira et al., 1995; Pereira et al., 1994). One apparent exception occurred in 1988 (see Figure 7-4) with a lower NUE in the irrigated-fertilized (IL) treatment. This occurred because trees in IL reached the phase of declining growth rates earlier than the other treatments. On the other hand, 1988 was a fairly "wet" year, with 70% more rainfall in spring–summer than in the previous years, and that may have contributed to improve NUE in the rain-fed treatments.

Interaction Between Water Deficits and Elevated CO_2

In recent years an increasing amount of data has been gathered on the effects of elevated CO_2 on carbon uptake and biomass formation, although often without the interaction with other environmental factors. Under ambient CO_2 conditions, oxygen competes with CO_2 for the active site of Rubisco, giving rise to the energy-spending process known as photorespiration. As CO_2 concentration increases, photorespiration is reduced and more carbon may be assimilated. In the short term, increases in the photosynthetic rate and in plant biomass in C_3 species have indeed been reported at elevated CO_2, as in the case of the CO_2 "fertilization" largely used by greenhouse growers. However, in the longer term and in some species, there is evidence for a "down-acclimation" to the enhanced atmospheric CO_2, expressed by a decline in photosynthetic rates to rates similar to or even lower than growth rates at ambient CO_2 (see Chapter 2 in this volume). It seems that the limitation by other resources may also reduce the intensity of plant response to elevated CO_2.

The short-term response of photosynthesis to enhanced CO_2 will depend on which process(es) control the rate of CO_2 fixation. According to Stitt (1991), the highest response will occur if Rubisco is limiting, an intermediate response if electron transport, ATP, or the regenerative phase of the Calvin cycle is limiting, and no response if end-product synthesis is limiting. In that case, accumulation of sugars following a "source-sink" imbalance may lead to a decline in photosynthesis by a direct feedback inhibition or by altering the expression of the genes encoding some of the Calvin Cycle enzymes (Sheen, 1989; Krapp et al., 1983). Under drought conditions (with stomata closed and low intercellular CO_2), high irradiances, and high temperature, we should expect that Rubisco activity is the main limiting factor for photosynthesis (Krapp et al. 1994), and consequently enhanced CO_2 will ameliorate plant performance. However, two features at the whole-plant level may counterbalance this positive effect of elevated CO_2: growth inhibition (and therefore sink limitation) due to water deficits, and faster development induced by higher temperature, resulting in shorter plant (or leaf) life cycles.

A survey of the literature concerning the interactive effects of high CO_2 and environmental stresses, such as drought and high temperature, reveals that only scant data are available (Sionit et al., 1980; Strain and Cure, 1985; Long, 1991; Chaves and Pereira, 1992; Guehl et al., 1994). From experimental evidence, it seems that elevated CO_2 protects against the effects of other factors at moderate stress levels, but the CO_2 enhancement effect is completely removed under severe stress conditions (Strain and Thomas, 1992).

One positive effect of elevated atmospheric CO_2 concentration is the increase in WUE as a consequence of higher rates of photosynthesis, often at decreased stomatal conductances. Enhancements of WUE by

40% to 100% have been reported for plants growing at high CO_2 (Wong, 1979; Morison and Gifford, 1984). The same is valid for water-stressed plants, e.g., in soybeans growing under elevated CO_2 where an increase of about 100% was recorded (Huber et al., 1984). The increased WUE, together with an increment in the root/shoot ratio promoted by elevated CO_2, may increase drought avoidance by plants in Mediterranean-type ecosystems. It is also possible that elevated CO_2 in the atmosphere enhances osmorregulation if photosynthesis increases proportionally more than growth, leading to an accumulation of carbon compounds.

The cooccurrence of high temperatures with elevated CO_2 may also be favorable, because the optimal temperatures for photosynthesis at light saturation and for maximum quantum yield seem to increase under high CO_2 (Long, 1991). To predict responses at canopy level, the rise in respiration rates and the shortening of the growing cycle of annuals by higher temperatures have to be taken into account. According to Long (1991), the interaction between temperature and CO_2 may invalidate conclusions based on their individual effects on photosynthetic productivity. It is therefore of major importance to get more information on the combined effects of high CO_2 concentration with water and temperature stress on the long-term growth of plants.

Drought Resistance and Climate Change in the Mediterranean Ecosystems

To predict the fate of Mediterranean vegetation under the influence of climate change, we need to understand its present characteristics. It should be noted that today's vegetation has been experiencing an increase in CO_2 in the atmosphere for the last century without apparent changes in distribution, i.e., anthropogenic impact may have overcome most natural changes. Likewise, the majority of plant species has survived the colder and more humid periods of the past (e.g., the end of the sixteenth and beginning of the seventeenth centuries) as well as the warmer and drier intervening periods, apparently without major changes. This was not necessarily so in the fringes of the Mediterranean, as demonstrated by the lowering of the upper altitudinal limits of human occupation in some mountainous areas in the early seventeenth century (Braudel, 1972). However the effects of climate change in modifying the arid-zone boundaries of Mediterranean vegetation are more difficult to illustrate because of the overwhelming impact of human activities in those areas.

The anticipated warming of the Mediterranean climate may make winter rainfall more useful in terms of plant productivity, as observed when precipitation falls in warmer periods of the year, e.g., when spring rain is higher than normal (see Interaction Between Water Deficits and Plant Nutrition). On the other hand, the effects of the increase in summer temperatures will depend upon the changes in the seasonality of rainfall. If summer rainfall remains low, the summers will be more arid because the ratio between actual and potential evapotranspiration will decrease. Higher temperatures and clear skies, together with drought, may drastically increase dehydration stress. On the other hand, if summer rainfall should increase, as some predict, then productivity of perennial woody plants might increase and spring–summer annuals would benefit as well. Under those circumstances, deep rooting for drought avoidance in perennials might not be so clearly advantageous as today. Plants with indeterminate growth might be better suited to summer rain than woody plants with determinate growth, which produce most of their foliage in spring, especially if combined with high root abundance in the upper soil horizons. These species are present in the native Mediterranean vegetation, but their growth period is often restricted by summer drought. On the contrary, an example of a woody species from a wetter climate being introduced in the Mediterranean region may be provided by *Eucalyptus globulus*, a native of regions with a higher and more equable rainfall (Hall et al., 1975). This species can be successfully cultivated in the Mediterranean provided that competition from native vegetation is artificially reduced during the first years after planting. Its indeterminate growth habit and the shallow root system allow the plants to capture water from any late rainfall whereas stomatal closure under drought (Pereira et al., 1986) and the presence of some sinker roots allow the plants to survive the present dry summers. The shallow and dense root system also permits a tight nutrient cycle.

The annual plants may be little affected by drought in the context of climate change. Higher temperatures in winter may allow these plants to develop faster and therefore reach fruit set before the onset of drought, if the seasonality of rainfall does not change. More precipitation in summer may favour some species that today are marginal and the introduced C_4 plants.

The role of stomata in plant adaptation may be twofold. Stomatal closure in part of the day during the early phase of acclimation to soil water deficits will avoid dehydration and therefore major tissue damage that would hamper recovery after soil rehydration. During the dehydration phase, stomatal closure may prevent the plants from reaching levels of water stress that may induce runway cavitation and embolism in the xylem and the irreversible loss of water flow capacity (Jones and Sutherland, 1991).

Some aspects remain unanswered and need further research. Does elevated CO_2 modify the resistance to xylem dysfunction and the tolerance to drought by increasing the amount of xylem produced each season or changing the dimensions of xylem elements? It is possible that species differ substantially in their responses, and some will not be capable of coping with greater potential evapotranspiration because stomatal control or hydraulic architecture does not provide enough protection from whole plant irreversible xylem disfunction. Finally, if climate does become warmer and more humid than in the present, it is possible that pathogens will increase the impact of embolism through increased water stress (e.g., through the occlusion of xylem or root killing or causing embolism, directly through "air seeding" as a consequence of the attack of xylem cell walls) (Tyree and Sperry, 1989). Differences in this respect may determine the persistence of certain species in the face of climate change. In any case, during the summer, plants have to withstand the combination of high temperatures, high light, and water deficits, and tolerance to these stresses may prove to be another discriminant of species persistence. In terms of species distribution, the effect of climate change on pathogens that may affect plant water relations cannot be neglected.

Acknowledgments. The authors gratefully acknowledge the collaboration of Julio and Leonor Osório, Maria-João Correia, J. Passarinho, and Tiago Faria in data collection and Isabel Leitão for typing the text.

References

Berry J, Björkman O. 1980. Photosynthetic response and adaptation to temperature in higher plants. Ann Rev Plant Physiol 31:491–543.

Braudel F. 1972. The Mediterranean and the Mediterranean World in the Age of Philip II. Vol. I. Fontana Press, London (English translation of 2nd French revised edition).

di Castri F, Mooney HA (eds). 1973. Mediterranean type Ecosystems. Origin and Structure. Springer-Verlag, Berlin, Heidelberg.

Chaves MM. 1991. Effects of water deficits on carbon assimilation. J Exp Bot 42:1–16.

Chaves MM, Pereira JS. 1992. Water stress, CO_2 and climate change. J Exp Bot 43:1131–1139.

Chaves MM, Osório ML, Osório J, Pereira JS. 1992a. Non stomatal limitation of photosynthesis under high temperature and water deficits in *Lupinus albus*. Photosynthetica 27:521–528.

Chaves MM, Wendler R, Osório ML, Osório J, Rodrigues ML, Faria T, Leegood RC, Stitt M, Pereira JS. 1992b. Down-regulation of photosynthesis under soil water deficits in grapevines. J Exp Bot 43 (suppl):35.

Cornic G, Briantais JM. 1991. Partitioning of photosynthetic e- flow between CO_2 and O_2 reduction in a C_3 leaf (*Phaseolus vulgaris* L.) at different CO_2 concentrations and during drought stress. Planta 183:178–184.

Cornic G, Ghashghaie J. 1991. Effect of temperature on net CO_2 assimilation and photosystem II quantum yield of electron transfer of French bean (*Phaseolus vulgaris* L.) leaves during drought stress. Planta 185:255–260.

Cornic G, Le Gouallec JL, Briantais JM, Hodges M. 1989. Effect of dehydration and high light on photosynthesis of two C_3 plants (*Phaseolus vulgaris* L. and *Elatostema repens* (Lour) Hall f.). Planta 177:84–90.

Correia MJ, Pereira JS, Chaves MM, Rodrigues ML, Pacheco CA. 1995. ABA xylem concentrations determine maximum daily leaf conductance of field-grown *vitis vinifera* L. plants. Plant Cell Environm.

Dancer J, David M, Stitt M. 1990. Water stress leads to a change in partitioning in favour of sucrose in heterotrophic cell suspension cultures of *Chenopodium*. Plant Cell Environm 13:957–963.

Davies WJ, Pereira JS. 1992. Plant growth and water use efficiency. In: Crop Photosynthesis: Spatial and Temporal determinants (NR Baker and H Thomas, eds). Elsevier Sci Publ BV, Amsterdam, pp. 213–233.

Davies WJ, Zhang J. 1991. Root signals and the regulation of growth and development of plants in drying soil. Ann Rev Plant Physiol Mol Biol 42:55–76.

Demmig-Adams B, Adams WW III, Winter K, Meyer A, Schreiber U, Pereira JS, Kruger A, Czygan FC, Lange OL. 1989. Photochemical efficiency of photosystem II, photon yield of CO_2 evolution, photosynthetic capacity and carotenoid composition during the midday depression of net CO_2 uptake in *Arbutus unedo* growing in Portugal. Planta 177:377–387.

Epron D, Dreyer E, Bréda N. 1992. Photosynthesis of oak trees [*Quercus petraea* (Matt.) Liebl.] during drought under field conditions: diurnal course of net CO_2 assimilation and photochemical efficiency of photosystem II. Plant Cell Environm 15:809–820.

Fredeen AL, Gamon JA, Field CB. 1991. Responses of photosynthesis and carbohydrate-partitioning to limitations in nitrogen and water availability in field-grown sunflower. Plant Cell Environm 14:963–970.

Gamon JA, Pearcy RW. 1989. Leaf movement, stress avoidance and photosynthesis in *Vitis californica*. Oecologia 79:475–481.

Gollan T, Schurr U, Schulze E-D. 1992. Stomatal response to drying soil in relation to changes in the xylem sap composition of *Helianthus annuus*. I. The concentration of cations, anions, amino-acids in, and pH of, the xylem sap. Plant Cell Environm 15:551–559.

Gowing DJG, Davies WJ, Jones HG. 1990. A positive root-sourced signal as an indicator of soil drying in apple, *Malus x domestica*, Borkh. J Exptl Bot 41:1535–1540.

Gowing DJG, Jones HG, Davies WJ. 1993. Xylem- transported abscisic acid: the relative importance of its mass and its concentration in the control of stomatal aperture. Plant Cell Environm 16:453–459.

Grime JP. 1979. Plant Strategies and Vegetation Processes. John Wiley & Sons. Chichester, UK.

Guehl JM, Picon C, Aussenac G, Gross P. 1994. Interactive effects of elevated CO_2 and soil drought on growth and transpiration efficiency in two European forest tree species. Tree Physiol 14:707–724.

Hall N, Johnston RD, Chippendale GM. 1975. Forest Trees of Australia. Australian Government Publication Service, Canberra.

Huber SC, Rogers HH, Mowry FL. 1984. Effects of water stress on photosynthesis and carbon partitioning in soybean (*Glycine max* L. Merr.) plants grown in the field at different CO_2 levels. Plant Physiol 76:244–249.

Jones HG, Sutherland RA. 1991. Stomatal control and xylem embolism. Plant Cell Environm 14:607–612.

Kaiser WM. 1987. Effects of water deficit on photosynthetic capacity. Physiol Plantarum 71:142–149.

Krapp A, Quick WP, Stitt M. 1991. Ribulose-1,5-bisphosphate carboxylase-oxygenase, other Calvin-cycle enzymes and chlorophyll decrease when glucose is supplied to mature spinach leaves via the transpiration stream. Planta 186: 58–69.

Krapp A, Hofman B, Schäfer C, Stitt M. 1993. Regulation of the expression of rbcs and other photosynthetic genes by carbohydrates: a mechanism for the sink regulation of photosynthesis. Plant J 3:817–828.

Krapp A, Chaves MM, David MM, Rodrigues ML, Pereira JS, Stitt M. 1994. Decreased Ribulose 1-5-Bisphosphate Carboxylase-Oxygenase in transgenic tobacco transformed with "antisense rbcS". VIII. Impact on photosynthesis and growth in tobacco growing under extreme high irradiance and high temperature. Plant Cell Environm 17:945–953.

Kuang JB, Turner NC, Henson IE. 1990. Influence of xylem water potential on leaf elongation and osmotic adjustment of wheat and lupin. J Exptl Bot 41:217–221

Lo Gullo MA, Salleo S. 1993. Different vulnerabilities of *Quercus ilex* L. to freeze- and summer drought-induced xylem embolism: an ecological interpretation. Plant Cell Environm 16:511–519.

Long S. 1991. Modification of the response of photosynthetic productivity to rising temperature by atmospheric CO_2 concentrations: Has its importance been underestimated? Plant Cell Environm 14:729–739.

Madeira M, Araújo MC, Pereira JS. 1995. Effects of water and nutrient supply on amount and on nutrient concentration of litterfall and forest floor litter in *Eucalyptus globulus* plantations. Plant Soil 168/169:287–295.

Masojidek J, Trivedi S, Halshaw L, Alexion A, Hall DO. 1991. The synergistic effect of drought and light stresses in sorghum and pearl millet. Plant Physiol 96:198–207.

Milburn JA. 1979. Water Flow in Plants. Longman, London.

Morgan JM. 1984. Osmoregulation and water stress in higher plants. Ann Rev Plant Physiol 35:299–319.

Morison JIL, Gifford RM. 1984. Plant growth and water use with limited water supply in high CO_2 concentrations II. Plant dry weight, partitioning and water use efficiency. Austral J Plant Physiol 11:375–384.

Osório J, Pereira JS. 1994. Genotypic differences in water use efficiency and ^{13}C discrimination in *Eucalyptus globulus*. Tree Physiol 14:871–882.

Osório ML, Chaves MM. 1992. Acclimation to high temperature at the photosynthetic level in *Lupinus albus*. Photosynthesis Res 34:215, P-631.

Pallardy SG. 1989. Hydraulic architecture and conductivity: an overview. In: Structural and Functional Responses to Environmental Stress (KH Kreeb, H Richter, and TM Hinckley, eds). SPB Academic Pub, The Hague, The Netherlands, pp. 3–19.

Passioura JB. 1988. Root signals control leaf expansion in wheat seedlings growing in drying soil. Aust J Plant Physiol 15:687–693.

Pereira JS. 1993. Gas exchange and growth. In: Ecophysiology of Photosynthesis, Ecol Studies Vol 100 (E-D Schulze and MM Caldwell, eds). Springer-Verlag, Berlin, Heidelberg, New York, pp. 148–181.

Pereira JS, Chaves MM. 1993. Plant water deficits in Mediterranean ecosystems. In: Plant Responses to Water Deficits – from Cell to Community (JAC Smith and H Griffiths, eds). BIOS Scientific Publ (Environmental Plant Biology Series), Oxford, UK, pp. 237–251.

Pereira JS, Pallardy S. 1989. Water stress limitations to tree productivity. In: Biomass Production by Fast-Growing Trees (JS Pereira and JJ Landsberg, eds). Kluwer, Dordrecht, The Netherlands, pp. 37–56.

Pereira JS, Beyschlag G, Lange OL, Beyschlag W, Tenhunen JD. 1987. Comparative phenology of four Mediterranean shrub species growing in Portugal. In: Plant Response to Stress. Functional Analysis in Mediterranean Ecosystems (JD Tenhunen, FM Catarino, OL Lange, and WC Oechel, eds). Springer-Verlag, Berlin, Heidelberg, New York, pp. 503–514.

Pereira JS, Chaves MM, Carvalho PO, Caldeira MC, Tomé J. 1994. Carbon assimilation growth and nitrogen supply in *Eucalyptus globulus* plants. In:

Whole-Plant Perspectives of Carbon Nitrogen Interactions (J Roy and E Granier, eds). SPB Publishing, pp. 79–89.

Pereira JS, Tenhunen JD, Lange OL, Beyschlag W, Meyer A, David MM. 1986. Seasonal and diurnal patterns in leaf gas exchange of *Eucalyptus globulus* trees growing in Portugal. Can J For Res 16:177–184.

Pereira JS, Linder S, Araújo MC, Pereira H, Ericsson T, Borralho N, Leal L. 1989. Optimization of biomass production in *Eucalyptus globulus* plantations. A case study. In: Biomass Production by Fast-Growing Trees (JS Pereira and JJ Landsberg, eds). Kluwer, Dordrecht, The Netherlands, pp. 101–121.

Quick WP, Chaves MM, Wendler R, David MM, Rodrigues ML, Passarinho JA, Pereira JS, Adcok MD, Leegood RC, Stitt M. 1992. The effect of water stress on photosynthetic carbon metabolism in four species grown under field conditions. Plant Cell Environ. 15:25–35.

Rodrigues ML, Chaves MM, Wendler R, David MM, Quick WP, Leegood RC, Stitt M, Pereira JS. 1993. Osmotic adjustment in water stressed grapevine leaves in relation to carbon assimilation. Aust J Plant Physiol 20:309–321.

Salleo S, Lo Gullo MA. 1989a. Xylem cavitation in nodes and internodes of *Vitis vinifera* L. plants subjected to water stress: limits of restoration of water conduction in cavitated xylem conduits. In: Structural and Functional Responses to Environmental Stresses (KH Kreeb, H Richter, and TM Hinckley, eds). SPB Academic Publishing, The Hague, pp. 33–42.

Salleo S, Lo Gullo MA. 1989b. Different aspects of cavitation resistance in *Ceratonia siliqua* L., a drought-avoiding Mediterranean tree. Ann Bot 64:325–336.

Samuelson ME, Eliasson L, Larsson CM. 1992. Nitrate-regulated growth and cytokinin responses in seminal roots of barley. Plant Physiol 98:309–315.

Schultz HR, Matthews MA. 1988a. Vegetative growth distribution during water deficits in *Vitis vinifera* L. Aust J Plant Physiol 15:641–656.

Schultz HR, Matthews MA. 1988b. Resistance to water transport in shoots of *Vitis vinifera* L. Relation to growth at low water potential. Plant Physiol 88:718–724.

Schulze E-D, Hall AE. 1982. Stomatal responses, water loss and CO_2 assimilation rates of plants in contrasting environments. In: Physiological Plant Ecology. II. Water Relations and Carbon Assimilation (OL Lange, PS Nobel, CB Osmond, and H Ziegler, eds). Encycl Plant Physiol New Series, Vol 12 B. Springer-Verlag, Berlin, pp. 181–229.

Schulze E-D, Turner NC, Golan T, Shakel KA. 1986. Stomatal responses to air humidity and to soil drought. In: Stomatal Function (E Zeiger, GD Farquhar, and IR Cowan, eds). Stanford University Press, Stanford, CA, pp. 311–321.

Sharp RE. 1990. Comparative sensitivity of root and shoot growth and physiology to low water potentials. In: (WJ Davies and B Jeffcoat, eds). Importance of Root-to-Shoot Communication in the Response to Environmental Stress. BPGRG Monograph 21-1990. British Society for Plant Growth Regulation, Bristol, pp. 29–44.

Sheen J. 1989. Metabolic repression of transcription in higher plants. Plant Cell 2:1027–1038.

Sinclair TR, Ludlow MM. 1985. Who taught plants thermodynamics? The unfulfilled potential of plant water potential. Aust J Plant Physiol 12:213–217.

Sionit N, Hellmers H, Strain BR. 1980. Growth and yield of wheat under CO_2 enrichment and water stress. Crop Sci 20:687–690.

Stitt M. 1991. Rising CO_2 levels and their potential significance for carbon flow in photosynthetic cells. Plant Cell Environm 14:741–762.

Strain BR, Cure JD. 1985. Direct effects of increasing carbon dioxide on vegetation. Nat Tech Infor Ser US Dept Comm Springfield, VA.

Strain BR, Thomas RB. 1992. Field measurements of CO_2 enhancement and climate change in natural vegetation. Water Air Soil Pollut 64:45–60.

Tegart WJ, McGsheldon GW, Griffiths DC (eds). 1990. Climate Change: The IPCC Impacts Assessment. Australian Government Publ Service.

Tenhunen JD, Pearcy RW, Lange OL. 1986. Diurnal variations in leaf conductance and gas exchange in natural environments. In: Stomatal Function (E Zeiger, GD Farquhar, and IR Cowan, eds). Stanford University Press, Stanford, CA, pp. 323–351.

Tenhunen JD, Catarino FM, Lange OL, Oechel WC (eds). 1987. Plant Response to Stress. Functional Analysis in Mediterranean Ecosystems. Springer-Verlag, Berlin, Heidelberg, New York.

Thomas H. 1992. Canopy survival. In: Crop Photosynthesis: Spatial and Temporal Determinants (NR Baker and H Thomas, eds). Elsevier Sci Publ BV, Amsterdam, pp. 11–41.

Tomé M, Pereira JS. 1991. Growth and management of eucalypt plantations in Portugal. In: Productivity in Perspective (PJ Ryan, ed). Third Australian Forest Soils and Nutrition Conference. Forestry Commission of NSW, Sydney, Australia, pp. 147–157.

Tourneux C, Peltier G. 1995. Effect of water deficit on photosynthetic oxygen exchange measured using $^{18}O_2$ and mass spectrometry in *Solanum tuberosum* L. leaf discs Planta 195:570–577.

Trejo CL, Davies WJ, Ruiz LMP. 1993. Sensitivity of stomata to abscisic acid. An effect of the mesophyll. Plant Physiol 102:497–502.

Turner NC, Jones HM. 1980. Turgor maintenance by osmotic adjustment: a review and evaluation. In: Adaptation of Plants to Water and High Temperature Stress (NC Turner and PJ Kramer, eds). John Wiley & Sons, New York, pp. 87–103.

Tyree MT, Sperry JS. 1989. Vulnerability of xylem to cavitation and embolism. Ann Rev Plant Physiol Mol Biol 40:19–38.

Wartinger A, Heilmeier H, Hartung W, Schulze ED. 1990. Daily and seasonal courses of leaf conductance and abscisic acid in the xylem sap of almond trees [*Prunus dulcis* (Miller) DA Webb] under desert conditions. New Phytol 116: 581–587.

Winkel T, Rambal S. 1993. Influence of water stress on grapevines growing in the field: from leaf to whole-plant response. Aust J Plant Physiol 20:143–57.

Wong SC. 1979. Elevated atmospheric partial pressure of CO_2 and plant growth. I. Interactions of nitrogen nutrition and photosynthetic capacity in C_3 and C_4 plants. Oecologia 44:68–74.

Zrenner R, Stitt M. 1991. Comparison of the effect of rapidly and gradually developing water-stress on carbohydrate metabolism in spinach leaves. Plant Cell Environm 14:939–946.

8. Functional Attributes of Angiosperm Hemiparasites and Their Hosts and Predictions of Possible Effects of Global Climate Change on Such Relationships

John S. Pate

The 3,000 or so species and almost 200 genera of parasitic angiosperms of the world flora collectively represent an extraordinarily broad assemblage of taxa from mostly distantly related families of dicotyledons and an equally profuse range of growth forms, morphologies, and life strategies (Kuijt, 1969; Press, 1989). All except certain diminutive subepidermal parasites (e.g., *Pilostyles* [Rafflesiaceae]) possess distinctive organs, so-called haustoria, through which water and solutes have been demonstrated or are presumed to be acquired from conducting tissues of a host. In some cases, haustoria are attached to above-ground parts of a host, in others they exploit host roots underground. Within both of these groupings, some species are noticeably chlorophyllous and accordingly are termed hemiparasites, while others which lack appreciable levels of chlorophyll are referred to as holoparasites. Under the latter two categories come herbaceous and woody species, species bearing only one or numerous haustoria, species usually parasitizing only one host species, and species simultaneously forming haustoria on a wide range of taxa. Parallel with this variation, parasitic species of one kind or other occur in all major land masses and ecosystems, including arctic vegetation, markedly arid habitats, heathlands, open woodlands, and temperate and tropical forests.

Although the general morphology, habitat typing, and taxonomy of angiosperm parasites are mostly well documented (Kuijt, 1969), only very few species have been examined physiologically. Especially favored for study of functional relationships with hosts have been certain hemiparasitic mistletoes of the families Loranthaceae and Viscaceae. A second grouping which has been extensively investigated consists of highly promiscuous root hemiparasites, including herbaceous species of Scrophulariaceae of temperate (*Odontites*, *Rhinanthus*, *Pedicularis*, *Bartsia*) or semitropical (*Striga*) regions, and woody shrub or tree genera of nutrient-poor heathlands and open woodlands (e.g., *Nuytsia* (Loranthaceae), *Santalum*, *Exocarpos*, and *Anthobolus* (Santalaceae) and *Olax* (Olacaceae) of SW Australian ecosystems). A third group receiving considerable attention comsists of three holoparasitic genera, two of these stem holoparasites (*Cassytha* [Lauraceae] and *Cuscuta* [Convolvulaceae]), the other a root parasite (*Orobanche* [Orobanchaceae]). This last group is not considered in this paper in view of the heterotrophy of its members.

The sections which follow summarize information of a functional nature on the two classes of hemiparasites described above, dealing in turn with the structure and absorptive qualities of the haustorium, photosynthetic CO_2 exchange, stomatal behavior and water relations, and finally with uptake, transport, partitioning, and utilization of mineral elements in relation to whole-plant growth.

Unfortunately, as far as the author is aware, no published experimental studies exist of the effects on hemiparasites of changing temperature, carbon dioxide concentration, or other variables relevant to global change. Comments on possible influences of such variables must therefore be highly conjectural, despite the excellent opportunities that the parasite–host relationships should ultimately provide as a model system for studying interaction of paired organisms to progressively changing environments. The final part of the chapter presents and discusses such predictions in relation to generalized schemes of functioning proposed for one class of hemiparasite.

Functional Biology of Hemiparasitic Mistletoes and Their Hosts

The Haustorial Connection of Mistletoes with Hosts

Mistletoes are essentially rootless plants which form large, anatomically complex perennial haustoria on the branches of their hosts. Cambial activity of the host is closely matched by increase in girth of opposed tissues of the parasite, resulting in a smooth external surface to the perimeter of the haustorial interface and a near-perfect graft junction between the two taxonomically unrelated partners. Mistletoes are generally regarded as xylem-tapping species, since the bulk of the haustorium faces xylem of the host, but dwarf mistletoes are suggested to be predominantly or even exclusively phloem feeders (Hull and Leonard, 1964), a feature consistent with their low photosynthetic capacities. In certain instances (e.g., *Viscum album* [Sallé, 1987], *Loranthus europaeus* [Buchleitner, 1983], and various species of *Amyema* and *Lysiana* [Kuo and Pate, 1990]), open lumen-to-lumen continuities exist between xylem conduits of parasite and host. These apparently form by parasite-mediated lateral dissolution of the walls of differentiating host xylem vessels by terminating tracheary elements of the parasite. Nevertheless, over 95% of the haustorium is composed of parenchymatous cells (Pate et al., 1991A).

Most host–mistletoe junctions display marked localized secondary thickening of the host close to the haustorium and continued generalized secondary thickening of the infested host branch even after host shoot parts distal to the haustorium may have deceased. Growth substances diffusing basipetally from mistletoe to host probably sponsor host secondary growth, but there is no evidence of accompanying bulk flow of phloem-borne photosynthates from parasite to host, at least judging from the $^{14}CO_2$ feeding experiments on *Arceuthobium* and *Phoradendron* reported by Hull and Leonard (1964).

Unimpeded flow of water across the haustorium into a transpiring mistletoe can be readily observed at the macroscopic level by feeding apoplast-mobile dyes (e.g., acid or basic fuchsin) or fluorescent markers through the cut base of a host branch to which a mistletoe is attached. Sectioning the haustorium then reveals a concentrated flow of dye through the low-resistance passages offered by the luminal continuities at the interface, alongside a much slower diffuse flow via the apoplast of surrounding parenchymatous tissue (J.S. Pate, unpublished observations). Detailed studies of the fate of host xylem-fed ^{3}H-lysine or electron-dense tracers in haustorial tissues of the dwarf mistletoe *Korthalsella lindsayi* (Coetzee and Fineran, 1987) have demonstrated effective apoplastic and symplastic pathways between and into the interface parenchyma of the parasite.

Water Relations, Transpiration, and CO_2 Exchange of Mistletoes and Their Hosts

The literature on water relationships of mistletoes is extensive in depth of investigation and number of species studied, thereby allowing one to extract a number of basic features crucial to functioning of this class of parasite. In virtually every case, the mistletoe exhibits substantially higher daytime rates of transpiration, higher stomatal conductances, and lower (more negative) leaf water potentials than its host (e.g., Fisher, 1983; Glatzel, 1983; Tocher et al., 1984; Ullmann et al., 1985; Ehleringer et al., 1986; Whittington and Sinclair, 1988; Davidson et al., 1989). Water flow is thus maintained continuously towards the parasite and at rates determined by the current leaf water potential gradient between parasite and host. Such gradients are maintained within carefully controlled limits by means of closely coordinated responses of the stomata of the parasite with those of its host (e.g., Glatzel, 1983; Davidson et al., 1989; Goldstein et al., 1989). Maintenance of such gradients involves earlier daily opening of stomata in the parasite, but, as water stress escalates through the day, lower (more negative) leaf water potential thresholds must be required to trigger partial or complete stomatal closure in the parasite than in the host (e.g., Glatzel, 1983; Whittington and Sinclair, 1988).

The studies of Glatzel (1983) on *Loranthus europaeus* parasitizing *Quercus robur* have drawn attention to the very high capacitance for water storage per unit area of mistletoe foliage in comparison with the host. For instance, in a heavily infested stand of oak, mistletoe capacitance was estimated at $1{,}400\,L\,ha^{-1}$ versus $300\,L\,ha^{-1}$ in the host, even though the mistletoe constituted a very much smaller proportion of the total biomass of the system. Glatzel (1983) has suggested that, due to its lower hydropassive threshold for stomatal closure, a partially wilted mistletoe would be likely to restore its capacitance preferentially on relief of stress and thus resume transpiration well in advance of its host.

The studies of Davidson et al. (1989) on the *Amyema linophyllum* (mistletoe)–*Casuarina obesa* partnership have shown that the haustorial junction accounts for some 80% of the resistance to water flow through the mistletoe during daytime and therefore is the major element responsible for affecting water potential gradients between the species. Pressure volume relationships for foliage of the mistletoe are found to be closely similar to those of the host, suggesting that tissue water relationships do not contribute appreciably to the promotion of gradients between partners. Both the above relationship and that between *Amyema fitzgeraldii* and *Acacia acuminata* (Davidson and Pate, 1992) are found to maintain essentially constant water potential gradients between host and parasite throughout the night. This, combined with evidence of much higher nocturnal water loss of the mistletoe than its host, implies that nighttime transpiration of the mistletoe must exercise a much greater influence on maintenance of water potential gradients with the host than might be caused by changes in capacitance of foliage(cf. studies of Glatzel, 1983, and Whittington and Sinclair, 1988). Experiments on *A. fitzgeraldii* (Davidson and Pate, 1992), using detached host branches supplied through their cut end with water, have shown that restoration of leaf water potential of the parasite lags behind that of the host, resulting in a transient buildup in leaf water potential gradient between partners. A phase of rapid flow into the parasite then ensues, presumably as a result of lowered haustorial resistance. This concept of highly variable haustorial resistance is consistent with the findings from a number of studies, including those mentioned above for *Amyema* spp., in which rates of transpiration of the mistletoe are found to vary a hundredfold or more between day and night, despite only slight concurrent changes in leaf water gradients with the host. Davidson and Pate (1992) have suggested that during the water stress of the day, the loose fibrillar wall components of parenchyma abutting the interface might contract as cell relative water contents fell, thereby creating a more effective apoplastic gateway into the parasite. Attendant lowering of the haustorial resistance would then promote greatly increased mistletoe transpiration. Conversely, at dusk, as relative water contents of host and parasite tissues become restored, the same pathways would become progressively obstructed and accordingly more resistant to flow. Stomatal conductances and attendant transpiration of the mistletoe would then fall drastically, albeit not to as great an extent as in the host.

The existence of luminal continuities between xylem conducting members of mistletoe and host implies that reverse flow from parasite to host should occur whenever the potential gradient of the partnership favors the host. Indeed, such a situation can be engineered artifically by enclosing and darkening a mistletoe clump just before dawn, thereby keeping its stomata closed and water loss to a minimium (Davidson et al., 1989). Then, once the host has commenced to transpire in early morning, the water potential gradient between partners reverses, thereby permitting water to be abstracted from the mistletoe by its host.

Where daytime CO_2 exchange rates of mistletoes have been studied alongside those of transpiration, net photosynthesis has been found to be slightly lower than or of similar magnitude to that of the host, or, if higher, never as great as to be expected from the much higher transpiration activity of the mistletoe. Water use efficiencies of mistletoes are accordingly in most instances noticeably poorer than those of hosts.

Mistletoe clumps are often well embedded in the tree canopy and therefore are less likely to experience the extremes of insolation and vapor pressure deficit encountered peripherally in the crown of the host. Mistletoes therefore tend to show better photosynthetic performance relative to adjacent host foliage at low than high light intensities (e.g., Davidson et al., 1989).

Carbon Isotopic Discrimination and Water Use Efficiencies of Mistletoes and Hosts and the Possibility of Partial Heterotrophy of Mistletoes for Carbon

According to the now well-substantiated relationships between WUE and $^{13}C/^{12}C$ natural isotopic discrimination originally proposed by Farquhar et al. (1982), the $\delta^{13}C$ value of dry matter of a plant deriving all of its carbon autotrophically by photosynthesis involving ribulose bisphosphate (RUBP) carboxylase can be defined as:

$$\delta^{13}\text{C plant} = \delta^{13}\text{C atmosphere} - a - (b - a)(C_i/C_a),$$

where a is the isotopic discrimination (4.4‰), due to slower diffusion of $^{13}CO_2$ than $^{12}CO_2$, b relates to isotopic discrimination by RUBP carboxylase (−27‰), and the $\delta^{13}C$ value of atmospheric CO_2 is −7.8‰. Merging this equation with that for WUE $\Delta C/\Delta W \times 1/1.56$), one finds a linear relationship between $\delta^{13}C$ of plant dry matter and WUE with a slope of ΔW $38.4/C_a$ and an intercept of −36.8 (e.g., Goldstein et al., 1989). Thus, since mistletoes generally display much poorer WUE than their hosts, parasite dry matter should show an appreciably lower (more negative) $\delta^{13}C$ value than that of its host. This differential in $\delta^{13}C$ value has in fact been consistently encountered across a broad taxonomic range of mistletoe species and geographical and ecological situations (e.g., Ehleringer et al., 1985; Pate, in press). However, only in a few instances have $\delta^{13}C$ measurements been accompanied by across-the-season measurements of WUE and C_i/C_a of the partners. However, Goldstein et al. (1989) have tested such relationships for the mistletoe *Pthirusa maritima* on the mangrove hosts *Conocarpus erectus* and *Coccoloba vinifera* and found that data for $\delta^{13}C$ and C_1/C_a or water use efficiency (WUE) of mistletoe and host do fit closely with the theoretical predictions alluded to above. However, in another study (Marshall and Ehleringer, 1990) involving the mistletoe *Phoradendron juniperinum* on *Juniperus osteosperma*, the measured $\delta^{13}C$ differential between host and parasite dry matter proved to be much less than that predicted from mean C_i values of the respective partners. Assuming the formula relating C_i to $\delta^{13}C$ to be correct, the authors conclude that the mistletoe must be partially heterotrophic on its host for carbon and, assuming there to be no isotopic discrimination during such transfer, they estimate the heterotrophic gain to be equivalent to 62 ± 2% of the parasite's requirements for carbon. A third example comes from studies by Schulze et al. (1991) on species of mistletoe (e.g., *Septulina glauca*) growing on hosts performing crassulacean acid metabolism (CAM). Despite their C_3 pattern of photosynthesis, such mistletoes show $\delta^{13}C$ values only 10‰ different from that of their hosts, compared with an expected difference of some 15–17‰ between a CAM and a C_3 plant behaving autotrophically. Again, the data suggest that a large fraction (60%) of the carbon of the mistletoe is being derived directly from its host.

The possibility of high levels of heterotrophy obviously challenges future use of $\delta^{13}C$ data as indices of relative WUE of parasites and hosts. In any event, significant amounts of carbon must accompany nitrogen whenever organic nitrogen is being derived from a host. For example, Pate, et al., 1991b have assessed heterotrophic intake of the mistletoe *Amyema linophyllum* from knowledge of the amino acid composition and C:N ratio of the xylem (tracheal) sap of its host (*Casuarina obesa*) and matching information on seasonal intake of nitrogen and net gain of and C:N ratio of dry matter of the mistletoe (Pate et al., 1991a). Nitrogen-linked heterotrophic intake via xylem is then shown to be equivalent to approximately one-quarter (24%) of the growth requirement of the mistletoe for carbon. The extent of heterotrophy would of course be even greater were solutes such as organic acids and sugars to be preferentially absorbed from host xylem or were the mistletoe to tap solutes from nonconducting tissues of the host. There might be occasions during the growing season, e.g., following bud burst in spring, when sugar levels in xylem are very high and benefit to the mistletoe increases accordingly. It is also highly likely that the extent of heterotrophy might vary between species of mistletoes, especially since some clearly possess phloem-tapping ability while others appear to be strictly xylem-tapping. In this connection it would be profitable to compare dark green species of mistletoes (e.g., *Lysiana, Loranthus* and *Amyema*) whose photosynthetic capacities approach those of their hosts with mistletoes yellow, brown, or light green in color and of proven or suspected abilities to exploit the phloem of their hosts (e.g., the dwarf mistletoes *Arceuthobium* spp. and certain hyperparasitic species [e.g., *Viscum capillatum*]).

Mineral Interrelationships of Mistletoes and Hosts

A great number of investigations spanning over 80 years have studied the mineral composition of mistletoes and their hosts (reviews of Lamont, 1983 and Pate, in press). In most of these cases, concentrations in dry matter of foliage or shoots of the mistletoe have been compared directly with those of the host, but in some studies comparisons have been extended to the composition of infested versus uninfested host branches to determine whether a significant drain of nutrients has occurred in the vicinity of the mistletoe lesion. Virtually all studies record significantly higher concentrations of most major nutrients in mistletoe than host. The concentration ratios for elements in the parasite to that in the host turn out to be by far the greatest for K and then decrease usually in the order Mg, Mn, Na, Ca, N, and Fe (Lamont, 1983), while comparisons of infested and unparasitized branches of the host mostly indicate that mistletoes have marked capacities to monopolize nutrients flowing into their branch of attachment (Lamont, 1983).

A particularly instructive series of comparisons has been made by Glatzel (1987) on tropical mistletoes from Sri Lanka, including a case where the composition of one species (*Dendrophtoe falcata*) has been examined on 48 different species of hosts. The data collectively show marked ability to adapt to extremely wide ranges of concentrations of specific nutrients in hosts (e.g., a 318-fold range difference for Cl levels in dry matter of *D. falcata* across the 48 hosts, 192-fold for Na, 11-fold for P, and 7- to 9-fold for K, Ca, and Mg). Despite such variation, levels of specific elements are generally well correlated with those of hosts, this being particularly well exemplified where the host of a partnership shows exceptional ability to accumulate specific elements, e.g., S, Na, or Mn. The principle that mistletoes are significantly more enriched than their hosts applies also to situations in which a mistletoe normally parasitizing nonmistletoe hosts is parasitizing an individual of the same species (e.g., in the cases of autoparasitism of *Phoradendron californicum* reported by Ehleringer and Schulze, 1985), in which one mistletoe species is occasionally parasitic on another (e.g., *Lysiana casuarinae* on *Amyema linophyllum*; J.S. Pate, unpublished data) or in which mistletoes are obligately hyperparasitic on other mistletoes (data for *Viscum capillatum* reported by Glatzel, 1987). In every instance, concentrations of most minerals, especially K, are found to increase outwards from nonmistletoe donor host to first and then to subsequent mistletoe(s).

There are a number of reasons why simple one-off comparisons of composition of dry matter of host and parasite are unreliable indicators of respective relative inputs from lower donor regions of the host (Pate, in press). First, the mineral composition of current shoot biomass may vary greatly in either species throughout the season. Second, leaching losses following interception of rain may occur differentially in host and parasite and probably to noticeably greater extents for certain elements (K^+, Na^+, and Cl^-) than others (divalent cations and organically bound nutrients). Third, basipetal loss from host shoots may occur by phloem-mediated cycling or mobilization from senescent organs, while probably not at all from the rootless mistletoe. Fourth, losses of minerals to reproductive parts or as shed leaves may occur differentially between elements or to differing degrees between host and parasite. Finally, measurements made on distally harvested host tissue may be complicated by monopolization of newly arriving minerals by the proximally sited mistletoe.

With the above complications in mind, two recent studies have attempted to model more adequately the uptake and partitioning of a range of mineral elements by a mistletoe and thereby shed light on the crucial issue of whether passive or active processes predominate in the highly complex uptake processes occurring within haustorial tissue. The first study by Glatzel (1983, 1987) has examined incorporation of N, P, K, Ca, and Mg into 3-year twigs of the European mistletoe *Loranthus europaeus* in a heavily infested coppice of oak (*Quercus petraea*). Ratios of seasonal uptake of each element on a leaf area basis into biomass of mistletoe and oak are estimated as 6.90 (uptake rate parasite: uptake rate host) for K, 2.38 for P, 1.84 for Mg, 1.31 for N, and 1.47 for Ca. Since the ratio value for Ca is commensurate with the ratio of measured rates of transpiration per unit leaf area of the partners, Glatzel (1987) suggests largely passive (mass) flow of Ca into the parasite, since, due to the phloem immobility of the element, no complications would be expected from basipetal transport and return flow of cycled Ca to the xylem stream. On the other hand, Glatzel (1987) interprets the higher uptake ratios for elements such as K and P on the supposition that the haustorial junction acquires xylem sap containing substantial recycling fractions of the host plant resource of that element alongside newly absorbed supplies of the element currently in transit to the shoot in the transpiration stream. He therefore sees no reason for invoking selective uptake by haustorial tissue to explain differential enrichment of the mistletoe with certain elements (e.g., K) versus others (Ca). He also cites the similar seasonal profiles of uptake of K and Ca in oak and mistletoe as favoring passive uptake across the haustorium.

The second case study of partitioning relates to the SW Australian mistletoe *Amyema linophyllum* on its host *Casuarina obesa* (Pate et al., 1991a). The uptake and utilization of a range of nutrients (Ca, Mg, K, N, P, Mn, Zn, and Ca) have been followed across a full reproductive season, portraying allocation of each element to secondary thickening of existing branches, investment into new shoots and leaves, losses as shed leaves, and prematurely lost reproductive units or successfully dispersed mature fruits. Proportional allotment to reproductive organs, stems, and leaves is then found to differ widely between elements; for example, particularly high proportions of host-derived Ca and Mn end up in leaf biomass of the mistletoe while P, K, Ca, and Zn are sequestered more strongly by reproductive organs than vegetative parts. Two further interesting characteristics of the partitioning system are the steep buildup of Ca, Mg, Na, and Cl with leaf age, and the very limited final retrieval of key limiting elements (P, N, and K) prior to leaf senescence. Thus, the foliage of the mistletoe abscises with almost full loads of nutrient elements, in marked contrast to its host *Casuarina* in which 70–90% of the N, P, and K is mobilized from old cladodes prior to their shedding (J.S. Pate, unpublished).

The study of Pate, True, and Kuo (1991) has included analyses of xylem sap collected at monthly intervals during the season of study. The data suggest that host xylem sap presents N, K, Ca, Mg, and P to the haustorium in proportions of 13.9:15.6:4.1:3.0:1, respectively. This compares with a ratio series of 21.5:20.2:6.6:1.6:1 (N:K:Ca:Mg:P) for net uptake by the mistletoe over the same season. The reasonably close mathematical similarity between the two sets of ratios has been taken to suggest that mass inflow of xylem fluid by the haustorium represents the major avenue of nutrient acquisition by the mistletoe.

Not all data on mistletoe:host associations suggest predominantly nonselective uptake from xylem of hosts. For example, Goldstein et al. (1989), analyzing pressure bomb-induced xylem sap of *Pthirusa maritima* on two mangrove hosts, have provided evidence of selective intake of K and Ca relative to Na and K. Similarly, in a series of comparisons involving eight associations of *Lysiana* spp. and six of *Amyema* spp. J.S. Pate, W.D. Jeschke, and W. Foulds (unpublished data) have used the ratio of K/Na of the parasite to K/Na ratio of the host as a comparative index of selectivity among the association. Ratio values are found to span a 90-fold range (9.9 to 0.11), suggesting that some mistletoes are particularly adept at accumulating K while excluding Na, whereas others are patently the reverse of this. Analyses of xylem exudates of these and other species are called for.

Functional Attributes of Root Hemiparasites

Rooting Characteristics of Root Hemiparasites

Members of this group are normally associated with species-rich communities in which selective advantage is clearly gained by forming effective haustoria on a multiplicity of species. This applies equally to the herbaceous hemiparasites (e.g., *Rhinanthis*, *Bartsia*, *Odontites*) of temperate grasslands of the Northern Hemisphere (e.g., Klaren and Dijk, 1976) and to the tree and shrub hemiparasites (e.g., *Nuytsia*, *Santalum*, *Anthobolus*, *Olax*, *Exocarpos*) of heathlands of SW Australia (e.g., Pate, et al., 1990c). In the latter instances, the parasitic species may be the only tall shrubs or trees species among a flora of small shrubby hosts, showing how effectively the parasitic members are able to monopolize resources of water and nutrients from their multiple hosts. Root systems of woody root hemiparasites (e.g., Pate, Dixon and Orshan, 1984; Pate et al., 1990c) are laterally extensive, exploiting a large number of host species across an area many times that of the foliar canopy of the parasite. In *Olax*, *Exocarpos*, and *Anthobolus* (J.S. Pate, G. Woodall, W.D. Jeschke, G.R. Stewart) roots do not penetrate below 30 cm, so attachment to deep-rooted hosts is essential to survival of the parasite through the dry summer season. In the Christmas tree (*Nuytsia floribunda*) deep "sinker" roots (Hocking and Fineran, 1983) tap underground water reserves, thereby supplementing supplies of water acquired by haustoria. This species forms clones of great extent through suckering.

Haustorial Structure and Functioning in Root Hemiparasites

There is great diversity in structure of haustoria among root hemiparasites, and this applies even within genera of the same family. However, in all cases reported so far, the anatomy of the haustorium clearly fosters close contact with xylem tissues of the host. Invasion of host tissues may lead to host phloem being disrupted and possibly still remaining in contact with outer parts of the haustorium, but uptake by the parasite from phloem tissue is unlikely to be significant, since the absorptive face of the haustorium is in contact solely with host xylem, and phloem tissue of the parasite does not extend into the endophyte. By contrast, files of tracheids usually extend through the haustorium to terminate against or in close contact with xylem vessels of the host (e.g., the three-dimensional representations of xylem connections shown for haustoria of *Olax* by Pate et al., 1990a).

Haustoria of a large number of parasitic species from the Santalaceae, Olacaceae, Opiliaceae, Loranthaceae, Krameriaceae, and Scrophulariaceae contain highly specialized graniferous tracheary elements. Tissue composed of such cells lies between and effectively connects to xylem elements of the endophytic part of the haustorium and the conventional xylem system of external parts of the root (Fineran, 1987; Sallé, 1987). The function of the cell type has not been properly defined, though it might act as loading point for solutes absorbed by interface parenchyma, or, as suggested by Fineran and Bulloch (1979), as a buffer against large fluctuations in water intake from the host.

Haustoria of most root hemiparasites are laterally attached, bell-shaped structures whose wedge-shaped sinker penetrates to the host cambium before splaying out laterally to form an absorptive interface against host xylem. In some cases, terminating xylem of the parasite contacts but does not penetrate host vessels (e.g., *Olax* [Pate et al. 1990a]); in other cases, direct luminal continuity is achieved (e.g., *Castilleja lutescens* [Dobbins and Kuijt, 1973], *Tozzia alpina* [Renaudin and Capdepon, 1979], and *Anthobolus foeviculatus* [Kuo and Pate, 1990]). In any case, as mentioned above for mistletoes, direct contacts between tracheary elements of the two partners comprise only a few percent of the area of the haustorial interface, the remainder on the parasite side of the junction consisting exclusively of specialized contact parenchyma (Fineran, 1987; Pate et al. 1990a). Transpirational (xylem) feeding of electron-dense tracers to roots parasitized by *Olax phyllanthi* (Kuo, Pate, and Davidson, 1989) has demonstrated an effective apoplastic pathway through the pits of host vessels into the apoplast of the haustorial interface and thence via tracheary elements or the cell walls of the contact parenchyma through to the xylem of the parasite root system.

The unusual "haustorial reticulum" of *Exocarpos sparteus* consists of branched ramifications encircling several centimeters of host root. A number of fine endophytic sinkers penetrate inwards from this network into the host xylem where they may even extend deeply into ray tissue (J.S. Pate and J. Kuo, unpublished).

The completely encircling haustorium of *Nuytsia floribunda* is highly unique in cutting the host root transversely by means of a secateur-shaped sclerenchymatous cutting device (i.e., the sclerenchymatous horn described by Fineran, 1983). Parenchymatous tissue of the endophyte abutting the still living proximal part of the host root then develops tubelike apical extensions into the cut host vessels, thereby possibly facilitating absorbtion of xylem solutes from host xylem sap (A. Calladine and J.S. Pate, unpublished). Conducting tissues of xylem of the *Nuytsia* haustorium terminate some distance from the interface, so absorbed solutes must traverse several layers of parenchyma before gaining access to the xylem stream of the parasite. Water intake may be minimal for, as mentioned earlier, *Nuytsia* also has deep sinker roots tapping underground reserves of water.

All of the above root hemiparasites possess the ability to form apparently effective haustorial contacts on a wide range of sizes and anatomies of host root. In *Olax phyllanthi* and *Anthobolus foeviculatus*, for example, the vast majority of dicotyledonous and some monocotyledonous species within the habitat are regularly parasitized (e.g., species list for *Olax phyllanthi* given by Pate et al., 1990a) and host roots from 0.5 to 2 cm in diameter can be successfully exploited. However, potential host species with an exfoliating periderm (e.g., Myrtaceae), sand-binding rhizosheaths (e.g., many species of monocotyledons), or thick fleshy roots (e.g., geophytes) show a general immunity to infection.

Photosynthesis, Transpiration, Water Use Efficiency, and Carbon Isotope Discrimination in Root Hemiparasites

A particularly well-documented picture of the carbon economy of a root hemiparasite is available for the economically-important *Striga hermonthica* when parasitizing cereals such as *Sorghum* spp. This herbaceous root hemiparasite shows low rates of photosynthesis but high rates of transpiration and night respiration (Stewart and Press, 1990). Unusual stomatal characteristics include high conductances, only partial closure at night, and relative insensitivity to abscisic acid (Shah et al., 1987). A carbon balance model of the *Striga–Sorghum* association, derived from growth and gas exchange measurements, has indicated that from 34% to as much as 74% of the carbon required by the parasite may be derived heterotrophically from the host (*Sorghum*) and that the amounts thus obtained are commensurate with the observed loss in grain yield when ravaged by the hemiparasite (Graves et al., 1989). Overall, host-derived carbon accounts for approximately one-third of the total carbon of the parasite, and this has been confirmed independently from measurements of the differentials in carbon isotope ratios of *Striga* (C_3 species) when parasitizing C_4 hosts such as *Sorghum* (Press et al., 1987). The means of acquisition of such large amounts of carbon are not clear, though a xylem origin is likely, especially since sugar levels in host xylem become particularly elevated when *Sorghum* is infected by *Striga* (Stewart and Press, 1990).

Photosynthesis and respiration rates, transpiration, and water use efficiencies of a range of northen European herbaceous root hemiparasites of the Scrophulariaceae have been studied by Press et al. (1988). Behavior generally resembles that of *Striga*, consistent with a general strategy of maximization of water loss to maximize heterotrophic gain of carbon from the host. Water use efficiencies are very low, ranging from 0.30 to 0.97 mmol $CO_2 \cdot mol\ H_2O^{-1}$. High nighttime water loss may well relate to massive guttation from leaf glands (e.g., study on *Odontites verna* by Govier et al., 1968). In the latter species, $^{14}CO_2$ feeding of hosts demonstrates substantial heterotrophic transfer carbon associated with intake of xylem-borne amino compounds (Govier et al., 1967). A study of the water relations of *Rhinanthus serotinus*, another herbaceous root hemiparasite of the Scrophulariaceae, has shown that shoot water potentials of the parasite become considerably more negative once parasitism has commenced (Klaren and Dijk, 1976). Study of the photosynthetic rates and ^{13}C discrimination values of a range of South African root hemiparasites by Harpe et al. (1981) have shown generally low levels of photosynthetic activity. However, $\delta^{13}C$ values for species of three C_3 parasitic genera (*Thesium*, *Cycnium*, *Striga*) prove to be close to those expected of their C_3 photosynthesis (values −23 to −26) despite being partnered with C_4 grasses (*Eragrostis* spp.) ($\delta^{13}C$ values −11.6 to −11.9). Thus, in contrast to the situation recorded above for mistletoes, *Striga*, and *Odontites*, these data suggest only minimal heterotrophy in respect of carbon.

Turning to woody root hemiparasites, studies on both *Nuytsia floribunda* (Hocking and Fineran, 1983) and *Olax phyllanthi* (Pate et al., 1990b) have shown less negative leaf water potentials in the parasite than the host throughout the day and night. In the last-named species, shallow-rooted hosts become severely water stressed during summer and haustoria associated with such hosts senesce once host leaf water potentials pass out of the absorptive capacity of the parasite. *O. phyllanthi* is thus forced to survive the summer using its long-lived haustorial contacts with deep-rooted hosts. The latter species tap the water table and exhibit water potentials consistently within the range of the parasite.

Unlike the herbaceous root hemiparasites described above, *O. phyllanthi* shows lower transpiration rates than its commonly exploited hosts. This, combined with commensurately lower rates of photosynthesis, results in water use efficiencies of the parasite (seasonal values spanning 1.7 to 6.6 mmol $CO_2 \cdot mol\, H_2O^{-1}$) being generally similar to those of its hosts (equivalent values 4 to 10 [same units]). As might be expected from their closely similar water use efficiencies, $\delta^{13}C$ values of the parasite (−26.3‰ for root dry matter and −27.8‰ shoot dry matter) turn out to resemble closely those of its common hosts (mean for 13 species, −26.5‰ for root dry matter and −27.8‰ for shoot dry matter) (Pate et al., 1990b). Unfortunately there are no parasitized C_4 or CAM species in the habitat of *Olax*, so it is not possible to test for heterotrophic intake of carbon as illustrated above for mistletoes and herbaceous root hemiparasites.

Mineral Relationships of Root Hemiparasites and Their Multiple Hosts

Mineral relationships of root hemiparasites are intrinsically difficult to study in view of possible independent uptake by the parasite's own root system alongside haustorial acquisition of nutrients. However, root hemiparasites are still substantially richer in certain elements, especially K, than their hosts, as shown by Klaren and Janssen (1978) for N, P, K, Mg, and Na in the herbaceous *Rhinanthus serotinus*. The same principle applies to woody root hemiparasites, according to studies on *Santalum spicatum* by Struthers, Lamont, Fox, Wijesuriya, and Crossland (1986) and on *Olax phyllanthi* by Pate, Kuo, and Davidson (1990c). In the latter species, dry matter becomes enriched differentially in N, P, and K as seedlings pass from their initial fully autotrophic stage to one of parasitism, implying that abstraction of these three elements becomes especially prominent once haustorial contacts have been made and commence to function effectively.

Predictions Regarding Possible Effects of Climatic Change on a Hemiparasite–Host Relationship

As mentioned earlier, there are no definitive experimental data on the likely effects on functioning of host–parasite relationships of the types of factors (e.g., increasing atmosphere CO_2, reduced rainfall, and elevated temperature) consequent upon global changes in climate. Added to this, while descriptions of the principal physiological and structural adaptations of hemiparasites are reasonably well documented, especially in relation to attraction of solutes and water from their hosts, matching information on how functioning of the partnership is regulated at cellular through to whole-plant levels is still grossly inadequate.

To simplify the issue, this section will concentrate on the single example of a xylem-tapping mistletoe (e.g., *Loranthus*, *Phoradendron*, or Amyema) and its corresponding woody host. The cardinal features of this system are set out in Figures 8-1 and 8-2, depicting possible interrelationships for phloem mobile minerals (e.g., N, P, or K) (Figure 8-1) as a basis for examining how mineral balance of the partnerships might respond to increased or decreased uptake by the host, and a summary of physiological characteristics of host and parasite (Figure 8-2) to portray possible interactive responses in terms of gas exchange, carbon balance, and water relationships. As a further simplification, only one climatic factor, elevated atmospheric CO_2, will be addressed, and the somewhat naive assumption made that net photosynthetic inputs of carbon will increase parallel with CO_2 increase and in proportionally equal amounts in both species. The primary effect is then assumed to be improved water use relative to carbon gain in the respective autotrophic activities of both species.

The scheme for mineral interrelationships (Figure 8-1) suggests that a mistletoe would have the capacity to regulate the size and composition of its mineral pool (MP) in three principal ways. First, intake by mass flow through the lumen-to-lumen pathway (item 7, Figure 8-1) with the host would be subject to increase or decrease in direct proportion to changes

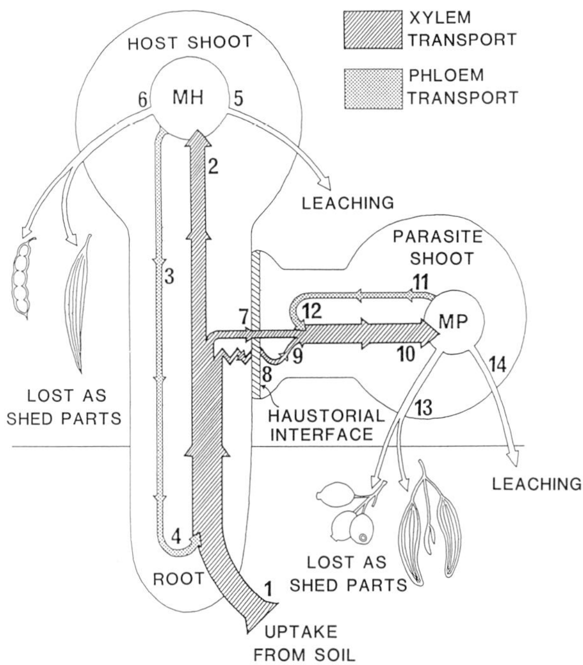

Figure 8-1. Principal features of the partitioning of minerals in a typical mistletoe–host partnership. The example applies to elements fully mobile in xylem and phloem. Xylem:phloem cycling is shown for both partners in relation to uptake by the host and losses through leaching and organ shedding of both species. Uptake across the haustorium is pictured to occur via the luminal continuities of xylem (7) or uptake by interface parenchyma (8). MH and MP are mineral pools of host and parasite, respectively.

in the same direction in transpiration rate. Second, selective uptake by interface parenchyma (item 8) would provide an obvious means of increasing or decreasing rates of uptake of certain elements relative to others. This might be regulated in turn by a feedback control system operating on the haustorium through the phloem–xylem circulatory system (items 9–12) of the parasite. Third, the shedding of foliage and reproductive structures and leaching of foliage in rain would offer a means of selectively excreting any mineral element building up to dangerous excess. Were this to be exercized in combination with effective retrieval of limiting nutrient elements prior to organ shedding, the element composition of the parasite biomass would then be altered radically relative to that of uptake from the host. Turning to the host, similar principles apply to regulation of composition through leaching and loss of shoot parts (processes 5 and 6, Figure 8-1). For example, on a liberal regime of mineral uptake, the recycling component of the pathway 2–4 would be expected to increase

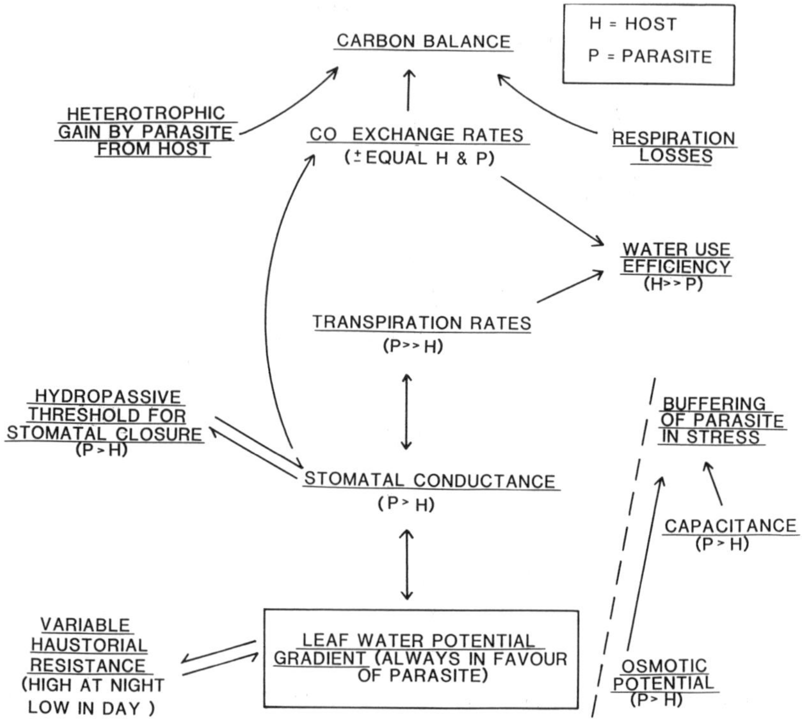

Figure 8-2. Scheme depicting the principal physiological characteristics of a hemiparasitic mistletoe and its host and the interactive effects of these features on carbon and water economies of the partner species.

for all phloem mobile elements. Proportional intake of circulating elements from host xylem by the parasite would then increase relative to that entering the host shoot and relative to that of phloem immobile elements such as Ca.

The matching set of attributes relating to gas exchange (Figure 8-2) and water relations highlights the pivotal role of coordinated stomatal responses of the mistletoe with its host in determining relative water use efficiencies, photosynthetic inputs, and mineral partitioning of the partners. For the mistletoe to survive, water flow must be continuously maintained in favor of the mistletoe. Higher transpiration rates, higher stomatal conductances, and lower thresholds for stomatal closure are essential to such an end, while the higher capacitance and osmotic potential of the mistletoe may be viewed as safeguards for retaining water effectively during times of severe stress, when stomata of both partners may be forced to close. The regulation of leaf water potential gradient between partners clearly consists of a delicate balance between the general level of stress and transpiration exhibited by the host, the parallel, but greater, transpiration activity of the parasite, and what appears to be an extremely large potential for structurally related changes in the impedance to flow at the haustorial junction. This apparently highly variable resistance of the haustorium may be critically affected by the respective turgor relations and relative water contents of opposing partner tissues and, through its effects on leaf water potential gradient, might well represent a signaling system for effecting the close tracking of day and night stomatal responses of parasite to those of its host. These and other possible control points within the system are illustrated in Figure 8-2.

Returning to the selected scenario of increased photosynthetic activity relative to water uptake within the partnerships, one encounters a number of equally conceivable outcomes, all of which might be profitably tested in future research. In the first, one would suppose that growth of the host remains strictly nutrient limited, say as in the predominantly N-driven or P-driven ecosystems recorded for certain pristine ecosystems of the world. Increases in photosynthetic efficiency under such circumstances might not necessarily result in significant growth increases of either partner, save possibly, say, in increased lignification or deposition of stored carbon in older organs. Such increments would not involve appreciable penalties in terms of additional requirement for mineral nutrients. However, one might also expect a diversion of some of the additional resource of carbon to root growth and to energizing of further nutrient absorption. This might then lead to limited enhancement of uptake of limiting nutrients with marginal benefits to the partners in terms of increased biomass.

A second series of outcomes might be envisaged for situations in which availability of water remains the limiting factor. In this case, net increases in size, and particularly in area of transpiring surface of both members of the partnership through more efficient gain of carbon and mineral nutrients, might well prove counterproductive because of increased subsequent demand for water and attendant vulnerability to drought and seasonal water stress. Under such conditions, one might expect relatively little long-term change in total biomass or balance of growth within the partnership or indeed in the ecosystem as a whole. Thus plants might be larger but their density less.

The third set of possible outcomes relates to situations in which water and nutrients remain at high, nonlimiting availability to the partnership. Then, with improved carbon economy, the host might be able to deploy greater proportions of its photosynthate to mineral uptake and thereby leading to xylem and phloem streams of the host becoming more enriched with nutrients. Assuming there to be no system of subsequent feedback control of mineral uptake by the host, the system outlined in Figure 8-1A would suggest that nutrients might accumulate to a greater extent in the parasite than in the host. Were this the case, the proportion of biomass of the partnership as parasite might increase, especially if increased mineral uptake were to be accompanied by commensurate increases in heterotrophic gain of carbon from the host. Experimental observations within such sets of circumstances might be particularly rewarding.

Acknowledgments. I am indebted to a number of colleagues who have collaborated in our experimental studies of parasitic plants. These persons include John Kuo, Neil Davidson, John Milburn and Dieter Jeschke and Bill Foulds, Ainslie Calladine, Geoff Woodall, Kathy True, and Edwin Rasins. All studies of our group referred to in this chapter have been most generously supported by the Australian Research Council.

References

Buchleitner E. 1983. Mikroskopische Untersuchungen über die Anatomie des Haustoriums und die Keimung der Eichenmistel (*Loranthus europaeus*). Dissertation Universität f. Bodenkultur, Wien. 170 pp.

Coetzee J, Fineran BA. 1987. The apoplastic continuum, nutrient absorption and plasma tubules in the dwarf mistletoe *Korthalsella lindsayi* (Viscaceae). *Protoplasma* 136:145–153.

Davidson NJ, Pate JS. 1992. Water relations of the mistletoe *Amyema fitzgeraldii* and its host *Acacia acuminata*. J Exp Bot 43:1549–1555.

Davidson NJ, True KC, Pate JS. 1989. Water relations of the parasite: host relationship between the mistletoe *Amyema linophyllum* (Fenzl) Tieghem and *Casuarina obesa* Miq. Oecologia 80:321–330.

Dobbins DR, Kuijt J. 1973. Studies in the haustorium of *Castilleja* (*Scrophulariaceae*). II. The endophyte. Can J Bot 51:923–931.

Ehleringer JR, Schulze E-D. 1985. Mineral nutrient concentrations in an epiparasitic *Phoradendron californicum* growing on *P. californicum* and its host *Cercidium floridum*. Amer J Bot 72:568–571.

Ehleringer JR, Cook CS, Tieszen LL. 1986. Comparative water use and nitrogen relationships in a mistletoe and its host. Oecologia 68:279–284.

Ehleringer JR, Ullmann I, Lange OL, Farquhar GD, Cowan IR. 1986. Mistletoes; a hypothesis concerning morphological and chemical avoidance of herbivory. Oecologia 70:234–237.

Ehleringer JR, Schulze E-D, Ziegler H, Lange OL, Farquhar GD, Cowan IR. 1985. Xylem-tapping mistletoes: water or nutrient parasites? Science 227: 1479–1481.

Farquhar GD, Ball MC, Von Caemmerer S, Roksandic Z. 1982. Effect of salinity and humidity on $\delta^{13}C$ value of halophytes–evidence for diffusional isotope fractionation determined by the ratio of intercellular/atmospheric partial pressure of CO_2 under different environmental conditions. Oecologia. 52:121–124.

Fineran BA. 1983. Ultrastructure of graniferous tracheary elements in the terrestrial mistletoe *Nuytsia floribunda* (Loranthaceae). Protoplasma 116:57–64

Fineran BA. 1987. Graniferous tracheary elements in haustoria of parasitic flowering plants. In: Parasitic Flowering Plants, eds H Chr Weber, W Forstreuter, pp. 197–200. Proceedings of the 4th ISPEP, Marburg.

Fineran BA, Bullock S. 1979. Ultrastructure of graniferous tracheary elements in the haustorium of *Exocarpus bidwillii*, a root hemi-parasite of the Santalaceae. Proc Roy Soc London B 204:329–343.

Fisher JT. 1983. Water relations of mistletoes and their hosts. In: The Biology of Mistletoes, eds DM Calder, P Bernhardt, pp. 161–183. Academic Press, New York.

Glatzel G. 1983. Mineral nutrition and water relations of hemiparasitic mistletoes: a question of partitioning. Experiments with *Loranthus europaeus* on *Quercus petraea* and *Quercus robur*. Oecologia 58:193–201.

Glatzel G. 1987. Haustorial resistance, foliage development and mineral nutrition in the hemiparastic mistletoe *Loranthus europaeus* Jacq. (Loranthaceae). In: Parasitic Flowering Plants, eds H Weber, W Forestreuter, pp. 253–262. Proceedings of the 4th ISPFP, Marburg, West Germany.

Goldstein G, Rada F, Sterngerg L, Burguera JL, Burguera M, Orozco A, Montilla M, Zabala O, Azocar A, Canales MO, Celis A. 1989. Gas exchange and water balance of a mistletoe species and its mangrove hosts. Oecologia 78:176–183.

Govier RN, Brown JG, Pate JS. 1968. Hemiparasitic nutrition in angiosperms II. Root haustoria and leaf glands of *Odontites verna* (Bell.) Dum. and their relevance to abstraction of solutes from the host. New Phytol 67:963–972.

Govier RN, Nelson MD, Pate JS. 1967. Hemiparasitic nutrition in angiosperms. I. The transfer of organic compounds from host to *Odontitles verna* (Bell.) Dum. (Scrophulariaceae). New Phytol 66:285–297.

Graves JD, Press MC, Stewart GR. 1989. A carbon balance model of the sorghum-*Striga hermonthica* host-parasite association. Plant Cell Environ 12:101–107.

Harpe AC, Visser JH, Grobbelaar N. 1981. Photosynthetic characteristics of some South African parasitic flowering plants. Z Pflanzenphysiol 103:265–275.

Hocking PJ, Fineran BA. 1983. Aspects of the nutrition of root parasitic Loranthaceae. In: The Biology of Mistletoes, eds DM Calder, P Bernhardt, pp. 185–257. Academic Press, Sydney.

Hull RJ, Leonard OA. 1964. Physiological aspects of parasitism in mistletoes (*Arceuthobium and Phoradendron*). I. The carbohydrate nutrition of the mistletoe. Plant Physiol 39:996–1007.

Klaren CH, van de Dijk SJ. 1976. Water relations of the hemiparasite *Rhinanthus serotinus* before and after attachment. Physiol Plant 38:121–125.

Klaren CH, Janssen G. 1978. Physiological changes in the hemiparasite *Rhinanthus serotinus* before and after attachment. Physiol Plant 42:151–155.

Kuijt J. 1969. The Biology of Parasitic Flowering Plants. University of California Press, Berkeley, CA.

Kuo J, Pate JS. 1990. Anatomy and ultrastructure of haustoria in selected West Australian parasitic angiosperms. Proceedings of the XIIth International Congress for Electron Microscopy, San Francisco Press, USA, pp. 690–691.

Kuo J, Pate JS, Davidson NJ. 1989. Ultrastructure of the haustorial interface and apoplastic continuum between host and the root hemiparasite *Olax phyllanthi* (Labill.) R.Br. (*Olacaceae*). Protoplasma 150:27–39.

Lamont B. 1983. Mineral nutrition of mistletoes. In: The Biology of Mistletoes, eds DM Calder, P. Bernhardt, pp. 185–204. Academic Press, New York.

Marshall JD, Ehleringer JR. 1990. Are xylem tapping mistletoes partially heterotrophic? Oecologia 84:244–248.

Pate JS. (in press). Mineral Relationships of Parasites and their Hosts. In: Parasitic Flowering Plants, eds MC Press, JD Graves. Chapman and Hall, London.

Pate JS, Dixon KW, Orshan G. 1984. Growth and life form Characteristics of Kwongan Species. In: Kwongan-Plant Life of the Sandplain, eds JS Pate, JS Beard, pp. 84–100. University of Western Australia Press Nedlands.

Pate JS, Kuo J, Davidson NJ. 1990a. Morphology and anatomy of the haustorium of the root hemiparasite *Olax phyllanthi* (Labill.) R.Br. (Olacaceae), with special reference to the haustorial interface. Ann Bot, 65:425–436.

Pate JS, Davidson NJ, Kuo J, Milburn JA. 1990b. Water relations of the root hemiparasite *Olax phyllanthi* (Labill) R.Br. (Olacaceae) and its multiple hosts. Oecologia 84:186–193.

Pate JS, Pate SR, Kuo K, Davidson NJ. 1990c. Growth resource allocation and haustorial biology of the root hemiparasite *Olax Phyllanthi* (Olacaceae). Ann Bot 65:437–449.

Pate JS, True KC, Kuo J. 1991a. Partitioning of dry matter and mineral nutrients during a reproductive cycle of the mistletoe *Amyema linophyllum* (Fenzl.) Tieghem parasitizing the swamp *Casuarina obesa* miq. J Exp Bot 42:427–439.

Pate JS, True KC, Rasins E. 1991b. Xylem transport and storage of amino acids by S.W. Australian mistletoes and their hosts. J Expt Bot 42:441–451.

Pate JS, Woodall G, Jeschke WD, Steward GR. 1994. Root xylem transport of amino acids in the root hemiparasitic Shrub *Olax phyllanthi* (Labill) R.Br. (Olacaceae) and its multiple hosts. Pl Cell Environ 17:1263–1273.

Press MC. 1989. Autotrophy and heterotrophy in root hemiparasites. Trends Ecol Evol 4:258–263.

Press MC, Graves JD, Stewart GR. 1988. Transpiration and carbon acquisition in root hemiparasites. J Exp Bot 39:1009–1014.

Press MC, Shah N, Tuohy JM, Stewart GR. 1987. Carbon isotope ratios demonstrate carbon flux from C_4 host to C_3 parasite. Plant Physiol 85:1143–1145.

Renaudin S, Capdepon M. 1979. Sur quelques aspects de l'ultrastructure des suçoirs de *Tozzia alpina* L. (Scrophulariacées). I–Etude des cellules absorbantes. Bull Soc Bot Fr Lett Bot 126:165–178.

Sallé G. 1977. Etude cytologique, cytochimique et histoautoradiographique du *Viscum album* L. (Loranthaces). Thèse, Univ. P. et M. Curie, 185 pp.

Sallé G. 1987. Originaliés des tissus conducteurs des phanérogames parasites. Bull Soc Bot Fr 134 Actual Bot 3/4:81–95.

Schulze E-D, Lange OL, Ziegler H, Gebauer G. 1991. Carbon and nitrogen isotope ratios of mistletoes growing on nitrogen and non-nitrogen fixing hosts and on CAM plants in the Namib desert confirm partial heterotrophy. Oecologia 88:457–462.

Shah N, Smirnoff N, Stewart GR. 1987. Photosynthesis and stomatal characteristics of *Striga hermonthica* in relation to its parasitic habit. Physiol Plant 69:699–703.

Stewart GR, Press MC. 1990. The physiology and biochemistry of parasitic angiosperms. Ann Rev Plant Physiol Plant Mol Biol 41:127–151.

Struthers R, Lamont BB, Fox JED, Wijesuriya S, Crossland T. 1986. Mineral nutrition of sandalwood (*Santalum Spicatum*). J Expt Brt 37:1274–1284.

Tocher RD, Gustafson SW, Knutson DM. 1984. Water metabolism and seedling photosynthesis in dwarf mistletoes. In: Biology of Dwarf Mistletoes, pp. 62–69. USDA Forest Science General Technical Report RM III.

Ullmann I, Lange OL, Ziegler H, Ehleringer J, Schulze E-D, Cowan IR. 1985. Diurnal courses of leaf conductance and transpiration of mistletoes and their hosts in Central Australia. Oecologia 67:577–587.

Whittington J, Sinclair R. 1988. Water relations of the mistletoe *Amyema miquelii* and its host *Eucalyptus fasciculosa*. Aust J Bot 36:239–256.

9. Effects of Global Change on Plant–Animal Synchrony: Implications for Pollination and Seed Dispersal in Mediterranean Habitats

W.J. Bond

Among the central problems of global change over the next few decades are increasing levels of atmospheric carbon dioxide, global warming, and the shrinking of natural habitats by human activities. The question addressed here is whether global change will influence the reproductive biology of plants by disrupting plant–animal mutualisms and whether such disruption could impact structure and composition of Mediterranean ecosystems.

Very little is known of the effects of increased atmospheric carbon dioxide on plant–animal interactions. The few available studies on insect herbivory suggest the net effect on herbivory levels will probably be slight (Nicolas and Sillans, 1989). Instead, the effects of possible climate change, especially temperature increase, will be considered here. Changes in precipitation, especially its seasonality, could have major effects on ecosystems, but the capacity to predict such change seems even less reliable than predictions of global warming. The problem of habitat fragmentation will not be considered since this is discussed elsewhere in this volume (Bond et al., 1988; Cowling and Bond, 1991, for fynbos studies).

The field of plant–animal interactions is filled with intriguing "special cases." Detailed studies on individual plants and their animal pollinators and dispersers abound, but community-wide studies are rare. Many studies have been directed at the identification of "syndromes," identifiable suites of traits associating particular plant features with particular animal partners (Faegri and van der Pijl, 1979; van der Pijl, 1982). More recently, interest has shifted to evolutionary aspects (e.g., Herrera, 1986). Studies on the importance of plant–animal mutualisms in determining distribution and abundance of species in communities are still in their infancy. The paucity of such studies is very apparent in exercises such as this requiring broad-scale predictions of vegetation change. Mutualisms are often unpredictable in time and space (e.g., Herrera, 1988; Horvitz and Schemske, 1990) and the generally reductionist nature of most studies means that predictions at the scale of whole floras or ecosystems are extremely difficult. Here the problem is addressed by making tentative generalizations on determinants of plant and insect phenology, the possibility of reproductive disruptions, and the general importance of seeds in regulating plant abundance. Examples are drawn, wherever possible, from Mediterranean-type vegetation, though inevitably biased towards the systems most familiar to the author. The main purpose of this contribution is to suggest relevant questions, mostly unanswered, and to offer tentative speculation on the broad effects of pollinator/disperser disruption on Mediterranean communities. The shaky foundations of both enterprises need emphasis! Hopefully, the scale of the problem, and the tentative predictions, will challenge students of pollination and dispersal to evaluate the broader context of their studies (it did for the author!).

Direct Effects of Climate on Plant Reproduction

Climate change can have direct effects on floral development which may disrupt the reproductive process. Microclimate can affect flower opening, anther extrusion and dehiscence, transfer of viable pollen, growth of the style, stigmatic receptivity, pollen germination, and growth of the pollen tube (Corbet, 1990, for a comprehensive review). Disruption of any of these processes in the flowering sequence can cause reproductive failure and has been shown to do so for some commercial species (Corbet, 1990). New threats may appear. Elevated levels of UV B radiation, equivalent to those anticipated as a result of ozone depletion, have been shown to cause sterility in a number of fynbos species (Musil and Wand, 1993).

Climate Cues and Reproductive Phenology of Plants

Seasonal Patterns of Flowering

Will climate change influence the seasonal phenology of plants and their interdependent pollinators and dispersers? If so, will plants and their reproductive partners become phenologically uncoupled?

There has been considerable interest in proximate and ultimate determinants of flowering phenology. Since climate change is likely to be much more rapid than the rate of evolutionary response, studies of proximate cues can best indicate possible reproductive shifts. Only three major environmental factors have been identified as cues that initiate flowering: photoperiod, temperature, and moisture (Wareing and Phillips, 1981; Rathcke and Lacey, 1985). Photoperiodic control is common in herbaceous, short-lived plants. Woody species usually flower in response to temperature cues, generally cumulative heat sums above some threshold level (for Mediterranean-type ecosystem examples see Jacobs and Honeyborne, 1979; Kummerow et al., 1981). In some tropical species flowering is triggered by rainfall, but no examples have been reported for Mediterranean regions. There are other cues besides photoperiod, temperature, and moisture. Fire triggers flowering in some petaloid monocots in South African fynbos ("fire lilies" *Cyrtanthus* spp. Amaryllidaceae; Keeley, 1993) and in *Xanthorrhea* in Australia (Gill and Ingwersen, 1976), possibly due to ethylene released in the smoke, so that the timing of flowering depends on the timing of fire.

The abundance of flowers, as opposed to their timing, has been linked to rainfall in the preceding rainy season in several woody chaparral species (Keeley, 1987) and South African Proteaceae (personal communication, J. Vlok, 1992). Keeley has suggested that the interdependence of flower number and previous season's rainfall is probably widespread in woody chaparral species.

From studies on physiological flowering cues one may speculate that the seasonal *timing* of flowering might shift to earlier in spring in woody plants as a result of global warming, whereas herbaceous species may show little change. The *quantity* of flowers produced on dominant woody shrubs will be sensitive to shifts in precipitation amount and seasonality. Finally, the proportion of individuals that flower or set fruit may also change, as shown in chaparral shrubs where the frequency of flowering or fruiting individuals varied greatly from year to year depending on rainfall (Keeley, 1977).

Topographic Effects on Flowering

Intraspecific differences in flowering phenology can be commonly observed along altitudinal gradients or on different aspects. For example, *Erica seriphiifolia*, a fynbos heath, flowers 3 months apart on adjacent level versus steep, south-facing aspects. The difference was maintained when the progeny were planted in a common garden (Bond, unpublished). Plant species in topographically diverse areas, with spatially variable flowering phenology, will likely be less vulnerable to global change than topographically homogeneous areas such as Western Australia.

Phylogenetic Determinants of Flowering Phenology

Flowering phenology has a strong phylogenetic component (Kochmer and Handel, 1986) so that flowering season is often similar among related species (Bell and Stephens, 1984, for Australian kwongan; Johnson, 1993, for fynbos). In fynbos, there is a clear sequence of flowering from Iridaceae in spring, through Orchidaceae in early summer, to Amaryllidaceae in autumn (Johnson, 1993) among monocotyledons. Among prominent woody families, Proteaceae show nearly exclusive spring flowering, in contrast to the year-round flowering of members of the Ericaceae. Petaloid monocotyledons have shorter flowering seasons than dicotyledons (Johnson, 1993), probably because the geophytic monocots have a limited number of flowering apices compared to the richly branched shoot systems of shrubs. Phylogenetic patterns in phenology may prove useful in attempting to generalize from single-species studies to whole floras. For example, Johnson's (1993) study suggests that monocots, wth their short flowering season, may be more susceptible to phenological disruption of plant pollinator interactions than shrubby dicots.

Seasonal Patterns of Fruiting

Fruiting and ripening follow sequentially from the flowering process and will therefore be affected to some extent by changes in flowering phenology. However, fruit ripening times may also be subject to climatic control, though determinants of fruiting time in natural habitats are poorly known. Fruiting phenology may be less dependent on specific climatic cues than on availability of resources, which influences the rate at which large fruits can be developed (Primack, 1985; Rathcke and Lacey, 1985). Experimental studies are needed which explore the potential for change in the period of fruit set along with changing climate. It would seem that woody plants would have greater capacity to vary flowering/fruiting interval than herbaceous or geophytic plants, simply because of larger resources, including more flowering apices (Johnson, 1992).

There is great variability in the period from flowering to viable seed or ripe fruit production and in the display period before a fruit begins to rot. This variability is seen in the variable relationship between flowering and fruiting among and within species. Thus Herrera (1984) found a broad range of fruit ripening rates and interindividual synchrony in a study of Mediterranean shrub phenology. Knight (1986, 1988), in a study of fleshy fruited species in the Mediterranean regions of South Africa, found all combinations of synchronous flowering followed by synchronous fruiting, synchronous flowering followed by variable fruiting period, and irregular flowering followed by synchronous fruiting. Some species displayed fruits for just a few weeks, others presented fruits throughout the year. Fruits vary too in how long they can persist before rotting, with some lasting less than a week and others for more than 3 months. Fruiting phenology can vary among populations, among individuals within populations, and among branches within individuals (Herrera, 1984; Pierce and Cowling, 1984; Knight, 1986, 1988). How does one predict the effects of climate change on fruiting asynchrony in the face of such complexity? Clearly there is scope for more studies on the relationship between flowering and fruiting phenology and variability in rates of fruit growth and development.

Climate Cues and Pollinator/Disperser Activity

The effect of climate on the timing of pollinator and disperser activity is complex and variable. Climate influences a variety of stages in insect life cycles and vertebrate behavior patterns, including migration. Climate also influences daily activity patterns, for example, diurnal movement periods of insects. Thus there is a host of possibilities determining possible macroclimatic effects on animal pollinators and dispersers.

Temperature has generally been considered a dominant factor controlling insect development and survival, with moisture playing a lesser role. There is a large literature on bioclimatic determinants of insect distribution related to their significance as pests (e.g., Messenger, 1959). Suffice it to say that climate change will almost certainly influence insect distribution and abundance. Climatic warming should allow the release of many insects from the seasonal constraint of low winter temperatures, possibly resulting in multivoltine life cycles. An extreme example of a potential consequence comes from the obligate mutualism between *Ficus carica* and its wasp pollinator. The fig can be grown from seed but does not reproduce north of the 46th parallel in France (Kjellberg and Valdeyron, 1990). Apparently the fig does not reproduce in the north because its wasp pollinator is absent due to conditions being too cool for completion of the two-generation cycle of the wasp. Climatic warming may lead to more generations in the wasp, allowing fig populations to spread northwards.

Determinants of insect phenology appear just as complex as for plants. Tauber and Tauber (1976) have reviewed climatic determinants of insect phenology. Diapause, the equivalent of dormancy in plants, is influenced by photoperiod and temperature. Photoperiod and/or low temperature serve to maintain early diapause, but the response to both low temperature and photoperiod decreases as the duration of diapause increases. Thus, early diapause is rigidly enforced by external cues, but diapause may end without a specific climatic signal (Tauber and Tauber, 1976). Postdiapause development typically depends on temperature regimes, but moisture is also required to break diapause in some species. Thus, temperature (both temperature thresholds and degree days), as in plants, is an important determinant of the phenology of insect emergence. Also as in plants, there is both within- and between-population variation in phenological response, indicating the capacity for selective change (Tauber and Tauber, 1976).

It is important to note that although plants and animals use similar external cues to time phenology, this does not necessarily mean they will respond synchronously if the timing of these cues change due to climatic shifts (see below).

Climate and Foraging Patterns

Foraging patterns of insects are strongly dependent on weather conditions. Ant foraging, for example, is markedly temperature dependent (Bernstein, 1979; Holldobler and Wilson, 1990), which may influence their effectiveness as dispersers (see below). The distribution of pollinator groups in California and Chile has been attributed to climatic variation (Mooney, 1977). In arid areas, small solitary bees, bombyliid flies, and wasps predominate, in contrast to foggy coastal areas dominated by large, endothermic bees, long-tongued flies, and hummingbirds. In the intermediate evergreen shrublands, all groups do well, with the addition of butterflies and hoverflies (Mooney, 1977). Along altitudinal gradients, bees and bee-pollinated flowers decline at high elevations in Chile and the European Alps. They are replaced as pollinators by butterflies, flies, and birds, depending on the geographic region (Arroyo et al., 1982). The implication is that climate change might lead to shifts in pollinator assemblages which might influence pollination and plant reproduction.

Evidence for Asynchrony Between Plants and Pollinators and Dispersers

Pollination

Plant–pollinator interdependence poses the classic "chicken or egg" paradox. Is the seasonal availability of pollinators and dispersers generally a cause or effect of plant phenology? Related plants often share similar flowering seasons (Bell and Stephens, 1984; Kochmer and Handel, 1986; Johnson, 1993), suggesting phylogenetic or developmental controls on flowering phenology so that in these groups insects follow plant flowering patterns rather than vice versa (Johnson, 1993). Many studies show a congruence between flowering time and the timing of maximum insect activity. In fynbos, the phenology of the Cape honey bee differs in the winter and all-year rainfall regions. Reproductive swarming takes place in September in the western Cape and November in the southern Cape, coinciding with differences in the peak flowering season in the two regions (Hepburn and Jacot-Guillarmod, 1991). Very few studies, especially from Mediterranean regions, explore the controls on phenology of both plants and their animal mutualists, so that it is difficult to evaluate the likelihood of reproductive failures if cues are mismatched.

Most work on the mismatching of plants and their dependent animals comes from the biological control literature, where failures in control due to climatic mismatching of insects and their plant hosts have long been recognized (DeBach, 1964). For example, St. John's Wort (*Hypericum perforatum*) has not been effectively controlled by chrysomelid beetles in moister, colder parts of British Columbia, though control is successful in areas with a Mediterranean-type climate matching the native habitat of the dominant beetle. Control is most effective where growth of the weed is curtailed by summer droughts. Where summer drought stress is relieved and winters are too cool for early insect emergence, *Hypericum* can escape its insect herbivores (Williams, 1985).

Evidence for pollinator–plant asynchrony is rare, especially asynchrony linked to climate. Waser (1979) found that seed set in ocotillo, *Fouquieria splendens*, a hummingbird-pollinated shrub in California arid shrublands, was greater for plants flowering coincidentally with the presence of hummingbirds than for plants flowering after hummingbirds had migrated. Several studies by Schemske and his colleagues (1978) in woodland herbs indicate that variation in seed set can be attributed to seasonal variation in the availability of pollinators. In this habitat, pollinator composition is very variable from year to year, implying frequent pollinator failures, some of which may be due to phenological mismatching. There are also examples of consistent mismatching of plant and pollinator phenologies. Zimmerman et al. (1989) studied an orchid pollinated by a single species of euglossine bee. Pollinator activity peaked in the dry season, but flowering peaked in the wet season. By far the best fruit set was from the few flowers produced earliest in the flowering season when the pollinators were still abundant. In this case, optimal pollinator activity coincides with leafing out of this deciduous species. The orchid, unlike its close relatives, has been unable to separate periods of growth from periods of flower and fruit production so that flowering is delayed. Many Mediterranean-type ecosystem (hysteranthous) geophytes have uncoupled flowering from growth and should be better able to track shifts in pollinator abundance (Dafni et al., 1981; Le Maitre and Midgley, 1992; Johnson, 1993). Geophytes that flower while foliage is on the plant (synanthous) may be more susceptible to pollinator asynchrony.

There seem to be rather few studies of plant–pollinator mismatching in Mediterranean shrublands. Keeley (1987) studied flowering and seed set in *Ceanothus crassifolius* in California over a 6-year period. Fruit production varied greatly from year to year, site to site, and individual to individual. On cool north-facing slopes, fruit production was higher in years with warmer March temperatures. Keeley attributed this pattern to temperature-dependent pollinator success, primarily because March temperatures were not correlated with fruit set on the warmer south slopes. However, other explanations are possible, for example, temperature-dependent changes in resource allocation to seed set.

Vaughton (1990, 1992) studied the effect of flowering season on pollinator visits and seed set in *Banksia spinulosa*. She showed that bird and honeybee visits changed through the flowering season, with fewest bird visits, most honeybee visits, and lowest seed set, presumably because bees are less effective pollinators, toward the end of the season (Table 9-1). Pollinator visits also changed from year to year, with three times more visits in the mid-flowering period in one year than the following. This difference was correlated with differences in follicle set (58.5% vs. 21.2%), but she was unable to separate the effects of resource limitation versus pollinator abundance on seed set.

To summarize, climate change may disrupt pollination systems in a number of ways and at a number of scales:

1. The microclimate of the flower can directly influence pollen viability and seed set, resulting in reproductive failure.
2. The daily activity of insect pollinators can be disrupted by microclimatic change.
3. Seasonal phenology of flowering and insect development are sensitive to climate cues and may change asynchronously.
4. The distribution of pollinator assemblages may change across altitudinal gradients.

Though there appears to be considerable scope for direct or indirect disruption of plant pollination by climate, there are very few well-substantiated examples from natural populations. There seems to be plenty of scope for experimental studies on climatic effects, for example, by exploiting the steep altitudinal gradients found in many Mediterranean regions. To be of relevance to global change predictions, studies should demonstrate effects of sufficient magnitude to influence population distribution or abundance.

Dispersal

The nature of dispersal problems related to global change will likely differ in different Mediterranean systems. In the Mediterranean Basin itself,

Table 9-1. Seasonal Variation in Pollinator Activity on *Banksia spinulosa*, from Vaughton, 1992

	Early	Mid	Late
No. of visits per inflorescence			
Birds (honeyeaters)	23.1	16.0	7.2
Honeybees	0.0	0.0	8.1
Flowers pollinated (%)	22.0	27.0	64.0
Fruit set (% inflorescences)	61.9	58.5	29.2

Early = May–June; Mid = July–August; Late = September.

frugivory by migrant birds may be disrupted if the timing of migrant visits becomes markedly uncoupled from fruiting periods (Herrera, 1982; Jordano, 1982; Izhaki and Safriel, 1985; see Jordano, 1992, for a wide-ranging review of vertebrate frugivory). Herrera (1984) suggested that the absence of summer and early autumn fruiting species in parts of Spain might be due to the local scarcity of autumn migrants. However, the fruiting season of Mediterranean shrubs lasts several months for most plant species; most dispersers are not migrants but winter visitors that stay in the region for several months; and most plant species are dispersed by many animal species, so that even a significant shift in plant or animal phenology might have little effect on the dispersal process (Herrera, 1984, 1986; Debussche and Isenmann, 1989). Plant species with large fruits, exceeding the gape width of local bird assemblages, are poorly dispersed at some Mediterranean sites (Herrera, 1984), mostly by mammals (Debussche and Isenmann, 1989). A comparison of the demography of the smaller-fruited species with large-fruited species might help indicate the demographic and distributional consequences of breakdown of bird dispersal.

In South Africa, fleshy fruits are rare in fynbos, though common in adjoining strandveld and forest (Siegfried, 1983; Knight, 1988). Fleshy-fruited species are generally forest precursors. Dispersal is the first stage of successional replacement of fynbos by forest and therefore is potentially important in triggering a complete switch in vegetation systems (Manders and Richardson, 1992). However, patterns of fruit dispersal are unlikely to be markedly affected by climate change, since seasonality of fruit display is extremely variable among species, within species, and even within individuals (Pierce and Cowling, 1984; Knight, 1986, 1988). Frugivory in fynbos is not strongly dependent on seasonal migrants, and it seems unlikely that frugivory will change greatly with global climate change. Frugivory in chaparral is also associated with later successional species that invade long-unburned chaparral (Keeley, 1991), so that altered dynamics of plant–frugivore interactions might have important consequences for chaparral composition. The effect of migrants on chaparral frugivory needs further study.

Climate change might impact myrmecochory, a common dispersal mode of many species in fynbos and Australian heathlands (Berg, 1975; Bond and Slingsby, 1983). Myrmecochory in fynbos is important not only for dispersal but also as a mechanism for reduction of predation by rodents, especially in large-seeded fynbos species (Bond and Slingsby, 1984; Bond and Breytenbach, 1985). Disruption of the dispersal process can lead to complete recruitment failure because of increased rodent predation (Bond and Slingsby, 1984; Slingsby and Bond, 1985). Seed predation by ants, rather than vertebrates, is more significant in Australia. Here the presence of elaiosomes increases the chances of seed discovery by seed-dispersing ants rather than seed-harvesting ants (Hughes and Westoby,

1992). Ants are very sensitive to temperature, and the diurnal activity of ants varies at different seasons (Steyn, 1954; Bernstein, 1979; Holldobler and Wilson, 1990). The most important seed-dispersing ant species in fynbos, *Anoplolepis custodiens*, is inactive in winter, forages only at midday in spring, but in the heat of midsummer rests at midday and forages only in the early morning and late afternoon (Steyn, 1954; personal observations). Ants are also very sensitive to moisture, and dispersal can fail if seeds are dispersed when the vegetation is wet (Bond and Breytenbach, 1985). Thus, seed dispersal by ants should be worst in winter and most effective in summer. Current flowering and fruiting patterns in *Leucospermum*, a myrmecochorous genus of Cape Proteaceae, result in seed release in midsummer when ants are most active (Pierce, 1984). The inferred seasonal variation in ant dispersal effectiveness is supported by recent experiments by Hughes and Westoby (1990) in Australia. They offered seeds of the same two species to ants throughout the year and found maximum dispersal rates in summer and minimum rates in winter (Figure 9-1).

Climate change could seriously disrupt myrmecochory. If seed release occurs too early, following temperature-cued early flowering, dispersal may be disrupted by rain. Changes in temperature-related foraging may expose seeds to greater risks of discovery by predatory vertebrates or ants. For some species, breakdown of myrmecochory could result in complete recruitment failure (Bond and Slingsby, 1984; Slingsby and Bond, 1985). In others the effects may be minor (Pierce and Cowling, 1991).

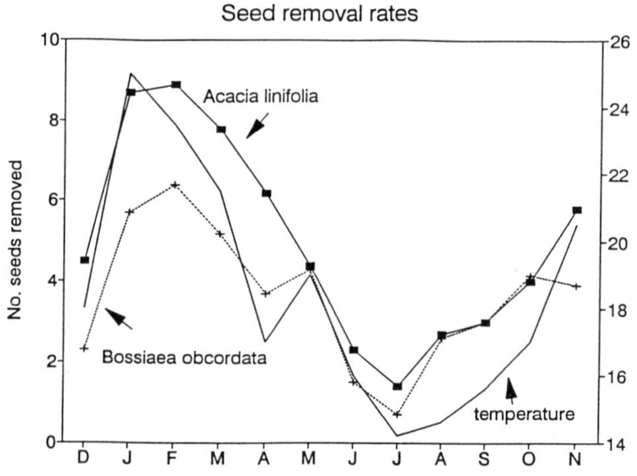

Figure 9-1. Seasonal variation in rate of seed dispersal by ants (Hughes and Westoby, 1990).

In summary, dispersal relationships are typically more diffuse than pollination and may be less sensitive to disruption. The phenology of fruiting is not as well explored as patterns of flowering. Are there, for example, phylogenetic patterns in fruiting season or ripening rate, or do local selective pressures play a greater role (cf. Herrera, 1984)? Even if climate change does alter fruiting phenology, the effects on vertebrate frugivory may be small because of long fruiting seasons and lack of disperser specificity. In this respect, it is worth noting that current disperser assemblages in Mediterranean shrublands evolved much later than the plants whose fruits they disperse, so that the flora has already survived changes in the disperser community (Herrera, 1984, 1986).

Myrmecochory seems much more vulnerable to climate-related disruption. Unlike vertebrate frugivory in Mediterranean shrublands, disruption of myrmecochory is known to have marked demographic effects in at least a few species, with potential system-wide consequences where myrmecochores are legumes (Shea et al., 1979). Ant activity also appears to be far more sensitive to microclimatic disruption than vertebrate activity.

Which Species and Communities Are Most at Risk?

Given that plant–mutualist disruptions may occur, what will be the likely impact on Mediterranean communities? Brief guidelines follow which may help focus attention on particular species or processes likely to be sensitive. The major impact of pollinators will be on seed numbers. Differences in seed quality due to different levels of pollinator movement or degree of outcrossing may also be influenced, but the effects will be less immediate than failures in seed set. Dispersal failures will influence the subsequent transition of seeds to surviving seedlings. The dependence of plants on these processes varies greatly. For example, plants may persist long after their putative dispersers have become extinct (Janzen and Martin, 1982; Herrera, 1984; Tredici, 1989).

Key attributes influencing which species may be at risk from climate change include plant breeding systems, pollination and dispersal mode, and reproductive dependence on seeds.

Breeding Systems

Many plants are facultative or obligate selfers and will be relatively independent of pollinators for seed production. In general, autogamy is most common in short-lived herbaceous plants, while xenogamy predominates in long-lived herbaceous woody perennials (Richards, 1986). There have been several surveys of breeding systems in Mediterranean regions (Moldenke, 1975; Keighery, 1982; Herrera, J., 1987; Steiner,

1987; Boaz et al., 1990), but self-compatibility levels in hermaphrodite plants are poorly known for most groups. There are many exceptions to the generalization that annuals are self-pollinated and perennials cross-pollinated. Crucifers in Mediterranean Israel, most of which are annuals, are predominantly xenogamous (Boaz et al., 1990). Some members of Proteaceae, all perennial shrubs, are self-compatible and can set seed without pollinators (Horn, 1962; Keighery, 1982; Vaughton, 1988). *Erica nabea* is a widespread cleistogamous fynbos shrub, while *Cliffortia ruscifolia* may produce seeds apomictically over much of its range. Nevertheless, at an initial guess, annuals, which can be important in postfire environments influencing nutrient uptake, are likely to be less sensitive to pollinator disruptions than longer-lived perennial woody or herbaceous plants.

Pollination Mode

It is tempting to assume that wind pollination will be less impacted by climate change than animal pollination because it does not involve animal partners. If this proves to be the case, wind-pollinated taxa may persist as a stable matrix in Mediterranean shrublands. Prominent wind-pollinated taxa include *Quercus*, *Garrya*, *Cliffortia*, *Anthospermum*, Cupressaceae, Pinaceae, Cyperacae, Poaceae, Restionaceae, etc. Given the importance of many of these taxa in Mediterranean shrublands, closer examination of the assumption that wind pollination is relatively immune from climate effects is warranted (e.g., Norton and Kelly, 1988).

Vertebrate pollinators are likely to be less influenced by climate change than insect pollinators because of body size and thermoregulatory capacity. Local movement patterns of sunbirds, sugarbirds, and hummingbirds may change, but these birds are likely to track changing resource patterns and still provide a good service as pollinators (Skead, 1967; Keast, 1968; Collins and Rebelo, 1987; Vaughton, 1990). Recent studies on Australian *Banksia* suggest that seed set may be influenced by flowering time in relation to bird pollinator activity (Vaughton, 1990, 1992). Nevertheless, given the roving nature of nectarivorous birds, it would seem that the dominant, bird-pollinated shrubs of many fynbos and Australian heathland communities may only be weakly affected by pollinator disruptions.

Among insect-pollinated groups, pollinator specificity varies greatly and will influence the likelihood of pollinator disruptions. Some plant species are visited by dozens of different insect species (e.g., *Lavandula latifolia*; Herrera, 1988) and some by both birds and insects (e.g., *Protea repens*; Coetzee and Giliomee, 1985). These will presumably be relatively unaffected by pollinator disruptions. Others are quite narrowly specialized. In fynbos, at least 15 species of spectacular red-flowered species are pollinated by a single butterfly, *Meneris tulbaghiae* (Johnson and Bond, 1994). The butterfly is only active in summer when all these plant species flower. Seed set is often low in *Meneris* plants, and climate-induced disruptions of plant–pollinator synchrony could quickly lead to species loss. Plants pollinated by oil-collecting bees are also highly specialized (Steiner, 1989), as are plants pollinated by long-tongued flies (Rebelo, 1987; Johnson, 1992). Clearly these species are potentially more vulnerable to pollinator shifts than those with a more generalist guild of pollinators. It seems that specialized pollination is most often associated with plant species that are relatively rare, at least in terms of biomass. Common species such as *Leucadendron* and *Protea*, which frequently form fynbos overstorys, and *Banksia* in Australian shrublands are pollinated by a variety of insects, birds (Coetzee and Giliomee, 1985; Collins and Rebelo, 1987; Ramsey, 1988; Hattingh and Giliomee, 1989; Vaughton, 1992), and even nonflying mammals (Goldingay et al., 1991). Thus, at least in fynbos, the basic structure of plant communities may be little influenced by pollinator disruption but biodiversity may be negatively impacted.

The same is true of dispersers. Myrmecochorous plants seldom dominate communities but include many rare and beautiful plant species. Only among the graminoid restios are species sufficiently common to seriously alter communities should dispersal be disrupted. Myrmecochores are also a relatively minor component in Australian heathlands measured in terms of biomass, though not in terms of species. However, many nitrogen-fixing Fabaceae are myrmecochorous, and examples have been reported of system disruption when legume reproduction fails (Shea et al., 1979).

Reproductive Dependence on Seeds: Are Seeds Limiting?

In chaparral, fynbos, and kwongan, with large populations of fire-stimulated nonsprouters, vegetation change is strongly related to fire. These areas have a higher potential population turnover rate than areas without fire and will be the first to change due to reproductive disruptions. Plants that resprout after fire have slower population turnover and should maintain their prominence in the face of reproductive failures longer than nonsprouting species that are killed by fire or that emerge only for brief periods after fire. Many shrublands of the Mediterranean basin are dominated by sprouters (Trabaud, 1984; Herrera, J., 1987) and may be slower to respond to reproductive disruption. Nonsprouting species, such as some of the *Cistus* species, or the legumes *Cytisus* and *Ulex*, may be the most sensitive indicators of climate-induced vegetation change.

Even among nonsprouters, the most abundant species may be little influenced by substantial reduction of seed supply. This is because many plant populations are essentially "seed saturated"; addition or subtraction of seeds would not impact population size because there is a surplus of seedlings over the space available for adults. High levels of seed predation, for example, may have negligible demographic effects (Anderson, 1989; Crawley, 1990, 1992). An example of a seed-saturated population is

Ceanothus megacarpus, where Schlesinger and Gill (1978) showed the elimination of many "surplus" plants by self-thinning in dense monospecific stands. Self-thinning is common in chaparral and possibly maquis and garrigue dominants and implies seed saturation. Large changes in seed numbers may have small effects on adult abundance. The heavy self-thinning characteristic of many chaparral nonsprouters is seldom seen in fynbos dominants. These seem to be seed limited, since seedling numbers are only one or two times more common than adults, and there is much less room for seedling depletion (Bond, 1989). However, the dominant Proteaceae are least likely to suffer from pollinator mismatching, since they have a mix of generalist insect and vertebrate pollinators. It should be relatively easy to set up experiments in which seeds are added to or subtracted from a postfire seed bed to test whether plant populations are seed saturated or seed limited. Such experiments would provide interesting insight into the importance of reproduction versus subsequent growth and survival as determinants of plant distribution and abundance.

In the less common species, densities may seldom reach high enough levels for self-thinning to act, and here seed supply may be more closely linked to population growth. Thus, the rare species again appear to be most vulnerable to disruptions of their reproductive biology. With the exception of systems where seed supply is low, such as fynbos Proteaceae and Australian heathlands, the basic structure of shrublands may be little affected.

Recently, seed supply has been implicated in patterns of coexistence (Chesson, 1986; Fagerstrom, 1988). The lottery for regeneration sites among species may underlie species abundance patterns more than subsequent interactions among plants. Preliminary studies indicate that lottery models might be applicable to fire-prone systems such as fynbos (Laurie, 1989; Bond et al., 1992). If recruitment overrides subsequent growth and survival in determining patterns of relative abundance, then shifts in species composition may occur due to differences in relative availability of seeds among co-occurring species. Again experiments are needed in which seed number is varied to determine whether competing populations are seed limited or seed saturated. If the latter, than climate-induced changes in plant growth or survival may prove more important in determining population size than changes in reproduction.

Though density dependence has been emphasized here as a process that uncouples seed number from population size, density-independent factors may also be important. *Banksia* seedlings in Australia, for example, can be severely impacted by drought in the first summer after a fire (Bradstock and O'Connell, 1988; Lamont et al., 1991), as are chamise seedlings after chaparral burns (Moreno and Oechel, 1992). Seedling survival is the key mortality factor here, and variation in seed supply may contribute little to population crashes. The important impact of climate change would be the probability of postfire drought rather than decline in seed availability through reproductive disruption. Again, experimental studies simulating realistic postburn climatic scenarios are needed.

Where Next?

This chapter has considered the potential for disruption of plant–pollinator/disperser interactions due to climate change in Mediterranean shrublands. Figure 9-2 summarizes the main effects of reproductive biology on survival through climate change. Vulnerability to pollinator disruption will be least in autogamous or apomictic species and possible wind-pollinated species, and most in pollen-limited outcrossing species. Vulnerability to dispersal breakdown will be least in abiotically dispersed species and most in species dispersed by migrant birds and especially by ants. The effects of both pollination and dispersal disruption on the persistence of vegetation will be most pronounced only where seed supply limits population size. Resprouters are the least sensitive, while nonsprouting species which experience little or no self-thinning and are therefore seed limited will be the most sensitive (see Bond, 1994, for approaches to screening vulnerability of plant populations faced with pollinator/disperser disrup-

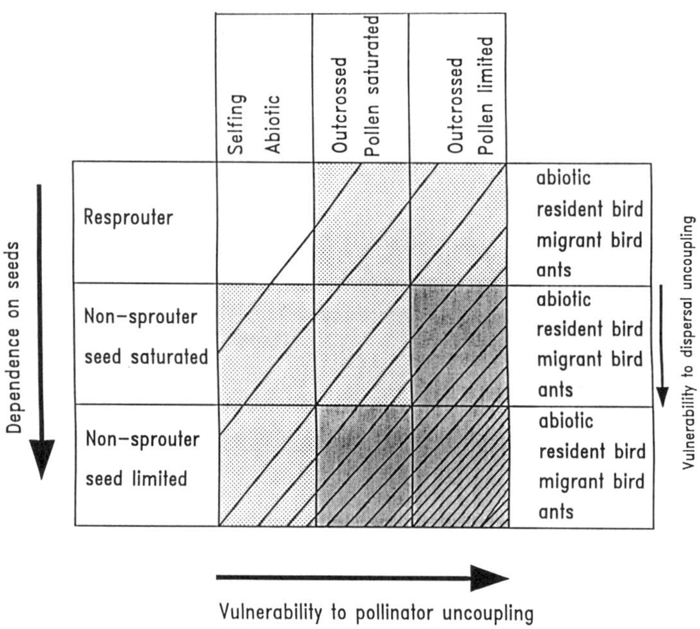

Figure 9-2. Assessing the risk of global warming in relation to plant reproductive traits. Density of diagonal hatching indicates increasing risk of reproductive failure.

tion). Predicting the spread of plant species, as opposed to their persistence, is a rather different problem, requiring consideration of dispersal and pollination under changing climates but with different emphasis on the demographic importance of seeds.

There is clearly much scope for research on a diversity of topics bearing on plant reproduction under changing climates. Major areas of ignorance where rapid progress could be made include the following:

1. Breeding systems of important species and the capacity to continue to produce seeds with diminished pollinator services.
2. The biology of key pollinators (and key plant resources on which many pollinators/dispersers depend) and their dependence on climate change. Obvious candidates for further study in fynbos include long-proboscid flies which may depend on marshy sites for their larval stages; bees, especially the large carpenter bees; and a better understanding of seasonal movements of birds in relation to nectar and fruit supplies.
3. The biology of key ant species implicated in myrmecochory and their likely activity patterns under changed temperature regimes. Similarly, the scale of dependence on migrant insects and vertebrate dispersers (and pollinators).
4. Identity and characteristics of seed-saturated versus seed-limited plant populations. The postfire seed bed offers unique opportunities for simple experimental manipulations of seed supply to test the impact of seed number on population size.
5. Direct climatic disruption of reproductive systems through, for example, increased UV radiation or microclimate-induced changes in floral development.

Mediterranean floras are notoriously diverse. Selecting the right species for detailed analysis of pollinator/disperser disruption is critically important for generalizing the results. There are many possible criteria for selecting species, but much effort could be wasted if they have little reproductive dependence on the mutualism or little demographic dependence on seeds. Regardless of whether or not climate change becomes a reality, studies that address the above areas will contribute significantly to our understanding of the functioning and management of Mediterranean shrublands.

Conclusions

There are so many unknowns that it might seem foolish to offer broad predictions. However, partly to summarize points made earlier in the paper, and partly to indicate their broader implications, the following speculative suggestions are offered:

1. Reproductive disruptions are unlikely to affect the basic structure of Mediterranean shrublands. The dominant shrubs and herbs are likely to persist either because there is little risk of pollinator failure or because the dominant species are not seed limited and undergo extensive self-thinning. Thus, essential system processes will probably be maintained. Even postfire annuals, which may be important in nutrient uptake, may be little affected because they are likely to include many selfing species.
2. The major impact is likely to be on the less common species with more specialized pollinators or dispersers where seed supply is more closely linked to population recruitment. Thus, biodiversity will likely be more impacted than system processes.
3. Of the Mediterranean regions, Australia, especially west Australia, appears most vulnerable, since topographic diversity is low, populations are often seed limited, pollinators are relatively scarce, ants are temperature sensitive dispersers, and turnover is rapid because of frequent fires. Chile and the Mediterranean itself may be most resilient because sprouting species often dominate communities, pollinators are diverse and abundant, and common plants are seed saturated (self-thinning is common). For the Cape fynbos, where as many as 50% of the flora are nonsprouters (Le Maitre and Midgley, 1992) and insect pollinators are often sparse (Johnson, 1992), there would seem to be a much higher potential for disruption.

Pollinator/disperser disruptions are among the more subtle consequences of climate change. Mediterranean shrublands offer model systems for assessing their significance in ecosystem change, both because of the diversity of reproductive mutualisms and because the demographic consequences of seed limitation can be comparatively easily assessed in fire-driven systems.

Acknowledgments. I would like to thank J. Moreno and W. Oechel for posing the question and inviting me to the Valencia meeting. A. Dafni, J. Donaldson, S. Johnson, J. Hoffmann, J. Keeley, A. Rebelo, M. Struck, and N. Allsopp provided useful advice and helped fill the many glaring areas of ignorance. The paper has benefited from reviews by two anonymous referees who were particularly helpful in drawing my attention to the extensive literature of the Mediterranean Basin.

References

Anderson AN. 1989. How important is seed predation to recruitment in stable populations of long-lived perennials? Oecologia 81:310–315.

Arroyo MTK, Primack R, Armesto J. 1982. Community studies in pollination ecology in the high temperate Andes of central Chile, I. Pollination mechanisms and altitudinal variation. Am J Bot 69:82–97.

Bell DT, Stephens LJ. 1984. Seasonality and phenology of Kwongan species. In: Pate JS, Beard JS (eds) Kwongan: Plant Life of the Sandplain. University of Australia Press, Nedlands, pp. 205–226.

Berg RY. 1975. Myrmecochorous plants in Australia and their dispersal by ants. Aust J Bot 23:475–508.

Bernstein RA. 1979. Schedules of foraging activity in species of ants. J Anim Ecol 48:921–930.

Boaz M, Plitmann U, Heyn CC. 1990. The ecogeographic distribution of breeding systems in the Cruciferae (Brassicaceae) of Israel. Israel J Bot 39:31–42.

Bond WJ. 1989. Describing and conserving biotic diversity. In: Huntley B (ed) Biotic Diversity in Southern Africa. Oxford University Press, Cape Town, pp. 2–18.

Bond WJ. 1994. Do mutualisms matter? Assessing the impact of pollinator and disperser disruption on plant extinction. Phil Trans Roy Soc Lond B 344:83–90.

Bond WJ, Breytenbach GJ. 1985. Ants, rodents and seed predation in Proteaceae. So Afr J Zool 20:150–154.

Bond WJ, Slingsby P. 1983. Seed dispersal by ants in shrublands of the Cape Province and its evolutionary implications. So Afr J Sci 79:231–233.

Bond WJ, Slingsby P. 1984. Collapse of an ant-plant mutualism: The Argentine ant, *Iridomyrmex humilis* and myrmecochorous Proteaceae. Ecology 65:1031–1037.

Bond WJ, Cowling RM, Richards MB. 1992. Competition and coexistence in proteoid shrubs. In: Cowling RM (ed) The Ecology of Fynbos: Divesity, Scarcity and Fire. Oxford University Press, Cape Town, pp. 206–225.

Bond WJ, Midgley J, Vlok J. 1988. When is an island not an island? Insular effects and their causes in fynbos shrublands. Oecologia 77:515–521.

Bradstock RA, O'Connell MA. 1988. Demography of woody plants in relation to fire: *Banksia ericifolia* L.f. and *Petrophile pulchella* (Schrad) R.Br. Aust J Ecol 13:505–518.

Chesson PL. 1986. Environmental variation and the coexistence of species. In: Diamond J, Case TJ (eds) Community Ecology. Harper and Row, Cambridge, pp. 240–256.

Coetzee JH, Giliomee JH. 1985. Insects in association with the inflorescences of *Protea repens* (L.) (Proteaceae) and their role in pollination. J Ent Soc So Afr 48:303–314.

Collins BG, Rebelo A. 1987. Pollination biology of the Proteaceae in Australia and southern Africa. Aust J Ecol 12:387–421.

Corbet SA. 1990. Pollination and the weather. Israel J Bot 39:13–30.

Cowling RM, Bond WJ. 1991. How small can reserves be? An empirical approach in Cape Fynbos, South Africa. Biol Conserv 58:243–256.

Crawley MJ. 1990. Plant population dynamics. Phil Trans Roy Soc Lond B 314:125–140.

Crawley MJ. 1992. Seed predators and plant population dynamics. In: Fenner M (ed) Seeds: The Ecology of Regeneration in Polnat Communities. CAB International, Wallingford, pp. 105–156.

Dafni A, Schmida A, Avishai M. 1981. Leafless autumnal flowering geophytes in the Mediterranean region—phytogeographical, ecological and evolutionary aspects. Plant Syst Evol 137:118–193.

DeBach P (ed). 1964. Biological Control of Insect Pests and Weeds. Chapman and Hall, London.

Debussche M, Isenmann P. 1989. Fleshy fruit characters and the choices of bird and mammal seed dispersers in a Mediterranean region. Oikos 56:327–338.

Faegri K, van der Pijl L. 1979. The Principles of Pollination Ecology. Pergamon Press, Oxford.

Fagerstrom T. 1988. Lotteries in communities of sessile organisms. Trends Ecol Evol 3:303–306.
Gill AM, Ingwersen F. 1976. Growth of *Xanthorrhea australis* R.Br. in relation to fire. J Appl Ecol 13:195–203.
Goldingay RL, Carthew SM, Whelan RJ. 1991. The importance of non-flying mammals in pollination. Oikos 61:79–87.
Hattingh V, Giliomee JH. 1989. Pollination of certain *Leucadendron* species (Proteaceae). So Afr J Bot 55:387–393.
Hepburn HR, Jacot-Guillarmod A. 1991. The Cape honey bee and the fynbos biome. So Afr J Sci 87:70–73.
Herrera CM. 1982. Seasonal variation in the quality of fruits and diffuse coevolution between plants and avian dispersers. Ecology 63:773–785.
Herrera CM. 1984. A study of avian frugivores, bird-dispersed plants, and their interaction in Mediterranean scrublands. Ecol Monogr 54:1–23.
Herrera CM. 1986. Vertebrate dispersed plants: why they don't behave the way they should. In: Estrada A, Fleming TH (eds) Frugivores and Seed Dispersal. Tasks for Vegetation Science 15. Junk, Dordrecht, pp. 5–20.
Herrera CM. 1988. Variation in mutualisms: the spatio-temporal mosaic of a pollinator assemblage. Biol J Linn Soc 35:95–125.
Herrera J. 1987. Flower and fruit biology in southern Spanish Mediterranean shrublands. Ann Missouri Bot Gard 74:69–78.
Holldobler B, Wilson EO. 1990. The Ants. Belknap Press, Cambridge, MA.
Horn W. 1962. Breeding research on South African plants: II Fertility of Proteaceae. So Afr J Bot 28:259–268.
Horvitz CC, Schemske DW. 1990. Spatiotemporal variation in insect mutualists of a neotropical herb. Ecology 71:1085–1097.
Hughes L, Westoby M. 1990. Removal rates of seed adapted for dispersal by ants. Ecology 71:138–148.
Hughes L, Westoby M. 1992. Effect of diaspore characteristics on removal of seeds adapted for dispersal by ants. Ecology 73:1300–1312.
Izhaki I, Safriel UN. 1985. Why do fleshy-fruit plants of the Mediterranean scrub intercept fall- but not spring-passage of seed-dispersing migratory birds? Oecologia 67:40–43.
Jacobs G, Honeyborne GE. 1979. The relationship between heat unit accumulation and the flowering date of *Leucospermum* cv. Golden Star. Agroplantae 11:83–85.
Janzen DH, Martin S. 1982. Neotropical anachronisms: the fruits the gomphotheres ate. Science 215:19–27.
Johnson SD. 1992. Plant-animal relationships. In: Cowling RM (ed) The Ecology of Fynbos: II Nutrients, Fire and Diversity. Oxford University Press, Cape Town, pp. 175–205.
Johnson SD, Bond WJ. 1994. Red flowers and butterfly pollination in the fynbos of South Africa. In: Arianoutsou M, Groves RH (eds) Plant animal interactions in mediterranean-type ecosystems. Kluwer, Dordrecht, pp. 137–150.
Jordano P. 1982. Migrant birds are the main seed dispersers of blackberries in southern Spain. Oikos 38:183–193.
Jordano P. 1992. Fruits and frugivory. In: Fenner M (ed) Seeds: The Ecology of Regeneration in Plant Communities. CAB International, Wallingford, pp. 105–156.
Keast JA. 1968. Seasonal movements in the Australian honeyeaters (Meliphagidae) and their ecological significance. Emu 67:159–209.
Keeley JE. 1977. Seed production, seed populations in soil, and seedling production after fire for two congeneric pairs of sprouting and non-sprouting shrubs. Ecology 58:820–829.

Keeley JE. 1987. Fruit production patterns in the chaparral shrub *Ceanothus crassifolius*. Madroño 34:273–282.

Keeley JE. 1991. Seed germination and life history syndromes in the Californian chaparral. Bot Rev 57:81–116.

Keeley JE. 1993. Smoke-induced flowering in the fire lily *Cyrtanthus ventricosus*. So Afr J Bot 59:638.

Keighery GJ. 1982. Bird pollinated plants in Western Australia and their breeding systems. In: Armstrong JA, Powell JM, Richards AJ (eds) Pollination and Evolution. Royal Botanic Gardens, Sydney, pp. 77–89.

Kjellberg F, Valdeyron G. 1990. Species-specific pollination: a help or a limitation to range extension? In: di Castri F, Hansen AJ, Debussche M (eds) Biological Invasion in Europe and the Mediterranean Basin. Kluwer, Dordrecht, pp. 371–378.

Knight RS. 1986. Fruit displays of indigenous and invasive alien plants in the south-western Cape. So Afr J Bot 52:249–255.

Knight RS. 1988. Aspects of plant dispersal in the south-western Cape with particular reference to the role of birds as dispersal agents. PhD thesis, University of Cape Town.

Kochmer JP, Handel SN. 1986. Constraints and competition in the evolution of flowering phenology. Ecol Monogr 56:303–325.

Kummerow J, Montenegro G, Krause D. 1981. Biomass, phenology and growth. In: Miller PC (ed) Resource Use by Chaparral and Matorral. Ecological Studies, Vol. 39, pp. 69–95. Springer-Verlag, New York.

Lamont B, Connell SW, Bergl SM. 1991. Seed bank and population dynamics of *Banksia cuneata*: the role of time, fire and moisture. Bot Gaz 152:114–122.

Laurie H. 1989. Can lottery models work for populations with discrete generations? Applications in Cape fynbos shrublands, University of Cape Town, Unpublished manuscript.

Le Maitre DC, Midgley JJ. 1992. Plant reproductive ecology. In: Cowling RM (ed) The Ecology of Fynbos: Nutrients, Fire and Diversity. Oxford University Press, Cape Town, pp. 135–174.

Manders PT, Richardson DM. 1992. Colonization of Cape fynbos communities by forest species. Forest Ecol Manage 48:277–293.

Messenger PS. 1959. Bioclimatic studies with insects. Ann Rev Entomol 4:183–206.

Moldenke AR. 1975. Niche specialization and species diversity along an altitudinal transect in California. Oecologia 21:219–242.

Mooney HA (ed). 1977. Convergent evolution of Chile and California—Mediterranean climate ecosystems. Dowden, Hutchinson and Ross, Stroudsburg, PA.

Moreno JM, Oechel WC. 1992. Post-fire controlling factors on the establishment of seedlings in southern California chaparral. Oecologia 90:50–60.

Musil CF, Wand SJE. 1993. Response of sclerophyllous Ericaceae to enhanced levels of ultra-violet B radiation. Environm Exp Bo 33:233–242.

Nicolas G, Sillans, D. 1989. Immediate and latent effects of carbon dioxide on insects. Ann Rev Entomol 34:97–116.

Norton DA, Kelly D. 1988. Mast seeding over 33 years by *Dacryidium cupressinum* Lamb. (rimu) (Podocarpaceae) in New Zealand: the importance of economies of scale. Funct Ecol 2:399–408.

Pierce SM. 1984. A synthesis of plant phenology in the fynbos biome. South African National Scientific Programmes Report 88, CSIR, Pretoria.

Pierce SM, Cowling RM. 1984. Phenology of fynbos, renosterveld and subtropical thicket in the south-eastern Cape. So Afr J Bot 3:1–16.

Pierce SM, Cowling RM. 1991. Dynamics of soil-stored seed banks of six shrubs in fire-prone dune fynbos. J Ecol 79:731–747.

van der Pijl L. 1982. Principles of Dispersal in Higher Plants. 3rd ed. Springer, Berlin.
Primack RB. 1985. Longevity of individual flowers. Ann Rev Ecol Syst 16:15–38.
Ramsey MW. 1988. Differences in pollinator effectiveness of birds and insects visiting *Banksia menziesii* (Proteaceae). Oecologia 76:119–124.
Rathcke B, Lacey EP. 1985. Phenological patterns of terrestrial plants. Ann Rev Ecol Syst 16:179–214.
Rebelo AG (ed). 1987. A preliminary synthesis of pollination biology in the Cape Flora. SANSP Report 141, CSIR, Pretoria.
Rebelo AG, Jarman ML. 1987. Pollination and community ecology. In: Rebelo AG (ed) A Preliminary Synthesis of Pollination Biology in the Cape Flora. SANSP Report 141, CSIR, Pretoria, pp. 155–192.
Richards AJ. 1986. Plant Breeding Systems. Allen and Unwin, London.
Schemske DW, Willson MF, Melampy MN, Miller LJ, Verner L, Schemske KM, Best LB. 1978. Flowering ecology of some spring woodland herbs. Ecology 59:351–366.
Schlesinger WH, Gill DS. 1978. Demographic studies of the chaparral shrub, *Ceanothus megacarpus*, in the Santa Ynez mountains, California. Ecology 59:1256–1263.
Shea SR, McCormick J, Portlock CC. 1979. The effects of fires on regeneration of leguminous species in the northern jarrah (*Eucalyptus marginata* SM) forest of Western Australia. Aust J Ecol 4:195–205.
Siegfried WR. 1983. Trophic structure of some communities of fynbos birds. So Afr J Bot 49:1–43.
Skead CJ. 1967. The sunbirds of southern Africa. Balkema, Cape Town.
Slingsby P, Bond WJ. 1985. The influence of ants on the dispersal distance and seedling recruitment of *Leucospermum conocarpodendron* (Proteaceae). So Afr J Bot 51:30–34.
Steiner KE. 1987. Breeding systems in the Cape flora. In: Rebelo AG (ed) A Preliminary Synthesis of Pollination Biology in the Cape Flora. SANSP Report 141, CSIR, Pretoria, pp. 22–51.
Steiner KE. 1989. The pollination of *Disperis* (Orchidaceae) by oil collecting bees in southern Africa. Lindleyana 4:164–183.
Steyn JJ. 1954. The pugnacious ant (*Anoplolepis custodiens* Smith) and its relation to the control of citrus scales at Letaba. Mem Ent Soc Sthrn Afr 3:1–96.
Tauber MJ, Tauber CA. 1976. Insect seasonality: diapause maintenance, termination and postdiapause development. Ann Rev Entomol 21:81–107.
Trabaud L. 1984. Fire adaptation strategies of plants in the French Mediterranean region. In: Margaris NS, Arianoutsou-Faraggitaki M, Oechel WC (eds) Being Alive on Land. Tasks for Vegetation Science 13:63–69.
Tredici PD. 1989. *Ginkgos* and multituberculates: evolutionary interactions in the Tertiary. Biosystems 22:327–339.
Vaughton G. 1988. Pollination and seed set of *Banksia spinulosa*: evidence for autogamy. Aust J Bot 36:633–642.
Vaughton G. 1990. Seasonal variation in honeyeater foraging behaviour, inflorescence abundance and fruit set in *Banksia spinulosa* (Proteaceae). Aust J Ecol 15:109–116.
Vaughton G. 1992. Effectiveness of nectarivorous birds and honeybees as pollinators of *Banksia spinulosa* (Proteaceae). Aust J Ecol 17:43–50.
Wareing PF, Phillips IDJ. 1981. Growth and Differentiation in Plants. 3rd ed. Pergamon, Oxford.
Waser NM. 1979. Pollinator availability as a determinant of flowering time in Ocotillo (*Fouquieria splendens*). Oecologia 36:223–236.

Williams KS. 1985. Climatic influences on weeds and their herbivores: biological control of St. Johns' Wort in British Columbia. Proc VI Int Symp Biol Contr Weeds. In: Delfosse ES (ed) Agric Can, pp. 127–132.
Zimmermann JK, Roubik DW, Ackerman JD. 1989. Asynchronous phenologies of a neotropical orchid and its euglossine bee pollinator. Ecology 70:1192–1195.

10. The Coastal Sage Scrub–Chaparral Boundary and Response to Global Climatic Change

George P. Malanson and John F. O'Leary

Boundaries of plant communities, ecosystems, and biomes have become a focus of attention in recent years. Structural or pattern aspects of edges have been reinvigorated by a focus on spatial processes at these locations on the landscape. Gosz (1991) described basic common ecological characteristics of such boundaries. A key insight is in their variability, even while an analogy to cell walls can be maintained. Forman and Moore (1992) listed aspects of membrane theory that may be relevant to the coastal sage scrub–chaparral boundary: intense disturbance destroys, relocates, or blurs boundaries; boundaries contain a record of interactions—strong interactions produce a structurally heterogeneous zone; and when boundaries move they increase the space of the system with higher variety. Boundaries may be particularly sensitive to changes in global climate and thus may be good indicators of climatic impacts, but boundary sensitivity is variable (Turner et al., 1991). Boundaries are conceptually important in ecology because they can be used to examine questions of scaling (Allen and Hoekstra, 1992) and system feedbacks (Wilson and Agnew, 1992).

The Coastal Sage–Chaparral Boundary

In the Mediterranean biome of California, four types of shrubland vegetation exist: chaparral, northern coastal scrub, coastal sage scrub, and coastal succulent scrub (Westman, 1983). Northern coastal scrub and coastal succulent scrub are spatially restricted to coastal areas of both the northernmost and southernmost parts of the Mediterranean biome in California. Evergreen sclerophyllous chaparral occupies the largest area and has extensive boundaries with coastal sage scrub (Figure 10-1). This boundary is found from central California south into Baja California in Mexico. Almost universally, the boundary occurs where chaparral occupies higher elevations on the coastal mountains, whereas coastal sage scrub is found as a band between chaparral and the sea, or in inland valleys between chaparral and inland grasslands or oak savannas.

Coastal sage scrub ranges in elevation from sea level to 1,000 m in more inland, southerly portions of its range. Unlike chaparral, sage scrub is characterized by malacophyllous subshrubs whose leaves absice during summer drought and are replaced by fewer, smaller leaves (Westman, 1981a). During moist winter and spring months, high transpiration and

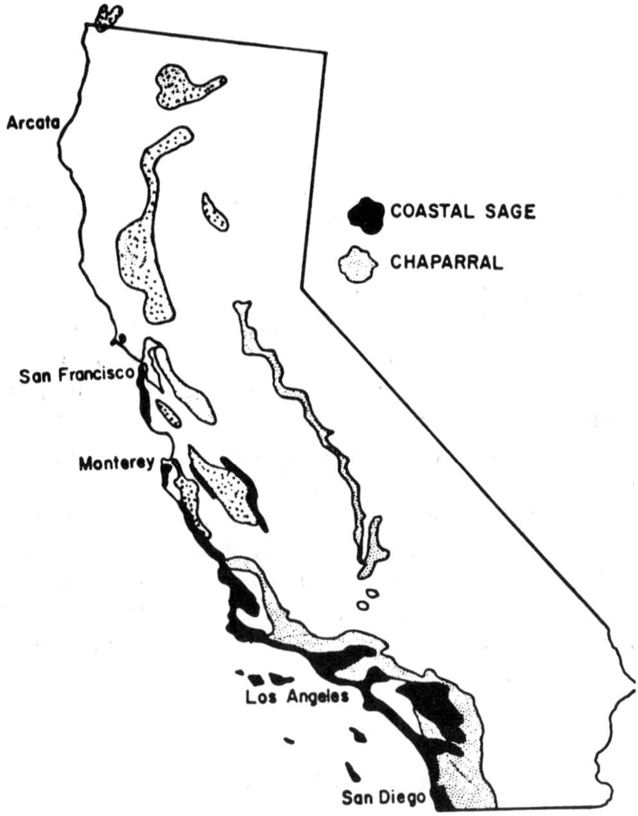

Figure 10-1. Distribution of the chaparral and coastal sage vegetation types in California. (Redrawn from Kuchler, 1964.)

carbon assimilation rates allow for rapid plant growth, flowering, and fruiting (Harrison et al., 1971). Most of the dominant species are drought evaders by virtue of their facultatively deciduous habit and are thus better adapted to prolonged summer-fall drought in areas of lower rainfall or in areas that are relatively xeric because of topographic position. Although sage scrub occurs on the full range of slope and aspect within its overall distribution, it occurs most frequently on steep equatorward-facing slopes (e.g., Kirkpatrick and Hutchinson, 1980). Chaparral occupies moister sites on poleward-facing slopes and in ravines and similar topographic locations which are likely to collect or retain greater soil moisture within areas of sage scrub, as well as higher on the slopes where orographic precipitation is likely to be greater. The physiognomy and total carbon gain in the two plant communities also indicate that coastal sage scrub has greater resource limitations than does chaparral (Mooney, 1988), and the only obvious gradient is in available moisture. At a more local scale, patches of sage scrub may replace chaparral on xeric sites underlain by shallow soils, or argillacious soils, on roadcuts, or in areas subject to chronic disturbance by grazing or subjected to very frequent fire (interval <10 yr) (Bradbury, 1978; Kirkpatrick and Hutchinson, 1980; Westman, 1981b; Zedler et al., 1983; Keeley and Keeley, 1988).

Sage scrub typically has distinct boundaries with chaparral at the scale of individual plants, but interdigitates at a landscape scale. This boundary appears to be quite stable. Bradbury (1978) found no change in a coastal sage scrub–chaparral pattern over a period of 41 years, as recorded in repeat photography. Gray (1983) found evidence for an ongoing invasion of coastal sage by chaparral, however. The invasion advances only in the immediate postfire years, however, and Bradbury's 41-year record apparently was fire-free.

The specific boundary between chaparral and coastal sage scrub has been variously attributed to climatic controls and to different fire regimes. Cooper (1922) and Howell (1929) speculated that lack of adequate winter moisture controlled the lower, dry limit of chaparral and therefore the moist upper limit of coastal sage scrub. Their hypothesis was later supported by both field observations and photosynthetic measurements made by Harrison et al. (1971) on plants of both types. A lack of difference in microclimate or soils has led to inference that the sharp boundary between the two types is maintained by fire (e.g., Bradbury, 1978). Moving beyond this inference, Oechel and Mustafa (1979) found no evidence that photosynthesis of chaparral species declined near the coastal sage scrub border. From these data they argued that factors other than growth, such as conditions for seedling establishment, might explain the spatial separation of the two shrubland community types.

The importance of fire regime is supported by observations of the sharpness of the boundary at the scale at which fires are a factor. The distinction in the boundary at different scales may be seen in comparison of maps of the vegetation types. The Weislander series (US Forest Service, 1945) shows considerable fine-scale detail in the boundaries in many locations. The Kuchler (1964) map shows broad-scale, smooth areas of contact between the two types. In contrast, the subassociations of coastal sage scrub which have distinct boundaries at a landscape scale are probably controlled by moisture differences, not fire regime (Malanson, 1984a). In San Diego County, the distinction mapped at 1:24,000 clearly shows an aspect-related difference with coastal sage scrub and chaparral interdigitating in east-west trending canyons with coastal sage scrub on the equatorward-facing slopes and chaparral on the poleward-facing slopes (Figures 10-2 and 10-3).

It is likely that both climate and fire are controlling factors of the boundary between coastal sage scrub and chaparral. The concern with climate change must address related changes in fire regime. Simple analysis of the potential effects of changing precipitation and temperature alone (e.g., Figure 10-4; Westman and Malanson, 1992) are insufficient. One approach to this boundary problem is to examine processes in a computer simulation model which can include both types of vegetation, and to use simulation experiments to analyze the relative importance of factors in determining the present and to project the future of the boundary.

Figure 10-2. Chaparral on poleward-facing slope (left) and coastal sage scrub on equatorward-facing slope (right) in San Diego County.

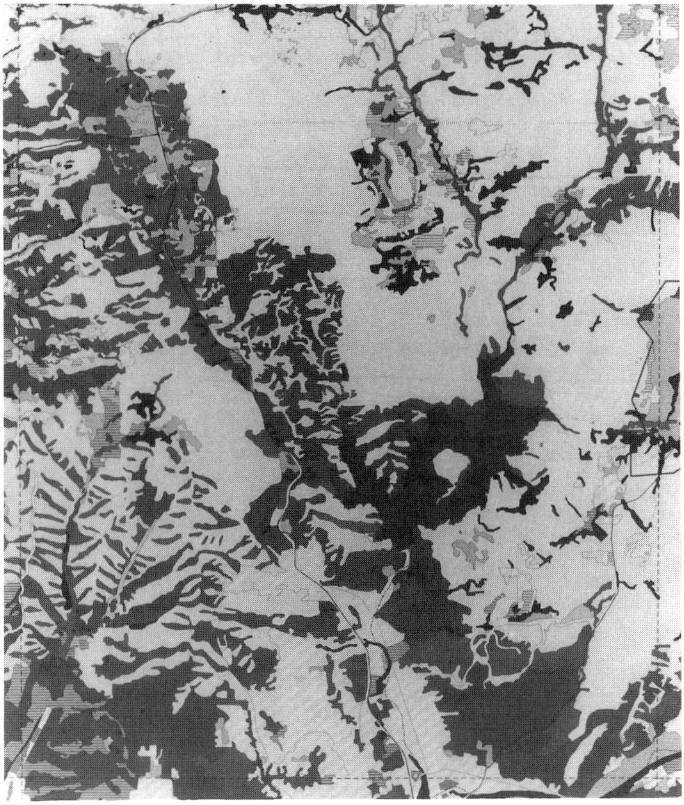

Figure 10-3. Boundary between coastal sage scrub and chaparral as mapped by San Diego County, 1992. The lower left corner illustrates the interdigitation of chaparral and coastal sage scrub on oppsite slopes of east-west trending valleys.

A Model of Californian Shrublands: FINICH

Three general types of models have been used to study the potential effects of global climatic change on ecosystems: transfer functions, stand (or gap) models, and physiological models. The stand models have been judged most versatile and appropriate for this purpose, given the criteria of filling societal information needs and including the responses demonstrated by species and ecosystems in the Holocene. Variations of the JABOWA-FORET model of forest dynamics are well known in this application (e.g., Solomon, 1986; Pastor and Post, 1988; Hanson et al., 1989).

Malanson and Westman (1991a) simulated the combined effects of climatic change and fire on coastal sage scrub. Although that model

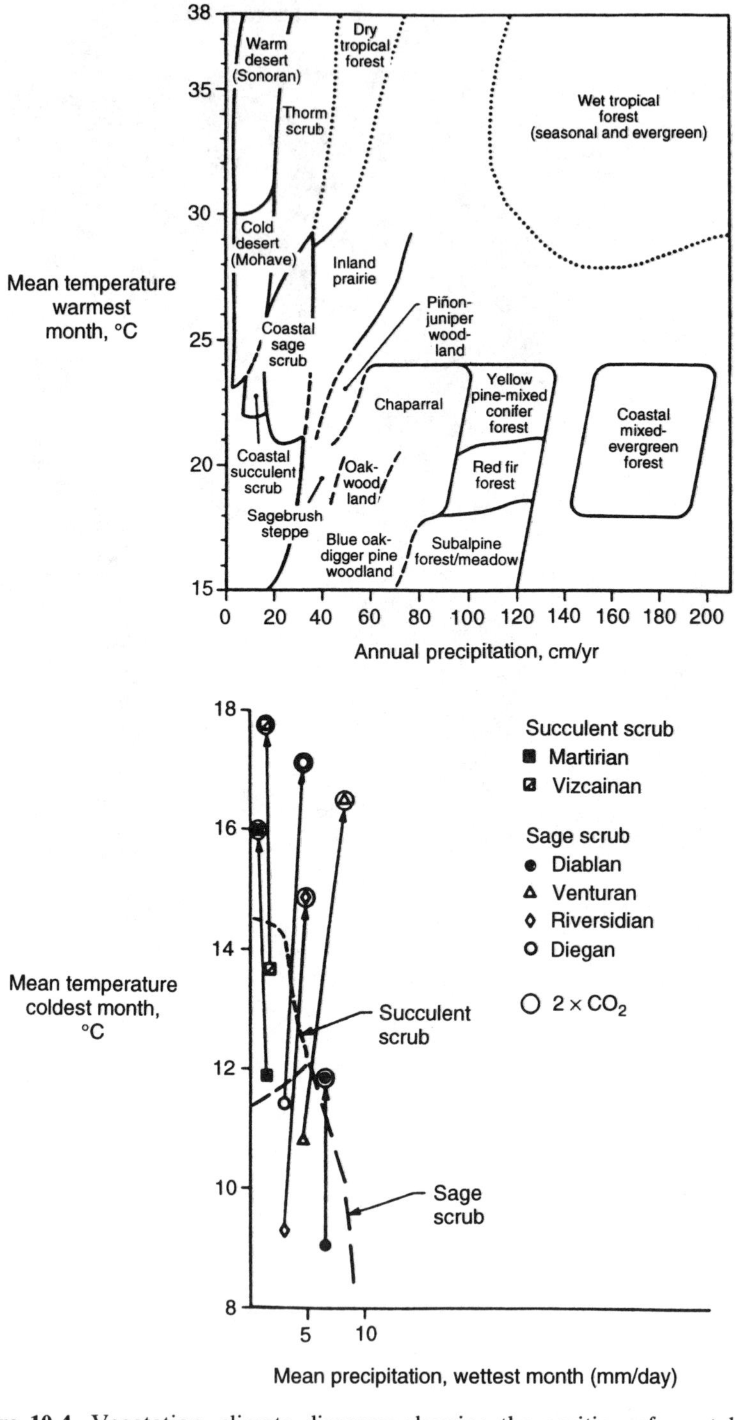

Figure 10-4. Vegetation–climate diagrams showing the position of coastal sage scrub and the change of specific locations following projected global climatic change; below, coastal sage scrub locations would change to positions occupied by cold desert and thorn scrub above.

projected some noteworthy changes, it was developed from an earlier model (FINICS, Malanson, 1984b) designed only to project the vegetation dynamics of coastal sage scrub under different fire regimes. It included only five species and was based on life history traits. A Leslie matrix format is used in which species in different age classes compete for space. It included no iterative feedback between vegetative growth and fire. Results suggested that altered fire regimes may result in greater changes in species composition than the direct effects of climatic change (Malanson and Westman, 1991a; Westman and Malanson, 1992).

Simulation Model Description

We developed a new model, FINICH (for a description see Malanson et al., 1992). The model differs from the JABOWA-FORET line models in using cohorts instead of individuals and in using more precisely defined climatic response functions. It differs from FINICS most notably in including the feedback between growth and fire intensity. The basic parts of the FINICS model that we continue to use are the type of resprouting function, in which the proportion of adults resprouting immediately after fire is a function of fire intensity, and competition for space, in which total foliar cover is used analogous to carrying capacity in a Lotka-Volterra demographic model. We made general changes in temperature alone or temperature and precipitation to reflect climatic changes projected by two different general circulation models (GCMs).

In FINICS, the populations of each species were processed as cohorts of individuals established in a given year; postfire resprouts and seedlings were kept separate. We retain this structure in FINICH. In the computations, we calculate growth for an individual and then multiply by the number of individuals in the cohort to obtain total growth for that cohort. The number of individuals in a cohort is determined in the year of establishment, and may be decreased in subsequent years through mortality. When mortality occurs, we reduce the total cover of the cohort proportional to the change in population. This structure is more economical of computations than are the individual-based models of JABOWA-FORET.

Growth Rates

We take the concept of a maximum growth rate from the JABOWA-FORET models. This maximum growth rate is then reduced by crowding and by climate, and we handle the process of continual basal sprouting in a new way. The equation for growth is:

$$\text{GROWT} = \text{MAX}_s * \text{F(PRECP, COLD, WARM)}_s * Z * ((150 - \text{TCOVER})/150)$$

where $\text{GROWT}_{s,c}$ is the total growth for an individual of cohort c of species s; MAX_s is the maximum rate for species s; F(PRECP, COLD,

$WARM)_s$ is the proportion of maximum growth found for species s at a given climate of annual precipitation PRECP, mean temperature of the coldest month COLD, and mean temperature of the warmest month WARM; Z is a factor to differentiate resprouts and seedlings, and TCOVER is the total foliar cover on the site, and 150 [%] is the limit to total foliar cover on the site (based on many previous studies).

We used observations of sizes reached during the first 2 to 5 years of growth following fire to estimate maximum growth rates. In these conditions, competition is reduced, nutrient levels are highest, and plants grow rapidly. For species for which we did not have data, we estimated this value from our observations in the field (e.g., Westman et al., 1981; Malanson and O'Leary, 1982, 1985; Malanson and Westman, 1985) and from observations on their size and form (Conrad, 1987). We defined seedlings establishing later to have maximum growth rates at 0.4 times those for postfire resprouts; we based this arbitrary value on observations by Malanson and O'Leary (1982) and adjustment in initial model trials.

Exponential growth is increasingly reduced as an upper limit, in our case a limit of total foliar cover, is approached. Because projected foliar cover can and does exceed 100% in this vegetation, we set an upper limit of 150%, and we then assumed that when cover exceeds 100%, an increasing proportion of the decreasing amount of growth is counted as replacement only.

Climatic Response

We use a direct measure of the success of a species in the climatic continuum defined by annual precipitation, temperature of the warmest month, and temperature of the coldest month. We derived response functions based on the present distribution of the species, i.e., their realized niche, and we compare these to hypothetical functions that we believe incorporate some aspects of their fundamental niche. Westman (1991) developed geographic distribution analysis to measure the direct response, or realized niche, of Californian shrubland species to climatic variables. In this research we used the same geographic distribution data. For a matrix of 120 possible combinations of temperature and precipitation classes, we calculated abundance at each point relative to the abundance at the locus of precipitation and temperatures where the species is at its maximum. The abundance of each species at every other combination of precipitation and temperatures is then proportional to this maximum (Figure 10-5a,b).

In California shrublands, climatic change could mean less stressful conditions, e.g., if precipitation were to increase. In some runs we altered the climatic response functions derived by geographic distribution analysis. We simply set the response at any precipitation value higher than the maximum value for a species equal to the maximum (1.0)

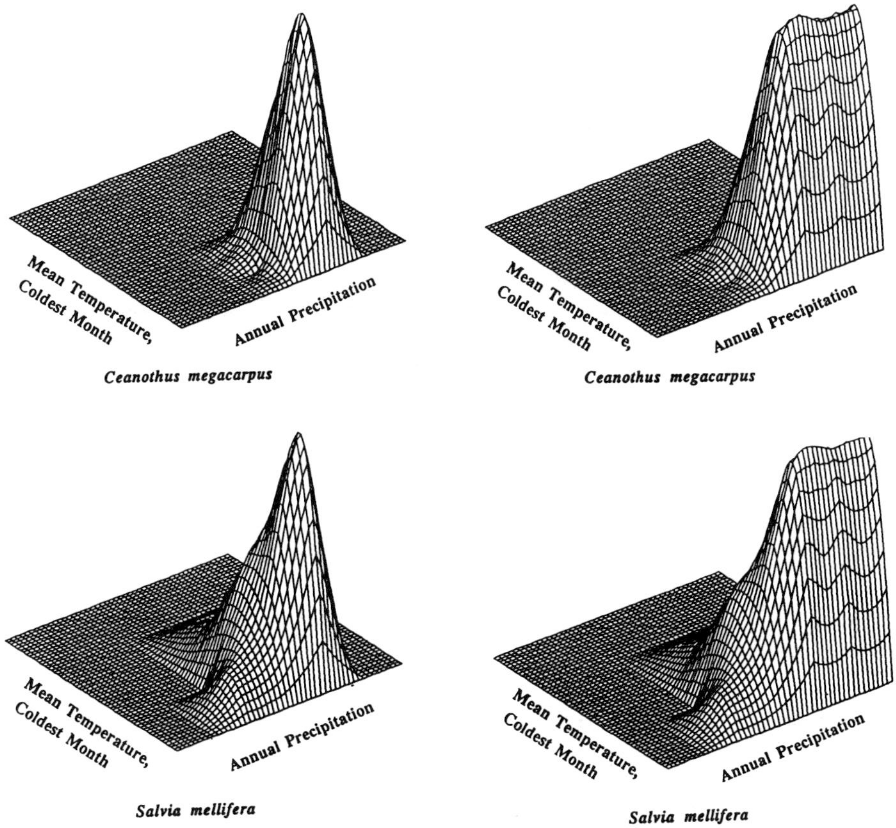

Figure 10-5. Geographic Distribution Analysis diagrams for two species showing both the calculated realized niche (a, b) and the assumed fundamental niche (c, d) responses.

(Figure 10-5c,d). Thus, if precipitation is greater than that found at the species maximum, its growth is not automatically reduced. Although it might be possible to assume that increasing winter temperatures or decreasing summer temperatures would also favor species, we felt less confident in including these concepts in the model at this time.

We included 11 species in the simulation. The parameterization for the species is approximate, because detailed data relative to the foliar cover of species in cohorts are not available.

Adenostoma fasciculatum has a broad and somewhat complex response. It has multiple peaks near the center of its distribution and a secondary area of abundance at the higher levels of precipitation. It performs best at the cold end of the winter temperature scale and in the middle of the summer temperature and precipitation scales.

Artemisia californica shows some irregularities and a broad distribution in the dimensions of summer temperatures and precipitation, and is more regular and limited by cold in winter.

Ceanothus cuneatus has a broad and irregular distribution with a secondary rise in the area of highest precipitation.

Ceanothus megacarpus has one of the most narrowly defined climatic responses. It is limited in every dimension.

Ceanothus spinosus also has narrow response functions. It is more abundant at the wetter end of the gradient than is *C. Megacarpus*.

Encelia californica has the same parameterization as *Salvia leucophylla*, and the two species mirror each other in the simulations.

Eriogonum fasciculatum has relatively broad and multimodal distributions.

Rhus integrifolia has a distinctly bimodal distribution along the precipitation axis when graphed with cold month temperature, and it becomes trimodal when graphed with warm month temperature. It occupies middle range areas of precipitation and warm month temperatures while being confined to the warmer end of the cold month temperatures.

Malosma laurina has a distinctly bimodal response along the warmest month axis, although it is bell shaped along the other two. It is limited by cold winter temperatures.

Rhus ovata is bimodal along the cold month axis, but otherwise bell shaped. It has a fairly narrow range in both temperature categories.

Salvia leucophylla has a response in the middle of the warm month temperatures and precipitation, and at the warmer end of the cold month temperatures. It is very regular and noticeably affected by the assumption of the fundamental niche shape.

Salvia mellifera has response peaks near those of *S. leucophylla*, but with greater breadth, including some very low abundance spread away from the peak area.

Life History Parameters

Seedling establishment is a different linear function of the foliar cover of the species. Resprouting success is calculated as a function of fire intensity. We used observations of resprouting rates for species of coastal sage scrub and chaparral for fires for which a measure of fire intensity was calculated (Malanson and O'Leary, 1982, 1985) using a computer simulation model of fire behavior (Albini, 1976). For mortality if growth is reduced below a threshold as climate changes, a proportion ($= 0.5^{1/6}$, for 6 iterations per year) of the cohort, which is the unit of record in our model, will die in an iteration, and that proportion of the biomass will be added to the dead fuel category (i.e., if the rate of growth is below the threshold for the whole year, 1/2 the cohort dies; because cover is reduced any time some die, the threshold is often crossed in only some iterations per year). This value is arbitrary, but in test runs with a constant climate produced communities similar to those observed now for that climate.

Fire Intensity Feedback

The development of a feedback incorporating fire intensity required the computation of a fuel load based on past growth (for a more detailed discussion, see Malanson and Westman, 1991b). The concept that the growth of plants in a simulation should determine the fuel load at the next fire seems straightforward. The fuel load, however, is more complex than a direct measure of foliar cover or of biomass would indicate. First, the distribution of wood in different branch sizes, with different surface-to-volume ratios and thus different rates of combustion, is known for present shrub species, but both the species and their growth forms may be altered under a different climate. Second, the heat content of the fuel, especially in Mediterranean-type ecosystems where the content of volatile oils is high, may change in a new climate. Third, the spatial distribution of fuels in three dimensions (e.g., grass-shrub proximity or fuel packing) could change with changing grass/shrub biomass ratio.

Both chaparral and coastal sage scrub can produce considerable amounts of dead fuel during long fire-free periods. Chaparral has been noted for this feature, which is often referred to as senescence (cf. Hilbert and Larigauderie, 1990). In coastal sage scrub, Malanson and Westman (1985) observed that individual shrubs continually produce new basal branches during the fire-free period, and Keeley and Keeley (1988) reported the same characteristic in certain chaparral species. A plant with this trait can replace its dead branches with live ones, and thus can produce new growth without expanding its area. Nonsprouting species may contribute to dead fuel loads differently (Montygierd-Loyba and Keeley, 1987). This sprouting ability is critical for the continued production of dead standing fuel. In the Mediterranean climate, standing dead fuel does not decompose rapidly, although fuel falling to the ground does not seem to accumulate, since litter loads are not heavy. This trait of coastal sage and chaparral shrubs, while indicating the importance of recording the change in fuel characteristics through time, may also indicate a pathway for assessing the potential effects of climatic change on standing dead fuel.

Although we set the upper limit of total foliar cover at 150% on a site, when cover exceeds 100% an increasing proportion of the decreasing amount of growth is considered to be replacement only. The increase in fuel will proceed as follows: following fire, live growth will begin to fill a site; as the site fills, the rate of growth will slow; once the canopy closes, the rate of growth continues to slow and much of the growth will be recorded as an accumulation of dead fuel; no upper limit for dead fuel is specified a priori. In this function, climate has no direct effect on the quantity of dead fuel, but affects it only indirectly by influencing growth. If growth is reduced below a threshold as climate changes, a proportion of the cohort, which is the unit of record in our model, will die and that proportion of the biomass will be added to the dead fuel category. Thus, if climates become more harsh, mortality will increase and add to dead fuel, while simultaneously releasing extant shrubs from competition and allowing continued growth.

Experiments

The present climate is a Mediterranean-type with cool, wet winters and warm, dry summers. Mean January temperatures are in the range of 5°–5°C; mean July temperatures are in the range of 20°–27°C; mean annual precipitation is in the range of 250–1500 mm. Our initial climate had annual precipitation and January and July temperatures of 400 mm, 5°C, 24°C. We ran simulations where we changed only temperature and where we changed both temperature and precipitation. We changed the January and July temperatures to 11°C and 30°C and we increased precipitation from 400 to 650 mm, while simultaneously changing the temperatures as noted. We ran the simulation for the initial climate for 100 yr and then changed the climate over the simulated period of 100–200 yr. Because of steps in the response functions of the species, this meant that the climate effectively switched at year 150. It is not clear to what extent climatic change will affect the fire frequency. Although warmer conditions might increase fire probabilities, wetter conditions would decrease them, and fire frequency in California also depends on human ignitions. We simulated fixed fire return intervals of 22, 50, and 100 yr. We report results here for all simulation experiments, but show graphs for only the most interesting cases.

Simulation Model Projections

Two general results are of interest. First, the simulation shows that climatic change can alter fire intensities. With the realized niche function, intensities reach a peak as climate changes, existing species dieback, and dead fuel accumulates. Then fire intensities drop to very low values in cases where not enough biomass has accumulated before the next fire, i.e., at 22-yr intervals, but at 50- and 100-yr intervals this drop is not seen because there is enough time after the first fire following dieback to rebuild a significant fuel load. Second, differences between simulations using the calculated realized niche and the assumed fundamental niche response functions are considerable (Malanson et al., 1992). Before climatic change, the cases are virtually identical, but when the climatic response function of a species is extended, species persist on the sites for a longer period, response to change is slower, and similarity is less.

Species Dominance

Temperature Change Only

Precipitation at 650 mm. Realized Niche. In this case the effect of the change in temperatures depends in part on the fire interval. In all cases total cover drops to zero at about year 175. At the 22-yr interval *Adenostema fasciculatum* is replaced by *Salvia mellifera*, and conditions change from one nearly monospecific stand to another. At the 50-yr interval, *A. fasciculatum* declines at 175 yr, then recovers some, but eventually disappears. *S. mellifera* increases during the period from 200 to 325 yr, then decreases. *Ceanothus spinosus* appears and maintains a steady but low abundance after 200 yr, whereas *C. megacarpus* increases and becomes dominant after 200 yr. At 100-yr intervals, the model's recovery of *A. fasciculatum* after 200 yr is attenuated and the abundance of *S. mellifera* reduced, while C. megacarpus becomes dominant.

Assumed Fundamamental Niche. In this case the primary difference at all three fire intervals is that *Adenostema fasciculatum* is able to survive longer. Where it once went quickly extinct with climatic change, it declines, recovers, but again dies out. *Salvia mellifera* becomes the sole dominant in the 22-yr case and greatly suppresses *Ceanothus megacarpus* in the 50- and 100-yr cases.

Precipitation at 400 mm. Realized and Assumed Fundamental Niche. In this case the drop to zero cover only briefly follows fire during the period of climatic change, and the differences between fire intervals are not pronounced, although the periods of dominance by different species are shifted slightly. *Adenostema fasciculatum* is dominant before climatic change, but disappears at year 150. *Salvia mellifera* and *Artemisia californica* then become codominant or exchange dominance in the period 150–300 yr. *S. leucophylla* and *Encelia californica*, after a slight rese just preceding year 150, disappear until about year 250, when they increase to become codominants. The effects are minimal for the assumed fundamental niche.

Precipitation at 150 mm. Realized and Assumed Fundamental Niche. In this driest case only the single species able to tolerate these conditions, *Eriogonum fasciculatum*, is able to exist, and without competition it does equally well in either temperature regime. Most notable is that although this species recovers after climatic change, there is a gap of close to 100 yr in which foliar cover is projected to be zero.

Both Temperature and Precipitation Change

Precipitation from 400 to 650 mm. Realized Niche. Major differences are seen among the three fire intervals (Figure 10-6a–c). At 22 yr, *Adenostema fasciculatum* dominates before climatic change, and is replaced by *Salvia mellifera*. At 50-yr intervals, *A. fasciculatum* disappears

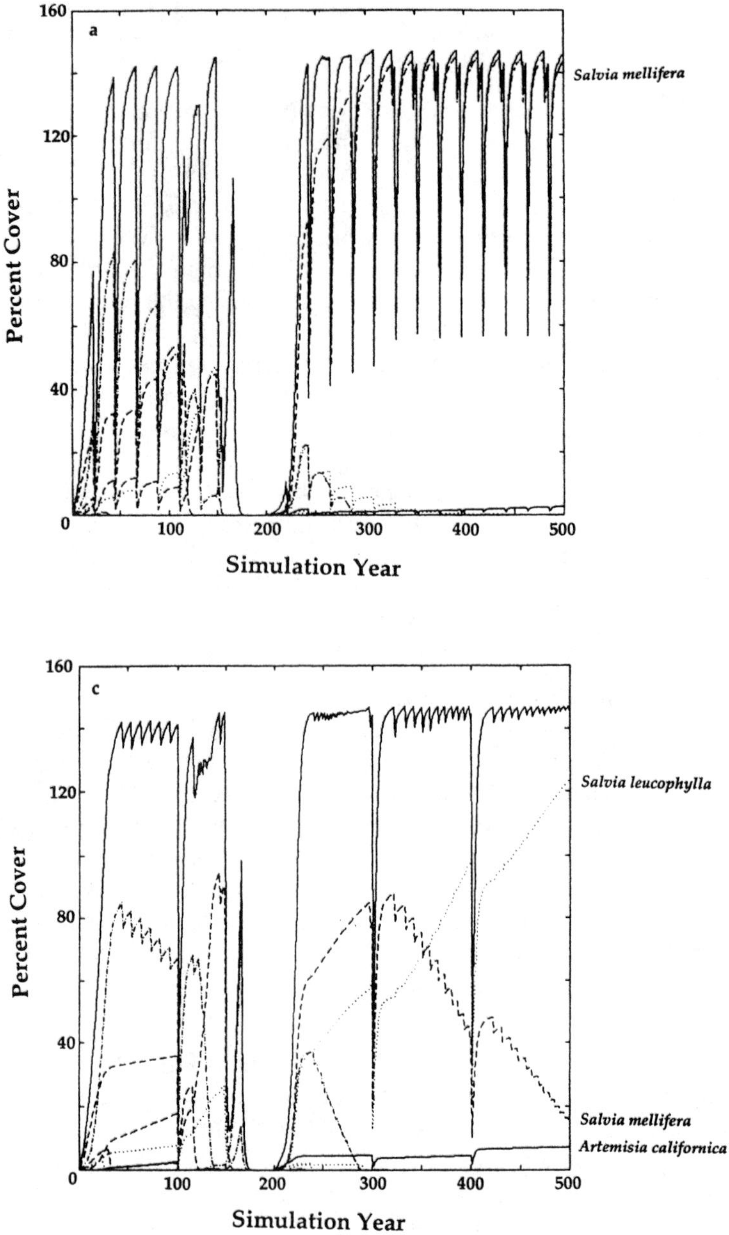

Figure 10-6. Projected responses to climate change at year 150. This case illustrates an increase in temperature and precipitation (from 400 to 650 mm) with the calculated realized niche function for three fire intervals (22, 50, 100 yr; a, b, c, respectively) and with the assumed fundamental niche function for a 50-year fire interval (d; the other intervals are generally similar). The topmost line is the sum of cover values.

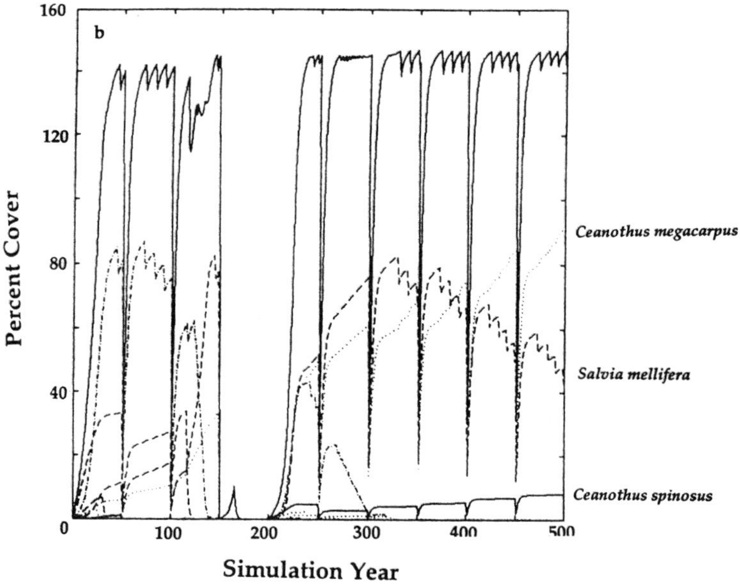

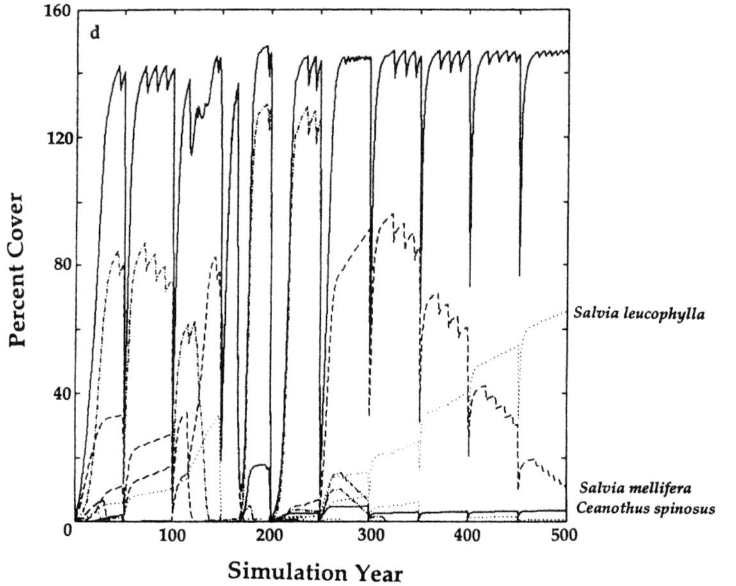

Figure 10-6. *Continued*

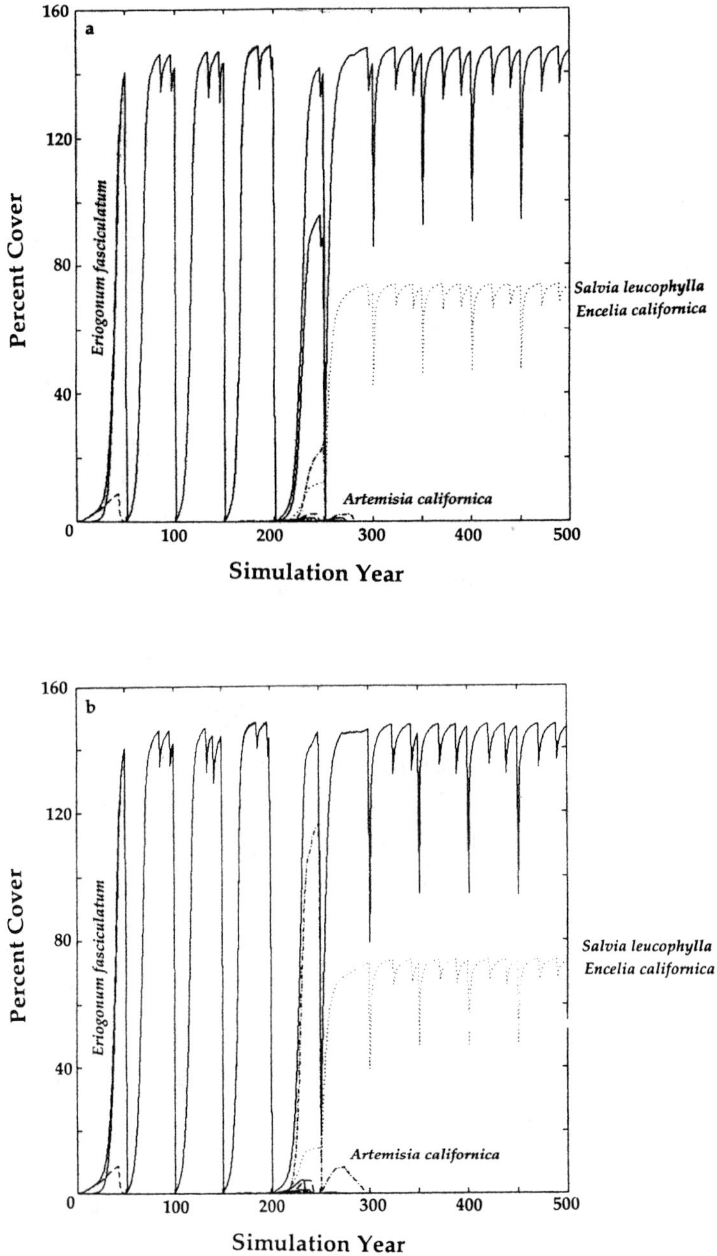

Figure 10-7. Projected responses to climate change at year 150. This case illustrates an increase in temperature and precipitation (from 150 to 400 mm) with the calculated realized niche function (a) and with the assumed fundamental niche function (b) for a 50-year fire interval (the other intervals are generally similar). The topmost line is the sum of cover values.

and is briefly replaced by *S. mellifera* until *Ceanothus megacarpus* rises to dominance, while *C. spinosus* maintains a low, steady abundance after climatic change. At 100-yr intervals *S. mellifera* and *Artemisia californica* are more successful, but *C. megacarpus* again dominates.

Assumed Fundamental Niche. *Adenostema fasciculatum* is able to survive longer in these cases, but *Salvia leucophylla* eventually becomes dominant at all three intervals (Figure 10-6d; patterns among fire intervals are similar).

Precipitation from 150 to 400 mm. Realized Niche. *Eriogonum fasciculatum* domintates in the years before climatic change, but the increase in precipitation allows new species to flourish (Figure 10-7a; patterns among fire intervals are similar). *E. fasciculatum* disappears in the 200- to 300-yr period. *Artemisia californica* then dominates for 50 to 100 years until it is replaced by *Salvia leucophylla* and *Encelia californica*. A notable contrast with the case without precipitation change is that the period with zero foliar cover is eliminated.

Assumed Fundamental Niche. The replacement process is delayed so that *Eriogonum fasciculatum* is abundant for an additional 50 to 100 yr, *Artemisia californica* reaches dominance later, and is eventually replaced by *Salvia leucophylla* and *Encelia californica* still later (Figure 10-7b; patterns among fire intervals are similar).

Discussion and Conclusions

Implications and Caveats

The projections indicate that changes in the relative abundance of species are to be expected in California shrublands. Although we have reported species-specific results, the parameterization of the model is not exact enough for us to claim that these are reliable projections for the future. The replacement of chaparral at its lower elevations by coastal sage scrub is a possibility, especially if precipitation does not increase while temperature does. It appears that even with increases in precipitation it is not likely that chaparral would replace existing coastal sage scrub. These projections depend, however, on the inadequate species response data, and should indicate the need for more research, rather than a position from which to base conservation policy.

Specifically, we do not claim that projections under the assumed fundamental niche, which show coastal sage scrub replacing chaparral in many scenarios, are reliable. These projections do have heuristic implications for the coastal sage scrub–chaparral boundary, however. Most importantly, it is clear that the direct response of species to climatic factors will need to be better understood in order to model stand level responses. One approach would be to use leaf and plant level models to build a foundation from which one can scale up to the stand and eventually the landscape. Crossing a range of model levels involves changing both temporal and spatial scales. Models running at fine spatial scales, i.e., a leaf, usually simulate processes with fast rates and so run for short periods. For single individuals, the models run at intermediate rates and longer periods, while stand models iterate slowly but run for centuries. The problem of scaling up has been addressed in the literature. Acock and Reynolds (1990) have argued that it is important to scale up from individual tree models to forests, but they propose extrapolating from individual tree models to the whole forest by aggregating information in similar cells based upon geographic information systems analysis. They wrote that higher-order models should not be considered because they represent approaches that are too empirical. Simple geographic aggregation is likely to produce error, however. Instead, the output of individual models should be used to parameterize stand models that can then incorporate stand level processes.

The literature on Mediterranean-type ecosystems is rich, and we believe that sources for improved parameterization are available. One area of needed improvement is in preburn demographic properties and the establishment phase. Resprouting and seedling establishment are very sensitive components in the model, and yet we do not have good information on the range of either in varying environments. Better parameterization of the relationship of fire behavior and intensity and resprouting is needed, in addition to more detailed information on seedling establishment. Recent work (e.g., Moreno and Oechel, 1991, and many of the papers in this volume) provides a global basis for this improvement. A second area is in the maximum growth of species and in variations away from the maximum. The literature on Mediterranean-type ecosystems is rich with physiological studies that need to be culled for the kind of information that can be used to parameterize a model of this type. This information is available, but the model needs to be organized so that available data for a global suite of species can be used to created parameter sets for many situations. Factors reducing growth, especially climatic ones, need to be better understood. We have illustrated the significance of understanding the fundamental response of species, and actual data of this type are hard to obtain, but needed. The geographic distribution analysis applied here is a black-box approach that may be useful for a first approximation, but the existing physiological studies may be usefully reinterpreted with model parameterization in mind.

In these simulations we show that differences in fire are extremely important because longer fire intervals lead to more intense fires which then limit resprouting. Although it has been argued that fire suppression efforts do not greatly alter mean fire firequency (Minnich, 1983), under suppression regimes the fires that do burn, even at the same interval, usually occur during dry, windy conditions and have a greater impact on the vegetation through intensity (cf. Malanson, 1987). Of all the changes that might occur as we face the future, fire behavior in response to both climatic change and management efforts may be the most significant. In these simulation runs, however, fire intervals are fixed. Fire ignitions and spread, and thus fire interval at places, would change with climate—a process that could be included by linking this model to a landscape-level fire behavior model.

Vulnerability of Coastal Sage Scrub

Historic losses of coastal sage scrub have been estimated from 66% to 90% (O'Leary et al., 1992). Nearly 100 of the plant and animal species associated with coastal sage scrub are classified as rare, sensitive, threatened, or endangered by state and federal agencies. To date, the greatest proportion of loss of sage scrub has occurred at lower to intermediate elevations within its distribution. Thus, much, if not most, of the remaining sage scrub occurs at or near its boundary with chaparral. The importance of the spatial relations of coastal sage scrub and chaparral has been magnified by direct human impact, and this boundary zone is likely to be increasingly important to the future of coastal sage scrub in response to the indirect impact of global climatic change.

Needed: A Unified Mediterranean-Type Ecosystem Simulator

Given the importance of understanding the potential response of coastal sage scrub and chaparral to global climatic change, and the analogous implications for other Mediterranean-type ecosystems, it seems desirable that a model be developed that corrects the shortcomings of FINICH. We propose that effort be aimed at the development of a unified Mediterranean-type ecosystem simulator. We propose that we take as an example the development of a unified boreal forest simulator, BOFORS. A recent book (Shugart et al., 1992) illustrates the development of this type of model. Based on conferences in 1985, 1987, and 1989, the contributors developed a simulation model based on the earlier JABOWA-FORET simulators that addressed specific concerns of boreal forests, e.g., soil temperature, nutrient cycling, and fire. This simulator is now being used to examine boundary questions at the boreal forest–tundra and boreal forest–northern hardwood forest boundaries and their response to global climatic change. A major attribute of this model is that it can be readily parameterized for species from the entire boreal biome, and in fact, parameter tables for the most important boreal species are included in the book.

We propose that the community of Mediterranean-type ecologists use this development as a guide. We believe that we will gain much more than a simulation model in the process. By organizing our knowledge around the common theme of a unified Mediterranean-type ecosystem simulator, we can bring together what seems to be a remarkably rich and diverse set of data and insights that is unfortunately fragmented. The FINICH model would need to be slightly deconstructed in order to serve as a starting point. What is needed is a consistent definition of maximum growth, a replicable set of multipliers reducing growth as simple functions of climate, nutrients, and other factors, and improved establishment parameters that recognize the diversity of fire and postfire conditions. Building up a hierarchical set of models, perhaps starting with MEDECS and its hierarchical submodel CAPS (Jacobson et al., 1981) or QUINTA (Tenhunen et al., 1989) and moving to the stand level with FINICH and perhaps to the landscape level with the REFIRES fire model presented by Davis (this volume), is a possibility for which we are well prepared. The physiological and life history avenues of research in our field can be usefully combined in this way, and still provide species specific responses at an appropriate and interesting temporal and spatial scale (Malanson, 1993).

Acknowledgments. The model development was funded by a grant to the late Walter E. Westman from the Environmental Protection Agency through Interagency Agreement DN-89933219-01-0 with the Department of Energy. Any views or opinions expressed here are those of the authors and do not necesarily reflect those of the sponsoring agencies. Z-J. Liu and Y-H. Yan assisted in model development. We owe a lasting debt to Walt Westman for initiating and guiding our studies in this area. We thank José Moreno and Walt Oechel for organizing the symposium and we thank the Universidad Internacional Menéndez Pelayo and the Centro de Estudios Ambientales del Mediterráneo for their sponsorship.

References

Acock B, Reynolds JF. 1990. Model structure and data base development. In: RK Dixon, RS Meldahe, GA Ruark, and WG Warren, eds. Process Modeling of Forest Growth Responses to Environmental Stress. Timber Press, Portland, OR, pp. 169–179.

Albini FA. 1976. Computer-Based Models of Wildland Fire Behavior: A User's Manual. US Forest Service, Intermountain Forest and Range Experiment Station, Ogden, UT.

Allen TFH, Hoekstra TW. 1992. Toward a Unified Ecology. Columbia University Press, New York.

Bradbury DE. 1978. The evolution and persistence of a local sage/chamise community pattern in southern California. Association of Pacific Coast Geographers Yearbook 40:39–56.

Conrad CE. 1987. Common shrubs of chaparral and associated ecosystems of southern California. US Forest Service General Technical Report PSW-99.

Cooper WL. 1922. The broad sclerophyll vegetation of California: an ecological study of the chaparral and its related communities. Carnegie Institute of Washington Publication 319.

Forman RTT, Moore PN. 1992. Theoretical foundations for understanding boundaries in landscape mosaics. In: AJ Hansen and F di Castri, eds. Landscape Boundaries. Springer-Verlag, New York, pp. 236–358.
Gosz JR. 1991. Fundamental ecological characteristics of landscape boundaries. In: MM Holland, PG Risser, and RJ Naiman, eds. Ecotones. Chapman and Hall, New York, pp. 8–30.
Gray JT. 1983. Competition for light and a dynamic boundary between chaparral and coastal sage scrub. Madrono 30:43–49.
Hanson JS, Malanson GP, Armstrong MP. 1989. Spatial constraints on the response of vegetation to climate change. In: GP Malanson, ed. Natural Areas Facing Climate Change. SPB Academic, The Hague, pp. 1–23.
Harrison AT, Small E, Mooney HA. 1971. Drought relationships and distribution of two Mediterranean-climate California plant communities. Ecology 52: 869–875.
Hilbert DW, Larigauderie A. 1990. The concept of stand senescence in chaparral and other mediterranean type ecosystems. Acta Oecologica 11:181–190.
Howell JT. 1929. The flora of the Santa Ana Canyon region. Madrono 1:243–253.
Jacobson MB, Stoner WA, Richards SP. 1981. Models of plant and soil processes. In: PC Miller, ed. Resource Use by Chaparral and Matorral. Springer-Verlag, New York, pp. 287–368.
Keeley JE, Keeley SC. 1988. Chaparral. In: MG Barbour and WD Billings, eds. North American Terrestrial Vegetation. Cambridge University Press, New York, pp. 165–208.
Kirkpatrick JB, Hutchinson CF. 1980. The environmental relationships of California coastal sage and some of its component communities and species. J Biogeog 7:23–28.
Kuchler AW. 1964. Natural Vegetation of California. Williams and Heintz Map Corp., Washington.
Malanson GP. 1984a. Fire history and patterns of Venturan subassociations of Californian coastal sage scrub. Vegetatio 57:121–128.
Malanson GP. 1984b. Linked Leslie matrices for the simulation of succession. Ecol Model 21:13–20.
Malanson GP. 1987. Diversity, stability, and resilience: effects of fire regime. In: L Trabaud, ed. Role of Fire in Ecological Systems. SPB Academic, The Hague, pp. 49–63.
Malanson GP. 1993. Comment on modeling ecological responses to climatic change. Climatic change 23:95–109.
Malanson GP, O'Leary JF. 1982. Post-fire regeneration strategies of Californian coastal sage shrubs. Oecoolgia 53:355–358.
Malanson GP, O'Leary JF. 1985. Effects of fire and habitat on regeneration in Mediterranean-type ecosystems: *Ceanothus spinosus* chaparral and coastal sage scrub. Oecol Plantar 6:183–195.
Malanson GP, Westman WE. 1985. Post-fire succession in Californian coastal sage scrub: the role of continual basal sprouting. Am Midl Nat 113:309–318.
Malanson GP, Westman WE. 1991a. Modeling interactive effects of climate change, air pollution, and fire on a California shrubland. Climatic Change 18:363–376.
Malanson GP, Westman WE. 1991b. Climate change and the modeling of fire effects in chaparral. In: SC Nodvin and TA Waldrop, eds. Fire and the Environment: Ecological and Cultural Perspectives. US Forest Service General Technical Report SE-69, pp. 91–96.
Malanson GP, Westman WE, Yan Y-L. 1992. Realized versus fundamental niche functions in a model of chaparral response to climatic change. Ecol Model 64:261–277.

Minnich RA. 1983. Fire mosaics in southern California and northern Baja California. Science 219:1287–1294.

Montygierd-Loyba TM, Keeley JE. 1987. Demographic structure of *Ceanothus megacarpus* chaparral in the long absence of fire. Ecology 68:211–213.

Mooney, HA. 1988. Southern coastal scrub. In: MG Barbour and J Majors, eds. Terrestrial Vegetation of California. John Wiley & Sons, New York, pp. 471–489.

Moreno JM, Oechel WC. 1991. Fire intensity and herbivory efffects on postfire resprouting of *Adenostema fasciculatum* in southern California chaparral. Oecologia 85:429–433.

Oechel WC, Mustafa J. 1979. Energy utilization and carbon metabolism in Mediterranean scrub vergetation of Chile and Califorria II. The relationship between photosynthesis and cover in chaparral evergreen shrubs. Oecologia 41:305–315.

O'Leary JF, Murphy D, Brussard P. 1992. An NCCP special report: the coastal sage scrub community conservation planning region. California Natural Community Conservation Planning Special Report No. 2.

Pastor J, Post WM. 1988. Response of northern forest to CO_2-induced climate change. Nature 334:55–58.

Shugart HH, Leemans R, Bonan GB (eds). 1992. The Global Boreal Forest Ecosystem. Cambridge University Press, Cambridge, UK.

Solomon AM. 1986. Transient response of forests to CO_2-induced climmate change: simulation modeling experiments in eastern North America. Oecologia 68:567–579.

Tenhunen JD, Reynolds JF, Rambal S, Dougherty R, Kummerow J. 1989. QUINTA: a physiologically-based growth simulator for drought adapted woody plant species. In: JS Pereira and J Landsberg, eds. Biomass Production for Fast-Growing Trees. Kluwer Academic, Dordrecht, pp. 135–168.

Turner MG, Gardner RH, O'Neill RV. 1991. Potential responses of landscape boundaries to global environmental change. In: MM Holland, PG Risser, and RJ Naiman, eds. Ecotones. Chapman and Hall, New York, pp. 52–75.

US Forest Service. 1945. Vegetation Types of California. California Forest and Range Experiment Station, Berkeley.

Westman WE. 1981a. Seasonal dimorphism of foliage in California coastal sage scrub Oecologia 51:385–388.

Westman WE. 1981b. Factors influencing the distribution of species of Californian coastal sage scrub. Ecology 62:439–455.

Westman WE. 1983. Xeric mediterranean-type shrubland associations of Alta and Baja California and the community/continuum debate. Vegetatio 52:3–19.

Westman WE. 1991. Measuring realized niche spaces: climatic response of chaparral and coastal sage scrub. Ecology 72:1678–1684.

Westman WE, Malanson GP. 1992. Effects of climate change on Mediterranean-type ecosystems in California and Baja California. In: RL Peters and T Lovejoy, eds. Global Warming and Biological Diversity. Yale University Press, New Haven, pp. 258–276.

Westman WE, O'Leary JF, Malanson GP. 1981. The effects of fire intensity, aspect, and substrate on postfire growth of Californian coastal sage scrub. In: NS Margaris and HA Mooney, eds. Components of Productivity of Mediterranean Regions. W. Junk, The Hague, pp. 151–179.

Wilson JP, Agnew ADQ. 1992. Positive-feedback switches in plant communities. Adv Ecol Res 23:263–336.

Zedler PH, Gautler CR, McMasher GS. 1983. Vegetation response to extreme events: the effect of a short time interval between fires in California chaparral and coastal scrub. Ecology 84:809–818.

11. Vegetation Response to Climatic Change in a Dune Ecosystem in Southern Spain

O. Merino, R. Villar, A. Martín, Diego García, and J. Merino

There is some controversy about the importance of the increase of atmospheric CO_2 as a determinant of climatic change (Croops and Schuurmans, 1986). Models available so far are very imprecise and unable to predict the direction and intensity of the change for a particular geographic location (Hulme et al., 1990).

Apparently for the Doñana area, the direct consequences of CO_2 increase should be an increase in the average temperature in summer of approximately 4° to 6°C and a decrease in annual rainfall of approximately 60 mm (Houghton et al., 1990). Considering the mean annual rainfall of the area (around 620 mm), this decrease could result in a significant decrease in water availability. These changes could be very fast, taking only about a hundred years at the present rate of CO_2 release into the atmosphere.

The effects of that decrease in water availability could cause the local extinction of the most sensitive species, resulting in a net loss in species diversity.

However, another possibility exists, since in response to the change in the climatic and soil water availability determined by the climatic change, the species could migrate northwards until they reached areas having temperature and evapotranspiration in equilibrium with their physiological characteristics. Simultaneously these species could be replaced in their places of origin by other immigrants coming from the south (Graham and Grimm, 1990). Such replacements have been reported to happen during the Holocene in several European localities from studies of pollen records (Birks, 1990). If this is the case, local communities should undergo changes in species composition, but species richness could be maintained by immigration.

In the present work we will discuss the sensitivity to decreases in water availability and the dispersion capacity of woody species representative of dry and mesic habitats in the configuration of the landscape of the Doñana Biological Reserve.

Area of Study

The area of study is located in the southwest of the Iberian Peninsula. The climate of the region is Mediterranean, modified in part by oceanic influence, with mild winters (mean daily temperature for January and December, the coldest months of the year, is 9.3°C), and hot summers (mean daily temperature for July, the warmest month, is 23.9°C). Average annual rainfall is around 620 mm, with a maximum in both early winter and spring, and with severe summer drought (Figure 11-1).

The area includes a large coastal plain covered by eolic sands. The topography consists of ridges separated by depressions. Because of the existence of a shallow and almost horizontal water table and high substrate permeability, the tops of the ridges are arid, while the depressions are mesic.

Scrub is the dominant vegetation type. Species composition appears to be controlled by topography through the depth of the water table (Allier

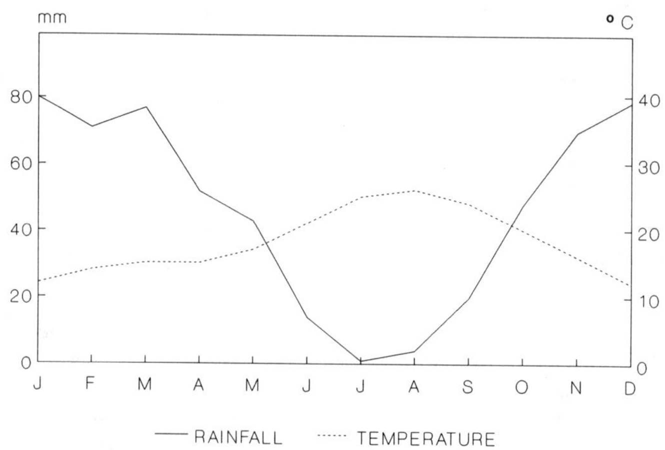

Figure 11-1. Ombrotermic diagram of the Doñana National Park area.

Figure 11-2. Idealized diagram of the stabilized sand dune system. At present the tree cover (*Juniperus oophora* on the ridges and *Quercus suber* on the depressions) is very narrow because of fire and human interference. (From Allier et al., 1974.)

et al., 1974). The composition changes from an open semidesertic scrub on the ridges, with mainly Cistaceae and Labiatae as dominant families, and scattered individuals such as trees of *Juniperus oophora*, to a more mesic scrub in the depressions, which includes *Erica scoparia* and *Quercus suber* (Figure 11-2).

Plant–Soil Water Relationships

For the analysis, two typical species from the very dry extreme of the gradient (top of the ridges), namely *Cistus libanotis* (a shrub) and *J. oophora* (a tree), and two species from the mesic extreme (depressions), namely *Erica scoparia* (a shrub) and *Quercus suber* (a tree), were selected.

Seasonal changes in water potential at dawn were estimated in a group of six individuals of each species with the aid of a Scholander-type pressure bomb (Scholander et al., 1965). The seasonal point of zero turgor pressure was determined in samples of the same individuals from pressure-volume curves (Tyree and Hammel, 1972).

Soil samples from both the dry and the mesic habitats were taken to the laboratory, and the relationship between soil water content and soil water potential at 0.03 and 1.5 MPa was established with the aid of a pressure plate (Hillel, 1980). From that relationship and from the seasonal evolution of the soil water content, the seasonal evolution of soil water potential was estimated.

Figure 11-3 shows the annual change in soil water potential in the dry and mesic extremes of the gradient. Figures 11-4 and 11-5 show the annual change in plant water potential at dawn and the evolution of the point of zero turgor in the four species considered.

Figure 11-4 shows that the two species representative of dry habitats, *Cistus libanotis* and *Juniperus oophora*, undergo high water stress during the summer months. The figure also shows that both species lose turgor (the line of zero turgor crosses the plant water potential line) during the

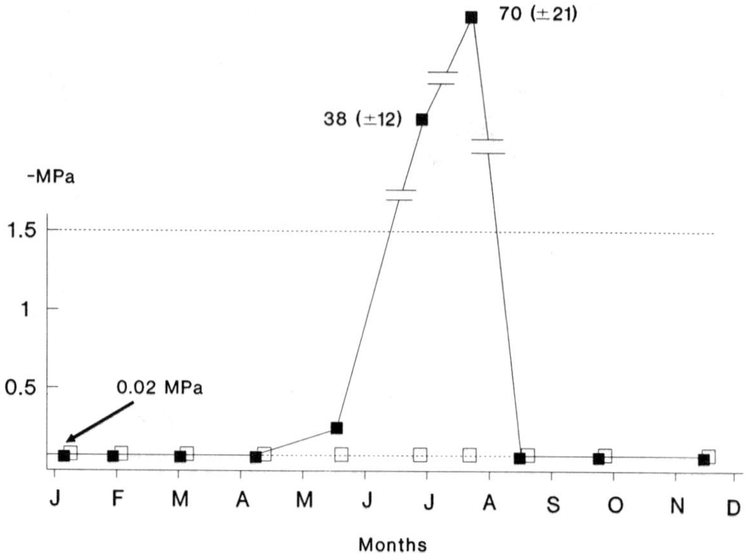

Figure 11-3. Annual change in soil water potential at 30 cm depth in the arid (■) and mesic (□) habitats considered in this study.

whole day for a significant part of the year (2 or 3 months). Considering the low water availability during the summer period (Figure 11-3), these results suggest that these species do not absorb any significant water during the summer months.

On the contrary, Figure 11-5 shows that the two typical species of the mesic habitat, *Erica scoparia* and *Quercus suber*, are turgid throughout the year (the line of zero turgor never crosses the plant water potential curve) so, at least apparently, they do not undergo any significant water stress.

In summary, and contrary to the expected, the species from the dry habitats have a high osmotic potential (positive) and thus a low capacity for water extraction. On the contrary, species from mesic habitats show a comparatively lower osmotic potential (more negative) and thus a higher capacity for water extraction.

At the top of the ridges, water availability during the summer months is extremely low (Figure 11-3), so it is probably not advantageous to decrease the osmotic potential for absorbing soil water. In this case plant strategy probably consists of developing an osmotic potential low enough to extract water when it is available in the soil (autumn, winter, and spring) combined with the ability to withstand a very high water stress during the summer months. The characteristics conferring this ability have not yet been studied.

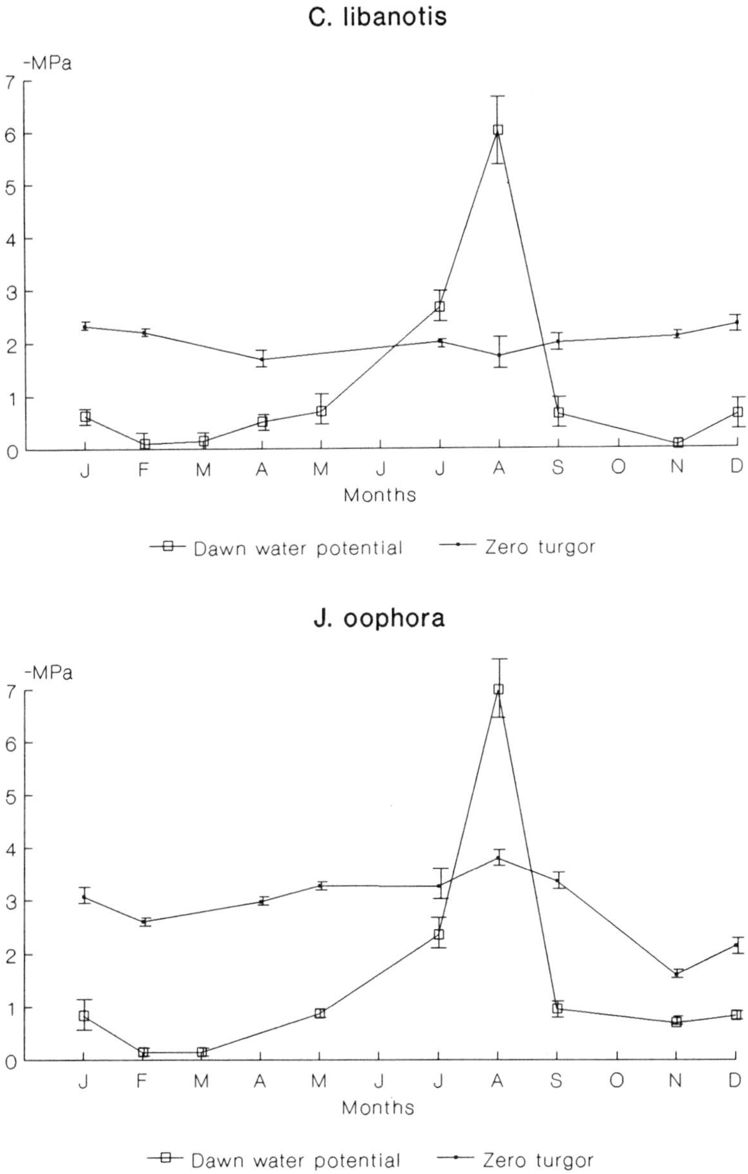

Figure 11-4. Annual change in xylem water potential and zero turgor point in two typical species of dry habitats.

In the mesic habitats, because of the proximity of the water table, water is available in the soil throughout the year. Thus a moderately low osmotic potential probably allows the plant to extract water throughout the whole year, including the summer months.

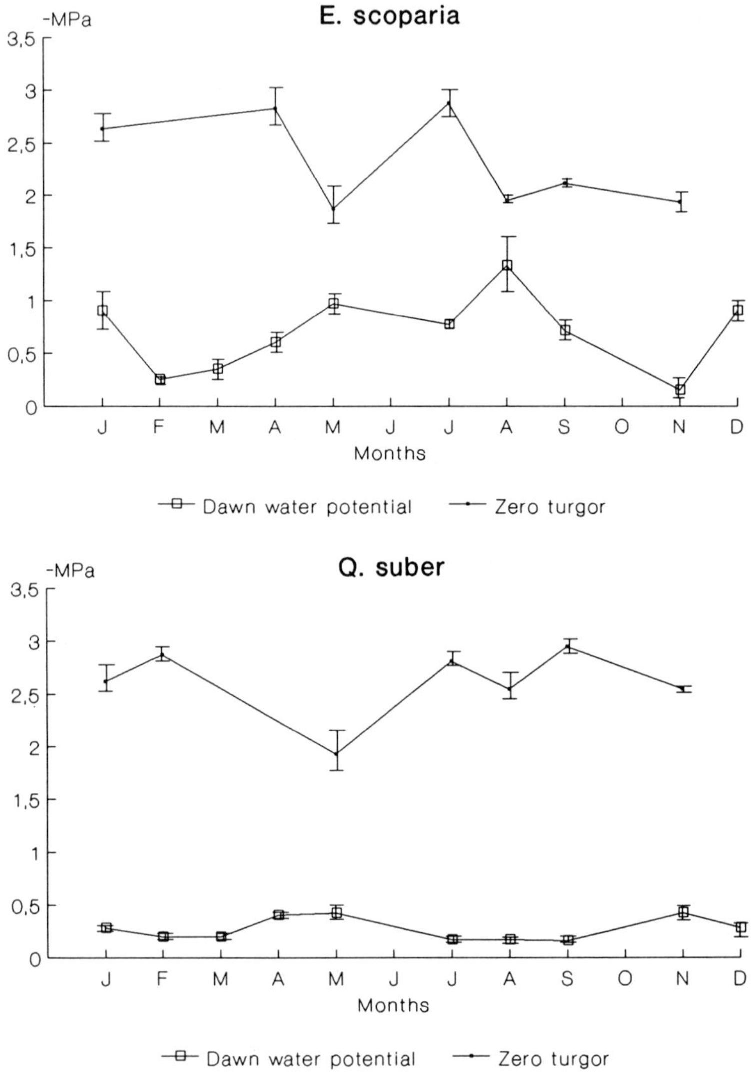

Figure 11-5. Annual change in xylem water potential and zero turgor point in two typical species of mesic habitats.

In summary, the species of very dry habitats are designed to withstand absence of water and should thus be considered endurers, while those from mesic habitats appear to be designed to keep an adequate water status (apparently by keeping an adequate osmotic potential) throughout the year, and so should be considered avoiders.

According to these considerations, the sensitivity of typical species from dry and wet habitats to unusually large changes in soil water accessibility should be different.

Litter Production

The annual litter production of five representative species of the community located at the dry extreme of the gradient, and one species typical of the mesic extreme (*Erica scoparia*), was estimated by setting containers for litter collection under individuals of the selected species. For the analysis, at least six individuals of each species were considered. Sampling was carried out for 3 consecutive years.

Figure 11-6 shows the changes observed in the annual litter production of the species from dry habitats and the species from mesic habitats in the 3-year period. During that period annual rainfall changed from 479 mm (a regular dry year) to 379 mm and to 266 mm (the driest year of the century).

Excluding *H. halimifolium*, a generalist able to colonize successfully dry and mesic habitats, Figure 11-6 shows that the litter production of the typical species from dry habitats is not very sensitive to strong changes in water accessibility. On the contrary, the litter production of *E. scoparia*, a representative species of the mesic habitats, decreased by around 50%.

If it is accepted that the annual litter production of a species is an indicator of its growth rate, the results suggest that the changes in water accessibility should affect the species of mesic habitats more intensely, while the typical species of dry sites should not undergo great variations

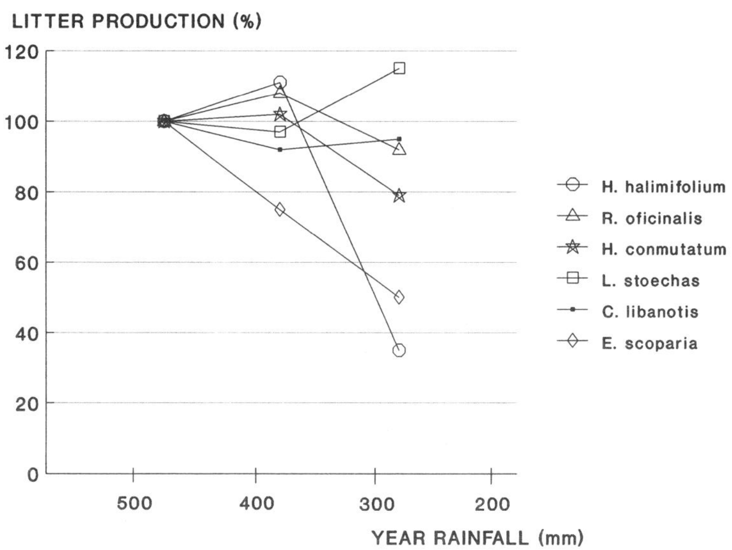

Figure 11-6. Annual litter production as a function of annual rainfall in species of mesic (Erica scoparia) and arid (Halimium halimifolium, H. commutatum, Rosmarinus officinalis, Lavandula stoechas, and Cistus libanotis) habitats. Points represent averages for at least six individuals.

in their growth rate and other related parameters (productivity, flower and fruit production, etc.).

These results are in good agreement with the results of other studies in the area. In effect, the first results of a study in progress on the relationship between the actual water stress on the individuals and their photosynthetic rate suggests that the species of dry habitats can maintain a much stronger positive carbon balance in situations of plant water stress than those of mesic habitats, which agrees with the results of litter production in this study.

Besides, in controlled growth conditions (pots), the lack of watering during relatively short periods induces a higher mortality in the species from mesic habitats. Circumstantial observations also showed that under natural conditions, populations of *Erica ciliaris* (a typical species of the humid habitats in the study area) displayed a high mortality following a period of intense drought in 1982, while no appreciable variations in mortality were detected in the typical species of dry habitats (Merino and Merino, 1988).

In summary, these results, taken together, suggest that the species from the dry habitats are quite insensitive to changes in water accessibility. On the contrary, low water accessibility during very dry years apparently overwhelms the control capacity of the typical species of mesic habitats, as reflected by both the strong decrease in growth rate and the strong increase in mortality they undergo.

Climatic Change and Species Behavior in the Doñana Area

The expected increase in average temperature and decrease in annual rainfall would necessarily result in an increase in water table depth (i.e., a decrease in water availability), which could cause the exclusion of the species typical of the mesic habitats (the most sensitive species in regard to the results discussed above). These habitats could then be invaded by the typical species from the drier habitats (the endurers), resulting in a net loss in species diversity in the Doñana area.

Such a change in the landscape as a consequence of a decrease in water availability has been reported for the area by Merino et al. (1990) by using indirect evidence from local historical records.

In fact, the historical records available in the archives of the Casa Ducal de Medina Sidonia, the owners of the Doñana area land since the twelfth century, indicate that by the start of the seventeenth century, the landscape of the area appeared similar to the present landscape (Figure 11-2), but with a wider cover of moist habitats, and a higher species diversity.

Because of low fertility of the soil, hunting was the main use of the area for many centuries. However, starting in 1628, the mode of land use changed drastically. Farming replaced gathering and several domesticated species were introduced in the area. In addition, prescribed fires were allowed in order to improve the productivity and palatability of the vegetation; settlements for shepherds appeared, and charcoal production and wood gathering increased considerably (Granados, 1987). These were all activities which, by their nature, removed the vegetation cover, suppressing its protective action on the soil.

Apparently, the low cohesion of the sandy soils under the direct effect of the usually strong sea winds started a dunification process which sprayed a sand layer over several kilometers far from the dune system, the thickness of the sand layer varying from several centimeters to some meters depending on the distance to the dune system and the topography (Gavala, 1952). The result of the sand layer deposition for the area was a general increase in the water table depth and, consequently, a decrease in water availability.

The historical records also show that by the end of the seventeenth century most of the lagoons of the region disappeared, along with the major part of the riparian forest (poplars, willows, tamarsk). Some conspicuous animal species became extinct (e.g., the swan) and some plant species underwent a strong decrease in numbers (e.g., the cork oak). In addition, all the attempts at reforestation with previously common species of economic interest (olive trees, cork oaks) failed. At the time, this failure was attributed to changes in the soil, which had become unstable and dry because of the deposition of the sand layer (Granados, 1987).

Historical records also reveal that the relative abundance of the species changed depending on their ecological characteristics. Thus, while species associated with the dry habitats increased in extent, those associated with the mesic habitats underwent a significant decrease in their abundance in the Doñana area.

The inability of the species typical of mesic areas to withstand strong water stress (see Plant–Soil Water Relationships) probably determined the exclusion of such species from the critical areas as water availability decreased. These areas were suitable then for invasion by the species typical of the dry habitats; both processes resulted in a net loss of species diversity in the Doñana area.

Species and Community Displacement

Global change models predict that the increase in temperature, combined with the decrease in water availability, will drive the species north of their present areas (Huntley and Webb, 1989; Graham and Grimm, 1990; Ozenda and Borel, 1990). So, in a global scenario, species excluded from the Doñana area because of the change in environmental conditions would be replaced by species coming from the surroundings, the efficiency of the process depending on the dispersion capacity of the species involved. If that is the case, the impact of climate change on the Doñana species diversity would be diminished.

Mediterranean woody species can be classified in two main groups according to their putative intrinsic dispersal capacity (Herrera, 1984). The first group (Type I, or "*r*" type plants) is made up mainly of common species of the first stages of succession (typically those of the Cistaceae and Labiatae families). They produce very small propagules, which are not easily blown around by the wind or transported by efficient dispersion vectors. Because of this, they have a very low intrinsic capacity for migrating and colonizing new areas.

The second dispersion group (Type II, or "*K*" type plants) is made up of species that produce berry-type or similar fruits, which have a high reward for the dispersers (for example, olive and cork oak trees), and therefore have a high intrinsic capacity for reaching new areas.

However, there are some proofs that the real dispersion capacity of both Type I and Type II plants is very low in the Doñana area. Thus, analysis of two series of aerial pictures of the Doñana area for the years 1956 and 1984 shows that many scrubland–grassland limits and many clearings in the dry shrubland have remained open for these 28 years. Animal activity and/or competition with grasses may play an important role in keeping these clearings (Davis and Mooney, 1985; Brown and Heske, 1990). In any case, the constancy in the shape of the clearing limits is striking, suggesting a low dispersal capacity of these species.

Also, the experience of farmers shows that after 10 years of continuous pulling up of the Cistaceae and Labiatae shrubs (i.e., after exhausting the seed banks), the shrubland does not regenerate for many years, unless there are nearby shrublands (Fernández Alés, personal communication).

Juniperus oophora (a Type II species) is the only treelike species growing in the dry habitats of the stabilized sand dunes of the Doñana area. It produces berry-like fruits which are basically dispersed by foxes (*Vulpes vulpes*) and a bird species (*Sturnus unicolor*) (Rau Acuña, 1987). This author reports that in the Doñana area the fox population processes more than 3×10^6 seeds per year, of which 10^5 germinate, which makes foxes the main dispersers of the *J. oophora* species in the area.

Until 40 years ago, *J. oophora* was harvested for building and fencing, and because of these activities, the forested area was strongly diminished, so that today, mature forests can be found forming only scattered patches. A careful exploration of the aerial pictures and field observations shows many stands of young (shrublike individuals) populations of *J. oophora* in the proximity (around 100 m) of the mature patches, indicating that, in the pacth borders, shrubs are replaced by *Juniperus* trees, which in turn suggests that the dispersion of this species has been quite efficient during recent years.

On the contrary, the substitution of shrubs by *J. oophora* trees in areas identical to those under colonization by this species, as regards their soil chemical and physical characteristics and water accessibility (Merino et al., 1976, 1990), but 3 or 4 km away from the forest patches, has been arrested. These areas bear no individuals of *J. oophora* even 100 years after the last harvest (Figure 11-7). It should be noted that in these areas, soil does not appear to pose particular troubles for germination, since, when transplanted to pots containing that type of soil, *J. oophora* seeds germinate and seedlings appear to progress quite well (unpublished data).

This incapacity of *J. oophora* to colonize new nondistant areas seems to be related to the behavior of its dispersers (*Vulpes* and *Sturnus*). Thus, during the fruit ripening period (summer–autumn), the activity of foxes is strongly associated with the forest patches of that species (Rau Acuña, 1987). A similar association effect seems to take place with *S. unicolor*, which uses the individuals of *J. oophora* as a perch as well as a food source, so that its whole activity remains restricted to the actual patches of *J. oophora* trees. The result is that, outside the immediate area around the forest patches, the value of these two species as dispersal agents is practically nil. Consequently, the population advance of *J. oophora* is very slow, as it only takes place in the limits of its patches.

These findings suggest that the real dispersion capacity of species, but particularly Type II species, may be very low and even considerably lower than that normally accepted (Dobson et al., 1989).

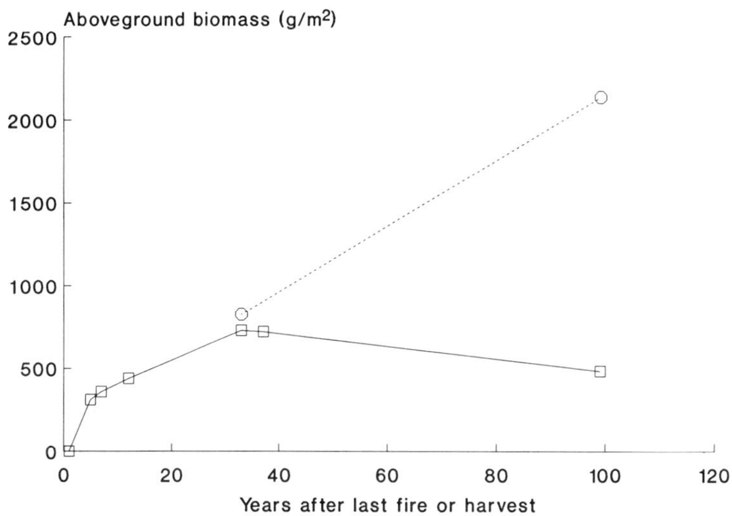

Figure 11-7. Change of biomass after fire in areas adjacent to the *J. oophora* forest patches (○) and in areas far (3 km) from the forest patches (□).

In a slowly changing scenario (for example, a smooth increase in temperature such as that reported for the first half of the Holocene), a low dispersal capacity would probably be harmless for the species. However, if it is assumed that an increase of 1°C in the average temperature is equivalent to an increase in latitude of some 100 km, and that the temperature change expected for the next 100 years is something around 4°–6°C, the results discussed in the present work suggest that the probability of successful latitude displacements for these species is very much lower than expected, which increases the probability that these species will become extinct because of the environmental change.

These observations contrast with the results obtained from pollen sediment analysis, according to which the species of the genus *Ulmus*, *Tilia*, *Corylus*, etc. showed strong northward displacements during the first half of the Holocene because of the climatic change. Some evergreen species of the genus *Quercus* displaced their area of distribution from southern Spain to Sweden (Birks, 1990), all which suggests a high capacity for migration.

However, the results discussed suggest that the considerations in respect of the real migration capacity of the species cannot be very profitable if they are only made at species level (the described strong association of *J. oophora* with *V. vulpes* and *S. unicolor* is a good proof of it), as the possibility of displacement and the success or failure of this displacement depend in many cases on how these changes affect the migrating capacity of the rest of the species and on the degree of integrity of the ecosystem.

Related to the last, it should be noted that during the last centuries, man has made substantial modifications to his environment, which have been translated into a considerable simplification and fragmentation of the landscape (Fernandez Alés et al., 1992). Present-day conditions therefore differ greatly from those which should have existed in the first half the Holocene, where the continuity of the landscape (communities and biotopes) probably allowed a much easier migration of species.

Because of all the above, the ability of the different woody species – and particularly those of *K* type—to migrate northwards, following both a rapid increase in the average temperature and a rapid decrease in water accessibility, seems to be very low because of their low real dispersion capacity. Therefore, the chance that these species will colonize distant areas seems rather improbable.

It follows that one of the general consequences of the predicted climatic change would be the extinction of many species (a global decrease in species diversity), and particularly, of those with more requirements for dispersion. For the Doñana area, the climatic change would result in a quite complex process, including both the massive extinction of the more labile species (those typical of the mesic habitats) and the partial substitution of local communities by others coming from the surroundings.

These would be integrated by species more resistant to water stress but impoverished in the *K*-type species.

Acknowledgments. We thank the Excelentísima Duquesa de Medina Sidonia and the Reserva Biológica de Doñana staff office for their support. This work was partially supported by projects PR84/1121 and PB90/0894.

References

Allier C, González Bernáldez F, Ramírez Díaz L. 1974. Reserva Biológica de Doñana. Mapa ecológico. Estación Biológica de Doñana. CSIC. Sevilla.

Birks HJ. 1990. Changes in vegetation and climate during the Holocene of Europe. In: Boer MM, de Groot RS (eds) Landscape Ecological Impact of Climatic Change. IOS Press, Amsterdam, pp. 133–158.

Brown JH, Heske EJ. 1990. Control of desert-grassland transition by a keystone rodent guild. Science 250:1705–1707.

Croops AJ, Schuurmans CJE. 1986. Detection of the CO_2 effect upon the climate of western Europe. Theor Appl Climatol 37:111–125.

Davis SD, Mooney HA. 1985. Comparative water relations of adjacent California shrub and grasslands communities. Oecologia 66:522–529.

Dobson A, Jolly A, Rubenstein D. 1989. The greenhouse effect and biological diversity. Tree 4:64–68.

Fernández Alés R, Martín A, Ortega F, Alés E. 1992. Recent changes in landscape structure and function in a Mediterranean region of SW Spain (1950–1984). Landscape Ecol 7:3–18.

Gavala J. 1952. Mapa Geológico de España, Hoja 1033, Palacio de Doñana, IGME, Madrid, Spain.

Graham WR, Grimm EC. 1990. Effects of global climate change on the patterns of terrestrial biological communities. Tree 5:289–292.

Granados M. 1987. Transformaciones históricas de los ecosistemas del Parque Nacional de Doñana. PhD. Thesis, University of Sevilla, Spain.

Herrera C. 1984. Tipos morfológicos y funcionales en plantas del matorral mediterráneo del Sur de España. Studia Oecol 5:7–34.

Hillel D. 1980. Fundamentals of Soil Physics. Academic Press, New York, 413 pp.

Houghton JT, Jenkins GJ, Ephrums JJ (eds). 1990. Climate Change. The IPCC Scientific Assessment. Cambridge University.

Hulme M, Wigley TML, Jones PD. 1990. Limitations of regional climate scenarios for impact analysis. In: Boer MM, de Groot RS (eds) Landscape Ecological Impact of Climatic Change. IOS Press, Amsterdam, pp. 111–129.

Huntley B, Webb T. 1989. Migration: species' responses to climatic variations by changes in the earth's orbit. J Biogeog 16:5–19.

Merino J, García Novo F, Sánchez Díaz M. 1976. Annual fluctuation of water potential in the xerophytic shrub of the Doñana Biological Reserve. Oecol Plant 11:1–11.

Merino J, Martín A, Granados M, Merino O. 1990. Desertification of coastal sands of south-west Spain. Agric, Ecosyst Environm 33:171–180.

Merino O, Merino J. 1988. Impacto potencial de la explotación del acuífero Almonte-Marismas en los ecosistemas del área de Doñana. Proceedings of the International Symposium on Hydrology of Wetlands in Semiarid and Arid Regions. Sevilla, Spain, pp. 123–126.

Ozenda P, Borel JL. 1990. The possible responses of vegetation to a global climate change. In: Boer MM, de Groot RS (eds) Landscape Ecological Impact of Climatic Change. IOS Press, Amsterdam, pp. 221–249.

Rau Acuña JR. 1987. Ecologia del Zorro, *Vulpes vulpes* (L.), en la Reserva Biológica de Doñana, S.O. de España. PhD Thesis, University of Seville, Spain.

Scholander PF, Hammel HT, Bradstreet ED, Hemmingsen E. 1965. Sap pressure in vascular plants. Science 148:339–346.

Tyree MT, Hammel HT. 1972. Measurement of turgor pressure and water relations of plant by the pressure bomb technique. J Exp Bot 23:267–282.

12. Monitoring Ecosystem Response to Global Change: High Spatial Resolution Digital Imagery

Allen S. Hope

Anthropogenically induced changes in the chemistry of the Earth's atmosphere and the hypothesized concomitant climatic shifts are distinguished from previous changes by the scale and pace with which they are expected to occur (Simon Silver and De Vries, 1990). A changing global environment is important for Mediterranean-type ecosystems because they generally occur in densely populated regions and often grade into arid zones which may be prone to expansion under hypothesized climatic shifts (Hope and Stow, 1993). Although the most direct effect of environmental change on ecosystems tends to be a physiological response of the organisms, the altered competitive interactions may be an important indirect effect leading to readjustments in the distributional limits of vegetation species (Ford, 1982). Other significant consequences of global change in Mediterranean-type ecosystems may include decreased species diversity and increased fire frequency or pest infestations.

Research projects that investigate the response of Mediterranean-type ecosystems to a changing global environment can benefit considerably from the use of remotely sensed data. These remote observations provide a consistent and cost-effective framework for monitoring ecosystems undergoing change and may produce information that is not revealed from ground studies. Furthermore, data collected from remote platforms may provide inputs for ecosystem models and can be a source of information to support field studies, including manipulative experiments.

The principal dimensions of remotely sensed data that can vary are their spatial, spectral, and temporal resolutions. The selection of appropriate resolutions is influenced by a large number of factors, including the study objectives, data availability, and costs. Remotely sensed data that are required for global circulation models or for global and regional monitoring need to be collected frequently and are required to cover large areas of the Earth's surface. With current satellite sensors and data storage and handling capabilities, observations made for these purposes generally have a ground resolution of at least a few square kilometers. Most high spatial resolution digital imagery collected using aircraft platforms tends to be collected for specific experimental objectives or for a detailed assessment of land cover type and condition. These data are, however, often overlooked as a source of information for the routine monitoring of ecosystems subject to a changing global environment. Much of the current global change research has a regional to global perspective, so remote sensing studies have generally focused on the acquisition and interpretation of low spatial/high temporal resolution satellite data.

This chapter explores the potential utility of high spatial resolution digital imagery for monitoring Mediterranean-type ecosystems that are subject to a changing global environment. High spatial resolution multispectral digital imagery is defined as imagery with picture elements not exceeding a ground resolution of 10 m. Given the spatial resolution of current satellite sensors, these data are by definition collected from aircraft platforms.

The specific objectives of ecosystem monitoring dictate the characteristics of a remote sensing system that will provide relevant and cost-effective information. The discussion begins with an examination of the role of ecosystem monitoring in global change and earth systems science studies. In reviewing the concept of spatial resolution in remote sensing studies, particular attention is given to the relationship between image information content and the spatial resolution. The unique remote sensing considerations associated with sensors and platforms used to collect high spatial resolution data are discussed briefly. Finally, the potential applications of high spatial resolution data in the monitoring of Mediterranean-type ecosystems are outlined.

Monitoring Ecosystem Response to Global Change

The term global change encompasses both climatic change and changes in the chemistry of the lower atmospheric boundary layer, particularly changes in CO_2 concentrations. The monitoring of ecosystem responses to global change has been identified as a crucial component of global change research programs. The scientific objectives of the US Global

Change Research Program are to "*monitor*, *understand*, and ultimately *predict* global change" (National Research Council, 1990, p. 3). The US Committee on Global Change (National Research Council, 1988, 1990) recognized the importance of long-term experiments for 1) monitoring the driving forces of climatic change, 2) monitoring the state variables or "vital signs" of the Earth where such changes are likely to manifest themselves, and 3) increasing the opportunity to observe and understand how unusual climatic events stress ecosystems. The two reports published by this Committee in 1988 and 1989 proposed that 1) systems should be examined along gradients where changes could be expected to occur; 2) sites need to be identified that can act primarily as "barometers" that will reflect the impact of climatic shifts on different ecosystems and that can be used for model and hypothesis testing; 3) sensitive ecosystems should be monitored for critical parameters such as species loss; and 4) there is a need to identify "indicator" species of ecosystem condition that can be observed easily (e.g., using remote sensing). This final proposal is consistent with the view that biological communities may be characterized by certain *keystone* species that are crucial to the functioning of the ecosystem (e.g., Paine, 1980; Westman, 1984).

Monitoring vegetation along ecosystem gradients or at specific barometer sites may include observations made from both satellite and aircraft platforms. However, high spatial resolution data collected from aircraft platforms are more compatible with the implied scales of observation for monitoring sensitive or indicator species than lower resolution data collected from conventional satellite platforms. The appropriate spatial resolution of remotely sensed data used in studies of ecosystem change is closely tied to the specific research questions under investigation, and these resolution requirements have to be balanced with the technical and cost constraints associated with the collection of high resolution data.

Spatial Resolution and Remotely Sensed Data

Spatial Resolution Defined

The amount of information contained in a remotely sensed image depends in large part on the image resolution (Simonett et al., 1983). As mentioned earlier, resolution can vary in three domains: spatial, temporal, and spectral/radiometric. The spatial resolution characteristics of a system or an image can be described in a variety of ways, including the geometric properties of the imaging system (Simonett et al., 1983). The geometric properties are referred to as the instantaneous field of view (IFOV) of the sensor, which is determined by the properties of the sensor and the altitude of the observing platform. The IFOV may also be termed "ground resolution," the size of the area on the ground from which the radiance measurements are derived (Townshend, 1980). Woodcock and Strahler (1987) regard this spatial resolution as analogous to the scale of the observations, while in the context of landscape models, King (1990) refers to it as the "grain" of the observation and the scene as the "extent" of the observation.

The selection of an appropriate grain and extent of an image for monitoring studies is complex, being a function of the type of environment under investigation and the information desired (Woodcock and Strahler, 1987). A framework for evaluating the potential information content of remotely sensed data at different spatial scales and the analytical approaches that can be employed at these scales is presented in the following sections.

A Remote Sensing Model

Image information content is resolution dependent, and tradeoffs exist between levels of spatial resolution. There tends to be a relationship between image grain and extent, finer-grained images tending to cover smaller areas than coarse-grained (low-resolution) images. Low-resolution picture elements (pixels) average the landscape detail but allow regional patterns to be readily observed. A remote sensing model described by Strahler et al. (1986) and summarized in this section provides a useful framework for evaluating levels of spatial resolution that are used in ecosystem monitoring studies.

The remote sensing model proposed by Strahler et al. (1986) has three basic components: a scene model, an atmospheric model, and a sensor model. The scene is real and exists on the ground, whereas the image is a collection of spatially arranged measurements drawn from the scene. These authors suggest that the remote sensing problem can be described as "inferring the order in the properties and distributions of matter and energy in the scene from the set of measurements comprising the image" (Strahler et al., 1986, p. 123). They also state that the problem of scene inference is a problem of model inversion where the order in the scene is reconstructed from the image and remote sensing model.

The scene model can be either discrete or continuous (Strahler et al., 1986). In the discrete model, objects have defined boundaries or discontinuities, whereas in the continuous model, changes in matter are continuous in space. In Mediterranean-type and many other ecosystems, scenes more often conform to the discrete rather than to the continuous model where the scene is made up of elements on a background. The elements have uniform properties or parameters, and in the case of vegetation, the leaf, branches, or entire plant can be treated as the elements in a scene. High spatial resolution data are well suited to monitoring landscapes made up of discrete rather than continuous features. The precise location and properties of specific elements, such as individual plants, can be derived directly from the image data.

Discrete scene models are divided into simple and complex models. The simple discrete models include one class of element in the scene, whereas in the complex models there is more than one class of element in the scene. Finally, there are two types of discrete scene models, H- and L-resolution models, which provide the link between the scene model and sensor spatial resolution. In the H-resolution model, elements in the scene are larger than the scene cells, whereas in the L-resolution model, the elements are smaller. Graetz (1990) incorporated the remote sensing model proposed by Strahler et al. (1986) into a schematic representation of data flow in remote sensing. This schematic contains the three model components identified by Strahler et al. (1986) and an image processing model (Figure 12-1). The "critical point" indicated in Figure 12-1 refers to the break between discrete and continuous scenes.

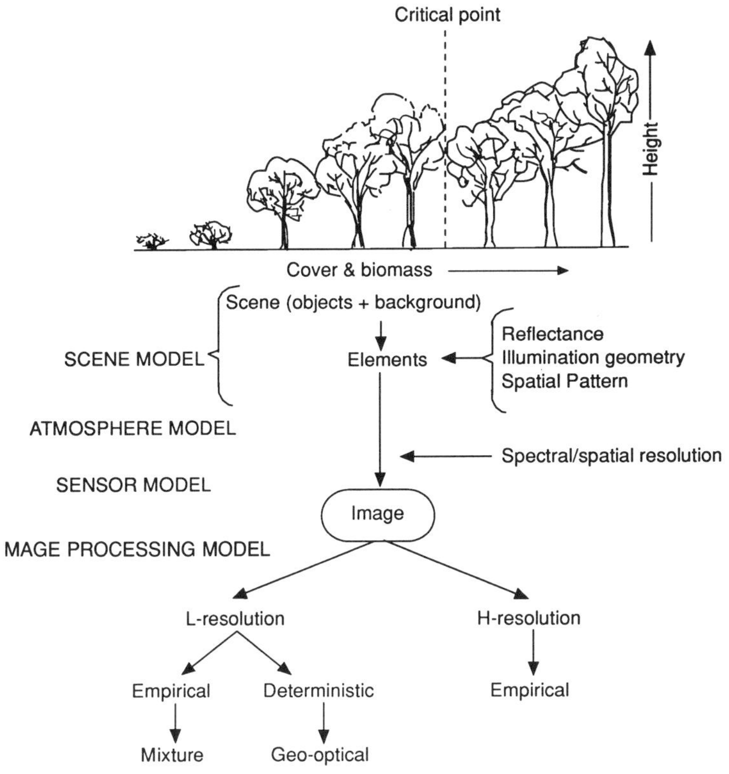

Figure 12-1. A schematic outline of the data flow in remote sensing including the remote sensing model of Strahler et al. (1986) and an image processing model. (Reprinted by permission of the publisher from "Remote sensing of terrestrial structure: An ecologist's pragmatic view." Graetz RD, In: Hobbs RJ and Mooney HA (eds) Remote Sensing of Biosphere Functioning. Ecological Studies 77:5–30. Copyright 1990 by Springer-Verlag.)

If a goal of remote sensing studies in Mediterranean-type ecosystems is to monitor species loss or indicator species along transects or within barometer sites, then the scene elements will be individual plants or uniform patches of vegetation. An appropriate scene model would be the H-resolution model using high spatial resolution data. If the scene conforms to the continuous rather than the discrete model, then some of the advantages of the high spatial resolution data and H-resolution model may not be as obvious. Coarser-resolution systems may be able to provide adequate area-averaged information for the investigator. However, it will be demonstrated later in this chapter that there may still be significant advantages to retaining high spatial resolution observations when the scene model is continuous.

Global Change Monitoring Scales and Remote Sensing

Global change studies are generally structured to address one or both of the following two fundamental questions (Hope and Stow, 1993):

1. How will global change affect the ecosystem (e.g., function and structure)?
2. How does the ecosystem or biome affect climate through exchanges of heat, mass, and momentum?

H-resolution scene models are more likely to be used for basic ecosystem monitoring at specific study sites, whereas L-resolution scenes are appropriate for deriving inputs for global circulation models (GCMs) and regional applications.

Studies in Mediterranean-type ecosystems that are designed to predict or describe the anticipated effects of a changing global environment and that include the collection of remotely sensed data can be 1) basic monitoring-type experiments that are conducted over long time periods (decades), 2) modeling-type experiments that focus on predicting the response of the ecosystems to the environmental changes, and 3) experiments that involve manipulations based on predicted changes or a range of changes in the climatic variables or atmospheric composition.

H-resolution scenes may benefit experiments in all three categories but are particularly suited to the monitoring of ecosystem responses to global change.

High Spatial Resolution Data: Sensors and Platforms

Scanned aerial photographs are a source of high spatial resolution data, but it is difficult to determine the absolute magnitude of spectral radiant exitance recorded on photographic film. However, these absolute values are routine data products from spectral radiometers or charge couple device (CCD) sensors. Although the following discussion will focus on data collected using spectral radiometers or CCD devices, it should be recognized that scanned data from photographs may suffice in applications where absolute radiometric quantities are not important.

The highest spatial resolution image data that are available from commercial satellites are provided by the French satellite system SPOT. These are panchromatic images with a ground resolution of 10 m. The usual platform for collecting high spatial resolution multispectral image data is an aircraft. In a general sense, any radiometer mounted on an aircraft platform can make high spatial resolution observations provided the altitude of the aircraft is low enough. Aircraft allow the investigator to exploit periods of clear sky conditions and to specify the time of day for data acquisition. It is also possible to collect data over selected or constructed calibration targets so that models can be developed to convert radiances to reflectances. Aircraft data can, however, be affected substantially by the instability of the platform. Ideally, concurrent pitch, roll, and yaw data are collected from the navigation system of the aircraft along with the radiometric data so that accurate geometric corrections can be applied to the image.

A second unique consideration for processing high spatial resolution imagery pertains to the corrections that are made for variations in terrain. Terrain corrections for low-resolution imagery are usually accomplished using a digital terrain model. These models may not be routinely available at the required scale to make precise corrections for high spatial resolution data. Consequently, these models would have to be developed specifically for the target sites if this correction was considered to be important.

Ecosystem Monitoring and High Spatial Resolution Data

Mediterranean-type ecosystem studies can include the collection of remotely sensed data to (1) classify a site's land cover, (2) quantify the state or condition of the vegetation cover, (3) determine biophysical quantities and rates, and (4) assist in the design of efficient field sampling schemes. These activities are not tied to the use of high spatial resolution data, but rather the high-resolution data offer unique possibilities regarding data analysis techniques and information extraction.

Land Cover Classification

Land cover classification is one of the earliest and best-developed applications of remote sensing. Remotely sensed data are generally used to classify a landscape when the area to be covered is large and it is not cost-effective to conduct a ground survey. High-resolution images tend to cover relatively small areas and are generally not used for conventional land use classification. High spatial resolution classification schemes are more likely to be implemented in studies where transects or barometer sites are monitored and there is interest in individual plants or small vegetation patches and their precise location (cf. Monitoring Ecosystem Response to Global Change). Changes in species abundance or patterns could be identified if the data were collected at an appropriate spatial resolution.

Classification is nearly always based on an H-resolution model where the objects in the scene are larger than the resolution cells (Strahler et al., 1986). Vegetation stands or fields are often regarded as the minimum resolution element in classification schemes based on satellite imagery so that information about individuals and their location in the landscape is lost. High spatial resolution imagery has the potential to reveal this locational information about individuals.

New analytical techniques for vegetation classification may be possible using high spatial resolution imagery. In a pilot study over a mixed forest in southern California, Hope and Stow (1993) analyzed data collected from the Airborne Data Acquisition and Registration (ADAR) system.

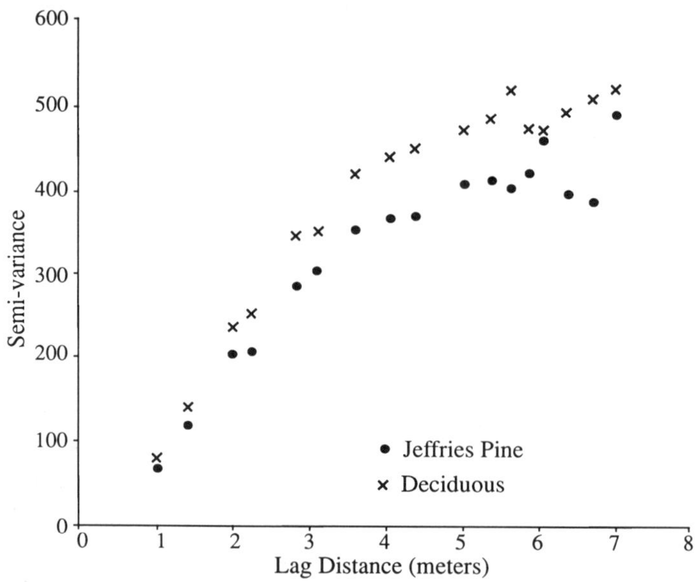

Figure 12-2. Semivariograms constructed using the simple spectral ratio (near infrared radiances divided by red radiances) obtained from ADAR 0.5 m resolution data for Jeffries pine and deciduous trees growing in southern California. (Reprinted by permission of the publisher from "An analysis of tree mortality in southern California using high spatial resolution remotely sensed spectral radiances: a climatic change scenario," Hope AS, Stow DA, Landscape and Urban Planning 24:87–94. Copyright 1993 by Elsevier Science Inc.)

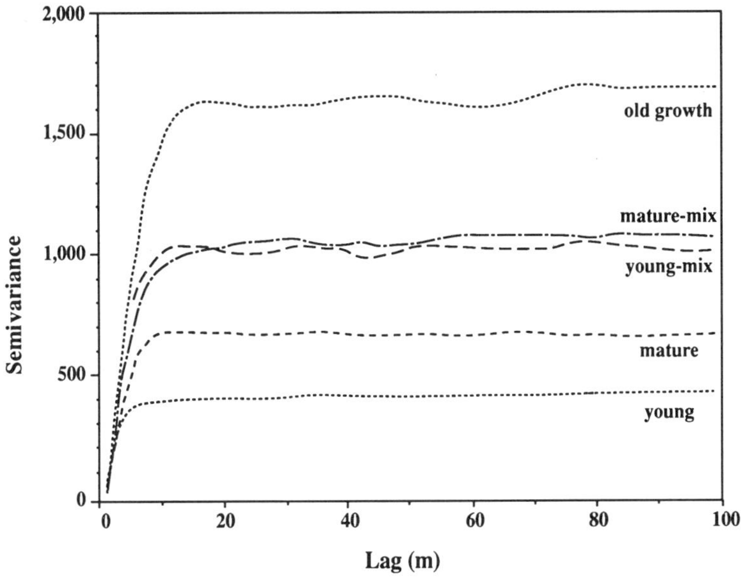

Figure 12-3. Semivariograms constructed using reflected red radiances (1.0-m resolution) for different-ageed Douglas fir stands growing in the Pacific Northwest region of the United States. (Reprinted by permission of the publisher from "Semivariograms of digital imagery for analysis of conifer canopy structure." Cohen WB, Spies TA, Bradshaw GA, Remote Sensing of Environment 34: 167–178. Copyright 1990 by Elsevier Science Inc.)

This system incorporates four full-frame CCD sensors covering visible and near infrared portions of the electromagnetic spectrum. The data were collected with a spatial resolution of approximately 0.5 m, and semivariograms were constructed for small uniform patches of deciduous and conifer trees using digital counts from the red, green, and near infrared spectral bands. The results suggested that differences in the semivariogram characteristics of deciduous and pine trees were different and these differences may be used to automatically discriminate between tree types (Figure 12-2). Although the results of Hope and Stow (1993) require further validation, a study of the canopy structure of Douglas fir trees in the northwest United States conducted by Cohen et al. (1990) had a similar outcome. In this study, the semivariograms of different-aged tree stands were examined using data with 1, 10, and 30 m ground resolution collected using a CCD detector. These authors found that semivariograms constructed using the 1 m resolution data were markedly different for the different-aged tree stands (Figure 12-3). These differences were not evident when the 10- and 30-m resolution data were used to construct the semivariograms.

The differentiation of plant types, age, or structural characteristics in Mediterranean-type ecosystems using high spatial resolution data and geostatistical techniques may be viable where detailed discrimination is required on a plant-by-plant or stand-by-stand basis. Cohen et al. (1990) suggest that application of the semivariogram technique to image data for within-stand structure analysis would require the calculation of separate semivariograms for each stand and this may not be practical for large scenes with many stands. A more likely application would be to use the semivariograms to monitor possible structural changes of selected stands over time.

Mediterranean-type ecosystems are often characterized by landscapes that conform to the discrete scene model where trees or shrubs are set against a background of soil or other vegetation. Classification of these scene components using high spatial resolution data allows investigators to determine directly the specific land cover fractions of each component without having to resort to the inversion of geometric-optical models as proposed by Li and Strahler (1985). Inversion of these models does not provide information regarding the specific location of each scene component but rather describes their abundance in statistical terms. Another unique advantage of high spatial resolution data for classification purposes pertains to image segmentation, which is intended to enhance classification accuracies. With high-resolution data, this segmentation can start at the level of the vegetation elements (background, shrubs, trees, etc.) and progress to the level of landscape-scale features, which is the common starting scale for segmentation using low-resolution data.

Vegetation Condition

The intensive monitoring of study sites (e.g., transects or barometer sites) to assess the impacts of global change is likely to include regular evaluations of vegetation condition or stress. A changing global environment may not have an equal impact on all vegetation species, so it is necessary to make these evaluations at the level of individual plants or species. Low-resolution remotely sensed data may have utility for such monitoring when the scene conforms to the continuous model. However, in landscapes with mixed vegetation species, the evaluation of vegetation at the level of individual plants (or uniform clusters) is necessary. With the individual plants as the scene element, only high-resolution imagery can provide the requisite data, as is demonstrated in the following example.

Events such as droughts provide an opportunity to observe and understand how unusual climatic occurrences stress ecosystems (National Research Council, 1988, 1990). These events can be treated as analog models of potential climatic change (Easterling et al., 1992). Hope and Stow (1993) invoked this climatic change analogy in a study of tree mortality in southern California. Following a prolonged drought, bark beetle (*Dendroctonus*) infestations increased in the forests of the Cuyamaca State Park, approximately 50 km east of San Diego. Since the infestations are initially at the scale of individual trees, high spatial resolution data (0.5 m) collected using the ADAR system were considered appropriate for identifying tree mortality and for relating to levels of infestation. It was shown by Hope and Stow (1993) that the identification of trees at different stages of bark beetle attack was possible using thcsc high-resolution data. The ratio of near infrared to red radiances and the blue and green reflected radiances were directly related to the level of infestation, even when there was no visible sign of vegetation stress or dieback (Figure 12-4).

While the results of Hope and Stow (1993) illustrated how high spatial resolution data can be used to identify infestation levels of individual

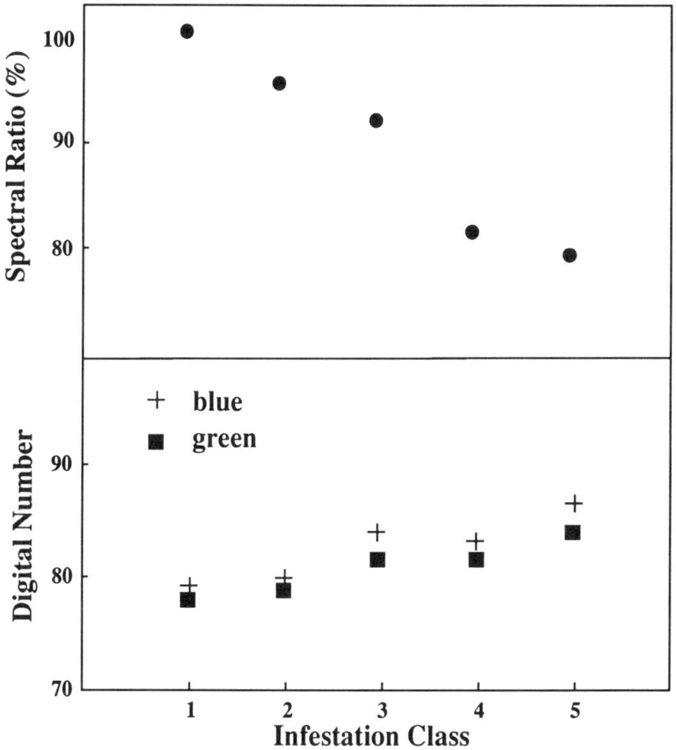

Figure 12-4. Relationship between the 0.5-m resolution ADAR near infrared: red spectral ratio, blue and green reflected radiances (digital numbers) and bark beetle infestation. (Reprinted by permission of the publisher from "An analysis of tree mortality in southern California using high spatial resolution remotely sensed spectral radiances: a climatic change scenario." Hope AS, Stow DA, Landscape and Urban Planning 24:87–94. Copyright 1993 by Elsevier Science Inc.)

trees or small patches on the basis of their spectral reflectance properties, Murtha and Wiart (1989) used high spatial resolution data to relate intraplant heterogeneity of the spectral radiances to levels of bark beetle infestation. These authors reported that trees under stress have a less variegated reflectance pattern than healthy trees (Murtha and Wiart, 1989). Consequently, the monitoring of Mediterranean-type ecosystems under stress may benefit from analyses of the spectral characteristics of high-resolution data as well as the characteristics of the intraplant variability in the reflected spectral radiances.

Predictions of climatic change based on the output from GCMs have a high degree of uncertainty. The uncertainty increases with a decrease in the size of the area being considered and is greatest for the prediction of changes in precipitation. Given these uncertainties, it still appears reasonable to anticipate a decrease in soil moisture over much of the continental area of the United States, including the Mediterranean-type ecosystems of California, because of the increase in evaporative losses associated with elevated air temperatures. Similar increases in aridity in other Mediterranean-type ecosystems may also be experienced.

The ability to monitor plant moisture stress in a timely and consistent manner at designated study sites is likely to be a key observational requirement of global change studies. Considerable research has been conducted into the use of remotely sensed thermal infrared radiances to detect water stress in plants (e.g., Clawson and Blad, 1982; Idso, 1982; Hope and Jackson, 1989) or for energy balance studies (e.g., Hatfield et al., 1984; Reginato et al., 1985). The use of these thermal infrared data has, however, been limited in part by the difficulty of assigning the observed thermal response to separate sources in the IFOV. Since water stress in a plant will show as an increase in the foliage temperature, thermal contributions from the soil background, dead plant material, and other nontranspiring surfaces will distort the interpretation of the radiometric temperature. With high spatial resolution thermal infrared data, these sources can be masked from the data set used to analyze plant water stress. Hope and Jackson (1989) demonstrated how the measurement of canopy temperature can be used to quantify moisture stress in a wheat crop, while Cliffe (1992) used a hand-held thermal infrared radiometer to isolate foliage temperatures from soil background contributions to detect moisture stress in grapevines growing in southern California. Similar approaches may be viable for monitoring water stress in natural ecosystems if the thermal emissions from the surface can be separated into soil and vegetation components.

Biophysical Quantities

A major effort in biophysical remote sensing over the past 15 years has been to relate quantities such as leaf area index, biomass, or net primary productivity (NPP) to the spectral reflectance properties of various vegetation communities (e.g., Drake, 1976; Tucker, 1979; Tucker et al., 1981; Aase et al., 1986). This effort has been motivated mainly by the prospect of making large area estimates of biophysical variables using data collected from satellite or aircraft platforms, which is attractive for regional ecosystem or agricultural studies. At a more local level where high spatial rcsolution data are collected, remotely sensed radiances may be used to obtain rapid and nondestructive estimates of canopy attributes (Asrar et al., 1989).

Data collected from low spatial resolution sensors often include scene components in the IFOV that have no relationship to the biophysical quantity under study. The reflectance from soil background or dead plant material is a major source of error in relationships between spectral reflectances and biophysical quantities (Sellers, 1985). This is of particular concern in semiarid and arid landscapes such as those found in Mediterranean-type ecosystems. As suggested earlier, these components can be isolated from the scene if high spatial resolution images are used in the study and if the preprocessing of the image includes classification and masking procedures.

The determination of some biophysical quantities from reflected spectral radiance data, such as net primary productivity or phenological development, requires multiple observations during a growing season. Since trajectories of the radiances over time are used in this type of analysis, the accurate coregistration of the images is imperative and high spatial resolution imagery is well suited to this task.

Field Sampling Design

The design of field sampling schemes to estimate area averaged vegetation quantities or the spatial distribution of these quantities requires the investigator to relate the sampling effort to the variation present in the variable of interest (Webster et al., 1989). Cost considerations may restrict the number of ground observations that can be collected in a preliminary survey, while a similar investment could provide a much greater density of measured radiances. Webster et al. (1989) suggest that the outlay for the reconnaissance stage in an experiment should be approximately 10% to 15% of the total sampling effort. Digital imagery can be used to assist with the design of an efficient strategy for sampling vegetation over large areas (Atkinson et al., 1992). Vegetation cover characteristics generally covary with the radiation that is reflected, so that vegetation properties and reflected radiation can be treated as coregionalized variables and cokriging can be used to develop an efficient sampling scheme (Atkinson et al., 1992). Since many biophysical variables are studied at scales less than 1 m, high spatial resolution data may be the appropriate digital imagery for designing preliminary surveys.

Conclusions

The potential use of remotely sensed data in earth systems science has been widely documented (e.g., Hobbs and Mooney, 1990). Most attention is generally given to the utilization of data from satellites, since global or regional scales often characterize global change research. If detailed ecosystem monitoring is accepted as part of the scheme to study global change, then there may be significant applications for high spatial resolution remotely sensed data. Studies along environmental gradients and monitoring small barometer sites for global change research are scales of observation that are consistent with the collection of these data.

Basic remote sensing research is required to investigate new information extraction techniques that exploit the unique characteristics of high spatial resolution data. Approaches that have been developed using lower-resolution data may not be appropriate for these images that can resolve individual plants or small uniform patches of vegetation. The image processing challenges associated with data acquisition from aircraft platforms will have to be addressed in these studies.

References

Aase JK, Millard JP, Brown BS. 1986. Spectral radiance estimates of leaf area and leaf phytomass of small grains and native vegetation. IEEE Trans Geosci Remote Sens GE-24:685–691.

Asrar G, Myneni RB, Kanemasu ET. 1989. Estimation of plant-canopy attributes from spectral reflectance measurements. In: Asrar G (ed) Theory and Applications of Optical Remote Sensing. John Wiley & Sons, New York, pp. 252–296.

Atkinson PM, Webster R, Curran PJ. 1992. Cokriging with ground-based radiometry. Remote Sens Environ 41:45–60.

Clawson KL, Blad BL. 1982. Infrared thermometry for scheduling irrigation of corn. Agron J 74:311–316.

Cliffe TW. 1992. Monitoring soil moisture deficit of grapevines under drip irrigation using the temperature-stress-day index. Thesis, San Diego State University, San Diego, CA.

Cohen WB, Spies TA, Bradshaw GA. 1990. Semivariograms of digital imagery for analysis of conifer canopy structure. Remote Sens Environ 34:167–178.

Drake BG. 1976. Seasonal changes in reflectance and standing crop biomass in three salt marsh communities. Plant Physiol 58:696–699.

Easterling WE, Rosenberg NJ, McKenney MS, Jones CA. 1992. An introduction to the methodology, the region of study, and a historical analog of climate change. Agric Forest Meteorol 59:3–15.

Ford MJ. 1982. The Changing Climate: Responses of Natural Fauna and Flora. George Allen & Unwin, London.

Graetz RD. 1990. Remote sensing of terrestrial structure: an ecologist's pragmatic view. In: Hobbs RJ, Mooney HA (eds) Remote Sensing of Biosphere Functioning. Ecological Studies 77. Springer-Verlag, New York, pp. 5–30.

Hatfield JL, Reginato RJ, Idso SB. 1984. Evaluation of canopy temperature-evapotranspiration models over various crops. Agric Forest Meteorol 32:41–53.

Hobbs RJ, Mooney HA (eds). 1990. Remote Sensing of Biosphere Functioning. Ecological Studies 79. Springer-Verlag, New York.

Hope AS, Jackson RD. 1989. Early morning canopy temperatures for evaluating water stress in a wheat crop. Water Resour Bull 25:1009–1014.

Hope AS, Stow DA. 1993. An analysis of tree mortality in southern California using high spatial resolution remotely sensed spectral radiances: a climatic change scenario. Landscape Urban Plann 24:87–94.

Idso SB. 1982. Non-water stressed baselines: a key to measuring and interpreting plant water stress. Agric Meteorol 27:59–70.

King AW. 1990. Translating models across scales in the landscape. In: Turner MG, Gardner RH (eds) Quantitative Methods in Landscape Ecology: The Analysis and Interpretation of Landscape Heterogeneity. Ecological Studies 82. Springer-Verlag, New York, pp. 479–517.

Li X, Strahler AH. 1985. Geometric-optical modeling of conifer forest canopy. IEEE Trans Geosci Remote Sens GE-23:705–721.

Murtha PA, Wiart RJ. 1989. Cluster analysis of pine foliage patterns aid identification of Mountain Pine beetle current-attack. Photogramm Eng Remote Sens 55:83–86.

National Research Council. 1988. Toward an understanding of global change: Initial priorities for U.S. contributions to the International Geosphere-Biosphere Program. Committee on Global Change, National Research Council, National Academy Press, Washington, DC, 213 pp.

National Research Council. 1990. Research strategies for the U.S. global change research program. Committee on Global Change, National Research Council, National Academy Press, Washington, DC, 291 pp.

Paine RT. 1980. Food webs: linkage interaction strength and community infrastructure. J Anim Ecol 49:667–685.

Reginato RJ, Jackson RD, Pinter PJ. 1985. Evapotranspiration calculated from remote multispectral and ground station meteorological data. Remote Sens Environ 18:75–89.

Sellers PJ. 1985. Canopy reflectance, photosynthesis and transpiration. Int J Remote Sens 21:143–183.

Simon Silver C, De Vries RS. 1990. One Earth One Future: Our Changing Global Environment. National Academy Press, Washington, DC.

Simonett DS, Reeves RG, Estes JE, Bertke SE, Sailer CT. 1983. The development and principles of remote sensing. In: Colwell (ed) Manual of Remote Sensing. American Society of Photogrammetry, Falls Church, pp. 1–36.

Strahler AH, Woodcock CE, Smith JA. 1986. On the nature of models in remote sensing. Remote Sens Environ 20:121–139.

Townshend JRG. 1980. The spatial resolving power of earth resources satellites: a review. Report TM 82020, NASA Goddard Space Flight Center, Greenbelt, MD.

Tucker CJ. 1979. Red and photographic infrared linear combinations for monitoring vegetation. Remote Sens Environ 8:127–150.

Tucker CJ, Holben BN, Elgin JH, McMurtrey J. 1981. Remote sensing of total dry matter accumulation in winter wheat. Remote Sens Environ 11:171–190.

Webster R, Curran PJ, Munden JW. 1989. Spatial correlation in reflected radiation from ground and its implications for sampling and mapping by ground-based radiometry. Remote Sens Environ 29:67–78.

Westman WE. 1984. Ecology, Impact Assessment, and Environmental Planning. John Wiley & Sons, New York.

Woodcock CE, Strahler AH. 1987. The factor of scale in remote sensing. Remote Sens Environ 21:311–332.

13. Monitoring Ecosystem Response to Global Change: Multitemporal Remote Sensing Analyses

Douglas A. Stow

If the impacts of global change on terrestrial ecosystems are to be monitored, remote sensing will likely play an essential role. However, the specific role that remote sensing will play and the value of the information that it can provide in the context of monitoring ecosystem response to global-scale disturbances is unknown. What is known is that the information content of data from a single remote sensing acquisition is insufficient, and that multitemporal data are required to discriminate important land surface features and to capture ecosystem dynamics. The major question that must be asked is: Can the signals associated with ecosystem responses to global-scale disturbances be resolved from the environmental and sensor noise associated with multitemporal remote sensing?

The focus of this chapter is on defining the potential utility of and requirements for monitoring responses of Mediterranean-type ecosystems to global change by multitemporal remote sensing. Global-scale changes that are likely to disturb these ecosystems are the certain increase in atmospheric CO_2 concentration and the uncertain changes in climate, particularly air temperature, precipitation, and humidity, in response to the increase of atmospheric CO_2 and other greenhouse gases (Watson et al., 1990). Note that changes in land use and other direct localized human disturbances are not within the scope of this chapter, but the sum of these localized changes is likely creating significant global change.

Optical imaging systems operating in the solar reflectance part of the electromagnetic spectrum will be emphasized because of the greater knowledge base that exists for land surface applications, and because Mediterranean climates tend to have many cloud-free days that enable remote sensing in these "optical" wavelengths, i.e., 0.3–14.0 μm micrometers. Finally, the context for much of this treatise and a majority of the examples are drawn from chaparral ecosystems of California (Hanes, 1977). Some of the approaches that are proposed for monitoring response to global change may be inappropriate for some Mediterranean-type landscapes, particularly those surrounding the Mediterranean Sea, where a long history of human disturbance may dominate the patterns of landscape change.

Multitemporal remote sensing of land surface features can be achieved for most spatial scales (from patch to global) and many temporal scales (from seconds to decades). However, for logistical and technical reasons, particular combinations of spatial and temporal scales are limited in an operational sense. In this chapter, an emphasis will be placed on remote sensing systems which capture data in image format, at landscape to regional spatial scales, and at temporal scales of weeks to decades. These scales of remote sensing fall within the domain of medium-resolution, polar-orbiting satellite systems such as Landsat and SPOT (Systeme Pour L'Observation de la Terre). Higher spatial resolution applications which are achievable by ground-level and aerial radiometry are emphasized in the chapter by Hope in this volume.

This chapter is organized into two major sections. The first section provides background on the technical requirements for capturing and processing multitemporal remote sensing, the radiative signal of landscape changes that can be extracted, and the sources of noise and error that can obscure this signal. The types of landscape-scale responses of Mediterranean-type ecosystems that are hypothesized to result from global-scale disturbances and that may have signals that can be remotely sensed are discussed in the second section. Elements of two pilot studies on the utility of intraannual (seasonal) satellite image data for mapping the distribution of vegetation types and fire fuels amount are incorporated into both sections.

Multitemporal Remote Sensing of Land Surface Phenomena

Remote sensing of environment entails the collection and analysis of data about surface, upper subsurface, and atmospheric features of the Earth by instruments that are not in direct contact with the features. Electromagnetic radiation (EMR) generally provides the link for remote sensing of environment, with aircraft and spacecraft usually acting as the remote observation platforms. Most optical remote sensing instruments capture multispectral radiances. "Multispectral" implies sensing in more than one waveband. Spectral radiance is the radiative flux (watts per square meter) per unit solid angle (steradian) per unit of waveband (micrometer). Hand-held measurements of EMR taken over targets of interest from the ground can be considered as close-range remote sensing.

Multitemporal remote sensing of land surface processes involves revisiting and collecting data for the same portion of the Earth at multiple points in time. For terrestrial ecosystem applications, multitemporal remote sensed data is usually exploited to capture intraannual (seasonal), or interannual (year-to-year) variations in surface features, as illustrated in Figure 13-1 (Hobbs, 1990). Multitemporal data provide added dimensionality for improved discrimination of surface features, e.g., vegetation classification based on phenological differences. Such data may also allow for the detection and quantification of ecosystem dynamics, e.g., seasonal plant production or postburn succession.

Global changes in atmospheric constituent concentrations and subsequent climatic responses that are likely to follow are expected to modify the function and structure of many ecosystem types, including Mediterranean ecosystems. If remote sensing is to play a role in monitoring changes in function and structure of terrestrial ecosystems, then some changing surficial condition(s) or feature(s) of the landscape that relate to ecosystem function or structure must be detectable, identifiable, or quantifiable. Thus, it is important to understand the relationship between remotely sensed radiances captured at a given spatial resolution and temporal repeat interval, and the changing landscape features that may potentially signal ecosystem responses to global-scale disturbances. It will also be important to understand the magnitude and implications of noise components in these remote measurements. Table 13-1 summarizes the scene and atmospheric contributions of radiance to a remote sensing instrument, which are discussed in more detail below. Isolating effects of global-scale disturbances from more localized disturbances and other transient variations of Mediterranean landscapes presents a significant challenge.

Factors Influencing At-Sensor Radiances

The canopies of shrubs and trees tend to dominate the vegetative component of reflected radiation received by an air-borne or space-borne sensor over Mediterranean-type landscapes because of their downward viewing perspective. This means that foliar elements most affect the received signal, except for sparse canopies where woody components are important (Lacaze and Debussche, 1984).

Mediterranean-type landscapes are dominated by shrub and tree plant forms, structurally characterized by projected foliar cover (PFC; ratio of the cover of foliage projected to the surface per unit surface area) ranging

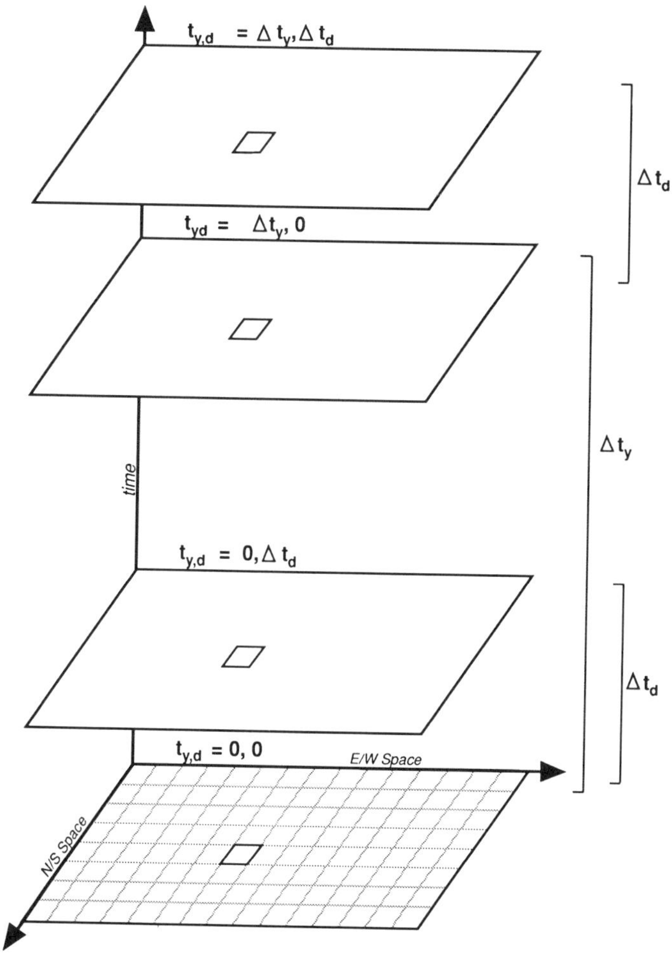

Figure 13-1. Conceptual model representing digital, multitemporal remotely sensed data. Intraanual image data acquired during a given growing season are separated by Δt_d days, and interannual comparisons are based on images acquired during growing seasons Δt_y years apart. All images are spatially registered such that, theoretically, the shaded pixel on all images represents the same ground resolution element on the earth's surface.

from 0.5 to 5.0 and tallest plant form heights of 0.5 to 5.0 m (Graetz, 1990). Bare soil and rock outcrops are significant contributors to satellite-sensed radiances from within and between shrub and tree canopies when the PFC < 1 (Graetz, 1990; Musick and Pelletier, 1986; Williamson, 1989). Annual and perennial forbs and grasses, and litter also contribute as noncanopy vegetative reflectors. When PFC > 1, e.g., old dense stands of *Ceanothus* shrubs, visible and near-infrared (NIR) canopy reflectance

Table 13-1. Scene, Atmospheric, and Sensor Variables Contributing to Spectral Radiance (in the Context of Remotely Sensing Vegetation Changes)

Factor	Effect
Soil composition	Modifies Lλ and SVIs for PFCs <1
Soil moisture	Modifies Lλ and SVIs for PFCs <1
Shadows	Modifies Lλ and SVIs, obscures some surface and canopy cover
Solar irradiance characteristics	Solar geometry influences illumination intensity, shadowing, canopy reflectance, atmospheric effects Magnitude of Lλ directly related to solar irradiance
Atmospheric path radiance	Reduces scence contrast; adds a temporally varying noise component, effects magnitude of SVIs
Atmospheric attenuation	Reduces target signal, affects magnitude of SVIs
Sensor calibration	Drift or errors results in imprecise determination of Lλ
Sensor view geometry	Variations (within scence or between dates may alter Lλ and SVIs for particular canopy types)

Lλ, Spectral radiance.
SVIs, Spectral vegetation indices.
PFCs, Projected foliar covers.

is most influenced by the depth of the foliar layer or essentially, leaf area index (LAI; Graetz, 1990).

Three general properties of vegetation tend to influence the spectral quality and quantity of solar reflected radiation received by satellite sensors, and therefore, through inversion, are the properties that are most likely to be accurately sensed. These vegetative properties are abundance, composition, and condition. Here, vegetation abundance refers to the relative or absolute amount of vegetation per some amount of ground area and does not include species abundance (Price, 1992). Abundance variables that have been shown to relate to reflected solar radiance are percent canopy and projected foliar cover, total and above-ground green biomass, and LAI (Aase et al., 1986). Essentially, the canopy reflectance is most directly indicative of the chlorophyll density, which is related to the abundance variables listed above (Wessman, 1990). Vegetation composition can be inferred from spectral and spatial, e.g., image texture, signatures of plants and their distributions. Differing species mixtures, plant physiognomy, and photosynthetic pigments most influence variations in multispectral radiances received by a satellite sensor. Vegetation conditions that may be remotely derived are photosynthetic rate (Sellers, 1985), leaf moisture and nitrogen content (Wessman et al., 1988), and plant stress (Cohen, 1991a) and mortality.

Besides the reflectance properties of land surface materials, other environmental factors influence the amount of radiance captured by a remote sensing instrument. The quantity and quality of solar illumination also determine the amount of remotely captured radiance. The quantity is defined as the solar irradiance, i.e., incoming radiation flux, which is determined by sun angle. The quality is determined by the ratio of direct solar to diffuse atmospheric, which depends on the sun angle and the optical properties of the atmosphere.

Spatial and temporal variability in atmospheric optical properties caused by changes in constituent concentrations also causes variations in the spectral radiance at the sensor. The effect of atmospheric constituents on radiation transfer is to attenuate the surface reflected signal and add a solar backscattered noise component. The result is a detrimental reduction in the ability to resolve or measure surface phenomena, and an equally detrimental change in the relative proportions of spectral radiances, since these atmospheric effects are wavelength specific.

The local terrain also affects the amount of reflected radiance by altering the actual solar incidence angle from that of a flat surface. In the extreme, certain terrain elements are not directly illuminated at all, resulting in shadowing (Hugli and Frei, 1983). Shadowed surfaces are only illuminated by diffusely scattered sky radiation, which limits the ability to observe or measure surface phenomena. Since shadowing varies with sun angle, which varies temporally, the degree of shadowing varies temporally. This can cause major variations in multitemporal radiances for a given terrain element, which may be confused as an apparent land surface change (Hugli and Frei, 1983). Another minor terrain effect for the solar reflected wavebands is the additive illumination caused by reflectance from adjacent terrain elements.

Thus, illumination and atmospheric conditions and the relative relationship of illumination and terrain varys temporally, and therefore, so do remotely sensed radiances (Duggin et al., 1985). Without properly accounting for or modeling these effects, the resultant temporal variability of multitemporal radiances can falsely signal or mask actual land surface changes.

In addition to artifacts and uncertainties due to environmental effects on radiation transfer are the effects of the sensor and its mode of sampling. Even when the same instrument is used, the earth-platform-sensor geometry relationship is different for each data acquisition. The resultant data are spatially autocorrelated because of optical diffraction and the scanning process. Atmospheric scattering also produces an adjacency effect which is a source of spatial autocorrelation in the image data (Singh, 1988).

Appropriate Satellite Systems

High-resolution satellite imaging systems such as Landsat and SPOT are likely candidates for monitoring ecosystem responses of global change for Mediterranean-type landscapes (Table 13-2). One justification for selecting such systems is that data have been acquired since July 1972 by Landsat. Although not all of the historical Landsat record is retrievable due to magnetic tape degradation and obsolete tape reading technology, Landsat Multispectral Scanner (MSS) data provide the longest historical record of any land satellite system. When selecting higher spatial resolution Landsat-type satellite data (10–100 m) over the coarse-resolution satellite data from NOAA Advanced Very High Resolution Radiometer (AVHRR; 1–10 km), one loses high temporal frequency of coverage (several weeks as opposed to daily, respectively). The actual repeat frequency depends on several factors, with cloud cover being the most important. An analysis of the Landsat data archive listing for the southern California region that contains Mediterranean-type ecosystems shows that four to eight cloud-free images have been acquired during the growing seasons of most years since 1972. If most of the archived tapes actually exist and are readable, this should provide sufficient temporal frequency for extracting information pertaining to plant growth and phenological variations through each growing season.

Another reason for advocating higher-resolution satellite data is the ability to spatially resolve many types of "edges" and individual vegetation stands that are small features relative to the pixel of an NOAA AVHRR-type sensor. Also, it is possible to stratify by vegetation, stand, terrain, and disturbance units, such that multiple pixels can be extracted for known surface conditions (Lacaze et al., 1984). NOAA AVHRR-type data will generally contain mixtures of vegetation stands, terrain slopes,

Table 13-2. High-Resolution Satellite Systems for Long-Term Monitoring

System	Start	Repeat (days)	Wavebands	Radiometric Resolution (bits)	Spatial Resolution (m)
Landsat MSS	1972	16*	G, R, NIR1, NIR2	7	80
Landsat TM	1982	16	BG, G, R, NIR MIR1, MIR2, TIR	8	30
SPOT XS	1986	26[a]	G, R, NIR	8	20
SPOT Pan	1986	26[a]	Visible	8	10

* Eighteen day repeat interval for Landsats 1–3.
[a] Pointable sensor increases effective frequency of image acquisitions.

and land use features. Finally, the Landsat and SPOT systems have reasonable radiometric calibration and normalization potential (Malila, 1985; Murphy et al., 1985) and very good geometric fidelity and registration potential (Nelson and Grebowsky, 1982).

Image Processing Requirements

Transforming multitemporal/multispectral radiances into useful information pertaining to landscape structure and change consists of several necessary and technically challenging image processing tasks. These can generally be grouped into geometric, radiometric, and information extraction processing procedures, the latter being based on models of the sensor, scene, and atmosphere (Strahler et al., 1986). Figure 13-2 is a flow diagram that summarizes the sequence of image processing procedures that are generally required to extract meaningful information on landscape change from multitemporal remotely sensed data.

Geometric processing entails correction and alignment of the digital image arrays such as to allow for precise image-to-image and image-to-ancillary data comparisons. The most critical of these operations for multitemporal monitoring is spatial registration, i.e., image-to-image matching (Nelson and Grebowsky, 1982). Small errors in alignment can translate into significant temporal artifacts, when comparing multispectral radiances, or spectral indices and thematic classes derived from such radiances. Georeferencing or geocoding is the process of aligning the image array to conform to an earth coordinate system (Novak, 1992). This allows pixels to be referenced to specific areas on the earth's surface, thus enabling direct comparison with field and GIS data. Spatial registration and georeferencing of multitemporal images is often achieved simultaneously, either by the application of deterministic earth-platform-sensor geometry models (Marvin et al., 1987) or by more empirical fit, "rubber sheeting" models (Novak, 1992). Geometric errors due to terrain relief displacement are significant in most Mediterranean-type landscapes, and can be corrected in the more deterministic models with the aid of digital terrain data (Marvin et al., 1987). Spatial registration and georeferencing accuracies are generally equivalent to ±1 pixel (Nelson and Grebowsky, 1982; Marvin et al., 1987) and ±0.5 pixel root mean square error (RMSE), respectively, with the highest precisions achieved with a hybrid of the deterministic and empirical geometric modeling approaches.

Spatial aggregation of image data can be a requisite geometric transformation when monitoring environmental change (Justice et al., 1989). By spatially aggregating radiance or transformed index values, one can minimize the misregistration and spatial autocorrelation effects (see below) and therefore, improve the signal:noise ratio and minimize locational uncertainty. This is achieved at the expense of reduced spatial

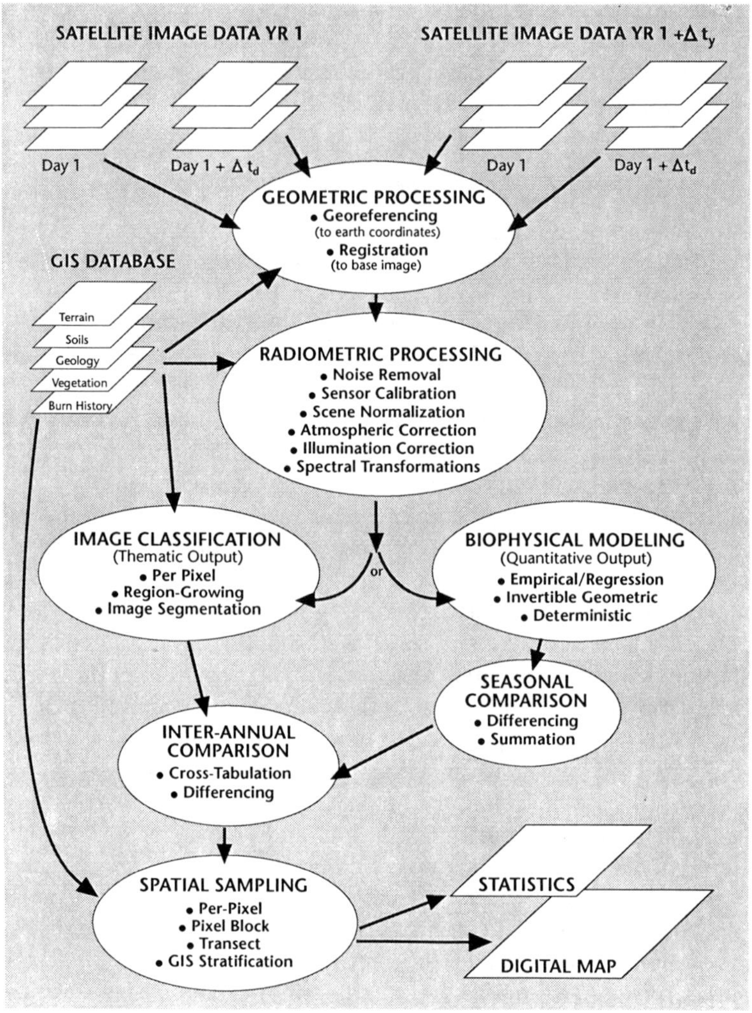

Figure 13-2. Flow diagram summarizing the likely sequence of image processing routines applied to intra- and interannual multitemporal data for monitoring ecosystem responses to global change.

resolution. Aggregating pixel data is usually achieved by averaging blocks of contiguous pixel values, but can be achieved by weighted averaging so as to minimize the spatial autocorrelation effects that result from the imaging process (Justice et al., 1989).

The other type of requisite image transformation is radiometric preprocessing, meaning that remotely sensed digital brightness values must be calibrated and/or converted so as to improve the relative spectral and temporal fidelity of the data. The essential first step is to radiometrically calibrate digital brightness values by converting them into spectral radiances (Robinove, 1982). Such calibration is based on published calibration coefficients derived from prelaunch laboratory calibration and postlaunch empirical corrections. Radiometric calibration serves to normalize spectral radiances between wavebands of the same image and between images, if the calibration coefficients are temporally stable or updated.

The most important requirement of radiometric correction is to ensure that changes in spectral radiances for corresponding pixels of a multitemporal image sequence are proportional to actual changes in spectral reflectance of the surface (Nelson, 1985). Differences in atmospheric optical properties and solar illumination between image dates influence the need to model their transient effects. More approximate, empirically based models may reduce these effects through scene normalization, but radiation transfer models offer a more precise, deterministic approach.

Scene normalization models are based on empirically matching multitemporal images based on scene features, e.g., features with stable reflectances such as rock outcrops or persistent shadows (Pech et al., 1986; Chavez, 1988; Schott et al., 1988; Laureau, 1991). A particularly promising example of this type of modeling is the "radiometric registration" approach developed by Hall et al. (1991). In an attempt to normalize atmospheric and illumination differences between seasonal Landsat TM images of chaparral and pine-oak forest in southern California, Stow et al. (1993) radiometrically "registered" an image data from a April 1987 scene to a scene acquired in November 1986 image using the Hall et al. (1991) method (Figure 13-3). TM brightness values for both dates were converted to "Brightness" (BR), "Greenness" (GR), and "Wetness" (WT) based on the "Tasseled-cap" transform (Crist and Cicone, 1984). Dark and bright control sets (pixel samples) were selected from BR-GR plots and used to develop a linear fit equation for normalizing TM spectral radiances for April 1987 relative to TM spectral radiances for November 1986. This enabled seasonal changes in BR, GR, and WT to be derived in an attempt to infer seasonal changes in fire fuels amount and condition (which is further described in a later section).

Multidate satellite data may also be normalized through more deterministic approaches that attempt to derive reflectance factors of land surfaces (Duggin, 1980; Moran et al., 1990). Reflectance factors are approximations of surface reflectances (and therefore should be temporally stable) that are based on an assumption of isotropically reflecting surfaces. Such approaches are based on radiative transfer modeling of solar illumination and atmospheric effects (Spanner et al., 1990). In the typically irregular terrain of many Mediterranean-type landscapes, the differential illumination effects from variable terrain slope facets should also be modeled, based on a digital elevation model (Holben and Justice,

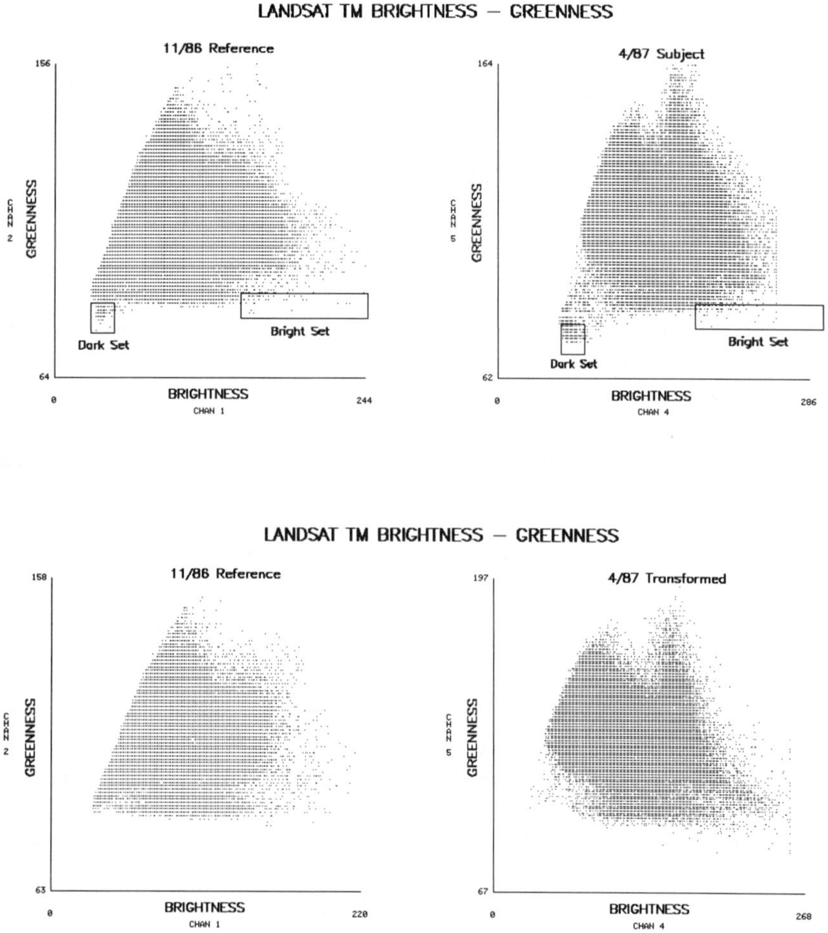

Figure 13-3. Illustration of the radiometric registration method developed by Hall et al. (1991). (a) Crist Transformed "Brightness" and "Greenness" scatterplot for Nov-86 and Apr-87 Landsat Thematic Mapper data, with "bright" and "dark" training sets outlined; (b) Crist Transformed "Brightness" and "Greenness" scatterplot for Nov-86 and radiometrically registered Apr-87 Landsat Thematic Mapper data.

1979). The greatest error and uncertainty in deriving reflectance factors results from the atmospheric modeling component. Particularly problematic is the lack of concurrent data on atmospheric constituent concentrations and their horizontal and vertical distributions (Caselles and Lopez-Garcia, 1989).

Once multidate image data are geometrically and radiometrically corrected and registered, another model must be applied to convert them into discrete thematic categories or continuous biophysical variables. If the information that is desired pertains to changes in vegetation composition or the movement of vegetation boundaries, then processed spectral radiance values (or reflectance factors) are transformed into thematic categories with the aid of statistical pattern recognition techniques. Such image classification can be achieved by categorizing single pixels (Stow et al., 1989), i.e., "per-pixel" classifier, or contiguous groups of pixels, i.e., image segmentation (Haralick, 1985). With either approach, the pattern recognition algorithm can be applied to each image date separately or to concatenated data from both dates. Nonimage data, i.e., ancillary or collateral data, such as digital terrain or climatic data, may also be incorporated into the classification procedure (Hutchinson, 1982; Frank, 1988). Conversion of processed radiance values into biophysical variables such as leaf area index or percent vegetation cover often involves a two-step process. First, calibrated spectral radiances are transformed into spectral indices using algebraic combinations, which are then related to the continuous (in both statistical and spatial contexts) biophysical variable of interest. The relationship may be established with an empirical model that relates a spectral index to field measurements of the landscape variable (Curran and Hay, 1986; Pech et al., 1986) by inverting a model of the structural and distributional characteristics of scene elements (Li and Strahler, 1985), or by "unmixing" subpixel elements (Rambal et al., 1990).

Scale Considerations

A further consideration when attempting to monitor ecosystem changes from multitemporal satellite image data is the choice of spatial unit and temporal interval. Theoretically, the finest changes that can be resolved are at spatial scales equal to the linear dimension of one to two pixels. However, multitemporal image comparisons of derived categorical or continuous data values at the pixel scale are prone to artifacts and other errors due to spatial misregistration effects. Such effects may be minimized by spatial aggregation, as was described previously. One can choose to aggregate a small block of pixels to preserve fine spatial details, or average a larger group of pixels that are extracted by some type of spatial sampling criterion or by landscape stratification. Examples of spatial sampling criteria are a linear transect, a rectangular block, or any arbitrarily shaped area that corresponds to a field experimental plot, which are illustrated in Figure 13-4. Aggregates of contiguous pixels associated with landscape strata may be defined by vegetation stands, terrain units, soil polygons, or land management units, all of whose boundaries may be extracted from a GIS data base. It is important to

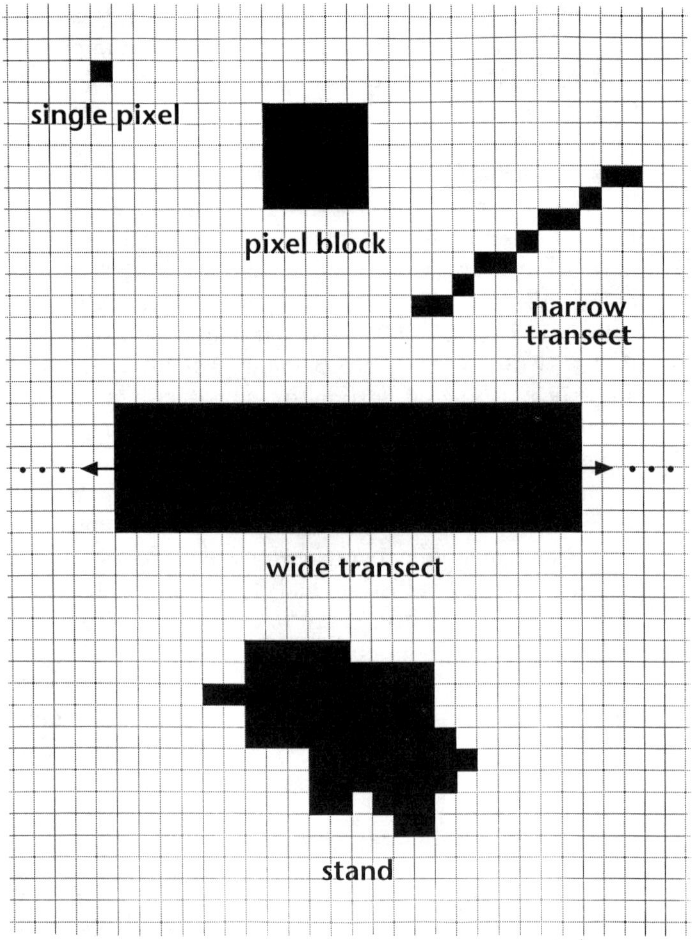

Figure 13-4. Examples of spatial sampling criteria for extracting specific pixel values from processed images, e.g., corresponding to thematic categories or biophysical properties. The dotted line on either end of the transect criteria implies that a transect can be very long relative to its width.

note that changes can be assessed at either the fine-scale or at the aggregate level for spatially or categorically defined sampling units.

The time interval for assessing change may depend on both intra- and interannual factors. Intraannual or seasonal signals of ecosystem functioning may be detected or quantified, and vegetation composition and ecotone boundaries may be better identified by incorporating satellite data acquired throughout the growing season (Allum and Dreisinger, 1987). The actual monitoring of signals associated with ecosystem response to global environmental change will necessarily occur at interannual intervals, although the signal may be based on intraannual phenomena (Adeniyi, 1985). A challenge is to identify the phenomena that are most responsive to the influence of "global changes" and whose associated remotely sensed signal can be detected in the shortest time interval possible. A related consideration is whether one is interested in detecting a global change signal in a hindcast mode by relying on archived satellite data, or in monitoring possible future change by comparing future satellite data to those from a baseline date (historical or present).

Possible Landscape and Regional-Level Responses of Mediterranean Ecosystems to Global Change

The purpose of this section is to identify landscape properties that are likely to be responsive to global-scale disturbances and can be remotely sensed. Once again, the two types of global-scale change that are likely to disturb Mediterranean-type ecosystems are: 1) the increase in atmospheric CO_2 and other "greenhouse gas" concentrations, and 2) changes in climate, particularly air temperature, precipitation, and humidity, in response to changes of greenhouse gas concentrations in the atmosphere. An emphasis will be placed on remotely sensing properties that may be responsive to increases in atmospheric CO_2 concentration, as this global change is already well documented (Keeling et al., 1982). Changes in the following ecological properties are candidates for remotely sensed indicators of global change in Mediterranean-type landscapes:

1. Plant abundance and growth
2. Stand species composition
3. "Edge" location
4. Fire regime

The actual signal of global change response will be the interannual change in the above properties, as estimated from multitemporal satellite data. Generally, the signal will be extracted from a sample of pixels by stratifying the image data based on environmental data layers, i.e., digital terrain or soils data from a GIS.

Changes in Vegetation Abundance, State, and Growth

Spectral vegetation indices (SVIs) derived from multispectral data have been shown to correlate most directly with illuminated chlorophyll density of terrestrial vegetation (Sellers, 1985). Indices such as the Simple Ratio (SR), Normalized Difference Vegetation Index (NDVI), and Greenness have also been shown to relate to plant abundance variables such as above-ground biomass (Everitt et al., 1989), leaf area index (LAI) (Badhwar et al., 1986; Curran and Williamson, 1987), and canopy cover, and with photosynthetic rate (Sellers, 1985) and absorbed (Asrar et al., 1992) and intercepted (Bartlett et al., 1990) photosynthetically active radiation. Net primary production has been related to the change or integral of SVIs over a growing season (Goward and Dye, 1987). Such relationships are not well established for Mediterranean-type vegetation, but some of these plant properties have been found to be correlated with spectroradiometric data for other semiarid vegetation types (Smith et al., 1990).

Let us assume that for Mediterranean-type vegetation, a satellite-derived SVI or its change through the growing season relate significantly and strongly to the amount of above-ground green plant material, to photosynthetic rate, and/or to seasonal growth. This still leaves two major questions to be answered:

1. How may such an index be exploited to infer differences in the amount or growth of above-ground vegetation matter that may be caused by changing CO_2 concentration or climate?
2. What environmental factors besides CO_2 concentration should be controlled that could also influence the variability of vegetative abundance or growth?

A promising approach for comparing above-ground green plant biomass for different years would be to derive SVI values from multitemporal satellite data acquired at the same time in the growing season and then extract samples (i.e., pixels) from even-age vegetation stands of nearly the same age (relative to each satellite image date). For example, fire history and vegetation stand maps could be used to extract SVI samples from areas corresponding to 20- to 25-year-old stands of mixed chaparral for Landsat MSS data acquired in 1973 and 1993. Presumably, a CO_2 enrichment effect would result in greater production and biomass accumulation for a 20- to 25-year-old stand in 1993, such that corresponding SVI values would be significantly higher. Results from CO_2 chamber enhancements applied to yearling chaparral shrubs showed a significant increase in biomass accumulation occurs at CO_2 levels that are twice current ambient concentrations (Oechel, this volume). However, it is not known whether older stands would accumulate green biomass at significantly higher rates or whether some type of homeostatic adjustment has occurred or would occur after a few years.

If SVIs for Mediterranean-type vegetation are more highly correlated with photosynthetic rate or fraction of absorbed photosynthetically active radiation than with above-ground abundance, as has been found in several studies (Sellers, 1985; Cihlar et al., 1991; Asrar et al., 1992) comparisons of single date, seasonally controlled SVIs from two different years may still provide evidence of CO_2 effects. Oechel (this volume) found that photosynthetic rate of yearling chaparral species increases at higher CO_2 concentrations. The potential problem with photosynthetic rate compared to abundance as an indicator of change is that it may be unstable and particularly responsive to short-term climatic variations. Thus, the short-term variability of an SVI that was more highly correlated with photosynthetic rate than with abundance could be as great as or greater than the interannual variability of vegetation stands.

Whereas above-ground abundance or photosynthetic rate can be estimated from SVIs extracted from a single satellite image, seasonal growth, i.e., net primary production, is estimated by differencing or integrating SVIs from multiple images acquired through the growing season (Goward and Dye, 1987; Running, 1990). Interannual variability due to CO_2 enrichment may be detectable as an increased growth signal in the differenced, e.g., end of season minus beginning of season, or integrated SVI. As with the abundance and photosynthetic rate properties, seasonal growth estimates for each interannual date should be extracted from stands of similar vegetation types and ages to minimize variations due to canopy architecture, size, and density. In studies of southern California chaparral, all three properties have been shown to vary by vegetation type and age (Oechel and Reid, 1984; Black, 1987).

Variations in cumulative insolation, moisture, and nutrients during different growing seasons could also account for differences in biomass accumulation during a given growing season and during the respective interannual intervals, e.g., 20–25 years, and between the differing stands, i.e., intersite variability. Some control for this variability could be achieved by stratifying according to slope/aspect and soil classes, and by selecting a large enough sample to represent most of the site variability between the two sample sets, i.e., between each date. Again, this should be the advantage of spatially integrated and regionally sampled satellite-derived measurements over in situ sampling. The most difficult variable to control would likely be differences in cumulative precipitation between the two time periods. This would suggest that an important criterion for choosing the length and specific years of the interannual intervals would be near equivalence in cumulative precipitation.

Changes in Species Composition

Other potential ecosystem responses to global change that may be detected using multitemporal remote sensing techniques are changes in species or plant form composition within vegetation stands. The assumption here is that through some type of physical control such as fire, terrain, or lithology, a generally persistent mosaic of vegetation stands exists across a Mediterranean-type landscape. However, changes in climate or CO_2 concentration could induce a change in the species or plant form composition within these stands.

Changes in composition as a result of climate change are more likely to occur than by increasing CO_2 concentration. Increases in woody shrub proportions (at the expense of herbaceous and tree cover) might result from a warmer and drier climate. Plants with greater water user efficiencies and deeper roots would be more competitive in such a changing climate. Higher levels of atmospheric CO_2 and/or higher temperatures might favor plants that fix carbon through C_4 over C_3 photosynthetic pathways, but C_3 plants tend to dominate Mediterranean-type ecosystems due to water limitations (Mooney, 1977).

Significant shifts in multispectral radiance signatures for contiguous groupings of pixels within stand boundaries over decadal time intervals might signal changes in the internal composition of vegetation stands. The challenge would be to distinguish compositional changes caused by global changes from those due to interannual differences in vegetation state and abundance, or due to successional vegetation change (Davis et al., 1988). Older vegetation stands would likely be the optimal sites for detecting global change response, since growth and succession are likely to account for most of the compositional change of younger vegetation stands.

Compositional changes within stands could be inferred by applying one of two general types of image processing routines that are generally referred to as "change detection" techniques (Price et al., 1992). In this case, change must be "identified" as well as "detected," meaning that some type of thematic classification routine must be invoked. Postclassification comparison approaches would entail the classification of single or multidate (Lacaze and Debussche, 1984) image data acquired many years apart (Singh, 1989). Thematic categories would be compared for each date on a per-pixel, pixel-block, or vegetation stand basis. The other general approach is multidate classification, where transition sequences (including no-change) are classified directly from combined multidimensional pixel vectors consisting of multispectral radiances from two satellite images. With either approach, the key to success is to identify categorical changes within the persistent and known polygonal boundaries of a vegetation stand (Lacaze et al., 1983). Also, the detail of the classification scheme, i.e., specificity of the vegetation types, used for the image classification must be more specific, e.g., community-level, than that used to delineate the vegetation stands.

An alternative approach to tracking compositional changes in the same vegetation stands through time would be to compare relative composition between different stands of the same age, based on classifications of multidate satellite data. This approach would be similar to that proposed for comparing abundance and growth based on SVIs from similar-aged stands. A large number of stands would need to be sampled to minimize the effects of intersite variability on vegetation type (Lepart and Debussche, 1980). The specific years for the interannual comparison would be chosen so as to minimize differences in cumulative precipitation prior to the specific image dates used for the classification.

The utility of multitemporal (seasonal) Landsat TM data for classifying chaparral vegetation types can be demonstrated with results from a pilot study conducted by Franklin et al. (1991). The objective of the study was to map vegetation composition and ecotones, which could provide a baseline for monitoring changes in composition and shifts in ecotones. The study was conducted at the Camp Pendleton military preserve area which is located 50 km north of San Diego, California (Figure 13-5). This study area consists of hilly to steep terrain covered by coastal sage scrub, chaparral, native and nonnative grasses, and diverse riparian vegetation types.

Multidate Landsat TM Digital data acquired April 1987, June 1990, and September 1988 (representing most of the phenological range for the plants in the study area) were analyzed. The spectral radiance signatures of the major vegetation types in the six reflected radiation wavebands of TM were compared to assess their relative separability on individual and combined dates. The spectral separability is important in that it partially influences whether particular vegetation classes can be distinguished from one another in either visual or computer-assisted classification processes.

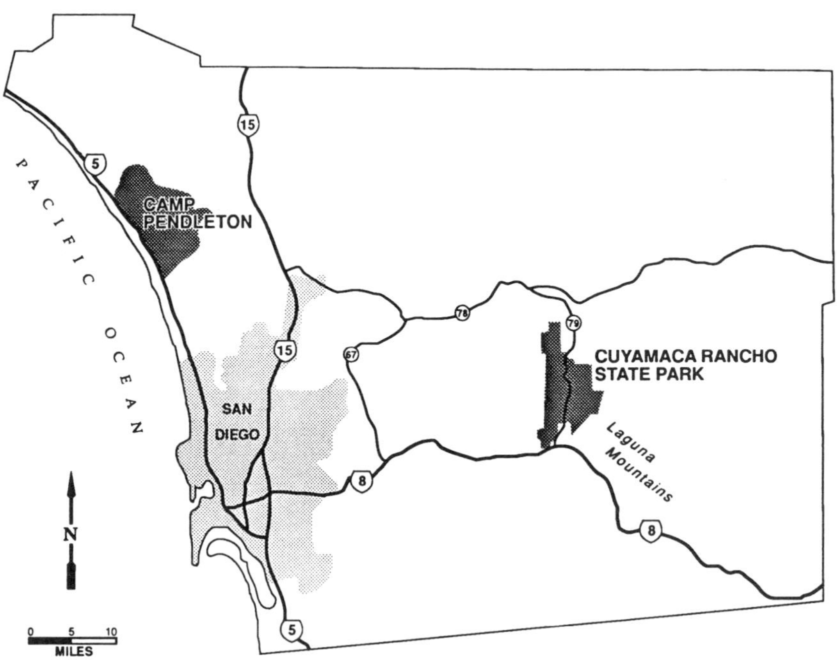

Figure 13-5. Locations of the Camp Pendleton and Cuyamaca State Park test sites in southern California.

Spectral signatures were analyzed for digital number values extracted from training sites, i.e., areas of known vegetation composition.

As expected, the TM4 (NIR) waveband provided the greatest information content on all four dates, and was least correlated with the other wavebands. TM1, TM2, and TM3 were highly correlated, as were TM5 and TM7. The overall separability of the signatures was greatest for the June 1990 image and the least for the April 1987 image. In early spring, most of the vegetation types that are water limited have "greened-up" and "leafed-out," which appears to increase the covariance (i.e., reduces separability) in the NIR (TM4) waveband. By June, many of the drought deciduous and annual vegetation types are dry and senescent and have lost leaves, which explains the minimal signature overlap for the June 1990 image data. Greater linearity and signature overlap relative to the June 1990 data were found for the April 1987 and September 1988 TM data. Evidently, drier and sparser grasses and shrub canopies in early fall result in many of the vegetation types having similar spectral reflectance properties, with much less band-to-band variability. The June 1990/September 1988 multitemporal composite image yielded the greatest separability between vegetation class pairs and was clearly superior to any single date image. Apparently the combination of late spring and early autumn images maximizes contrast between coastal sage scrub and chaparral types, as well as between the chamise and mixed chaparral vegetation classes.

Ongoing research efforts in comparing "per-pixel" and image segmentation approaches to image classification of vegetation stands and habitat types are exploiting the added discrimination capabilities of multitemporal (seasonal) Landsat TM data (Shandley et al., 1992). This should improve the ability to detect changes in within-stand composition and possibly movement of "edges" for interannual assessments of landscape response to global change.

Shifts in "Edge" Location

In ecology and biogeography, "edges" are high-gradient zones that represent a sharp boundary or rapid transition between vegetation, soil, rock, or disturbance units. Similarly, in image processing, "edges" are high-gradient zones of image brightness values that may represent linear features or boundaries between land surface units in a given scene. In scenes containing natural vegetation, image edges commonly correspond to vegetation composition edges. If vegetation edges shift in response to climate change or anthropogenic disturbances, then it may be feasible to identify and determine the rate of movement of shifting edges from multitemporal image data.

Several types of vegetation edges can be defined, based primarily on differences in spatial scale and exogenous controls. At the finest scales are edges between plant community units that may be controlled by localized edaphic and moisture conditions. Moving up in scale, the boundaries between vegetation stands are edges which generally reflect abrupt changes in topography or lithology. Ecotones occur where major changes in topography or climate cause high-gradient transition zones between distinctly different vegetation types or forms. At the coarsest scale are the relatively broad boundaries between biomes.

Besides localized anthropogenic disturbances, changes in temperature and precipitation are most likely to cause vegetation edges to shift through time. Increasing CO_2 could influence some edge shift, but most likely in conjunction with changing climatic variables. Edges associated with the boundaries of plant communities and vegetation stands would likely occur because of differences in species competiveness associated with rising air temperatures and greater moisture variability (Gates, 1990). Ecotone movement in chaparral ecosystems could result from advancing xeric vegetation types with increasing summer air temperatures, or possibly from the advance of mesic vegetation types with increasing CO_2.

Two conditions must be satisfied for movements of vegetation edges to be successfully detected with multitemporal remotely sensed data. First, edges of interest must be detectable with minimum confusion from other scene or sensor phenomena that create high-gradient zones. Second, the movement of the edge must be greater than the ground resolution element and the positional accuracy and uncertainty of the image-to-image registration (Townshend and Justice, 1988).

Edges may be directly detected by applying image segmentation algorithms to digital image data (Corr et al., 1989). For example, Shandley et al. (1992) applied an image segmentation algorithm to Landsat TM data to delineate stands of chaparral and pine-oak forest vegetation. Another approach for delineating vegetation edges is to apply moving window filters or raster-vector conversion routines to the categorical data output from image classifications (Johnston and Bonde, 1989).

Once edges are delineated on each image of a registered, multidate pair, movements could be inferred by manual or automated edge tracking procedures. The detection of false edges and false edge movements could be minimized by comparing adjacent vegetation categories on either side of an edge for consistency between image dates. Another criterion for determining acutal edge shifts is that the edge movements should be *at least* twice the size of the ground resolution element (GRE) of the imagery, e.g., >160 m for Landsat MSS imagery. Otherwise, errors in spatial registration could yield false detections (Townshend and Justice, 1988). Given this minimal displacement criterion and the GRE size of land satellite data, one may conclude that: (1) the utility of multitemporal satellite data for monitoring edge shifts will likely be restricted to movements of major vegetation stands and ecotones, and (2) a long time interval between interannual image acquisitions may be required.

A completely different and potentially more viable approach for detecting edge shifts is to select narrow swaths of image data that are oriented so as to maximize environmental gradients that would likely enhance movements of vegetation boundaries. Swaths or transects of image data could be located along moisture gradients that contain distinctive boundaries between mesic and xeric vegetation types. Rather than classifying multispectral data into vegetation categories, SVI values could be computed, averaged in the across-transect dimension to reduce local variability, and then plotted along transect, as shown in Figure 13-6. The position of inflexion points along the SVI transects could be tracked through time to infer and quantify edge movements pertaining to climate change.

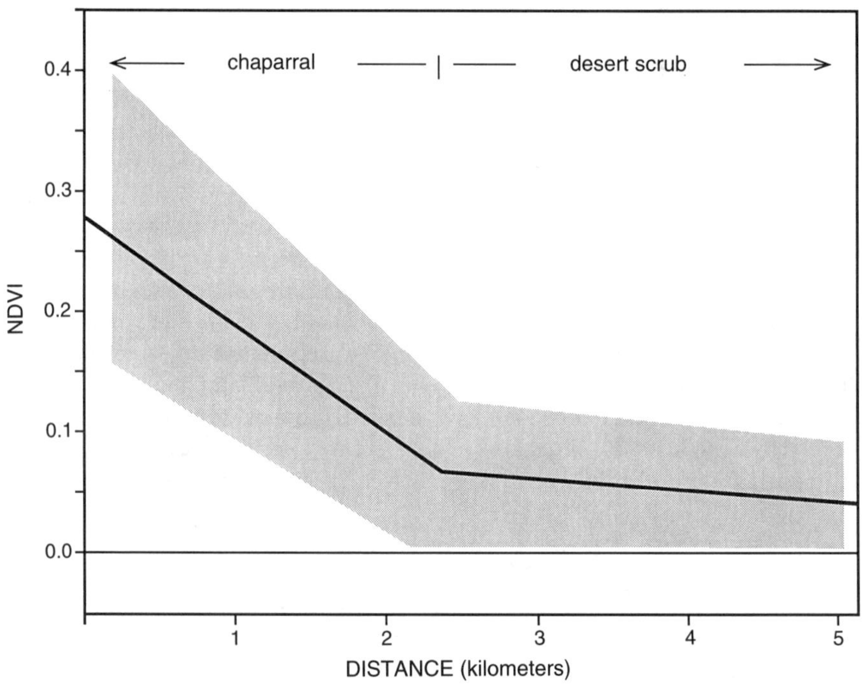

Figure 13-6. NDVI values derived from a Landsat TM image of southern California, east of San Diego. Samples were extracted from a 50-pixel-wide transect trending east-west along a strong aridity gradient. The two line segments represent the least-squares lines on either side of a vegetation "edge," and the envelope around the line segments represents the range for the 50 pixels (that trend north-south) for each point on the transect. (Unpublished data from Allen Hope.)

Changes in Fire Regime

The last of the general types of landscape change that may signal an ecosystem response to global change are temporal variations in fire regime. Since fire is one of the dominant features of Mediterranean-type ecosystems, it is reasonable to expect that changes in CO_2 and/or climate would likely alter the recurrence frequency and behavior of fires, and the nature of recovery from fire (Torn and Fried, 1992; Davis and Michaelsen, this volume). Many aspects of fire can be remotely sensed by satellites, including the location and extent of recently burned areas, amount and condition of fire fuels, and successional state following fires (Cosentino et al., 1981; Burgan and Shasby, 1984; Chuvieco and Congalton, 1989).

The direct effects of global changes pertaining to fire regime in Mediterranean-type ecosystems are likely to pertain to the amount and condition of fuels, i.e., live and dead vegetation, and to stochastic aspects of pertinent climatic variables. Increasing atmospheric CO_2 may be currently altering present or future fire regimes by increasing net primary production and therefore the amount of fuels, by altering water use efficiency and therefore plant moisture, and by modifying species competitiveness of postburn recovery. Anticipated climatic change would also likely have an effect on fule mass and moisture and on postburn plant succession. In addition, an unknown change in humidity, wind velocities, surface moisture, and frequency of lightning strikes during the fire season (whose timing and length could also change) are also likely.

Four aspects of a changing fire regime could be monitored through multitemporal remote sensing. The first two, changes in amount and condition of fuels and post burn succession, could essentially be considered as plant abundance/condition/growth and compositional differences within vegetation stands, respectively (Yool et al., 1985; Lopez et al., 1991).

Stow et al. (1993) conducted a pilot study to examine the information content of multitemporal Landsat TM data for representing the distribution of fire fuels properties of a chaparral and Mediterranean-type forest landscape in southern California. The study was based on archived Landsat and GIS data sets covering the Cuyamaca Rancho State Park in San Diego, County, California (McKinsey, 1988; Figure 13-5). Mixed chaparral shrubland (*Adenostoma* and *Ceanothus* species), pine-oak forested woodland (*Pinus* and *Quercus* species), and grasslands cover the granite soils of the park. The recent fire history of the park has been mapped and digitized. Geometrically and radiometrically registered (Figure 13-3) Landsat TM digital image data acquired in November 1986 (Nov-86) and April 1987 (Apr-87) were "tasseled-cap" transformed (Crist and Cicone, 1984) to create "Brightness" (BR), "Greenness" (GR), and "Wetness" (WT) images for both dates. Seasonal difference images of

BR, GR, and WT were generated by subtracting Nov-86 index values from the Apr-87 values on a pixel-by-pixel basis. Vegetation-stand age and vegetation-aspect classes (from GIS files) were used to stratify BR-GR-WT images for subsequent statistical analysis, as shown in Figure 13-7.

The Stow et al. (1993) study found that BR and GR images provided unique spatial information pertaining to vegetation conditions. Distributions of WT were similar to GR and inverse to BR. Seasonal change in BR was highest for burn and meadow areas, whereas high values of seasonal change in GR corresponded to deciduous oak woodlands, grasslands, and vegetated north-facing slopes. Wetness appeared to be influenced by the effects of shadowing and by recent burns. End-of-season GR for Mixed Chaparral varied with stand age in a manner which was similar to field measurements of total and live standing biomass reported by Black (1987) in mixed chaparral stands less than 50 km away from Cuyamaca State Park. Seasonal changes in GR were generally equal to or slightly greater than beginning of season GR, while BR changes were much smaller than beginning of season BR. Coefficients of variation for BR and GR generally decreased with stand age, which may be explained by the closing of canopies as chaparral stands mature. Differential illumination effects were apparent, as seasonal changes in all three image transforms were highest for northern aspects, due to differential shadowing for the two image dates. Illumination variations must be normalized in future efforts if useable information on vegetation dynamics and global change effects is to be derived. Field-based fuels and canopy property measurements should be compared with concurrent image-derived spectral indices.

A third aspect of changing fire regime could pertain to fire behavior, where variables such as fire intensity and rate of spread could be determined from real-time thermal infrared or passive microwave imaging (Ambrosia and Brass, 1988). Comparisons of remotely sensed fire behavior variables could be made between historical and current or future fire events. Data from numerous events would be required in order to separate situation-specific behavior from longer-term changes in fire regime.

Figure 13-7. Examples of scatterplots (±1 S.D. from mean value) for "Tasseled cap" transformed "Brightness," "Greenness," and "Wetness" and their seasonal changes from Landsat TM data (November, 1986 and April 1987). Samples extracted from representative stands of various chaparral and pine-oak vegetation types. Transformed data from (a) November 1986, (b) April 1987, and differences between the two dates (c) stratified by stand ages. Transformed data from (d) November 1986, (e) April 1987, and differences between the two dates (f) stratified by aspect classes. Labels of the ordinate axis represent stand type (po = pine-oak forest, mxc = mixed chaparral; mc = montane chaparral, cc = *Ceanothus chaparral*), stand age or age interval in years (number; gt = greater than), or aspect (n = north facing, f = flat, s = south facing).

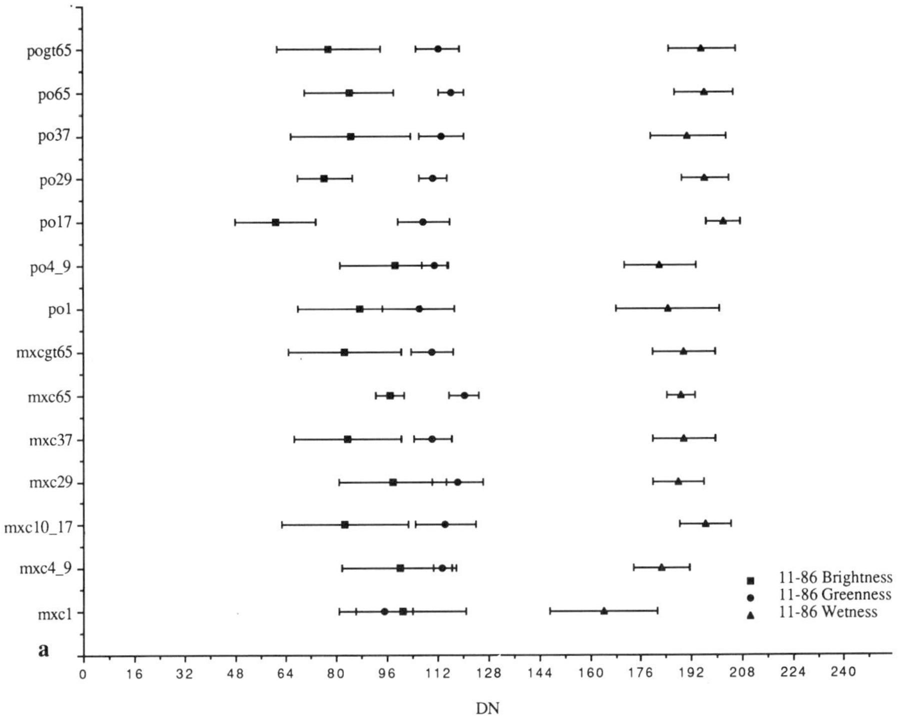
pogt65
po65
po37
po29
po17
po4_9
po1
mxcgt65
mxc65
mxc37
mxc29
mxc10_17
mxc4_9
mxc1
a
0 16 32 48 64 80 96 112 128 144 160 176 192 208 224 240
DN
11-86 Brightness
11-86 Greenness
11-86 Wetness

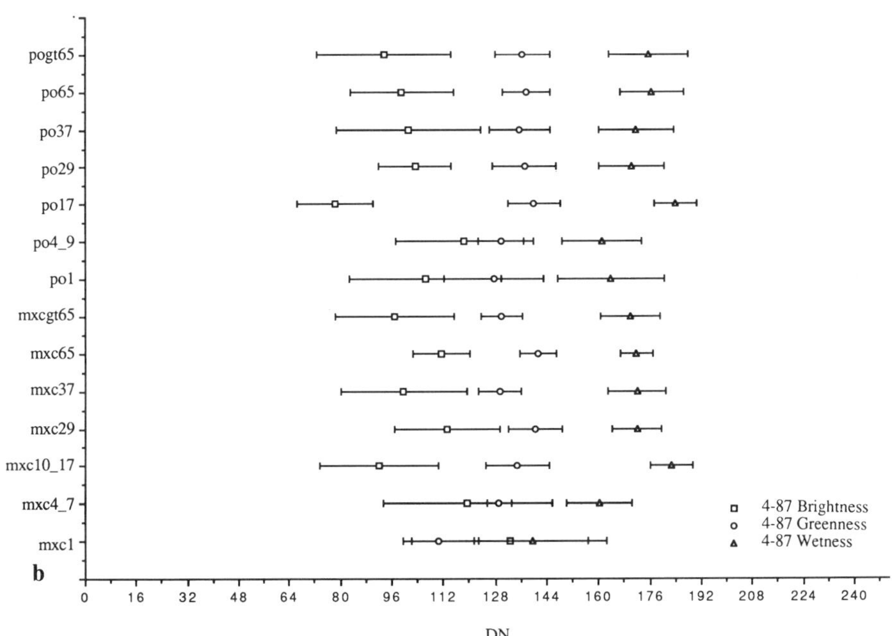
pogt65
po65
po37
po29
po17
po4_9
po1
mxcgt65
mxc65
mxc37
mxc29
mxc10_17
mxc4_7
mxc1
b
0 16 32 48 64 80 96 112 128 144 160 176 192 208 224 240
DN
4-87 Brightness
4-87 Greenness
4-87 Wetness

Continued

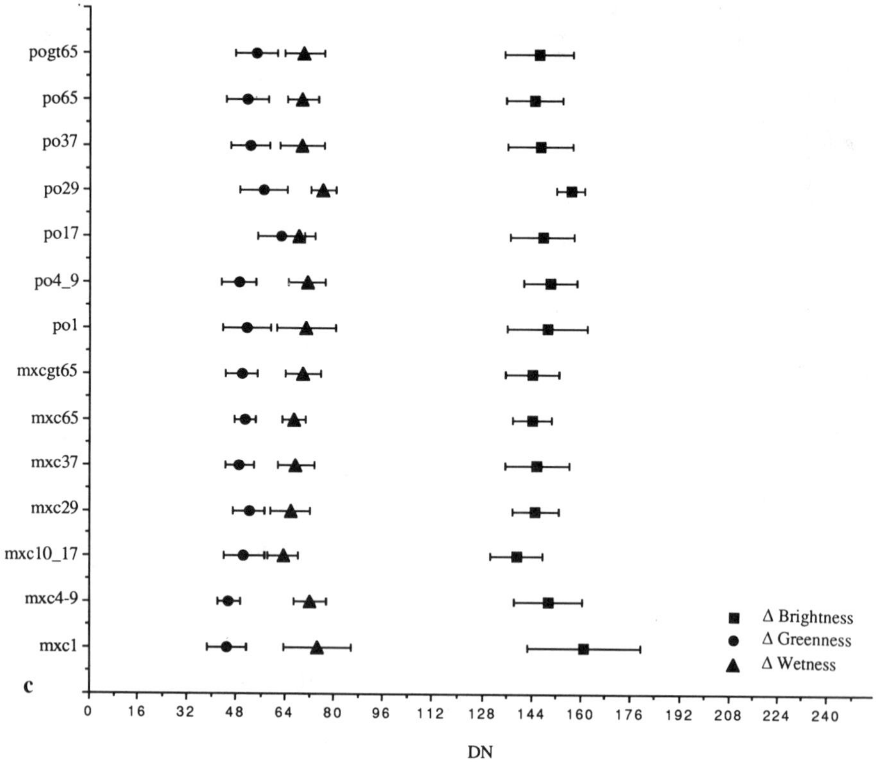
pogt65
po65
po37
po29
po17
po4_9
po1
mxcgt65
mxc65
mxc37
mxc29
mxc10_17
mxc4-9
mxc1
Δ Brightness
Δ Greenness
Δ Wetness
c
0 16 32 48 64 80 96 112 128 144 160 176 192 208 224 240
DN

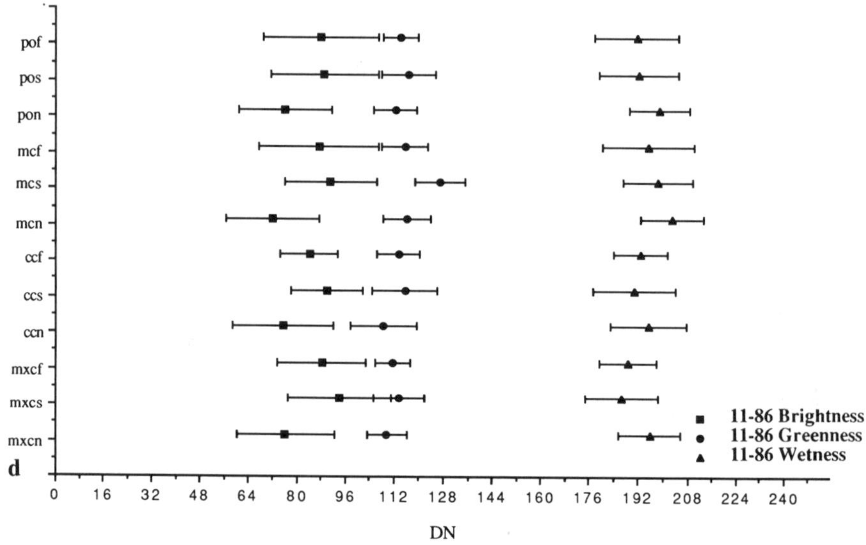
pof
pos
pon
mcf
mcs
mcn
ccf
ccs
ccn
mxcf
mxcs
mxcn
11-86 Brightness
11-86 Greenness
11-86 Wetness
d
0 16 32 48 64 80 96 112 128 144 160 176 192 208 224 240
DN

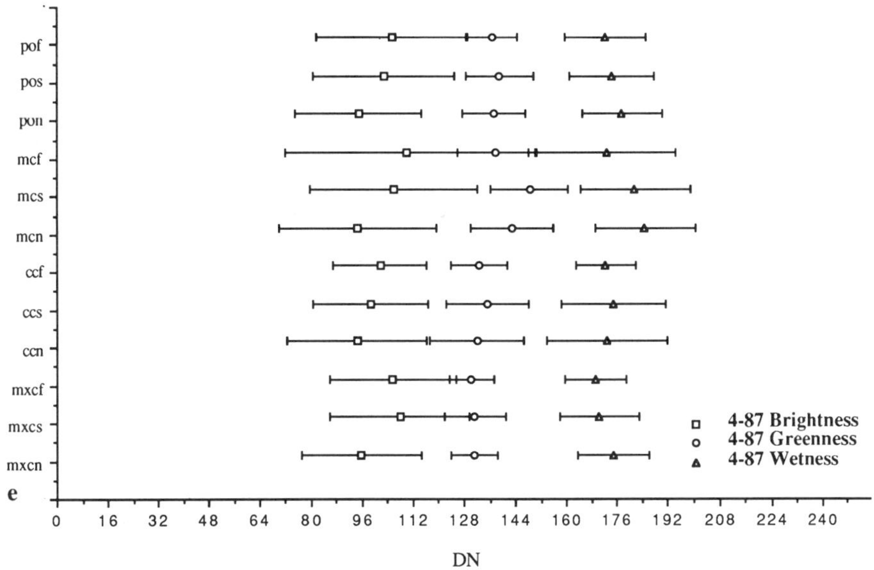

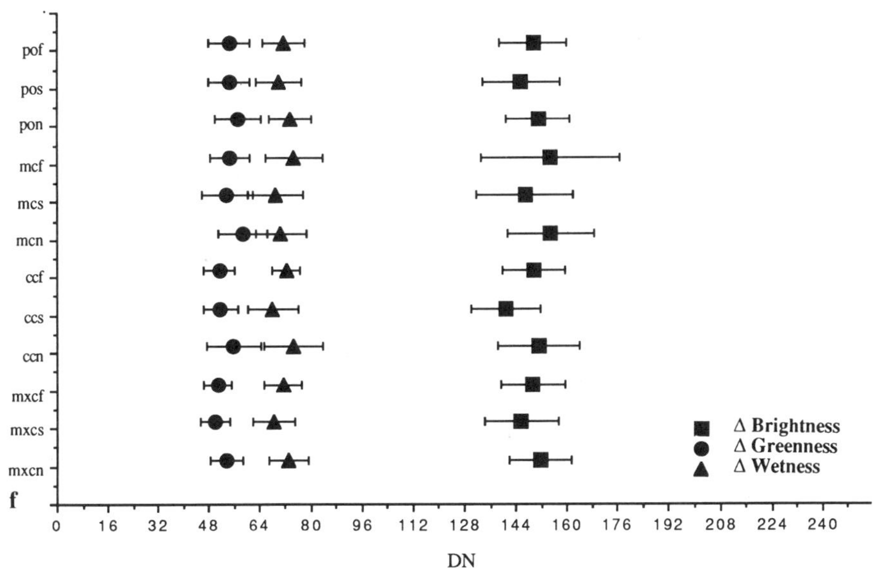

Figure 13-7. *Continued*

The fourth aspect would be the size and recurrence frequency of burn units across the landscape. The boundaries of specific burn events can be mapped from remotely sensed image data acquired within days to a few years of the fire (Chuvieco and Congalton, 1988). Landsat-type image data acquired annually at the end of the fire season should provide an effective monitoring capability. Changes in the size distribution of burn units for historical and more recent time periods could indicate changes in fire regime (Tanaka et al., 1983).

Conclusions

Detecting past effects and monitoring future responses of Mediterranean-type ecosystems to global change via multitemporal remote sensing is a worthwhile, promising, and challenging endeavor. Although the potential effects of global change and the approaches to monitoring such effects are not necessarily unique to Mediterranean-type ecosystems, such ecosystems may be the most responsive and therefore exhibit the earliest signs of an impact (see Oechel in this volume). For example, a CO_2 enrichment effect might be detectable first in Mediterranean-type ecosystems because of the synchronic winter maximum of the seasonal atmospheric CO_2 oscillation and the winter peak in plant production that is fairly unique to these ecosystems.

The major challenge to successfully monitoring global change effects with multitemporal remote sensing is to detect a potentially weak signal related to ecosystem responses relative to possible noise associated with the environment, sensors, and image processing. The signals that have been identified in this chapter as having some potential to be remotely sensed are changes in plant abundance and growth, differences or changes in species composition within vegetation stands, shifts in vegetation "edges," and changes in fire regime. Other potential signals that were not addressed might include changes in plant properties such as moisture and nutrient content, or detection of plant stress due to direct (moisture and temperature; Cohen, 1991b) or indirect (insect infestation; Hope and Stow, 1993) effects. Examples of possible environmental noise are the temporal variability in landscape properties caused by factors other than global change and the temporal variability in atmospheric optical properties. Sensor-related noise could result from drift in radiometric calibration, differences in spectral sensitivity, and inconsistencies in spatial sampling. Artifacts resulting from image processing techniques such as geocoding, spatial registration, resampling, atmospheric correction, spectral transformation, and image classification can also produce noise for multitemporal image analyses.

The major advantages of a remote sensing approach to monitoring ecosystem responses to global change are the ability to efficiently draw a large sample over a large areal extent, and the fact that fine-scale variability can be effectively integrated within a ground resolution element or by pixel aggregation. In relation to in situ measurements, the first advantage means that many more samples can be extracted to capture intersite variability, while the second advantage means that much of the extraneous intrasite variability at the plot level is minimized. In addition, other spatial technologies, such as GIS and spatial statistics, can be integrated with remote sensing techniques to help reduce more transient temporal and/or landscape-scale spatial influences on signals that could relate to global scale effects. Of course there can be a significant cost associated with the production and maintenance of GIS data bases which are rarely extant for Mediterranean-type regions.

Three factors that may limit the viability of detecting historical global change signals in the archived satellite data record, and which must be paid attention to for future monitoring, are sensor calibration, measurements to support atmospheric correction, and data degradation related to specific computer storage media. Even if a CO_2 enrichment effect has already influenced changes in Mediterranean-type vegetation over the last 20 years, it may be undetectable with data in the Landsat archives. Failure to document Landsat MSS calibration coefficients and their temporal drift, and the inability to properly correct for atmospheric effects, may significantly affect uncertainties in measured spectro-radiometric change. The loss of image data stored originally on magnetic, computer-compatible tapes may mean that cloud-free Landsat MSS scenes are not available for times that are critical to capturing important seasonal patterns. Even if historical changes are not retrievable, an effort should be made immediately to establish baseline images that are properly calibrated and archived. Recent programmatic efforts pertaining to the US National Aeronautics and Space Administration's (NASA) Earth Observing System (EOS) and Mission to Planet Earth are steps in the right direction.

The other recommended research thrust is to continue to test the relationship between vegetation properties (as indicators of ecosystem change) and spectro-radiometric measurements/transformations. Multitemporal remote sensing of properties pertaining to any or all of the four proposed indicators of global change should be explored. Spatial variations in these properties should be used as analogs for temporal variations at fixed locations, to test the sensitivity of spectro-radiometric indices and classifiers. Also, the magnitude of intraannual (seasonal) variations in the properties and their spectro-radiometric surrogates needs to be quantified so as to determine the feasibility of isolating interannual fluctuations.

Acknowledgments. I am grateful to Walt Oechel, Allen Hope, Frank Davis, and José Moreno for sharing their insights and knowledge. Janet

Franklin, Heather Pray, Kathy Prosser, and David McKinsey provided image processing support and/or data for the two pilot studies.

References

Aase JK, Millard JP, Brown BS. 1986. Spectral radiance estimates of leaf area and leaf phytomass of small grains and native vegetation. IEEE Trans Geosci Remote Sens 24:685–691.

Adeniyi PO. 1985. Digital analysis of multitemporal Landsat data for land-use/land-cover classification in a semi-arid area of Nigeria. Photogrammetric Eng Remote Sens 51:1761–1774.

Allum JAE, Dreisinger BR. 1987. Remote sensing of vegetation change near Inco's Sudbury mining complexes. Int J Remote Sens 8:399–416.

Ambrosia V, Brass J. 1988. Thermal analysis of wildfire effects on global ecosystem cycling. Geocarto Int 3:29–40.

Asrar G, Myneni RB, Choundhury BJ. 1992. Spatial heterogeneity in vegetation canopies and remote sensing of absorbed photosynthetically active radiation: a modeling study. Remote Sens Environm 41:85–103.

Badhwar G, MacDonald RB, Mehta NC. 1986. Satellite-derived leaf-area-index and vegetation maps as input to global carbon cycle models—a hierarchical approach. Int J Remote Sens 7:265–281.

Bartlett D, Whiting GJ, Hartman JM. 1990. Use of vegetation indices to estimate intercepted solar radiation and net carbon dioxide exchange of a grass canopy. Remote Sens Environm 30:115–128.

Black CH. 1987. Biomass, nitrogen, and phosphorus accumulation over a southern California fire cycle chronosequence. Plant Response to Stress-Functional Analysis in Mediterranean Ecosystems, J Tenhuenen, F Catarino, O Lange, and W Oechel, eds. NATO ASI Series, G15:445–458.

Burgan RE, Shasby MB. 1984. Mapping broad-area fire potential from digital fuel, terrain, and weather data. J Forestry 82:228–231.

Caselles V, López-García MJ. 1989. An alternative approach to estimate atmospheric correction in multitemporal studies. Int J Remote Sens 10: 1127–1134.

Chavez P. 1988. An improved dark-object subtraction technique for atmospheric scattering correction of multispectral data. Photogrammetric Eng Remote Sens 24:459–479.

Chuvieco E, Congalton RG. 1988. Mapping and inventory of forest fires from digital processing of TM data. Geocarto Int 4:41–53.

Chuvieco E, Congalton RG. 1989. Application of remote sensing and geographic information systems to forest fire hazard mapping. Remote Sens Environm 29:147–159.

Cihlar J, St-Laurent L, Dyer, JA. 1991. Relation between the normalized difference and ecological variables. Remote Sens Environm 35:279–298.

Cohen WB. 1991a. Response of vegetation indices to changes in three measures of leaf water stress. Photogrammetric Eng Remote Sens 57:195–202.

Cohen WB. 1991b. Chaparral vegetation reflectance and its potential utility for assessment of fire hazard. Photogrammetric Eng Remote Sens 57:203–207.

Corr DG, Tailor AM, Cross A, Hogg DC, Lawrence DH, Mason DC, Petrou M. 1989. Progress in automatic analysis of multitemporal remotely-sensed data. Int J Remote Sens 10:1175–1195.

Cosentino M, Woodcock C, Franklin J. 1981. Scene analysis for wildland fire-fuel characteristics in a mediterranean climate. Proceedings 15th International Symposium on Remote Sensing of Environment. ERIM, Ann Arbor, MI, pp. 635–646.

Crist EP, Cicone RC. 1984. A physically-based transformation of Thematic Mapper data—the TM tasseled cap. IEEE Trans Geosci Remote Sens GE22: 256–263.

Curran PJ, Hay AM. 1986. The importance of measurement error for certain procedures in remote sensing at optical wavelengths. Photogrammetric Eng Remote Sens 52:229–241.

Curran PJ, Williamson HD. 1987. Airborne MSS to estimate GLAI. Int J Remote Sens 8:57–74.

Davis FW, Michaelsen J. 1995. Sensitivity of fire regime in chaparral ecosystems to global climate change. This volume.

Davis FW, Hickson DE, Odion DC. 1988. Composition of maritime chaparral related to fire history and soil, Burton Mesa, Santa Barbara County, CA. Morono 35(3):69–195.

Duggin MJ. 1980. The field measurement of reflectance factors. Photogrammetric Eng Remote Sens 46:229–241.

Duggin MJ, Sakhavat H, Lindsay J. 1985. Systematic and random variations in Thematic Mapper digital radiance data. Photogrammetric Eng Remote Sens 51:1427–1434.

Everitt JH, Escobar DE, Richardson AJ. 1989. Estimating grassland phytomass production with near-infrared and mid-infrared spectral variables. Remote Sens Environm 30:257–261.

Frank TD. 1988. Mapping dominant vegetation communities in the Colorado Rocky Mountain front range with Landsat TM and digital terrain data. Photogrammetric Eng Remote Sens 54:1727–1734.

Franklin J, Stow D, Prosser K, Soreco M. 1991. An examination of the utility of satellite imagery for mapping vegetation in San Diego County. Final Report to the San Diego Association of Governments, San Diego, CA, p. 27.

Gates DM. 1990. Climate change and the response of forests. Int J Remote Sens 11:1095–1107.

Goward SN, Dye DG. 1987. Evaluating North American net primary productivity with satellite data. Adv Space Res 7:165–174.

Graetz RD. 1990. Remote sensing of terrestrial ecosystem structure; an ecologist's pragmatic view. In: Remote Sensing of Biosphere Functioning, RJ Hobbs and HA Mooney, eds. Springer-Verlag, New York, pp. 5–30.

Hall FG, Strebel DE, Nickeson JE, Goetz SJ. 1991. Radiometric rectification: toward a common radiometric response among multi-date multisensor images. Remote Sens Environm 35:11–27.

Hanes TL. 1977. California chaparral. In: Terrestrial Vegetation of California, MG Barbour and J Majors, eds. John Wiley & Sons, New York, pp. 417–469.

Haralick RM. 1985. Image Segmentation Techniques. Comput Vision Graphics Image Proc 29:100–132.

Hobbs RJ. 1990. Remote sensing of spatial and temporal dynamics of vegetation. In: Remote Sensing of Biosphere Functioning, RJ Hobbs and HA Mooney, eds. Springer-Verlag, New York, pp. 203–220.

Holben BN, Justice CO. 1979. Evaluation and Modeling of the Topographic Effect on the Spectral Response from Nadir Pointing Sensors. NASA Technical Memorandum 80305, p. 19.

Hope AS, Stow DA. 1993. An analysis of tree mortality in southern California using high spatial resolution remotely sensed spectral radiances: a climatic change scenario. Landscape Urban Plan 24:87–94.

Hugli J, Frei W. 1983. Understanding anisotropic reflectance in mountainous terrain. Photogrammetric Eng Remote Sens 49:671–683.

Hutchinson C. 1982. Techniques for combining Landsat and ancillary data for digital classification improvement. Photogrammetric Eng Remote Sens 48: 123–130.

Johnston C, Bonde J. 1989. Quantitative analysis of ecotones using a GIS. Photogrammetric Eng Remote Sens 56:1643–1647.
Justice CO, Markham BL, Townshend JRG, Kennard RL. 1989. Spatial degradation of satellite data. Int J Remote Sens 10:1539–1561.
Keeling CD, Bacastow RB, Whorf TP. 1982. Measurements of the concentration of carbon dioxide at Mauna Loa Observatory, Hawaii. In: Carbon Dioxide Review, WC Clark, ed. Oxford University Press, New York, pp. 377–385.
Lacaze B, Debussche G. 1984. Integration of multiple thematic data with Landsat data: some result about the feasibility of Mediterranean land cover inventories. Proceedings Integrated Approaches in Remote Sensing, European Space Agency, Guilford, UK, pp. 31–39.
Lacaze B, Debussche G, Jardel J. 1983. Spatial variability of Mediterranean woodlands as deduced from Landsat and ground measurements. Proceedings of the International Geoscience and Remote Sensing Symposium II(4):1–5.
Lacaze B, Debussche G, Khelfa A, Mazurek H, Rambal S. 1984. A comparison of Landsat-MSS, HCMM and NOAA-AVHRR features for identifying dominant vegetation patterns in Mediterranean areas. Proceedings of the Eighteenth International Symposium on Remote Sensing of Environment, Paris, France, October 1–5, pp. 787–802.
Laureau J. 1991. De-hazing Landsat Thematic Mapper images. Photogrammetric Eng Remote Sen 57:1297–1302.
Lepart J, Debussche M. 1980. Information efficiency and regional constellation of environmental variables. Vegetatio 42:85–91.
Li X, Strahler A. 1985. Geometrical-optical modeling of a conifer forest canopy. IEEE Trans Geosci Remot Sens 23:705–721.
López S, González F, Llop R, Cuevas JM. 1991. An evaluation of the utility of NOAA AVHRR images for monitoring forest fire risks in Spain. Int J Remote Sens 12:1841–1851.
Malila WA. 1985. Comparison of the information contents of Landsat TM and MSS Data. Photogrammetric Eng Remote Sens 51:1449–1457.
Marvin J, Labovitz M, Wolfe R. 1987. Derivation of a fast algorithm to account for distortions due to terrain in earth-viewing satellie sensor images. IEEE Trans Geosci Remote Sens 25:2.
McKinsey D. 1988. The priority ranking for prescribed burning in the Cuyamaca Rancho State Park using a geographic information system. Proceedings of the GIS/LIS Conference, San Antonio, TX 2:961–970.
Mooney HA. 1977. Primary production in Mediterranean-climate regions. In: Terrestrial Vegetation of California, MG Barbour and J Majors, eds. John Wiley & Sons, New York, pp. 249–255.
Moran MS, Jackson RD, Hart GF, Slater PN, Bartell RJ, Biggar SF, Gellman DI, Santer RP. 1990. Obtaining surface reflectance factors from atmospheric and view angle corrected SPOT-1 HRV data. Remote Sens Environ 32: 203–214.
Murphy JM, Ahern FJ, Duff PF, Fitzgerald AJ. 1985. Assessment of radiometric accuracy of Landsat-4 and Landsat-5 Thematic Mapper data products from Canadian production systems. Photogrammetric Eng Remote Sens 51: 1359–1369.
Musick BH, Pelletier RE. 1986. Response of some Thematic Mapper band ratios to variation in soil water content. Photogrammetric Eng Remote Sens 52: 1661–1668.
Nelson R. 1985. Reducing Landsat MSS scene variability. Photogrammetric Eng Remote Sens 51:583–593.
Nelson R, Grebowsky G. 1982. Evaluations of temporal registration of Landsat scenes. Int J Remote Sens 3:45–50.

Novak K. 1992. Rectification of digital imagery. Photogrammetric Eng Remote Sens 58:339–344.
Oechel WC, Reid CD. 1984. Photosynthesis variability along an elevational gradient in the chaparral. Proceedings of the Mediterranean Bioclimatology Symposium, Bull Soc Bot France, Actual Bot 131:399–409.
Pech RP, Davis AW, Lamacraft RR, Graetz RD. 1986. Reflectance modelling and the derivation of vegetation indices for an Australian semi-aird shrubland. Int J Remote Sens 7:1729–1750.
Price JC. 1992. Estimating vegetation amount from visible and near infrared reflectance. Remote Sens Environm 41:29–34.
Price KP, Pyke M, Mendes DAL. 1992. Shrub dieback in a semiarid ecosystem: the integration of remote sensing and geographic information systems for detecting vegetation change. Photogrammetric Eng Remote Sens 58:455–463.
Rambal S, Lacaze B, Winkel T. 1990. Testing an area-weighted model for albedo or surface temperature of mixed pixels in Mediterranean woodlands. Int J Remote Sens 11:1495–1499.
Ripple WJ. 1986. Spectral reflectance relationships to leaf water stress. Photogrammetric Eng Remote Sens 52:1669–1675.
Robinove CJ. 1982. Computation with physical values from Landsat digital data. Photogrammetric Eng Remote Sens 48:781–784.
Running SW. 1990. Estimating terrestrial primary productivity by combining remote sensing and ecosystem simulation. In: Remote Sensing of Biosphere Functioning, RJ Hobbs and HA Mooney, eds. Springer-Verlag, New York, pp. 65–86.
Schott J, Salvaggio C, Volchok WJ. 1988. Radiometric scene normalization using pseudoinvariant features. Remote Sens Environm 26:1–16.
Sellers PJ. 1985. Canopy reflectance, photosynthesis and transpiration. Int J Remote Sens 6:1335–1372.
Shandley J, Franklin J, White T, Rechel J. 1992. Testing a segmentation procedure for defining regions in a digital image corresponding to natural vegetation stands. Proceedings of the GIS/LIS Annual Conference and Exposition, San Jose, CA, November 10–12, pp. 700–708.
Singh A. 1989. Digital change detection techniques using remotely-sensed data. Int J Remote Sens 10:989–1003.
Singh SM. 1988. Estimation of multiple reflection and lowest order adjacency effects on remotely-sensed data. Int J Remote Sens 9:1433–1450.
Smith MO, Ustin SL, Adams JB, Gillespie AR. 1990. Vegetation in deserts: I. A regional measure of abundance from multispectral images. Remote Sens Environm 31:1–26.
Spanner MA, Pierce LL, Peterson DL, Running SW. 1990. Remote sensing of temperate coniferous forest leaf area index. The influence of canopy closure, understory vegetation and background reflectance. Int J Remote Sens 11: 95–111.
Stow D, Burns B, Hope A. 1989. Mapping arctic tundra vegetation types using digital SPOT/HRV-XS Data: a preliminary assessment. Int J Remote Sens 10:1451–1457.
Stow D, Hope A, McKinsey D, Pray H. 1993. Deriving dynamic information on fire fuel distributions in southern California chaparral from remotely sensed data. Landscape Urban Plan 24:113–127.
Strahler AH, Woodcock CE, Smith JA. 1986. On the nature of models in remote sensing. Remote Sens Environm 20:121–139.
Tanaka S, Kimuram H, Suga Y. 1983. Preparation of a 1:25,000-scale Landsat map for assessment of burnt area on Etijma Island. Int J Remote Sens 4(1): 17–31.

Torn MS, Fried JS. 1992. Predicting the Impacts of global warming on wildland fire. Climatic Change 21:257–274.

Townshend JRG, Justice CO. 1988. Selecting the spatial resolution of satellite sensors required for global monitoring of land transformations. Int J Remote Sens 9:187–236.

Watson R, Rohde H, Oeschenger H, Siegenthaler U. 1990. In: Climate Change the IPCC Scientific Assessment, JT Houghton, GJ Jenkins, and JJ Ephraums, eds. Cambridge University Press, Cambridge, UK, p. 365.

Wessman C. 1990. Evaluation of canopy biochemistry. In: Remote Sensing of Biosphere Functioning, RJ Hobbs and HA Mooney, eds. Springer-Verlag, New York, pp. 135–156.

Wessman C, Aber J, Peterson D, Melilo JM. 1988. Remote sensing of canopy chemistry and nitrogen cycling in temperate forest ecosystems. Nature 335: 154–156.

Williamson HD. 1989. Reflectance from shrubs and under-shrub soil in a semiarid environment. Remote Sens Environm 29:263–271.

Yool SR, Eckhardt DW, Estes JE, Cosentino MJ. 1985. Describing the brushfire hazard in southern California. Ann Assoc Am Geographers 75(3):417–430.

14. Microbe–Plant Interactions in Mediterranean-Type Habitats: Shifts in Fungal Symbiotic and Saprophytic Functioning in Response to Global Change

Michael F. Allen, Sherri J. Morris, Fred Edwards, and Edith B. Allen

In any terrestrial ecosystem, the major allocation of carbon and the largest carbon sink is into components in soils responsible for the acquisition of nutrients and water. Although generally unappreciated, soil microbes are the dominant consumers of carbon. Mycorrhizal fungi are estimated to be the largest consumer group because of their large mass and direct access to the host carbon (Figure 14-1). Both these organisms and all others ultimately end up in decomposer mass, with most plant mass going directly to decomposers without passing through animals. Mediterranean-type habitats are semiarid regions that accumulate significant quantities of carbon below ground (e.g., Kummerow et al., 1978). Because of the arid conditions and the sclerophyllous nature of much of the plant tissue, decomposition tends to be very low. Therefore, the dynamics of microbes and their responses to change in the global environment are critical to predicting changes in ecosystem processes that will affect the regions of interest.

CO_2 is both a prerequisite for primary production and the product of decomposition. The fixation, utilization, and release of carbon provide the common currency that binds all components of the ecosystem together. As any one pool of CO_2 changes, the results will be noticed throughout the system. We project three scenarios (not mutually exclusive) that could result with increasing atmospheric CO_2. For example, if photosynthesis is increased, the carbon/nutrient ratio increases resulting in

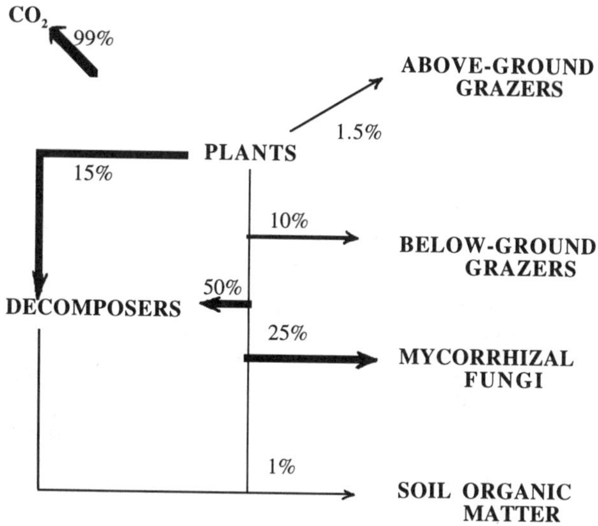

Figure 14-1. A generalized model of carbon flow in ecosystems (from Allen, 1991).

potential decreases in decomposition due to a nitrogen deficiency, but potentially increasing mycorrhizal nutrient acquisition because of the nutrient deficiency. CO_2 may not stimulate production but may rather, as a greenhouse gas, increase temperature and alter precipitation patterns. This could increase decomposition and reduce mycorrhizal development. Then again, because of a homeostatic adjustment, no measurable change in ecosystem processes may result. The only attempt to assess an entire system clearly demonstrated that there were important feedback interactions between populations that resulted in complex interactions not predicted by scaling from physiological responses of plants (Korner and Arnone, 1992). Clearly, whatever the outcome, we have little understanding of the ecosystem processes that can be affected by global environmental change.

Plants are the first component directly impacted by global change in most terrestrial ecosystems (e.g., Oechel and Strain, 1985). However, because plants are dependent on soil organisms to extract most of the critical resources, a feedback loop rapidly forms that alters the resource base and accessibility (e.g., Schlesinger et al., 1991). For this reason, Field et al. (1992) suggested that understanding the resource base was critical to predicting the responses of primary producers. Thus, indirect effects of microbial responses to changing plant dynamics may be the critical alterations associated with changing global conditions.

Microbial–Plant Interactions in Mediterranean-Type Habitats

Saprophytic Activity

When attempting to understand the dynamics of soil microbes in Mediterranean-type (or any type) habitats, it is important to characterize the types of organisms catalyzing critical ecosystem processes. Decomposition is really two separate processes that together regulate nutrient retention and availability, immobilization, and mineralization. The first process in decomposition is the mineralization of carbon and the immobilization of nutrients; the second process is the mineralization of nutrients. The first process is primarily catalyzed by soil fungi and bacteria, the second by bacteria and soil animals. Ideally, carbon should be mineralized and nutrients immobilized until the plants either directly or via their mycorrhizal associates are able to take the mineralized elements up. Thus, the importance of efficient carbon mineralization while immobilizing scarce elements in decomposer mass followed by uptake into plant mass regulates the ability of the system to retain nutrients in a closed system or lose them resulting in system degradation or retrogressive succession.

Mediterranean-type habitats are characterized by plant communities ranging from highly sclerophyllous shrublands to savannas. What this means is that the saprophytic organisms must deal with substrates ranging from material with a high proportion of lignified material (shrubs) to those with a low lignin content and higher cellulose and hemicellulose content (grasses). The higher lignin contents generally tend to be decomposed by fungi that have a slower growth rate, complex enzyme activity, and a high capacity for nutrient immobilization. Simpler materials tend to be readily decomposed by bacteria and "sugar fungi," organisms with rapid growth rates and a low capacity for immobilizing nutrients. Decomposer organisms also are influenced by the accumulation of organic matter. In systems with higher organic material, fungi tend to predominate, whereas when organic matter is reduced, decomposition is predominantly bacterial. Therefore, the ratio of fungal to bacterial predominance is dependent on the substrate input and, in turn, affects the immobilization of critical nutrients (Ingham et al., 1989).

The role of species diversity in below-ground processes has been virtually ignored in the ecosystem literature. However, the data on the ability of microbes to adjust to the prevailing conditions and substrates are rather remarkable. Two factors determine the relative adaptability of the soil biota: diversity and plasticity. Neither has been adequately assessed in Mediterranean-type ecosystems. The closest relevant information may be for cold desert shrublands. Christensen (1981) noted that only after 1,500 isolates did the fungal species increment curve begin to level out in a sagebrush community, and we found that after 375 isolates, no leveling had occurred in 25 soil samples taken within a 4 m^2 area of an *Artemisia-Agropyron* site (Allen, 1985). Estimates of 15,000 to 40,000 species of prokaryotes per gram soil have been made based on the DNA diversity (Atlas and Bartha, 1993). No such attempts at estimating the microbial diversity have been made for Mediterranean-type ecosystems. No attempts at studying the ability of saprophytic microbes to utilize differing substrates have been performed for Mediterranean-type ecosystems. However, studies of other biomes have shown remarkable plasticity. Flanagan and Scarborough (1974) demonstrated that the microfungi from the Alaskan tundra were capable of utilizing a wide diversity of carbon compounds, and simple lists of the carbon that can be used by any number of fungi show wide variability (Domsch et al., 1980).

The amount of saprophytic hyphae in soil can be important. We have found generally 50 to 300 m of living hyphae per gram soil in both coastal sage and chaparral sites. This represents about 4 to 20 mg of living fungal tissue per gram of soil.

Mediterranean-type habitats are also characterized by relatively small seasonal variation in temperature coupled with large seasonal variations in precipitation. This results in rather stable conditions in soil temperature and large variations in soil moisture. As a general rule, fungi are more tolerant of dry conditions than are bacteria. Moreover, when moisture is adequate, bacterial turnover is very rapid, resulting in nutrient loss. Alternatively, fungi tend to grow slower and immobilize nutrients into the hyphal mass.

The high lignin activity coupled with low moisture results in extremely slow decomposition and nutrient release. Phosphorus turnovers estimated in hundreds of years may occur, and nitrogen turnovers of decades are possible. Nevertheless, decomposition provides a large fraction of the regrowth needs. Schlesinger (1985) suggested that in a California chaparral, most of the nitrogen needs were met by decomposition but that phosphorus may be limiting. We have found that in an oak-annual grass savannah, soil phosphorus levels tend to be quite high and most of the phosphorus is inorganic; there is little indication that phosphorus is immobilized in dead tissue (Cannon, 1993). In a northern California annual grassland, virtually all of the bound nitrogen turned over within a single growing season (Jackson et al., 1989). In the arid Cape region of South Africa, Mitchell et al. (1986) noted that fire probably accounts for the majority of nutrient release.

Mycorrhizal Activity

Mycorrhizal activity in Mediterranean-type habitats is highly variable. Unlike above-ground activity, the few data available suggest that there is no convergence in below-ground structure, although the largest fraction of plants form vesicular-arbuscular (VA) mycorrhizae in all regions. For example, in South Africa there appear to be no ectomycorrhizal plants and there is an unusually high proportion of nonmycotrophic species in the Proteaceae, along with a few ericoid mycorrhizae (e.g., Allsopp, 1992). The vast majority of species form VA mycorrhizae. In Australia, there is a high fraction of ectomycorrhizal plants with plants that form ericoid, orchid and nonmycotrophic perennials (Lamont, 1984; McGee, 1986). However, there are reports that the "nonmycotrophic" plants, primarily the Proteaceae, can form mycorrhizae under appropriate conditions (Bellgard, 1991). Interestingly, in these extremely nutrient deficient soils, Epacridaceae form ericoid mycorrhizae, virtually identical morphologically to the mycorrhizae of plants in the Ericaceae formed in bogs in northern latitudes where nutrients are immobilized in organic matter (Read, 1992). No arbutoid mycorrhizae have been reported from these areas. In California, the pattern appears quite different. At the SDSU Sky Oaks and Santa Margarita Ecological Reserve, there is a relatively high proportion of ectomycorrhizal oaks and pines and the ericaceous plants form arbutoid mycorrhizae (Kummerow and Borth, 1986; Allen, 1991). (Molina et al. [1992] consider arbutoid mycorrhizae to be a form of ectomycorrhizal association.) No ericoid mycorrhizae appear to be found in the American Mediterranean-type habitats (The early report of ericoid mycorrhizae in *Arctostaphylos* by Kummerow and Borth [1986] was based on the old distinctions. It is now classified as arbutoid because of the hyphal mantle and only limited cellular penetration). The only nonmycotrophic species appear to be exotic annual weeds (Table 14-1). Interestingly, our observations suggest that all of the roseaceous plants form VA mycorrhizae (Allen, 1991), whereas in northern California, ectomycorrhizal *Adenostoma* have been found (Horton and Parker, 1992). All of the remaining shrubs that have been observed appear to be VA mycorrhizal (Allen, 1991). Mediterranean ecosystems are known for their high incidence of ectomycorrhizal species such as oaks and pines as well as common VA mycorrhizal grasses and shrubs similar to those of southern California (e.g., Puppi and Tartaglini, 1991). One difference may be the appearance of ericoid mycorrhizae in *Erica* in burned (but not unburned) sites in Italy. This possible pattern deserves further study.

This pattern is probably related to soil P dynamics and soil organic matter. In South Africa, the Proteaceae appear to predominate in soils with higher phosphorus and high fire frequency which may preclude the accumulation of an organic base inadequate to support ectomycorrhizal fungi (Allsopp, 1992). In western Australia, the precipitation is relatively high. There is little organic matter buildup, as decomposition probably proceeds rather rapidly. In southern California, soil phosphorus tends to be high in the newer soils and the organic matter accumulates rather rapidly due to the rapid growth in the cool precipitation period and little

Table 14-1. Mycorrhizal Status of Plants of Mediterranean-Type Vegetation

Location	Community	Ecto M[a] % of sp	VAM % of sp	NM % of sp	Ericoid % of sp	Other % of sp	Avail P µg/g so:
[b]S. Africa	Lowland Fynbos	0	72	23	<1	3	1.4
	Renoster-veld	0	77	18	0	5	0.8
	Strand-veld	0	64	27	0	9	40
[c]Australia	Semiarid	33	83	9	0	1	ND
[d]USA	SMER	6	77	14	0	2	15–40
[d]USA	Sky Oaks	9	84	8	0	1	15–30

[a] Many species can form both ectomycorrhizal and VA mycorrhizal associations.
[b] Data from N. Allsopp (1992).
[c] Data from P. McGee (1986).
[d] Data compiled by M. Allen.

decomposition during the warmer dry summer. This may support the greater ectomycorrhizal activity (Table 14-1). Unfortunately, no soils data were presented in the Australian paper, but those soils also tend to be extremely phosphorus deficient. It is clear, however, that despite the observed "convergence" in above-ground morphology and function of plants in Mediterranean-type ecosystems, few similarities exist below ground. This is probably because of the high diversity in soils ranging from the very old and phosphorus-deficient South African and Australian soils to the younger and relatively high-phosphorus Mediterranean and California soils.

Plants in Mediterranean-type habitats also form mycorrhizal associations with a wide variety of fungi, the extent of which is poorly known. Truffles have been prized mycorrhizal fungi in Europe since the early Greek and Roman mycologists. Only recently has the taxonomy of the truffle fungi in Australia been studied, with a surprising diversity being found (J. Trappe, personal communication). In two southern California field sites encompassing shrub and savanna vegetation, we have found all major genera of VA mycorrhizal fungi and more than 15 species. We have also found numerous ectomycorrhizal fungi including *Pisolithus tinctorius*, *Amanita ocreata*, *Boletus satanus*, an undescribed species of *Hydnotryopsis*, and *Hysterangium separabile* associated with native oaks. *Hydnangium carneum* (that probably immigrated from Tasmania) and *Laccaria fragilis* (possibly from Australia) were found with the eucalyptus (M. Allen et al., 1994). Because of the wide diversity in soils, microsites, plants, and mycorrhizal types, Mediterranean-type habitats may have the greatest diversity in mycorrhizal fungi of any vegetation type in the world.

The distribution of fungi is a function of each individual species. For example, we surveyed the VA mycorrhizal fungi associated with *Artemisia tridentata* ssp. *tridentata* at 89 sites around the Great Basin of western North America (E. Allen et al., 1994). An ordination of the sites based on the species composition showed a distinct grouping of the fungi (E. Allen et al., in preparation). There was a distinct north-south gradient and samples within sites were highly clustered. These fungi are considered promiscuous, forming mycorrhizae with any genetically compatible plant. Within the ectomycorrhizal plants, some have broad host ranges (such as *P. tinctorius*, which we have found associated with oaks, pines, and eucalyptus at our southern California sites), while others appear to be relatively specialized (e.g., *B. satanus* and a new species of *Hydnotryopsis* that we have found only associated with valley oak in southern California) (Molina et al., 1992).

Functionally, the mycorrhizal fungi in Mediterranean-type habitats is also highly diverse. Although there is a suite of mycorrhizal benefits, including enhanced water and nutrient status as well as enhanced photosynthesis and production (Allen, 1991), the differing mycorrhizae have widely varying aspects to that enhancement. One fungus can have different effects on different hosts, and different fungi can have different effects on the same host. For example, *Artemisia tridentata* ssp. *tridentata* will grow tall and narrow with *Acaulospora elegans* and short and broad with *Scutellospora calospora* (Allen et al., 1992). This effect has been retained through three growing seasons and has resulted in very different whole-plant carbon, water, and nutrient balances (Hickson, 1993). When grown with a mix of *Glomus* spp., *Stipa pulchra* increases root growth whereas *Avena barbata* increases leaf production (Nelson and Allen, 1993). The ectomycorrhizal fungus *Pisolithus tinctorius* was found in association with oaks and with eucalyptus in highly inorganic soil, whereas *Boletus satanus*, *Hysterangium separabile*, and *Hydnotryopsis sp. nov.* were found under oaks in areas with a large leaf litter accumulation and a relatively highly organic soil. *Hydnangium carneum* was abundant in a stand of planted eucalyptus with a relatively thick surface litter layer but little soil organic matter. These patterns imply very differing ecologies of the different species and the potential for very different functional responses to changing environmental conditions.

Interactions Among Microbial Groups

Most studies on soil fungi concentrate on studying either mycorrhizal or saprophytic fungi. However, the differing soil organisms interact directly in important ways beyond their individual ecosystem responses (Azcon-Aguilar and Barea, 1992). Little is known about these interactions in native habitats and little in agricultural systems. The interactions between mycorrhizal fungi and nitrogen-fixers or mycorrhizae and plant pathogens is rather well documented (Azcon-Aguilar and Barea, 1992), but the interactions with saprophytes are also important. Immobilization can limit production, especially with phosphorus (Schlesinger, 1985). Mycorrhizal fungi compete with saprophytic fungi for mineralized phosphorus and nitrogen; they represent an intensive microscale mechanism for absorbing nutrients and water (Allen, 1991). A single VA mycorrhizal hyphal network can provide over 100 cm of hyphae per centimeter of root length (Friese and Allen, 1991), providing up to 30 m of hyphae per gram soil of living absorbing hyphae (Allen and Allen, 1986). We have seen extensive mat formation associated with *Quercus dumosa* formed by *Hysterangium separabile*. These mats form dense hyphal networks wherein most of the total soil mass is fungal. At Sky Oaks, we have found fruiting bodies of *Pisolithus tinctorius* located over 10 m from the host oak canopy presumably with a large hyphal network competing with the VA mycorrhizal hyphae associated with surrounding grasses and shrubs and with the saprophytic hyphae.

The interactions beyond direct competition can be very important. The *Hysterangium* mats under *Q. dumosa* have extensive oxalate activity on the surface (M. Allen et al., in preparation) similar to that found in conifer forests (Cromack et al., 1979). The VA mycorrhizal fungi appear not to have surface oxalate crystals (M. Allen et al., in preparation). The oxalate alters both phosphorus and cation cycling by preferentially binding the cations, releasing the phosphorus for uptake (Jurinak et al., 1986). Subsequently, a large variety of saprophytic fungi, bacteria, and actinomycetes mineralize those oxalates (Morris and Allen, 1994). This alters the soil pH as well as releasing the cations for uptake (after the phosphorus has previously been taken from that location). This may account for the accumulation of cations (especially Ca and Fe) associated with the hyphae of ectomycorrhizal fungi, whereas the VA mycorrhizal fungi have a more even distribution of soil elements along the hyphal surface (M. Allen et al., in preparation).

Projected Changes in the Global Environment

CO_2

Certainly the best-documented global change in the environment is the increase in CO_2. This change may drive increased photosynthesis and production. This may have the effect of creating greater nutrient and water limitations because of the higher nutrient demand associated with the greater carbon fixation rates (e.g., Harley, Hilbert, and Canadell, this volume). Norby et al. (1992) noted that fine root tissue increased significantly with elevated CO_2. Presumably this will place a greater demand on the mycorrhizae to return and take up those added needed resources (Field et al., 1992).

Temperature

Global warming is hypothesized to occur as a result of the greenhouse effect associated with increasing CO_2 and other greenhouse gases. The worldwide increase has been hypothesized to range from 1° to 5°C for the next 50 years. However, at the level of the individual soils, given the daily and seasonal variation, the rise in temperature per se will probably be undetectable by the soil organisms.

Precipitation Timing

Precipitation amount and timing are also hypothesized to change as a response to global warming. Currently, Mediterranean-type ecosystems are characterized by winter/spring precipitation and relatively mild temperatures. For example, the sand plains of western Australia are characterized by 400–800 mm of precipitation, peaking in June and July (Beard, 1984). San Diego receives approximately 300 mm of precipitation, peaking between January and March. Ely et al. (1993) noted that increasing late summer and autumn storms appear to be one manifestation of the El Niño conditions that occur as a result of the warming of the southern Pacific Ocean. The increases in these conditions since industrialization are rather dramatic, and it has been hypothesized that they will continue to increase with increasing CO_2. This could change the major season of moisture inputs in Mediterranean-type habitats, which could dramatically alter microbial processes. Specifically, now both decomposition and mycorrhizal activity are correlated in time. Both occur predominantly during the winter/spring rains. However, if precipitation shifts substantially to summer, we would predict that decomposition would shift rapidly but mycorrhizal activity would shift slowly as plants change. This could result in leaching of nutrients mineralized during decomposition but not taken up because of the lack of mycorrhizal activity.

Plant Species

By altering plant species composition via competition, a changing global environment probably could have the greatest effect on soil fungi. Soil fungi are extremely sensitive to the plant composition, and the types of mycorrhizal associations are genetically controlled. For example, if weedy annuals are favored over woody perennials in a global change scenario, many of those weeds actively reject the mycorrhizal fungi (Allen et al., 1992). We would expect a loss in mycorrhizal fungi and the benefits that they provide. If VA mycorrhizal plants such as grasses are favored over ectomycorrhizal oaks, the enzymatic activity and oxalate production by the oak mycorrhizae would be lost and nutrient cycling patterns shifted as a result.

Human Population Pressure

Certainly one change that is occurring at a dramatic rate is the disturbance associated with habitat fractionation due to human development in Mediterranean-type habitats. This is occurring and will continue to occur with or without the other changes. Urbanization and agriculture have dramatic effects on both saprophytic and mycorrhizal associations. We have documented shifts from ectomycorrhizal plants to VA mycorrhizal plants in riparian areas subjected to urbanization and a decrease in fungal/bacterial ratios in human-disturbed shrublands. Both types of changes result in a more open nutrient cycle.

Influence of Global Change on Saprophytic Activity

Timing

Timing of activity could present a major change if increasing summer temperatures result from global change or if the timing of the precipitation is altered. Moisture is probably a major factor affecting decomposition in Mediterranean-type habitats (e.g., Meentemeyer, 1978; Stock and Midgley, this volume), and precipitation patterns are highly likely to be affected by increasing CO_2. Mycorrhizal activity is concentrated in the spring in mediterranean-type habitats (M. Allen, unpublished observations). However, summer and autumn storms are one expression of El Niño events that are hypothesized to increase with increasing CO_2 (Ely et al., 1993). Unless there are corresponding rapid changes in rooting phenology and mycorrhizal fungal species, with summer precipitation and the associated increased decomposition will come nutrient mineralization and loss due to an inadequate root and mycorrhizal surface area for retention. Particularly, this change coupled with other shifts due to habitat changes described below could have major impacts on Mediterranean-type habitats.

Evidence from the cold desert suggests that an altered precipitation pattern associated with El Niño events could have dramatic effects on the microbial activity and nutrient dynamics. During the El Niño years of the early 1980s, there was a dramatic increase in late summer and early autumn precipitation. Total precipitation increased up to four times normal. The decomposer mass was significantly higher in the autumn than at other time periods, there were high levels of shrub mortality associated with snow mold fungi, and there was a crash in the mycorrhizal fungal activity associated with a large number of dead and punctured spores. The total soil P concentrations declined by as much as 40% on a stand level basis (Allen et al., 1987; Allen, 1993).

Changes in Microbial Processes

Our scenario suggests that decomposition rates per se by fungi and bacteria probably will not be affected by global change. However, superimposing global change on anthropogenic perturbations could be devastating for Mediterranean-type habitats. Weedy annuals appear to be invading many coastal shrub and grassland Mediterranean-type habitats (O'Leary and Westman, 1988; Nelson and Allen, 1993; T. Zink and M. Allen, unpublished data). We surveyed a slope undergoing conversion from a coastal sage community to an annual community composed of mixed patches of introduced annual grasses and forbs. At a nearby slope, where patches of annual species (predominantly *Centarea*, *Avena*, and *Brassica*) were present within the same coastal-sage community, the density of fungi was affected dramatically by the vegetation shifts. The coarse organic matter associated with shrub litter disappeared, particularly under the annual grasses, and so did the fungi (Table 14-2). Fungal colony sources declined by 95% to 99%, richness by 85%, and hyphal length by 35% (unpublished data). An even greater change was associated with the establishment of an adjacent avocado grove (Table 14-2).

Table 14-2. Fungi Associated With Differing Vegetation Patches in an Undisturbed Coastal-Sage Site Undergoing Invasion by Weedy Annuals (Morris and Allen, 1994)

	No. Colonies		Richness		Hyphal Length
Plant	Autumn	Spring	Autumn	Spring	Spring
Native					
Salvia	26	66	11	18	168
Artemisia	52	27	11	22	163
Stipa	13	51	9	20	228
Weedy					
Centaurea	9	n.d.	9	n.d.	n.d.
Brassica	18	41	9	19	147
Avena	1	56	2	21	99
Agricultural					
Avocado	n.d.	17		19	38

Colony sources are expressed as colony numbers $\times 10^3$/g soil. Richness is the taxon richness per plate. Hyphal length is in meters per gram soil.

Influence of Global Change on Mycorrhizal Activity

One hypothesis is that the increase in CO_2 will cause increasing nutrient deficiencies which could be alleviated by increasing mycorrhizal activity. Chamber studies in sterilized and inoculated soils seem to confirm this hypothesis (O'Neill and Norby, 1988). In a phytotron study, C. Monz and W. Hunt (personal communications) found that *Agropyron smithii* and *Bouteloua gracilis* had increased mycorrhizal infection in high CO_2. We have taken a different approach. Using a set of developmental rules (Friese and Allen, 1991) we developed a model describing the hyphal activity of VA mycorrhizal fungi that fits the observed field hyphal lengths (M. Allen, in preparation). The model is based on the number of hyphal entry points, or infection units, and the external development of a fan-shaped external mycelial network. The external matrix increases in bifurcating branches at regular intervals in space and time (Friese and Allen, 1991) with regular die-back and disappearance based on the glass-plate observations. The model appears to describe the mass of mycorrhizal development in successional sites accurately (Allen and Allen, 1992). Using this model, we assessed the *external* mycorrhizal hyphal activity in undisturbed soils (Figure 14-2). This model predicts that as a rooting front passes through a soil zone, there will be approximately 50 m of living hyphae in every cm^3 associated with 4 cm of root for that same soil. This means that there is approximately a 3 to 5 cm depletion zone for every root, indicating a virtual saturation of the existing soil with mycorrhizal hyphae that search the soil. The roots die back annually and the process is repeated. Based on this scenario, we suggest an alternative hypothesis: these systems are already supporting the maximum amount of mycorrhizae, and the VA mycorrhizae, at least, are taking up the maximum resource load possible to support the existing vegetation.

This scenario is supported by two sets of observations: observations of the amount of mycorrhizal activity under conditions of altered root growth, and the preliminary results from the Jasper Ridge project. O'Neill et al. (1991) found that the percentage VA mycorrhizal infection did not change in yellow poplar with elevated CO_2 but the fine root activity was greater, suggesting a greater total mycorrhizal mass per plant. We (Allen et al., 1989) observed the total amount of mycorrhizae per unit soil volume and the percentage of root infection in response to differing moisture and clipping (carbon input) regimes over several years. Under optimum conditions of carbon gain, the roots appeared to outgrow the mycorrhizal fungi in that the percentage of roots with mycorrhizae declined significantly but the total number of infection units remained constant. In a set of open-chamber experiments, Julie Whitbeck assessed the infection frequency of the root systems of native plants and found that there was no change in mycorrhizal activity with elevated CO_2 (personal communication).

As a possible carbon sink, VA mycorrhizae may also serve a major role. The respiration estimated by the above hyphal model peaks in the

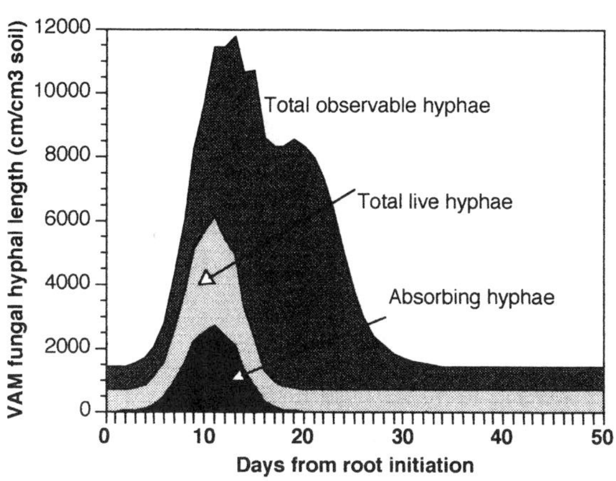

Figure 14-2. External VA mycorrhizal fungal hyphal lengths with time in an expanding rooting network. The data presented are live absorbing hyphae (dark), total live hyphae (absorbing plus runner hyphae) (light shading), and total hyphae (gray) assuming dieback of the absorbing network as the network passes through and depletes the soil volume of extractable resources, a steady-state value for the runner hyphae, and a 5-day disappearance of dead hyphae (see text for details).

micromolar range, whereas root respiration is in the millimolar range (Williams and Kemp, 1978). Our estimates of intensive internal VA mycorrhizal activity (arbuscular) show that even highly infected roots have only ~20% arbuscular infection, and in those regions, using image analysis, we estimated that heavily infected roots contain ~20% fungus by volume (M. Allen, S. Morris, B. Weinbaum, and S. Bennett, unpublished data). Thus, only a peak of 4% of the rooting volume is fungal. The maximum carbon sink provided by the mycorrhizal fungi should not exceed 5% to 10% of that allocated below ground.

Importantly, these data only relate to a situation in which the existing fungal and vegetation composition and the phenology do not change. Current estimates suggest that as much as 30 kg/ha of nitrogen is deposited on the coastal sage and chaparral communities of the Los Angeles Basin (Fenn and Bytnerowicz, 1993). If there are changes in the soil nutrient status due to anthropogenic nutrient inputs, the fungal composition may change, probably to a less efficient set of isolates (e.g., Johnson, 1993). In addition, even processes such as habitat fractionation appear to alter both plant and mycorrhizal fungal species composition, which could exacerbate CO_2 effects. We conducted an experiment by observing plant and mycorrhizal responses to anthropogenic disturbances in 10 m^2 plots in undisturbed coastal sage scrub. Importantly, in this study we did not directly disturb plots but assessed the undisturbed plots surrounded by anthropogenic disturbance. Annual grasses (*Bromus* and *Avena*) invaded plots manipulated by fractionating the habitats with cement blocks and tillage even during the first growing season. By the second growing season, there was a significant reduction in the diversity of mycorrhizal fungi (Edwards, 1995). Finally, if the timing or amount of precipitation changes, presumably the root and fungal phenology will also change. We have no way currently to predict how the interactions of these climatic factors affect the mycorrhizal species composition or activity.

Ectomycorrhizal associations probably provide a larger carbon sink and certainly provide much more external hyphae (Finlay and Soderstrom, 1992; Read, 1992). If a CO_2 increase were the only likely global change, these associations might expand and enhance uptake of limiting resources. However, work in many areas shows that the ectomycorrhizal and arbutoid mycorrhizae (which are often the same fungi; Molina et al., 1992) are probably the most sensitive to pollution and the organic matter losses associated with increasing human population pressure (Kowalski, 1987; Vosatka et al., 1990). VA mycorrhizal associations are relatively well adapted to mineral soils (Read, 1983) but are reduced or eliminated when the sites are dominated by many annual weedy forbs (Allen and Allen, 1980). Even though annual grasses form mycorrhizae, the fungal species composition can be altered to a less effective group of species for maintenance of a perennial community (Nelson and Allen, 1993).

In summary, we hypothesize that in a fully developed Mediterranean-type habitat, mycorrhizal activity is probably already acting at levels approaching the maximum efficiency. The most likely scenario is an increasing CO_2 superimposed on a biome increasingly pressured by habitat fractionation and pollution. We suggest that the most likely scenario is that the decomposition process will be decoupled with the mycorrhizal and fine root activity. It is likely that mycorrhizae will decrease in effectiveness, not increase, to meet the increasing nutrient demands of the invading plants. This decoupling will only enhance the invasion of exotic weedy species and reduce the diversity and stability of these critical ecosystems.

Conclusions

Our conclusion does not paint a rosy picture. We suggest that saprophytic activity probably cannot meet the needs of a potential increasing carbon input. Fire is the most prevalent mineralization agent in Mediterranean-type habitats. As most decomposition of complex carbon substrates is probably undertaken by fungi, a shift toward a bacterial-dominated

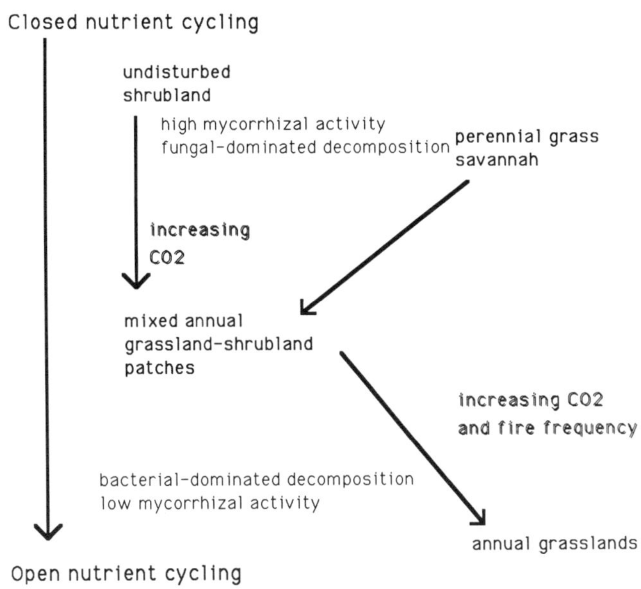

Figure 14-3. A possible scenario of changes in soil fungal functioning with multiple global changes, including primarily increasing human perturbation, weed invasions, and increasing CO_2.

decomposition process likely means that fire will become of even greater importance. Further, since there is a high turnover associated with bacteria compared with fungi, upon leaf decomposition by bacteria, those nutrients that are released via decomposition probably will not be immobilized by the hyphal mass but will be lost by leaching or erosion. Finally, mycorrhizal activity probably will be uncoupled from decomposition and therefore will be unable to increase resource acquisition. As Mediterranean-type habitats are systems already on the edge of deserts, slight changes in the nutrient retention capacity could lead to increasing desertification (Figure 14-3).

These scenarios are, of course, only hypotheses. Few data are available to make accurate predictions of the dynamics of below-ground processes even with no change. The wide array of perturbations facing Mediterranean-type habitats are nonlinear and highly interactive. A better understanding of the organisms and their control of critical processes such as decomposition and nutrient uptake is essential before appropriate management strategies can be formulated to face the future.

Acknowledgments. We thank E.G. O'Neill for her helpful comments on this manuscript. This work was funded by grants from the Department of Energy, the National Science Foundation, and the Environmental Protection Agency, with additional support from the California Department of Transportation.

References

Allen EB, Allen MF. 1980. Natural re-establishment of vesicular-arbuscular mycorrhizae following stripmine reclamation in Wyoming. J Appl Ecol 17: 139–147.

Allen EB, Allen MF. 1986. Water relations of xeric grasses in the field: interactions of mycorrhizae and competition. New Phytol 104:559–571.

Allen EB, Allen MF, Helm DJ, Trappe JM, Molina R, Rincon E. 1994. Patterns and regulation of arbuscular and ectomycorrhizal plant and fungal diversity: a hypothesis. Plant Soil 161:1–16.

Allen MF. 1985. Spatial patterning and soil saprophytic microbiota: impacts of strip mining, importance, and management strategies. Pages 322–326 in Proceedings of the American Society for Surface Mining and Reclamation, Symposium on "How Microorganisms Improve Reclamation: Their Importance and Management in Mine Soils."

Allen MF. 1991. The Ecology of Mycorrhizae. Cambridge University Press, Cambridge, UK.

Allen MF. 1993. Microbial and phosphate dynamics in a restored shrub steppe in southwestern Wyoming. Restoration Ecol 1:196–205.

Allen MF, Allen EB. 1992. Development of mycorrhizal patches in a successional arid ecosystem. Pages 164–170 in DJ Read, A Fitter, and I Alexander, eds. Mycorrhizas in Ecosystems. International Press, Oxford, UK.

Allen MF, Allen EB, West NE. 1987. Influence of parasitic and mutualistic fungi on *Artemisia tridentata* during high precipitation years. Bull Torrey Bot Club 114:272–279.

Allen MF, Richards JA, Busso CA. 1989. Influence of clipping and water status on vesicular-arbuscular mycorrhizae of two semiarid tussock grasses. Biol Fertil Soils 8:285–289.

Allen MF, Allen EB, Dahm CN, Edwards FS. 1993. Preservation of biological diversity in mycorrhizal fungi: importance and human impacts. In: G Sundnes, ed. Human Impacts on Self-Recruiting Populations. Third International Kongsvoll Symposium. Tapir Press, Trondheim, Norway, pp. 81–108.

Allen MF, Clouse SD, Weinbaum BS, Jeakins S, Friese CF, Allen EB. 1992. Mycorrhizae and the integration of scales: from molecules to ecosystems. Pages 488–515 in MF Allen, ed. Mycorrhizal Functioning. Chapman and Hall, New York.

Allsopp N. 1992. The occurrence and ecophysiology of vesicular-arbuscular mycorrhizal plants in the Cape Floristic region. PhD Thesis, University of Cape Town, South Africa.

Atlas RM, Bartha R. 1993. Microbial ecology: fundamentals and applications. Benjamin Cummings Publishing Co. Redwood City, CA.

Azcon-Aguilar C, Barea JM. 1992. Interactions between mycorrhizal fungi and other rhizosphere microorganisms. Pages 163–198 in MF Allen, ed. Mycorrhizal Functioning. Chapman and Hall, London.

Beard JS. 1984. Biogeography of the Kwongan. Pages 1–26 in JS Pate and JS Beard, eds. Kwongan: Plant Life of the Sand Plains. University of Western Australia Press, Nedlands WA.

Bellgard SE. 1991. Mycorrhizal associations of plant species in Hawkesbury Sandstone vegetation. Aust J Bot 39:357–365.

Cannon JP. 1993. The effects of oxalic acid produced by *Salsola kali* on phosphorus uptake by *Stipa pulchra*. MS Thesis, San Diego State University, San Diego, CA.

Christensen M. 1981. Species diversity and dominance in fungal communities. Pages 201–232 in DT Wicklow and GC Carroll, eds. The Fungal Community: Its Organization and Role in the Ecosystem. Marcel Dekker, New York.

Cromack KJ, Sollins P, Graustein WC, Speidel K, Todd AW, Spycher G, Liu CY, Todd RL. 1979. Calcium oxalate accumulation and soil weathering in mates of the hypogeous fungus *Hysterangium crassum*. Soil Biol Biochem 11:463–468.

Domsch KH, Gams W, Anderson T-H. 1980. Compendium of Soil Fungi. Academic Press, London.

Edwards FS. 1995. Indicators of low intensity ecosystem perturbation: shifts in herbaceous plants and arbuscular mycorrhizal fungi in two semi-arid ecosystems. MS Thesis, San Diego State University, San Diego, CA.

Ely LL, Enzel Y, Baker VR, Cayan DR. 1993. A 5000-year record of extreme floods and climate change in the Southwestern United States. Science 262: 410–412.

Fenn ME, Bytnerowicz A. 1993. Dry deposition of nitrogen and sulfur to Ponderosa and Jeffrey pine in the San Bernardino National Forest in southern California. Environ Pollut 81:277–285.

Field CB, Chapin FS, Matson PA, Mooney HA. 1992. Responses of terrestrial ecosystems to the changing atmosphere: a resource based approach. Ann Rev Ecol Syst 23:201–235.

Finlay R, Söderstrom B. 1992. Mycorrhiza and carbon flow to the soil. Pages 134–162 in MF Allen, ed. Mycorrhizal Functioning. Chapman and Hall, New York.

Flanagan PW, Scarborough A. 1974. Physiological groups of decomposer fungi on tundra plant remains. Pages 159–182 in AJ Holding, OW Heal, SF MacLean, and PW Flanagan, eds. Soil Organisms and Decomposition in Tundra. IBP Tundra Biome Steering Committee, Stockholm.

Friese CF, Allen MF. 1991. The spread of VA mycorrhizal fungal hyphae in the soil: inoculum type and external hyphal architecture. Mycologia 83:409–418.

Hickson LE. 1993. The Effects of Vesicular-Arbuscular Mycorrhizae on Morphology, Light Harvesting, and Photosynthesis of *Artemisia tridentata* ssp. *tridentata*. MS Thesis, San Diego State University, San Diego, CA.

Horton TR, Parker VT. 1992. *Pseudotseuga menziesii* invasion into *Arctostaphylos gladulosa*: assessment of mycorrhizal facilitation and other mechanisms. Bull Ecol Soc Am 73:212.

Ingham ER, Coleman DC, Moore JC. 1989. An analysis of food-web structure and function in a shortgrass prairie, a mountain meadow, and a lodgepole pine forest. Biol Fertil Soils 8:29–37.

Jackson LE, Schimel JP, Firestone MK. 1989. Short-term partitioning of ammonium and nitrate between plants and microbes in an annual grassland. Soil Biol Biochem 21:409–415.

Johnson NC. 1993. Can fertilization of soil select less mutualistic mycorrhizae? Ecol Applis 3:749–757.

Jurinak JJ, Dudley LM, Allen MF, Knight WG. 1986. The role of calcium oxalate in the availability of phosphorus in soils of semiarid regions: a thermodynamic study. Soil Sci 142:255–261.

Korner C, Arnone JA. 1992. Responses to elevated carbon dioxide in artificial tropical ecosystems. Science 257:1672–1675.

Kowalski S. 1987. Mycotrophy of trees in converted stands remaining under strong pressure of industrial pollution. Angew Botanik 61:65–83.

Kummerow J, Borth W. 1986. Mycorrhizal associations in chaparral. Fremontia 16:11–13.

Kummerow J, Krause D, Jow W. 1978. Seasonal changes of fine root density in the southern California chaparral. Oecologia 37:201–212.

Lamont BB. 1984. Specialized modes of nutrition. Pages 126–145 in JS Pate and JS Beard, eds. Kwongan: Plant Life of the Sand Plains. University of Western Australia Press, Nedlands WA.

McGee P. 1986. Mycorrhizal associations of plant species in a semiarid community. Aust J Bot 34:585–593.

Meentemeyer V. 1978. Macroclimate and lignin control of litter decomposition rates. Ecology 59:465–472.

Mitchell DT, Coley PGF, Webb S, Allsopp N. 1986. Litterfall and decomposition processes in the coastal fynbos vegetation, South-western Cape, South Africa. J Ecol 74:963–976.

Molina R, Massicotte H, Trappe JM. 1992. Specificity phenomena in mycorrhizal symbiosis: community-ecological consequences and practical implications. Pages 357–423 in MF Allen, ed. Mycorrhizal Functioning. Chapman and Hall, New York.

Morris SJ, Allen MF. 1994. Oxalate metabolizing microorganisms in sagebrush steppe soils. Biol Fertil Soils 18:255–259.

Nelson LL, Allen EB. 1993. Restoration of *Stipa pulchra* grasslands: effects of mycorrhizae and competition from *Avena barbata*. Restoration Ecol.

Norby RJ, Gunderson CA, Wullschleger SD, O'Neill EG, McCracken MK. 1992. Productivity and compensatory responses of yellow-poplar trees in elevated CO_2. Nature 357:322–324.

Oechel WC, Strain BR. 1985. Native species responses to increased atmospheric carbon dioxide concentration. In: BR Strain and JD Cure, eds. Direct Effects of Increasing Carbon Dioxide on Vegetation. United States Department of Energy, Washington, DC.

O'Leary JF, Westman WE. 1988. Regional distrubance effects on herb succession patterns in coastal sage scrub. J Biogeog 15:775–786.

O'Neill EG, Norby RJ. 1988. Differential responses of ecto- and endomycorrhizae to elevated atmospheric CO_2. Bull Ecol Soc Am (supplement) 69:248–249.

O'Neill EG, O'Neill RV, Norby RJ. 1991. Hierarchy theory as a guide to mycorrhizal research on large-scale problems. Environ Pollut 73:271–284.

Puppi G, Tartaglini N. 1991. Mycorrhizal types in three different communities affected by fire to different extents. Acta Oecol 12:295–304.

Read DJ. 1983. The biology of mycorrhiza in the Ericales. Can J Bot 61: 985–1004.

Read DJ. 1992. The mycorrhizal mycelium. Pages 102–133 in MF Allen, ed. Mycorrhizal Functioning. Chapman and Hall, London.

Schlesinger WH. 1985. Decomposition of chaparral shrub foliage. Ecology 66: 1353–1359.

Schlesinger WH, Reynolds JF, Cunningham GL, Huenneke LF, Jarrell WM, Virginia RA, Whitford WG. 1991. Biological feedbacks in global desertification. Science 247:1043–1048.

Vosatka M, Cudlin P, Mejstrik V. 1990. Establishment of VAM association in grass cover in process of spruce forest decline under different pollution stress. Pages 298 in MF Allen and SE Williams, eds. Eighth North American Conference on Mycorrhizae. University of Wyoming Agricultural Experiment Station, Laramie, WY.

Williams GJ III, Kemp PR. 1978. Simultaneous measurement of leaf and root gas exchange of shortgrass prairie species. Bot Gaz 139:150–157.

15. Soil Organic Matter in Mediterranean-Type Ecosystems and Global Climatic Changes: A Case Study—The Soils of the Mediterranean Basin

P. Bottner, M.M. Coûteaux, and V.R. Vallejo

On a worldwide basis, the soil organic carbon pool is around two to three times the amount of carbon stored in the atmosphere and two to three times that stored in the terrestrial biomass (Bohn, 1982; Rozanov, 1990; Scharpenseel and Becker-Heidmann, 1990). The sensitivity of Soil organic matter (SOM) to climate change raises two essential questions. Are the soils going to be a source or a sink of atmospheric CO_2 in direct or indirect responses to the predicted physical or chemical climate change? The SOM is the major regulating factor of nutrient availability in the ecosystem; therefore the second question is to know to what extent the climate changes will modify the stabilization-mineralization balance of the nutrients by modifying the SOM turnover rates.

To provide some insight on these questions, this review will first try to analyze the specific nature of the SOM and the parameters affecting its dynamic in MTEs and afterwards discuss the mechanisms of responses to global change, foreseeable from the present basis of knowledge.

An important feature of the Mediterranean Basin is the long-term intensive and extensive land use which probably has strongly modified landforms, soils, and vegetation (Naveh and Lieberman, 1984). This feature sets the Mediterranean Basin apart from the other Mediterranean climate regions of the world.

Climatic Characteristics of the Mediterranean Basin

In the Mediterranean Basin, precipitation ranges from 20 to 2,000 mm within a range of altitudes of −400 to 4,000 m and distances from the sea from 0 to 1,000 km; mean minimum air temperatures in January range from −15° to +15°C and mean maxima in June from 15° to 45°C. Annual potential evapotranspiration ranges between 500 and 2,500 mm.

The Mediterranean climate is characterized by wet winters and warm, dry summers. A dry month is empirically defined as having precipitation lower than one-third of potential evapotranspiration (Le Houerou and Popov, 1981). Summer drought can last 3 to 11 months. From the length of the rainy and dry seasons and the mean yearly precipitation, different bioclimatic Mediterranean zones are defined: hyperarid, arid, semiarid, subhumid, humid, and hyperhumid. Water stress may range from moderate and of short duration in summer to extremely severe and prolonged. Winter cold stress may be nonexistent along the seashore, whereas hard frost and/or snow cover lasts over 6 months in high mountains or continental zones.

The Organic Matter in Mediterranean Soils

The size of the carbon reservoir in soils results from the balance between inputs of plant material and losses through decomposition and, ultimately, heterotrophic respiration. The flow of carbon in the soil mainly depends on two environmental factors. 1) Climatic conditions influence production and decomposition at specific rates, which in turn determine the equilibrium between the input and output flow of carbon, hence the soil carbon reservoir (Franz, 1990). Comparing major world forest biomes, litterfall ranges differ by one order of magnitude (Cannell, 1982) whereas litter decomposition rates differ by more than two orders of magnitude (Meentemeyer, 1984); therefore, probably factors other than macroclimatic conditions introduce more variability in decomposition rates than in litter production. Special attention must be paid to the sclerophyllous nature of the Mediterranean vegetation. 2) Soil properties affect the mineralization rates of the organic compounds, through physical and chemical interactions with the soil minerals and by conditioning the structure and activity of the microbial community.

Based on 2,700 soil profiles from "natural ecosystems" distributed in the major world life zones, Post et al. (1982) tried to establish the correlation of soil carbon density ($kg\,C\,m^{-2}$) with climatic variables on a world scale (Figure 15-1). They organized the world distribution of carbon according to the Holdridge Life Zone Classification System (Holdrige climagram), which is based on biotemperature, annual precipitation, and potential evapotranspiration. Some zones, such as warm and hot deserts,

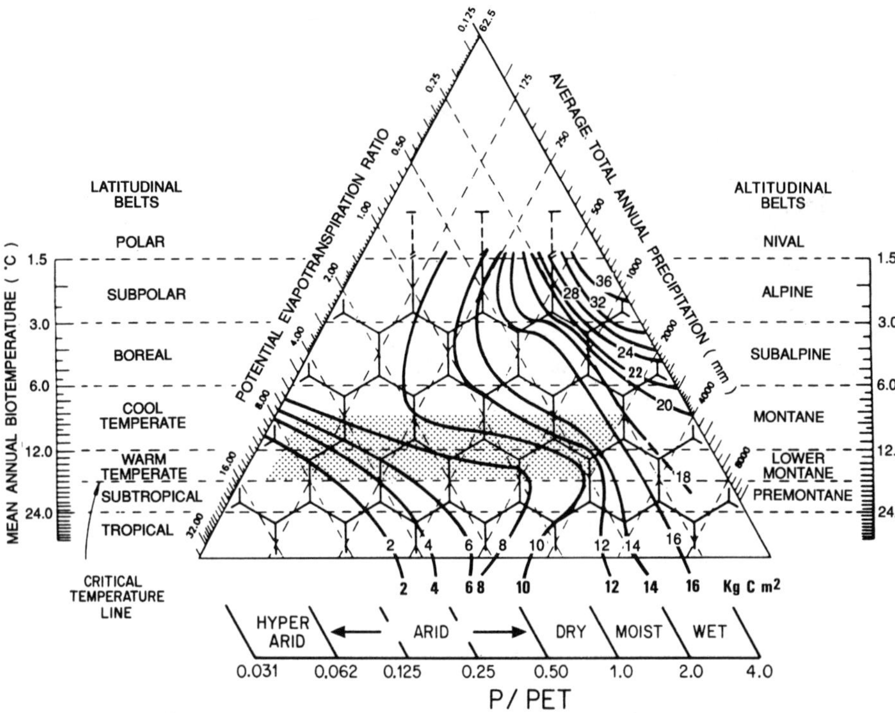

Figure 15-1. Soil organic carbon densities ($kg\,m^{-2}$) plotted as contours on the Holdridge life-zone chart (source: Post et al., 1982). The Mediterranean bioclimatic zones are indicated by shading.

are poorly represented, but their carbon content is low. Climatic Mediterranean zones are indicated in the figure. Minimum and maximum biotemperatures have been used for that delimitation. The general pattern of carbon distribution shows: (1) a increase in carbon density from left to right, reflecting greater production and soil carbon accumulation with higher annual precipitations; (2) in most of the zones, a soil carbon increase with decreasing biotemperature for any particular annual precipitation, reflecting a limitation of decomposition by temperature. Most life zones bordering the PET/P = 1.0 line have a carbon density of around 10–12 $kg \cdot m^{-2}$, but in the warm temperate belt, including the Mediterranean climate, lower soil carbon densities are found (8–10 $kg \cdot m^{-2}$). This could be explained by a strong climatic seasonality, limiting summer production but with decomposition conditions remaining favorable for a longer period of the year.

In semiarid conditions in the Ebro Basin (Catalonia, NE Spain), the carbon densities of shallow forest soils range from 3.5 to 16 $kg \cdot m^{-2}$ (Vallejo, 1983). For colluvial holm-oak forest soils in subhumid conditions

(Montseny Mountains, Catalonia), carbon pools (including the forest floor) range from 5 to 14 kg carbon m^{-2} (Hereter, 1990). The mean values of the latter observations, about 10 kg carbon m^{-2}, are close to the total biomass in these forests (Canadell and Rodá, 1989; Figure 15-2). The large variability found within a single climatic zone stresses the effect of factors other than macroclimate on soil organic carbon accumulation. Vogt et al. (1989) stressed the lack of correlation between climate and forest floor mass in a worldwide analysis, because local conditions contributed to wide variation within latitudes or forest types. Figure 15-3 shows some estimates of the organic carbon content in Mediterranean soils, developed on limestone parent material, in some major bioclimatic zones of the Mediterranean Basin, ranging from the subhumid zone in southern France, to the Sahara hyperarid zone, through the semiarid and arid zones in Algeria (Bottner, 1981; Pouget, 1980). The values are presented in carbon percentage of dry soil for the upper mineral A horizon. At a mean bulk density of 1.5 $kg \cdot dm^{-3}$, an exponential decrease of SOM with soil depth, and on limestone material a stoniness of 10–20%, the maximum values of 7% carbon in Figure 15-3 correspond to a carbon pool of about 16–20 $kg \cdot m^{-2}$. These carbon densities correspond in the diagram of Figure 15-1 to the cool or temperate wet forests. The lowest values of 0.5% carbon in the arid zone, with 100–200 mm of precipitation, correspond to 1.5 kg carbon m^{-2} (the desert bush life zone in Figure

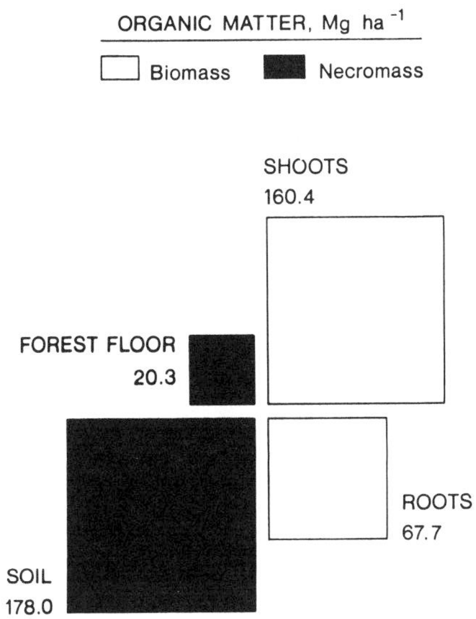

Figure 15-2. Organic matter distribution in a holm oak forest ecosystem from the Montseny Mountains (Barcelona). Source: Canadell and Rodá (1989) and Hereter (1990).

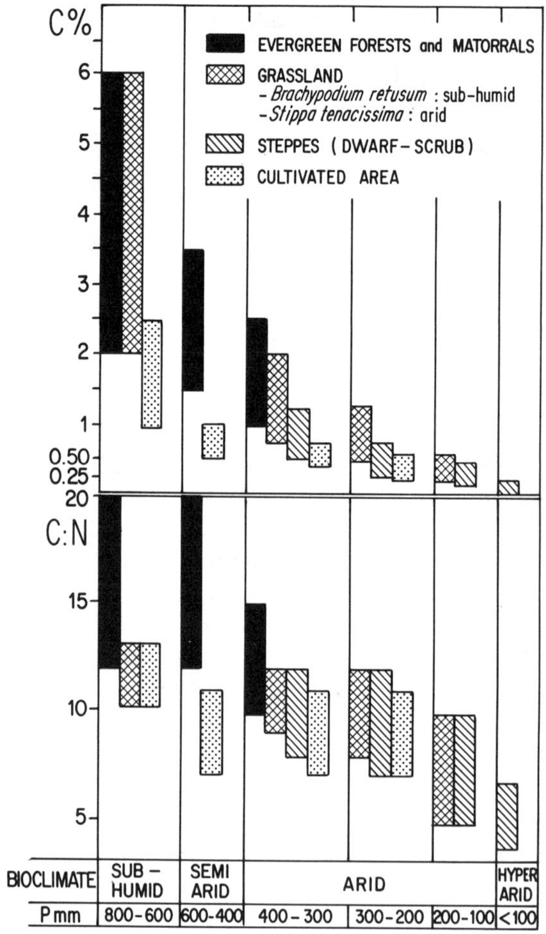

Figure 15-3. Organic carbon concentrations (in % dry soil) and C:N ratios in soils on limestone of some major life zones in southern France (source: Bottner, 1981) and Algeria (source: Pouget, 1980).

15-1). In the subhumid zone, the carbon contents in the evergreen *Quercus ilex* forest are not different from the carbon contents in the degraded *Brachypodium retusum* grassland, indicating that although the vegetation is continuously changing in a degradation-recovery dynamic, the soil organic matter remains rather stable. In drier zones, the influence of land use and vegetation types seems to be more important.

The recent estimations of the soil carbon on a global basis range from 1,400 to 2,000 Gt C (Ajtay et al., 1979; Post et al., 1982; Buringh, 1984; Schlesinger, 1984; Bolin, 1986). On a worldwide basis, the carbon pool ranges from less than 1 kg carbon m^{-2} in desert to 30 kg m^{-2} in rain tundra. According to Figure 15-3, in the MTEs the values range from approximately 2 to 20 kg m^{-2}. Nevertheless, taking into account the relatively small areal extent of the Mediterranean climatic zones, the Mediterranean carbon reservoir obviously represents a small part of the world soil carbon pool. Thus, the effect of global environmental changes on Mediterranean soils, as source or sink of CO_2, will probably be negligible.

Diversity of Soil Properties and Quality of Soil Organic Matter

The "natural ecosystems" have a large variability in SOM contents within the life zones (Anderson, 1991), and their distribution pattern related to climate will probably be improved, if other factors, like mineral soil properties, soil age, land use, and the history of the vegetation are taken into account. The high diversity of the soil types within relatively small areas is one of the essential features of the soils developed in the Mediterranean Basin. In the northern cold temperate climatic zones, the soils and the soils' parent material were vigorously disturbed and mixed by active glacial geomorphological processes during the last glaciation. During the more recent and warmer periods (Holocene), with low geomorphological disturbances, soil development took place, giving relatively young soils, related to recent or present vegetation and climatic conditions. In contrast, the soils in the Mediterranean Basin were less affected by glacial perturbations; the organization of the soil profile and its mineralogy reflects the subtropical paleoclimatic conditions preceding the last glacial period. Thus, the coexistence of paleosols and recent soils explains the high variety of Mediterranean soils on relatively small areas. In conclusion, many Mediterranean soils are paleosoils in their profile organization and in their mineral part and modern soils in the organic part, which is influenced by the recent and present climatic conditions. The recent soil perturbations related to human activity affected essentially the upper horizons and the soil organic matter profile.

The interdependency of pedogenetic profile and organic profile is illustrated in the diagram of Figure 15-4. In the humid to subhumid zone of the Mediterranean Basin, two main processes dominate soil formation and evolution: the leaching of carbonates and the leaching of mineral clays. In the soil succession presented in Figure 15-4, the carbonate soils at the left of the diagram are generally young soils with high $CaCO_3$ contents still remaining in the top soil. The highly differentiated luvisols at the right of the figure, with deep accumulations of carbonates and clay alluviated and compacted Bt horizons and with thick leached E horizons, are generally old paleosols, formed under different climatic conditions, probably since the earlier Quaternary periods.

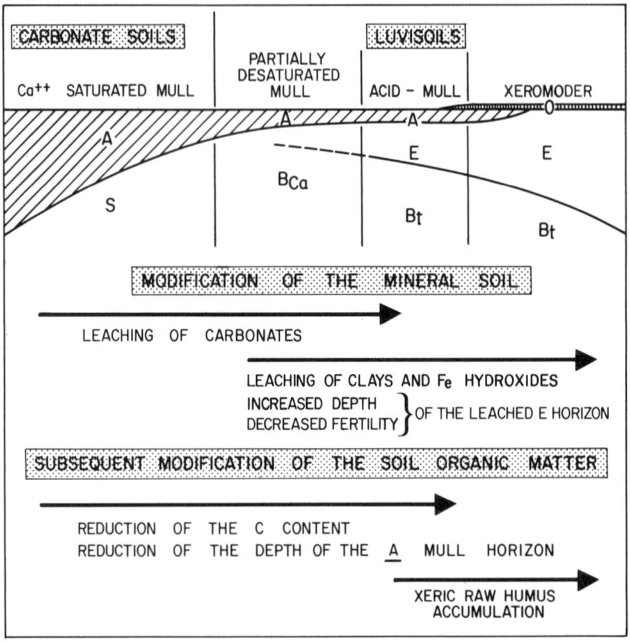

Figure 15-4. Depth distribution and density of SOM are controlled by physicochemical properties of the mineral components of the soil: paleopedogenesis conditioning the present SOM profile. A: Mull humus form; O: Raw humus horizon; C: Parent material; B_{Ca}: Calcium carbonate accumulation; B_t: clay accumulation; E: carbonate and clay leached horizon.

The present SOM profile is closely controlled by the soil mineral environment (Figure 15-4). In calcareous soils, the organic carbon is deeply distributed and the carbon content is relatively high, since the clay minerals remaining in the topsoil and the high Ca^{2+} concentration protect the organic matter against microbial degradation (Van Veen and Paul, 1981), forming clay-Ca^{2+}-humus complexes, which are characteristic of the calcareous or calcic mull humus types. Bottner and Peyronel (1977) found a mean residence time of 3,000 years for this type of organic matter. From the carbonate soils at the left, through the range of luvisols in the center of the diagram, the organic A horizon becomes more and more shallow, with decreasing organic carbon concentrations. The lowering of the SOM is essentially due to the fact that the carbonates and clays are leached out from the upper horizons. The turnover time and the mean age of the SOM which is not protected any more by a clayey environment are shortened and the carbon contents are lowered.

In the oldest and most leached soils (at the right), most of the fine mineral fractions are eliminated from the upper horizons; only sandy material remains in the topsoil, resulting in a lowering of the water retention capacity and in a subsequent reduction of the activity and diversity of the soil fauna. Elimination of the earthworms (an essential factor of the organic matter incorporation) results in accumulation at the surface of the soil of a shallow raw humus layer: xeromoder or xeromor, with C:N ratios ranging from 20 to 30, indicative of decreased litter decomposition rates, due to a reduction of faunal activity. Eighty percent of the organic matter of the soil is then concentrated in a 2- to 5-cm thick superficial layer.

Figure 15-5 illustrates similar relationships between the organic profile and the properties of the mineral soil. In a set of forest soils formed from different types of calcareous parent materials in semiarid conditions in NE Spain, organic carbon densities are related to the combined effects of the particle size, lime content, and degree of soil profile differentiation (Vallero, 1983).

Thus, the density, quality, and dynamics of SOM, its distribution through the soil, and its landscape distribution cannot be solely related to the climate, but must also be explained by the mineral soil environment. In the Mediterranean countries, the sometimes very complex paleopedogenetic evolution of the mineral soil will be an essential feature to understand the present organic matter dynamic.

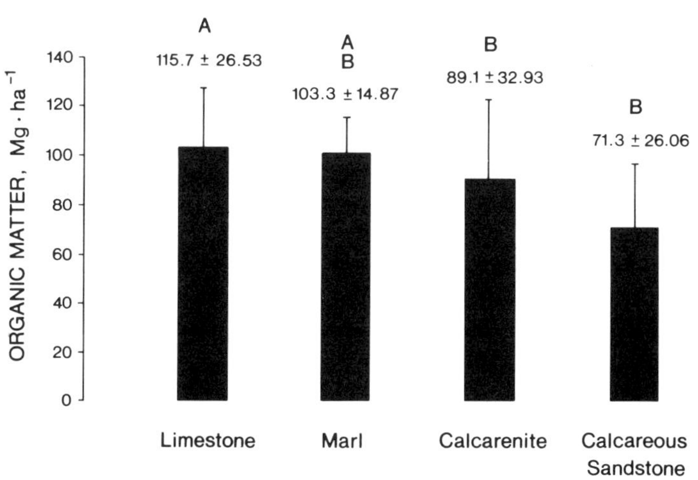

Figure 15-5. Organic matter density in the forest floor plus the 0–15 cm top mineral soil of different types of soils grouped by rock type in the Ebro Basin (Spain), including *Quercus rotundifolia*, *Q. coccifera*, *Q. faginea*, and *Pinus halepensis* forests. Soil characteristics were closely dependent on lithology: noncalcareous silty clay loam on limestones ($n = 15$); carbonate-rich silty clay loam on marls ($n = 18$); carbonate-rich sandy loam on calcarenites ($n = 7$); noncalcareous to calcareous sandy loam on calcareous sandstones ($n = 10$). Different letters indicate significant differences with $p < 0.05$ (Fisher's LSD multiple comparison). Bars show confidence limits at $p < 0.05$.

Mediterranean Soil Organic Matter and Global Climate Change

Global climate change is expected to affect SOM indirectly through modifications of the litter quality in response to CO_2 increase and directly by modifying the decomposition rates, through temperature increase. A third path for SOM pool modification is expected through change of land use. Since it has a strong sociological component, the effect of land use will not be discussed in this review.

Indirect Effects of Atmospheric CO_2 Increase on SOM Decomposition

The basic assumptions are (1) a rise in the atmospheric CO_2 concentration, reaching $700\,\mu L \cdot L^{-1}$ during the next century and (2) a predicted concomitant increase in annual temperature of $3 \pm 1.5°C$ in the Mediterranean Basin (Dickinson, 1986; Bach, 1988; Bouwman, 1990). The response of the Mediterranean vegetation to CO_2 fertilization must be linked to soil nitrogen availability. The high stability of the organic compounds is a major characteristic in many Mediterranean soils, and is generally explained by simultaneous effects of high turnover rates and of repeated seasonal alternation of dry and wet soil conditions, resulting in low C:N ratios, in high condensations of the humified compounds, and ultimately in the formation of stable humus–clay associations (Bottner, 1985). A high proportion of organic nitrogen is generally in protected organic structures with low mineralization rates. Despite the relatively high amounts of total nitrogen in some Mediterranean forest soils, most of the natural Mediterranean ecosystems are nitrogen stressed. Therefore, the response of the vegetation to CO_2 increase cannot be predicted as a single-factor response but must be simultaneously analyzed at least by considering the nitrogen availability.

In 1981, Luxmoore hypothesized that the CO_2 rise, by increasing the below-ground translocation of carbon in the plants, may stimulate rhizosphere activity and hence stimulate soil nutrient mobilization, through nitrogen fixation, mycorrhizal activity, and soil nitrogen mineralization. Billes et al. (1993) investigated the rhizosphere activity of wheat by doubling the CO_2 concentration; they demonstrated an increased exudation and a light stimulation of the soil nitrogen net mineralization. Thus, the essential question is to know to what extent this strategy may improve the ecosystem nitrogen availability, in response to CO_2 fertilization.

From numerous laboratory and greenhouse experiments investigating the response of the plants to increased CO_2 and using very different cultivated or natural C_3 species, it appears that in nitrogen-limited soils, the nitrogen concentration of the plant material is generally lowered. Table 15-1 presents the C:N ratios of 3-year-old chestnut trees cultivated

Table 15-1. Carbon and Nitrogen Distribution in Plant Parts of 3-Year-Old Sweet Chestnut Trees (*Castanea sativa* Mill.) Exposed Over 2 Years to 350 $\mu L \cdot L^{-1}$ and 700 $\mu L \cdot L^{-1}$ CO_2 Concentrations

		Carbon g $Plant^{-1}$		Nitrogen g $Plant^{-1}$		Nitrogen % Dry w.		C:N	
CO_2 $\mu L \cdot L^{-1}$		350	700	350	700	350	700	350	700
Leaves	a	6.14	7.66	0.11	0.06	0.80	0.39	54.4	120
	b	1.85	2.13	0.03	0.01	(1)		(1)	
	c	ns		**					
	d	1.25		0.56		0.49		2.20	
Stem and branch		14.5	16.7	0.18	0.12	0.58	0.32	80.1	145
		4.10	4.60	0.04	0.03	0.07	0.04	9.99	19.6
		ns		**		**		**	
		1.16		0.65		0.55		1.81	
Tap root		8.92	9.62	0.12	0.08	0.59	0.36	75.7	127
		2.81	2.38	0.03	0.02	0.05	0.06	6.64	22.4
		ns		**		**		**	
		1.08		0.66		0.61		1.67	
Roots > 1 mm		5.07	7.49	0.11	0.10	1.00	0.61	45.6	76.4
		1.67	1.12	0.03	0.03	0.16	0.14	7.5	19.8
		*		ns		**		**	
		1.48		0.93		0.61		1.68	
Roots 1 − 0.5		11.0	15.1	0.28	0.32	1.20	1.03	40.2	46.8
		2.19	2.53	0.05	0.06	0.11	0.12	4.2	5.4
		**		ns		*		*	
		1.36		1.18		0.86		1.16	
Roots < 0.5 mm		1.52	2.17	0.05	0.06	1.55	1.32	30.8	36.2
		0.37	0.52	0.012	0.016	0.11	0.17	2.59	4.59
		*		ns		**		*	
		1.43		1.23		0.85		1.17	
Whole plant		47.1	58.7	0.84	0.74	0.82	0.59	55.3	79.0
		11.9	12.6	0.16	0.12	0.08	0.08	5.4	11.9
		ns		ns		**		**	
		1.24		0.88		0.72		1.43	

(1): pooled samples.
[a] Mean values ($n = 7$ for 350 ppm and $n = 8$ for 700 ppm).
[b] Standard deviation.
[c] $** p < 0.01.99$; $*$ $0.01 < p < 0.05$; ns: $0.05 < p$.
[d] Values at 700 $\mu L\, CO_2 \cdot L^{-1}$/values at 350 $\mu L\, CO_2 \cdot L^{-1}$.

for 2 years in 350 and 700 $\mu L \cdot L^{-1}$ CO_2 concentrations respectively, in the greenhouse, on natural chestnut forest soil. The nitrogen concentration is lowered in all the plant parts except in the fine roots. The highest reduction of the nitrogen concentration appears in the inactive plant parts (the stem, the tap root, and the coarse roots) and in the leaves. The lower reduction appears in the active fine roots. Thus, from these experiments, the general assumption is that in nitrogen-stressed soils, the quality of the plant material is modified in response to the doubling of the atmospheric CO_2. Stimulation of root exudation and rhizosphere activity by increased CO_2 is likely inefficient or poorly efficient for mobilizing an extra amount of nitrogen from the stabilized organic compounds to compensate for the increased carbon assimilation. Therefore, many CO_2-controlled experiments showed that in soils with low nitrogen availability, CO_2 fertilization stimulated plant production but did not affect or only slowly affected nitrogen mineralization from the soil. The hypothesis is that in MTEs nitrogen stress and the stability of nitrogen organic compounds will probably limit the CO_2 fertilization effect on plant growth.

Thus, from our present knowledge, it appears that for most plants growing in natural soil conditions, the plant residues returning annually to the soil will have lower nitrogen concentrations in response to increased atmospheric CO_2 concentration. No evident trends have been found in the literature with respect to changes in the concentrations of other nutrients. In some cases, a parallel increase of lignin content has been found (Anderson, 1991). C:N ratio or lignin:N ratio are common indicators of litter quality. Therefore, it is fundamental to understand to what extent these sometimes dramatic quality modifications may affect the decomposition processes and rates. The task of the soil biologists is to built up models able to predict how these modifications may affect (1) SOM turnover, reservoir, and residence time and (2) nutrient and, especially, nitrogen supply for plants.

It is generally assumed that the expected decrease in nitrogen concentrations of the plant material will increase the residence time of carbon in soils and hence increase carbon sequestration in the soil reservoir. However, some recent examples show that the response of the soils will probably be more complex. In an experiment in progress, litter bags of different C:N ratios, incubated in the same site under Mediterranean climate conditions, showed the same mass loss pattern during the first year (Vallejo, personal communication). Berg and Ekbohm (1991), incubating different litter types in bags in Swedish coniferous sites, found that at the initial steps of decomposition, mass loss was positively related to the initial nitrogen content, but after incubation over 3 years, the situation was reversed, and finally the litter with the lowest initial nitrogen content decomposed faster. The hypothesis posed by these authors is that nitrogen in the litter that is richer in nitrogen may inhibit lignin decomposition during the late decomposition stages. Coûteaux et al. (1991) incubated chestnut litters (with C:N ratios = 40 and 75, obtained at CO_2 concentrations of 350 and 700 $\mu L \cdot L^{-1}$ respectively), for 11 months in microcosms with controlled temperature and moisture, and controlled soil animal groups. In the incubation systems with the simplest food web, the litter with the lowest nitrogen content decomposed more slowly than the one with higher nitrogen content. In contrast, in the most complex food webs, the litter having the lowest nitrogen concentration showed surprisingly faster CO_2 release than the litter with high nitrogen concentrations. Thus, the change in quality of the plant material in response to CO_2 increase results in complex and partly unknown effects on the decomposition rates. The results were explained by direct effects of the substrate quality on the microbial activity, through interactions between soil fauna and microflora. In a recent review, Anderson (1991) concluded that initial litter quality appears to be a reasonable descriptor of decomposition rates, only during the initial periods when 40–50% of original mass is lost; afterwards, the accumulation of recalcitrant products increasingly determines the process.

These illustrations of the indirect effects of atmospheric CO_2 increase point out the necessity to link plant production models to decomposition models. In MTEs, it seems essential to analyze the plant response to CO_2 increase in relation to water potentials and to nitrogen availability. The carbon and nitrogen allocation in the plant determines the quality of the plant material. The leaves and root litter quality and the exudates determine the decomposition rates and finally the nitrogen mineralization rates. The processes are regulated by the soil properties.

Direct Effects of the Predicted Temperature Increase

The mean annual temperature increase in the Mediterranean Basin is estimated at 3 ± 1.5°C (Bach, 1988; Brinkman and Brammer, 1990). There are few indications predicting a change in temperature seasonality, which is an essential feature of the Mediterranean conditions. In addition, GCMs do not provide reliable indications of changes in the amount and the seasonal distribution patterns of rainfall. A slow decrease or little change is expected in mid-latitudes, e.g., in Mediterranean areas (Brinkman and Brammer, 1990). A 3°C increase would cause an average altitudinal upward shift of the vegetation belts of 500 m, the tree line would reach 2,500 m under 46°N and 3,500 m under 30°N, and each bioclimatic latitudinal vegetation zone would move one step upwards, the xerothermo Mediterranean climate zone, for example, replacing the actual thermo-Mediterranean zone (Le Houerou, 1990). In the case of high rainfall at higher elevation and higher latitudes, the productivity would increase as a result of a longer growing season and reduced winter cold stress. In the cold arid zone, the growing season would be likewise extended. In most of the arid zones, where winter temperature is not a major constraint to primary productivity, the temperature increase would adversely affect the length of the growing season by lowering the critical P/PET ratio.

Figure 15-6A shows the geographic distribution of a combined coefficient of temperature × moisture, controlling the soil organic matter

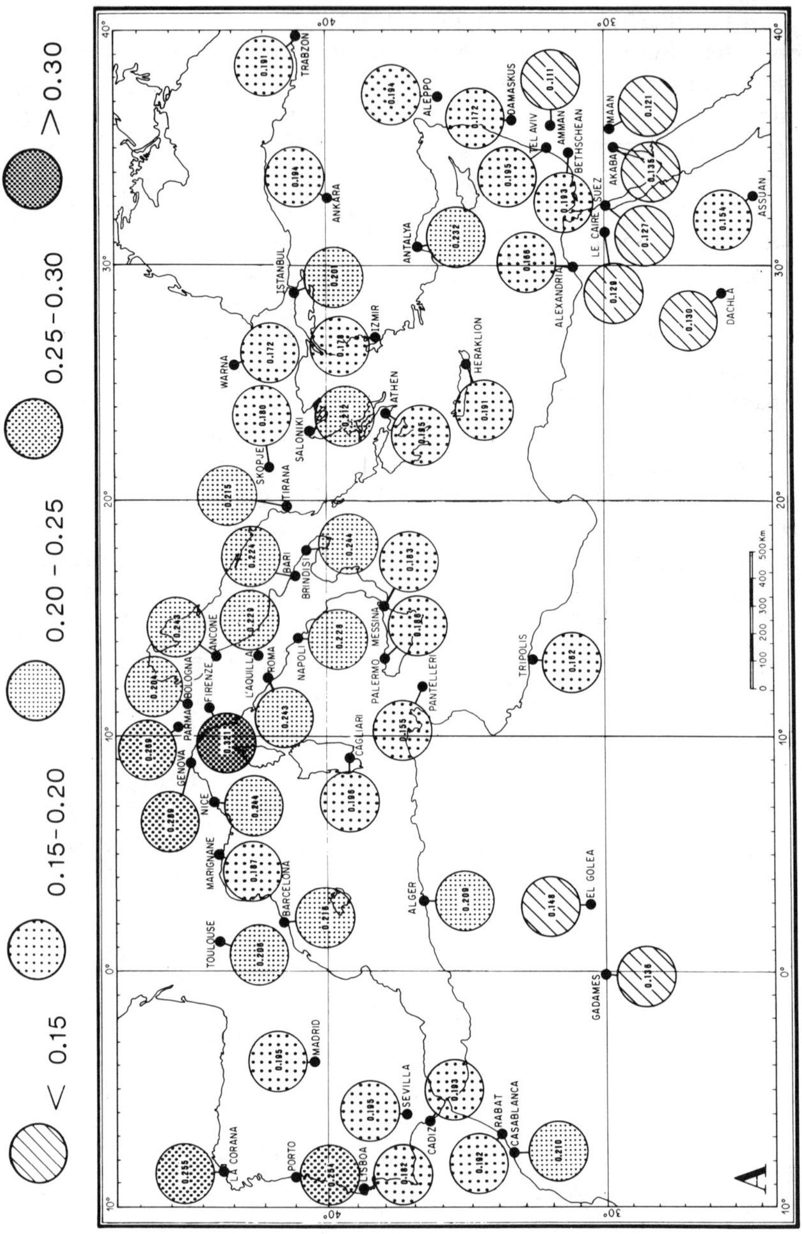
< 0.15
0.15-0.20
0.20 - 0.25
0.25-0.30
> 0.30
A

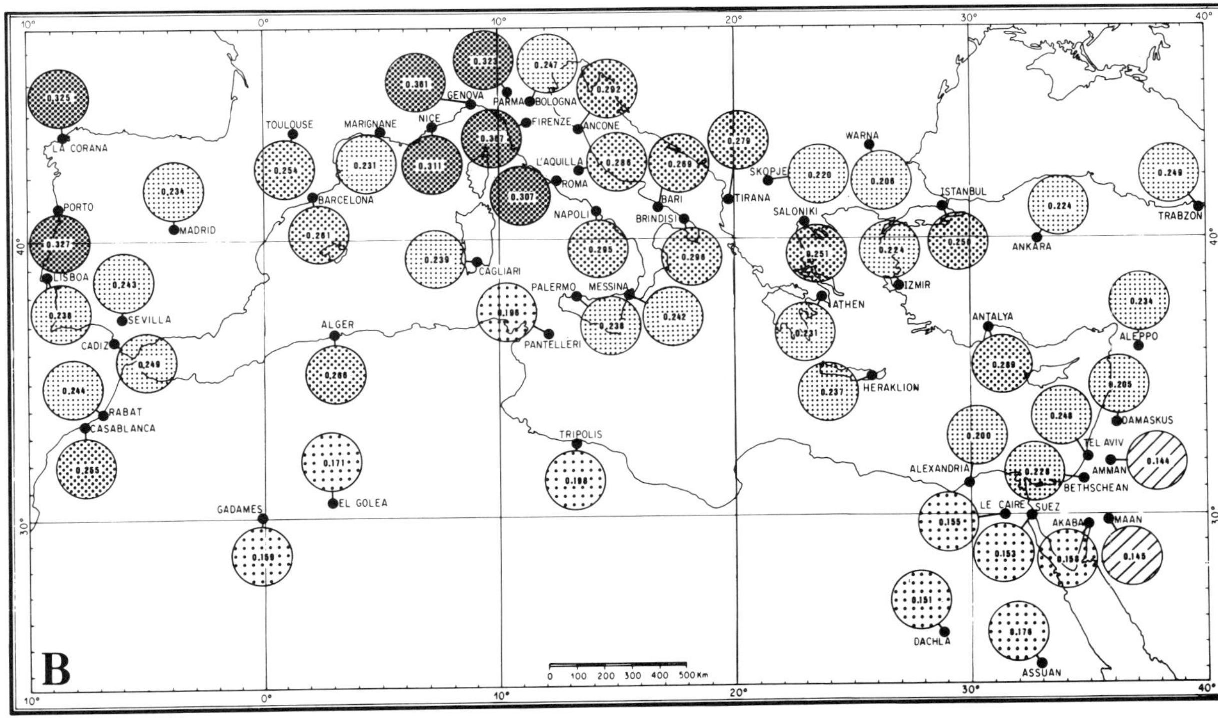

Figure 15-6. (A) Geographic distribution of the climate-related coefficient controlling the decomposition rate. The coefficient is calculated for some essential meteorological stations of the Mediterranean Basin, using the temperature and PET/P response functions of Parton et al. (1987). (B) Simulation of 3°C temperature increase, without changing the precipitation conditions. In A and B, PET is calculated according Le Houérou (1990).

decomposition rates, derived from data from representative meteorological stations throughout the Mediterranean Basin. The coefficient, ranging from 0 to 1 (the theoretical optimum climatic conditions), is calculated, on a monthly basis, according to the response functions of the decomposition rates (1) to temperatures and (2) to precipitation/potential evapotranspiration ratios (P/PET), according to Parton et al. (1987). The map shows that the coefficient increases from south to north, indicating that in the Mediterranean area the effect of moisture generally prevails over the effect of temperature. Figure 15-6B results from a simulation of a 3°C rise of the temperature without changing the precipitation patterns: the simulation increases the temperature component of the response function and decreases its P/PET component. The decomposition coefficients (temperature × P/PET) increase in all the bioclimate zones, including the most moisture-stressed zones. Nevertheless, the major effect of the temperature increase is expected to occur in the Mediterranean humid areas.

This is illustrated in Table 15-2, showing the stimulated effect of a 3°C temperature rise—without changing the precipitation patterns—on the SOM pool in four major bioclimatic zones of the Mediterranean Basin, according to the Century model (Parton et al., 1987). The carbon losses range from 15% to 28% of the initial contents and the loss is higher in the humid than in the drier zones, demonstrating again the prevailing role of moisture conditions. Also, in the arid zones the carbon stock is low (about 10 times less than in humid areas) and any change of its size is of little importance. Thus, the soil warming will have greater effect in the

Table 15-2. Loss of Soil Organic Matter (SOM) from Some Climatic Zones, Expressed in % of Initial Content, in Response to a Simulated 3°C Temperature Increase

Climatic Zones	Winter Conditions	Meteorological Stations	Mean Temperature °C (1)	Annual Precipitation mm (2)	SOM Loss of Initial %
Hyperhumid	Temperate	Ain Draham (Tunisia)	15°0	1,534	28
Subhumid	Cool	Montpellier (France)	14°2	699	20
Hyperarid	Warm	Cairo (Egypt)	21°9	26	15
Hyperarid	Cool	Bechar (Algeria)	20°2	90	15

(1) average of the temperatures (30 years).
(2) average of the annual precipitations (30 years).

wet Mediterranean zones, when both the carbon pools and the percentage loss of carbon are higher. The specific response time was calculated for the site of Montpellier: the half-times needed to reach equilibrium are 2, 20, and 515 years for the active, slow and passive fractions of the Century model (Parton et al., 1987), respectively. Thus, taking into account the size of the organic pools, one may conclude that the slow fraction will be the main source of soil CO_2 in response to a possible temperature increase during the next century.

The organic carbon content usually decreases exponentially with depth, but the carbon density can follow very different types of distribution. In Mediterranean forest soils having approximately 1 m of depth, Hereter (1990) found that 50% of the total soil carbon reserve is accumulated in the upper 25 cm. The foreseen increase in air temperature is not going to affect organic matter at all depths in the same way. Owing to the prevailing dry conditions in the Mediterranean areas, a low thermal conductivity and high warming effect on the top soil layer are expected. Therefore, the upper soil layer would be mostly affected by any warming effect.

In the decomposition process models, the control by temperature is described as a single response function: the temperature dependency (i.e., the Q_{10} values) is similar for all the organisms, regardless of the diversity of microbial function and populations and regardless of the high diversity of organic substrates. Schleser (1982), investigating the respiration response to temperature increase in soils coming from a large range of climate conditions, showed that Q_{10} values actually ranged from 2 to 8. Thus, in terms of effects of temperature increase on the SOM reservoir, there is need to improve the description of the temperature dependency of the major decomposition processes.

Conclusions and Perspectives

The aim of this review was to analyze the biological component of the Mediterranean soils and to stress some essential characteristics, with regard to the sensitivity of the soils to global environmental changes and possible significant feedbacks on global system functioning. The fundamental functions and processes governing the transformation of the plant residues to SOM, and resulting ultimately in carbon and nutrient mineralization, are probably comparable in most of the bioclimatic zones and especially under temperate, Mediterranean, and subtropical climates. Nevertheless, despite a large diversity of physical and chemical abiotic soil properties (mineralogy, geomorphology, history, soil use and management, etc.), the global spatial distribution pattern of the SOM is qualitatively (humus types) and quantitatively (see Figure 15-1) governed by bioclimatic conditions. The bioclimatic factor acts directly or indirectly through plant production, litter quality, and decomposition rates.

In the Mediterranean climate, the relatively low and moisture-stressed productivity of the ecosystems explains the relatively low SOM densities. The deep root system and the deep below-ground seasonal activity of the soil fauna, both related to dry summer conditions, cause appreciable SOM content into the soil profile. The Mediterranean climate conditions also shape some major soil chemical features, e.g., moderately acidic to basic soil pH, resulting in a soil chemical environment favorable to high biological activities and large population diversity and leading to high decomposition rates. The characteristic stability of the organic compounds and of clay–humus complexes in Mediterranean soils is due to the combination of (1) a high biological activity and (2) the alternation of wet winter conditions and strong dry summer conditions. Thus, in the MTEs, some major SOM properties result from direct or indirect effects of the climate conditions.

Other SOM properties are related to the mineral soil component. In the Mediterranean Basin, many properties of the mineral soil are inherited from paleoclimatic conditions, giving a juxtaposition of a high diversity of soils, with different ages and with different properties. Some qualitative and quantitative properties of organic matter are closely related to the mineral environment of these diversified soils. The intensity of the leaching of carbonates and of clays, the mineralogy and degree of weathering, are some important factors governing organic matter distribution through the soil. Therefore, the high diversity of the Mediterranean soil types explains the high diversity of the types of SOM.

Owing to their limited areal extent and their low standing crop density of carbon in plant and soils, the MTEs will probably have a small impact on the global balance of CO_2 and other trace gases. In contrast, the increasing atmospheric CO_2 and the subsequent expected temperature rise will probably have some major direct and indirect effects on the functioning of the Mediterranean ecosystems. Their understanding requires the definition of some specific research priorities and recommendations. Among them, five crucial questions are of special interest in the Mediterranean countries.

1. The chemical and physical specific properties of the plant material, resulting essentially from the sclerophyllous nature of the MTEs, seems to require high activity and diversity of the soil fauna (Bertrand et al., 1987), which is an essential step of the initial litter decomposition process, preceding and conditioning the microbial activity. Thus it seems important to understand the shift in activity and structure of the soil fauna, as a result of modifications of the litter quality in response to CO_2 fertilization and increased sclerophylly.

2. A high proportion of the soil organic nitrogen is in condensed and rather stable and inactive forms, due to seasonal alternation of wet and dry soil conditions. The question is to understand and to predict to what extent (1) the effect of increased mineralization rates, related to climate change, and (2) the effect of modifications in the litter quality, related to CO_2 fertilization, will modify, within the nitrogen cycle, the balance between stable and active nitrogen. It is essential to predict if the active fraction of the nitrogen will increase or decrease in response to physical and chemical climate change. The answer to this question is rather difficult, since temperature rise and CO_2 increase will probably have opposite effects on the mobility of nitrogen. In addition, the nitrogen dynamic in the soil is driven by carbon, which is the microbial energy support. Thus, the effect of environmental change on nitrogen balance cannot be understood without analyzing its link with the carbon cycle.

3. It also seems important to clarify the effect of CO_2 increase on the Mediterranean vegetation, in terms of production, vegetation structure (sensitivity to fire), and competition, with a special focus on the carbon and nitrogen allocation governing the quality of leaf and root litter. Litter quality is essential in predicting modifications in the nitrogen cycle, and the response of plants cannot be explained without describing nitrogen availability. Thus, for MTEs it seems essential to link the response to CO_2, to nitrogen cycling, and to water use efficiency. The effects of climate change must be considered in terms of a multiple factor response. The highly diversified plant species composition of the MTEs is another difficulty when answering these questions.

4. The stimulation of rhizosphere activity by doubling CO_2 and the consequences on nutrient availability have recently been investigated. The nitrogen supply seems to be a major factor controlling the exudation response of the plants and the microbial utilization of the exudates (Merckx et al., 1987), and the rhizosphere activity seems to partly control nitrogen availability (Billes et al., 1993). In MTEs, an important topic that requires further research is to understand to what extent water limitation may affect root activity and rhizospheric microbial activity. The interest of these investigations is to analyse the response of the plant–soil system as an interactive response to CO_2, nutrient availability, and water potential.

5. This review analyzed the effects of global climate changes, excluding the impact of land use, which is another essential cause of profound transformations of Mediterranean soils. MTEs in the Mediterranean Basin have been affected by intensive human activities for centuries; great changes in land use are expected in the near future (Le Houerou, 1990). Further research is needed to investigate the effect of land use on SOM in the context of fire regime modification, cropland abandonment in the European Mediterranean, and desertification by increased cultivation in North Africa.

References

Ajtay GL, Ketner P, Duvigneaud P. 1979. Terrestrial primary production and phytomass. In: Bolin B, Degens ET, Kempe S, Ketner P, eds. The Global Carbon Cycle, SCOPE 13, 491p:129–171.

Anderson JM. 1991. The effects of climate change on decomposition processes in grassland and coniferous forests. Ecol Appl 1(3):326–347.

Bach W. 1988. Development of climatic scenarios from general circulation models. In: Parry ML, Carter TR, Konijn NT, eds. The Impact of Climatic Variations on Agriculture, Sect 4:125–158. Kluwer, Dordrecht, The Netherlands.

Berg B, Ekbohm G. 1991. Litter mass-loss rates and decomposition patterns in some needle and leaf litter types. Long-term decomposition in Scots pine forest. VII. Can J Bot 69:1449–1456.

Bertrand M, Janati-Idrissi A, Lumaret JP. 1987. Etude expérimentale des facteurs de variation de la consommation de la litière de *Quercus ilex* L. et *Q. pubescens* Willd. par *Glomeris marginata* (V.) (Diplopoda. Glomeridae). Rev Ecol Biol Sol 24:359–368.

Billes G, Rouhier H, Bottner P. 1993. Modifications of the carbon and nitrogen allocations in the plant-soil system, in response to increased atmospheric CO_2 concentration. Plant Soil 157:215–225.

Bohn HL. 1982. Estimate of organic carbon in world soils: II. Soil Sci Soc Am J 46:1118–1119.

Bolin B. 1986. The carbon cycle and predictions for the future. In: Bolin B, Döös BR, Jäger J, Waerwick RA, eds. The Greenhouse Effect, Climate Change and Ecosystems. SCOPE 29:93–156. John Wiley & Sons, Chichester, UK.

Bottner P. 1981. Evolutions des sols et conditions bioclimatiques méditerranéennes. Ecol Mediterr 8:115–134.

Bottner P. 1985. Response of microbial biomass to alternate moist and dry conditions in a soil incubated with ^{14}C and ^{15}N-labelled plant material. Soil Biol Biochem 17:329–337.

Bottner P, Peyronel A. 1977. Dynamique de la matière organique dans deux sols méditerranéens étudiée à partir de techniques de datation par le radiocarbone. Rev Ecol Biol Sol 14:385–393.

Bouwman AF. 1990. Exchange of greenhouse gases between terrestrial ecosystems and the atmosphere. In: Bouwman AF, ed. Soils and the Greenhouse Effect. John Wiley & Sons, Chichester, UK, pp. 62–191.

Brinkman R, Brammer H. 1990. The influence of a changing climate on soil properties. Transactions 14th International Congress of Soil Science (Kyoto) Vol V:283–288.

Buringh P. 1984. Organic carbon in soils of the world. In: Woodwell GW, ed. The Role of Terrestrial Vegetation in the Global Cycle, SCOPE 23:91–109. John Wiley & Sons, Chichester, UK.

Canadell J, Rodá F. 1989. Biomasa y mineralomasa subterránea del encinar de la Castanya, Montseny. Options Méditerranéennes; Série Séminaire 3:13–18.

Cannell MGR. 1982. World Forest Biomass and Primary Production Data. Academic Press, London.

Coûteaux MM, Mousseau M, Célérier ML, Bottner P. 1991. Increased atmospheric CO_2 and litter quality: decomposition of sweet chestnut leaf litter with animal food webs of different complexities. Oikos 61:54–64.

Dickinson RE. 1986. The climate system and modelling of future climate. In: Bolin BR, Döös BR, Jäger J, Waerwick RA, eds. The Greenhouse Effect,

Climate Change and Ecosystems. SCOPE 29:207–271. John Wiley & Sons, Chichester, UK.

Franz EH. 1990. Potential influence of climatic change on soil organic matter and tropical agroforestry. In: Scharpenseel HW, Shomaker M, Ayoub A, eds. Soil on a Warmer Earth. Elsevier, p. 274.

Hereter A. 1990. Els sols forestals del Massis Montseny. Tesi doctoral, Universitat de Barcelona, Barcelona, Spain.

Le Houérou HN. 1990. Global change: vegetation, ecosystems and land use in the Southern Mediterranean Basin by the mid twenty-first century. Israel J Bot 39:481–508.

Le Houérou HN, Popov GF. 1981. An ecoclimatic classification of inter-tropical Africa. Plant production and protection paper No. 31. FAO, Rome.

Luxmoore RJ. 1981. CO_2 and phytomass. Bioscience 31:626.

Meentemeyer V. 1984. The geography of organic matter decomposition rates. Ann Assoc Am Geographers 74:551–560.

Merckx R, Dijkastra A, Den Hartog A, Van Veen JA. 1987. Production of root-derived material and associated microbial growth in soil at different nutrient levels Biol Fert Soils 5:126–132.

Naveh Z, Lieberman AS. 1984. Landscape ecology. Springer-Verlag, New York.

Parton WJ, Schimel DS, Cole CV, Ajima DS. 1987. Analysis of factors controlling soil organic matter levels in Great Plains grasslands. Soil Sci Soc Am J 51:1173–1179.

Post WM, Emanuel WR, Zinke PJ, Stangenberger AG. 1982. Soil carbon pools and world life zones. Nature 298:156–159.

Pouget M. 1980. Les relations sol-vegetation dans les steppes sud-algéroises. Trav et doc ORSTOM, F-93140 BONDY, 553 p.

Rozanov BG. 1990. Human impacts on evolution of soils under various conditions of the world. Transactions 14th ICSS Plenary Papers, 53-62 Kyoto.

Scharpenseel HW, Becker-Heidmann P. 1990. Overview of the greenhouse effect. Global change syndrome; general outlook. In: Scharpenseel HW, Schomaker M, Ayoub A, eds. Soil on a Warmer Earth. Elsevier, 274 p.

Schleser GH. 1982. The response of CO_2 evolution from soils to global temperature changes. Z Naturforsch 37a:287–291.

Schlesinger WH. 1984. Soil organic carbon: a source of atmospheric CO_2. In: Woodwell GM, ed. The Role of Terrestrial Vegetation in the Global Carbon cycle: Measurement by remote sensing. SCOPE 23:111–131. John Wiley & Sons, Chichester, UK.

Vallejo VR. 1983. Los suelos forestales de la depresion Central Catalana. Thesis doctoral, Univesitat de Barcelona, Barcelona, Spain.

Van Veen JA, Paul A. 1981. Organic carbon dynamics in grassland soils. I. Background information and computer simulation. Can J Soil Sci 61:185–201.

Vogt KA, Grier CC, Vogt DJ. 1989. Production turnover, and nutrient dynamics of above- and below-ground detritus of world forests. Adv Ecol Res 15:303–307.

16. Ecosystem Response to Elevated CO_2: Nutrient Availability and Nutrient Cycling

W.D. Stock and G.F. Midgley

Atmospheric CO_2 levels have risen from an estimated preindustrial concentration of $280\,\mu L\,L^{-1}$ (Friedli et al., 1986; Neftel et al., 1985) to $350\,\mu L\,L^{-1}$ today (Boden et al., 1990) and, even using conservative assumptions regarding future energy sources, could increase to $600\,\mu L\,L^{-1}$ by the end of the next century (Ausubel et al., 1988). The increase in agricultural crop yield which could result from a doubling of preindustrial CO_2 levels has been suggested to be in the range of 30–40% (Kimball, 1983; Cure and Acock, 1986). However, natural ecosystem responses to increasing atmospheric CO_2 levels are more difficult to predict, as their mechanistic basis remains poorly understood (Bazzaz, 1990), especially in terms of system processes and interactions between system components (Morison, 1990). Such an understanding depends heavily on the elucidation of the linkages between plant carbon uptake, carbon allocation, and nutrient cycling. Because these processes interact, disruption of any one often induces changes in others, which then leads to either positive reinforcement or negative feedback at the ecosystem level. This interaction is critical for the ultimate structural, functional, and floristic nature of the altered ecosystem. Natural ecosystems of certain Mediterranean climate regions have been shown to be altered substantially by changes in nutrient availability (Specht, 1963) in terms of species composition and system functioning, and it appears that rising atmospheric CO_2 levels may hold a similar threat, but at a global scale.

The nutrient-limited environments of the world that have a Mediterranean-type climate have species-rich floras and show interesting convergence in many aspects of plant growth form and ecosystem function (Cody and Mooney, 1978; Cowling and Campbell, 1980). Species of nutrient-poor environments have low nutrient requirements (Chapin, 1980; Berendse et al., 1992) and have developed mechanisms for increasing their efficiency of acquiring nutrients (through specialized root systems or microbial associations) and nutrient use (evergreenness, internal recycling). Low relative growth rates appear to be an inherent component of these adaptations (Chapin et al., 1986; Chapin, 1988). It is possible elevated atmospheric CO_2 will disrupt the functioning of these systems, with severe consequences for the maintenance of their biotic diversity (Mooney et al., 1991).

Ecosystem processes potentially affected by elevated CO_2 (Figure 16-1) include those responsible for the cycling of nutrients and hence control of nutrient availability. The impacts of environmental stresses on ecosystems are generally evident as changes in either the total nutrient pools within

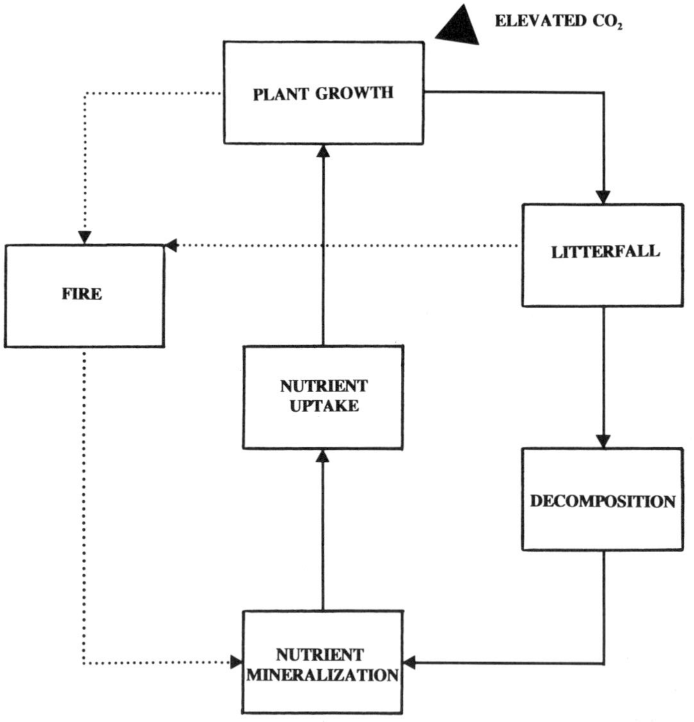

Figure 16-1. Diagrammatic representation of the main processes involved in nutrient cycling in Mediterranean ecosystems which may be affected by increased atmospheric CO_2 levels.

the vegetation or in the rates of nutrient turnover. Alien nitrogen-fixing invaders on acid fynbos soils, for instance, have been shown to substantially increase the nutrient content of the vegetation and soils of the southwest Cape, while having little impact on the rates of nutrient release from decomposition (Stock and Allsopp, 1992). Without an attempt to understand possible interactions and feedbacks at the ecosystem level, elevated CO_2 studies at the individual plant level are poor predictors of potential changes in our terrestrial environment.

The objectives of this chapter are, first, to investigate whether nutrient availability affects plant photosynthetic and carbon allocation response to elevated CO_2 directly, and second, to determine whether an interaction between elevated CO_2 levels and nutrient availability has implications for patterns and rates of nutrient cycling in Mediterranean-type ecosystems occurring on soils of differing nutrient status.

Nutrient Availability and Plant Response to Elevated CO_2

Plant responses to elevated CO_2 possibly mediated by low nutrient availability that could impact processes at the ecosystem level (Figure 16-1) fall into three basic groups. These are potential changes in (1) plant relative growth rates and productivity (and hence biomass structure and accumulation rates) through enhanced photosynthetic CO_2 uptake and altered carbon allocation patterns at the whole-plant level; (2) leaf-level carbon allocation patterns (altered C:N ratios, C:P ratios, and secondary compound production and composition); and (3) rhizosphere interactions including the allocation of carbon to root symbionts.

Interactions between nutrient availability and elevated CO_2 have been considered in agronomic and forestry studies, which include mainly fast-growing species with high nutrient requirements. Although these studies are useful for deriving a general understanding of the relationship between these two parameters, it is premature to extrapolate to species which are adapted to nutrient-poor environments.

Growth Rates and CO_2 Response

Plant response to elevated CO_2 in C_3 plants, which dominate Mediterranean-type climate regions, appears to be mediated by carbon source:sink relationships within the plant (Arp, 1991) which may be extremely complex (Cure et al., 1991). For instance, it is possible that initial widespread reports of photosynthetic acclimation after a treatment period of atmospheric CO_2 enrichment could have been due to limited root expansion in small pot volumes, which has been shown to reduce sink demand from the root system (Arp, 1991; Thomas and Strain, 1991). Restricted demands for carbon could also result from low nutrient availability, or the evolutionary adaptation of species to low-nutrient environments (i.e., genetically determined low plant growth rates). Either mechanism would inhibit plant response to elevated CO_2 in the short term but would have different implications in the long term. For instance, additional carbon fixed by fast-growing plants under nutrient-limiting conditions could be invested in root growth or the production of exudates, both of which could alleviate the nutrient limitation, thereby allowing increased growth and potentially enhancing rates of nutrient cycling. Genetically limited, slow-growing plants, on the other hand, have few options for increasing growth rates under rising atmospheric CO_2 conditions, and thus minor changes in ecosystem processes would be predicted.

Photosynthetic carbon uptake rates of C_3 plants are initially increased by elevated CO_2 concentrations; in particular, the rate of leaf photorespiration is reduced and Rubisco activity increased (Bazzaz, 1990; Arp, 1991; Bowes, 1991). Such enhancement effects may disappear with time of exposure to enriched CO_2, reflecting acclimation processes which are not only physiological and biochemical, but may also be anatomical or morphological in nature (Bowes, 1991). The nutrient dependence of the acclimation process is poorly understood in agricultural and wild species, and this topic requires further study before predictions of long-term plant photosynthetic and growth responses to elevated CO_2 in nutrient-limited environments can be made (Bazzaz, 1990).

In agricultural and forestry situations, plant growth is limited by nutrient supply, particularly of nitrogen, more than any other resource except water (Kramer and Kozolowsi, 1979). The largest nitrogen investment in a plant is in the leaf, and Field and Mooney (1986) have demonstrated a strong positive relationship between leaf nitrogen content and photosynthetic rate in a wide range of plant species. Thus, suggestions such as that of Bazzaz (1990) that plant response to elevated CO_2 is more pronounced under high levels of resources such as water and nutrients are appealing. The notion of resource-constrained responses implies that under low nutrient availability, decreased photosynthetic responses to CO_2 would be expected; however, conflicting reports appear in the literature. Some species, such as *Sinapis alba* (a C_3 annual crop plant), produced more biomass under elevated CO_2, only when nutrients were not limiting, while in the same experiment, production of *Abutilon theophrasti* (a wild C_3 annual) was independent of nutrient supply (Coleman et al., 1991). Norby and O'Neill's (1989, 1991) comparison of white oak (*Quercus alba*), a species which responds relatively little to increased nutrient supply, and yellow poplar (*Liriodendron tulipifera*), a species which benefits from greater nutrient supply, showed little change in plant relative photosynthetic enhancement or biomass accumulation under elevated CO_2. These studies suggest that the increase in plant growth under elevated CO_2 is related to the maximum growth rate of the study species under ambient conditions. Species with low growth rates and therefore low sink demands would respond less to nutrient or CO_2 fertilization. Hunt et al. (1991) also found that plants with the slow-growing, stress-tolerant strategy (*sensu* Grime, 1977) showed the lowest response to elevated CO_2 conditions (Figure 16-2).

At the community level, responses to elevated CO_2 depend on the nutrient requirements and growth characteristics of the vegetation type involved. Productivity of both C_3 and C_4 communities in a nutrient-rich saltmarsh system showed continued stimulation of carbon uptake and growth, even after 2 years of CO_2 enrichment (Drake and Leadley, 1991). By contrast, the carbon gain of nutrient-limited tundra vegetation under elevated CO_2 rapidly decreased, after initial stimulation, to the same level as that prior to CO_2 fertilization (Tissue and Oechel, 1987). A controlled-environment study of three important tundra plant species showed that their production was not increased by elevated atmospheric CO_2, even with nutrient enrichment (Oberbauer et al., 1986). In a greenhouse-based study of two sclerophyll Proteaceae species from the nutrient-poor fynbos region, preliminary data on plant photosynthetic responses

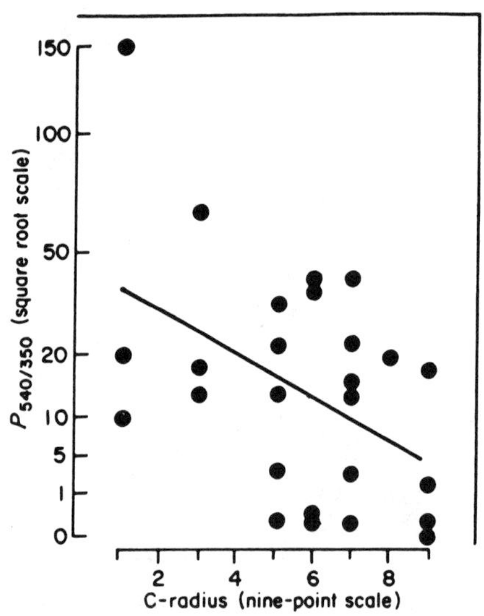

Figure 16-2. Relation between CO_2 responsiveness and C-S-R plant life history strategy in 26 species of C_3 metabolism. $P_{540/350}$ is the percentage increase in plant yield predicted from the experiments, measured between the current nominal CO_2 concentration (350 $\mu L\,L^{-1}$) and that predicted for the year 2050 (540 $\mu L\,L^{-1}$), plotted on a square root scale. Plant strategy is expressed as distance from the pure competitive strategy on a nine-point scale, such that C = 1 and R = S = 9. The fitted line is significant at $p < 0.02$ ($\sqrt{P_{540/350}} = 0.73 - 0.55$ (C-radius)), $r = -0.46$. (From Hunt et al., 1991.)

show conflicting results. *Leucadendron laureolum* showed photosynthetic down regulation under low-nutrient supply conditions while *L. xanthoconus* (Figure 16-3) showed no such response. Such conflicting results among sister species from the same low-nutrient environment suggest that responses to elevated CO_2 cannot be predicted with any degree of certainty from current agricultural nutrition paradigms.

The application of the "limiting factor" paradigm, first proposed by Liebig, which forms the underlying basis of agricultural research, appears inadequate when attempting to predict plant photosynthetic and growth response to the interaction between nutrient availability and elevated CO_2 supply (Sinclair, 1992) for plants from both nutrient-rich and nutrient-poor environments. The assumption that all plants are limited by resource availability throughout their lifespan ignores evidence obtained from natural ecosystems where nutrient addition experiments (Witkowski et al., 1990) have elicited little or no increase in growth. Alternative paradigms, such as the carbon/nutrient balance hypothesis (Bryant et al., 1983; Baas, 1989) and the economic analogy (Bloom et al., 1985), consider the ecological and evolutionary consequences of patterns of carbon allocation to various organs and functions, and as such have a conceptual framework in which source-sink relationships (already shown to be of importance in interpretation of elevated CO_2 studies [Arp, 1991; Thomas and Strain, 1991]) are an integral part.

Plant Carbon Allocation

Patterns of carbon allocation to growth versus carbon-rich secondary compounds at the plant and leaf level under elevated CO_2 may be considered in the light of the carbon/nutrient balance hypothesis (Bryant et al., 1983). According to this paradigm, low nutrient supply and adequate light conditions are suggested to result in an excess carbon condition in plants, even under ambient CO_2 levels. This excess carbon may be stored as starch or allocated to carbon-based secondary compounds which have a role in plant defense or which regulate a plant's interaction with the biotic environment (allelopathy). Several studies have shown that elevated CO_2 does not boost the production of structural carbon compounds (fiber) or secondary compounds (Bazzaz, 1990; Overdieck, 1990; Fajer et al., 1992), which suggests that their rate of production is demand-limited. In Mediterranean climate regions characterized by slow-growing plants, the situation may be very different. Poorter and Bergkotte (1992) have shown that the carbon-based compounds accumulated by fast- and slow-growing plants are very different (Figure 16-4). Slow-growing species accumulated more hemicellulose, insoluble sugars, and lignin than did fast-growing herbaceous species. Thus we could expect the slow-growing species to respond very differently to elevated CO_2, with the often-reported increase

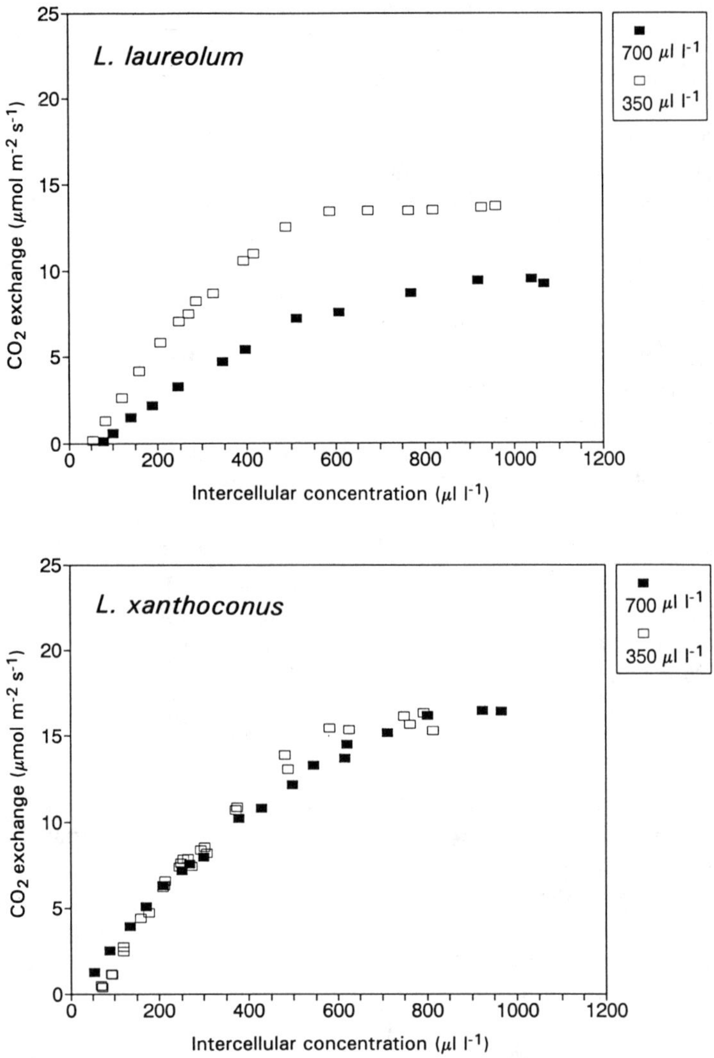

Figure 16-3. Photosynthetic response curves for single leaves of 9-month-old *Leucadendron laureolum* and *L. xanthoconus* (Proteaceae) individuals grown under ambient atmospheric CO_2 concentration (roughly 350 μL L^{-1} CO_2) and elevated CO_2 concentration (ambient + 350 μL L^{-1} CO_2). Plants had been grown under these treatment condition for 2 months, in open-top chambers of polycarbonate construction in a polycarbonate-covered greenhouse. Plants were grown in a 3:1 acid-washed sand to sterile perlite mix in 2.75-L pots (0.5 m deep), and were fed weekly with 100 mL of a 5% Long Ashton nutrient solution containing 0.05 mM nitrogen in ammonium and nitrate form.

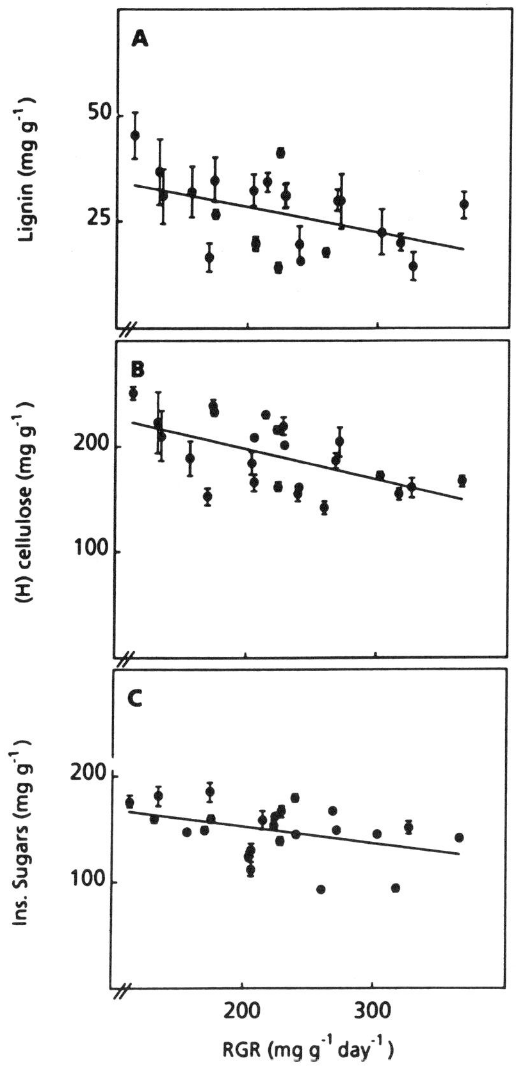

Figure 16-4. Concentration of (A) lignin, (B) (hemi)cellulose, and (C) insoluble sugars of whole plants of 24 species differing in relative growth rate: mean values ±SE (determinations of two independent bulk samples). (From Poorter and Bergkotte, 1992.)

in starch storage noted for fast-growing agricultural species (Wong, 1990; Korner and Arnone, 1992) being replaced by greater structural carbon allocation.

Whole-plant carbon allocation changes in response to CO_2 enrichment result in an increased investment in roots, especially if soil resources are limiting (Bazzaz, 1990). For instance, white oak seedlings showed increased production of fine roots under elevated CO_2 (Norby et al., 1986). The implications of the greater below-ground investment are substantial, as this investment could fund further above-ground growth through greater nutrient and water uptake.

Respiratory maintenance costs of increased root production could be met by increased above-ground carbon fixation, although predictions of the impact of elevated CO_2 on plant respiration rates are equivocal, as both increases and decreases have been reported (Amthor, 1991). Whole-plant carbon allocation patterns, the balance between carbon fixation and respiration, and possible impacts of altered allocation patterns on nutrient acquisition under elevated CO_2 are topics that need to be addressed in a range of ecologically significant species before reasonable predictions of the effects of elevated CO_2 can be made.

Rhizosphere Interactions and Root Symbionts

The potential CO_2-fertilization response of species adapted to nutrient-poor environments may also be enhanced by increased activity in the rhizosphere. Factors that alter photosynthetic activity have been shown to alter microbial activity in the rhizosphere and hence nutrient availability at the root surface (Scaglia et al., 1985). A study by O'Neill et al. (1987), however, on yellow poplar seedlings (*Liriodendron tulipifera*) was unable to demonstrate a positive contribution of rhizosphere bacteria to the enhanced performance of the seedlings under CO_2 enrichment. It was found that in the rhizosphere of plants under elevated CO_2, there were significantly fewer nitrite-oxidizing and phosphate-dissolving bacteria, and the ability of this species to show growth responses under elevated CO_2 was attributed to increased soil exploration and improved nutrient use efficiency (O'Neill et al., 1987). Although total rhizosphere activity might increase under elevated CO_2 conditions, competition between roots and microbes for nutrients and changes in bacterial species composition could lead to a decreased availability of nutrients to plants.

Evidence indicating a positive interaction between CO_2 enrichment and greater activity of root symbionts is clear. Norby (1987) demonstrated enhanced nodulation (but no enhancement of nitrogenase activity) in three woody species, which, under nutrient limitation, contributed substantially to increased plant growth. However, Thomas et al. (1991) found that a similar response in another woody species, *Gliricidium sepium*, was dependent on soil nutrient status. Root mycorrhizal colonization and development has also been shown to be enhanced under elevated CO_2 (Norby et al., 1986). The implications of changes in nutrient acquisition capabilities on plant performance are considerable, particularly for elements such as phosphorus which are relatively immobile in soils.

Patterns and Rates of Nutrient Cycling in Response to Elevated CO_2

Studies of ecosystem nutrient cycling processes have shown that they invariably change in both magnitude and rate in response to addition of nutrients (Heddle and Specht, 1975; Witkowski et al., 1990); however, the impact of increased CO_2 levels on these processes is unknown. Enhanced CO_2 could probably induce the reverse effect of nutrient fertilization, as the extra biomass that might be produced would dilute existing nutrient capital and thereby restrict rates of nutrient turnover. Processes crucial in controlling rates of nutrient cycling in Mediterranean ecosystems are litterfall, decomposition, and fire (Schlesinger and Hasey, 1981; Mitchell et al., 1986; Stock and Allsopp, 1992), and the impact of elevated CO_2 should be most evident at these stages.

Litterfall

The return of nutrients to the soil via litterfall is a vital process in nutrient cycling because the nutrients can be made available to plants by mineralization. Elevated CO_2 could affect the magnitude of litterfall inputs as well as the nutritional composition or litter quality, which, together with climate, are the major controls on decomposition and subsequent nutrient release (Witkamp, 1966; Meentemever, 1978).

Litter inputs to the various Mediterranean-type ecosystems are low, ranging from 71 to 598 $g\,m^{-2}$ (Maggs and Pearson, 1977; Mitchell et al., 1986; Arianoutsou, 1989), and litter quality is generally poor, having a high organic polymer and low mineral content (Schlesinger and Hasey, 1981; Mitchell et al., 1986). If acclimation of carbon fixation is not a common occurrence and net primary production of C_3-dominated Mediterranean systems does increase in response to elevated CO_2, then the larger standing biomass could lead to increased litter inputs. The fate of the increased litter inputs will depend upon decomposition processes in the various ecosystems. In low-nutrient fynbos ecosystems, where fire is the main mineralizing agent (Stock and Allsopp, 1992), litter could accumulate more quickly and lead to more regular and intense fires. Possible increased fire frequencies and intensities have profound implications for the maintenance of nutrient cycling processes in these low-nutrient ecosystems, although these are the low-resource habitats where Williams et al. (1988) suggest the effects of elevated CO_2 on productivity are highly attenuated.

In other, more nutrient-rich Mediterranean ecosystems, where decomposition processes are responsible for nutrient turnover, the quality of the litter under elevated CO_2 conditions becomes critical. Changes in foliage and subsequent litter quality under elevated CO_2 may have fundamental effects on ecosystem processes such as decomposition, nutrient cycling, fire frequency, and herbivory. Therefore, leaf carbon allocation patterns are important in linking individual plant responses with ecosystem processes, and may provide critical feedback effects which ultimately determine the characteristics of the changing ecosystem.

Foliage quality responses to elevated CO_2 involve changes in foliar C:N and C:P ratios, and allocation to various carbon compounds, including phenolics and lignin, which are strong determinants of decomposition rates in various ecosystems (Meentemeyer, 1978; Schlesinger and Hasey, 1981; Melillo et al., 1982; Mitchell et al., 1986). C_3 plants, in particular, show higher foliar C:N ratios under elevated CO_2, which may vary with stage of tissue development (Curtis et al., 1989; Conroy, 1992). However, litter quality may not be affected by elevated CO_2, due to nitrogen retranslocation processes before senescence and abscission (Larigauderie et al., 1988). Penuelas and Matamala (1990) suggest that C:N ratios response to elevated CO_2 is asymptotic, with progressively smaller C:N changes occurring with progressively higher atmospheric CO_2 concentration.

Depending on the species concerned, the concentrations of several other foliar nutrient constituents have been shown to decrease with rising atmospheric CO_2 supply to a similar degree to nitrogen (Overdieck, 1990), or to decrease somewhat less than nitrogen (O'Neill et al., 1987; Midgley, unpublished results). Decreases in nitrogen in response to elevated CO_2 have been attributed to a reduced flux of nitrogen through the photorespiratory cycle as well as to a decline in concentrations of Rubisco or other enzymes of the photoreductive cycle (Conroy, 1992). On the other hand, nutrients, such as phosphorus, may be required in the same or larger quantities under elevated CO_2 conditions because the foliar nutrient requirements necessary to sustain maximum photosynthetic rates increase (Conroy et al., 1990a). Phosphorus may be needed for photosynthate transport because inorganic phosphorus fluxes into the chloroplast increase with the higher flux of carbon through the photoreductive cycle, or it may be needed for sink generation (Conroy et al., 1990b, 1992). Although vague at present, a knowledge of what precise changes in nutrient concentrations of plants exposed to elevated CO_2 occur is essential for the development of successful decomposition models which will allow for a better understanding of nutritional effects at the ecosystem level.

In systems where nutrient concentrations in the litter are not greatly altered, substantial increases in the amount of nutrient per unit ground area could be expected if productivity and phytomass of communities increase with elevated CO_2 (Overdieck, 1990). Nutrient turnover per unit of ground area would also increase with larger quantities of nutrients becoming available. Thus, despite no change in litter quality, the increased quantities could have a profound influence on nutrient dynamics.

Decomposition

Decomposition models suitable for predicting regional variations in decay rates generally include both climate and litter characteristics (Witkamp, 1966; Meentemeyer, 1978; Schlesinger and Hasey, 1981; Berendse et al., 1987; Horner et al., 1988). Climate changes during the period of CO_2 increase will vary among each of the five Mediterranean regions depending upon local climate features, and impacts on decomposition will be specific to each area. Litter characteristics important in controlling the decomposition rate include moisture content, nutrient content (in particular nitrogen content), polyphenolic content (tannins and lignin), and the ratios between carbon and certain nutrients (C:N and C:P ratios), all of which could, as mentioned earlier, be affected by elevated CO_2 conditions.

In nutrient-rich ecosystems, rates of microbial decomposition appear to be predicted reasonably well from the C:N ratio of the litter when it is 30:1 or less (Fog, 1988; Taylor et al., 1989). Addition of nitrogen to high C:N organic material such as wood and bark samples, however, does not enhance decomposition (Allison and Klein, 1961), and hence hypotheses which include a consideration of the influence of recalcitrant organic compounds are more applicable to most natural ecosystems. In agricultural ecosystems, the effect of elevated CO_2 should be to reduce decom-

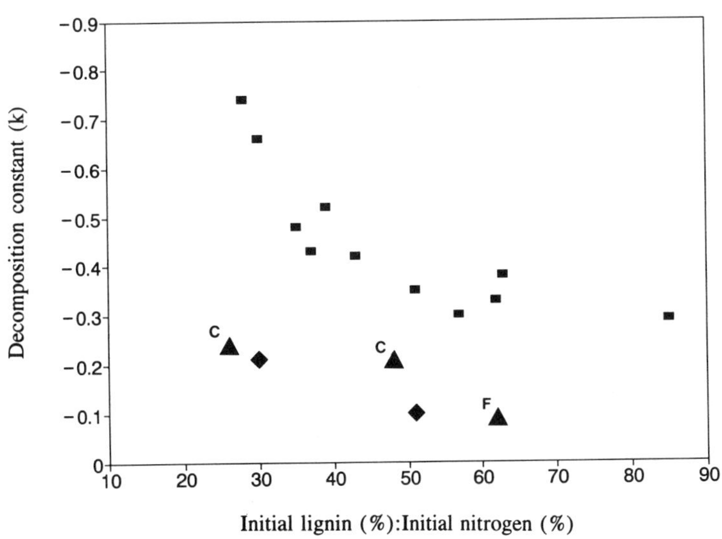

Figure 16-5. Relationships derived for deciduous forest (■), evergreen forest (◆) and Mediterranean ecosystems (▲, F–fynbos [*Leucospermum parile*] and C–chaparral [*Salvia mellifera* and *Ceonothus megacarpus*]) relating the patterns of loss of mass in decomposition to parameters of initial lignin and nitrogen content. (Modified from Melillo et al., 1982; Schlesinger and Hasey, 1991; and Mitchell et al., 1986.)

position rates in parallel to the observed increase in C:N ratios of the leaves, provided retranslocation of nitrogen compounds out of the leaf at senescence does not increase proportionally (Larigauderie et al., 1988; Curtis et al., 1989; Conroy, 1992).

In the natural ecosystems of the Mediterranean regions, predictions of decomposition rates from models such as those of Meentemeyer (1978) and Melillo et al. (1982) have been attempted (Schlesinger and Hasey, 1981; Mitchell et al., 1986) and found to be only partly applicable. Decay constants (*k*; Olson, 1963) are much lower than would be predicted from any of the models, including those which consider the initial lignin: nitrogen ratio which has been successfully applied in hardwood forests but failed for coniferous systems (Figure 16-5, modified from Melillo et al., 1982). Therefore predicting the impacts of elevated CO_2 is problematic, particularly where the phytochemical composition of the community and the persistence of the various secondary metabolites in litter remain unknown. If, however, the response of slow-growing plants of the low-nutrient environments is to accumulate starch or other nonstructural carbohydrates under elevated CO_2, then patterns of decomposition should be less affected than if the plants increase lignin, tannin, or other carbon polymer concentrations.

Conclusions

A dichotomy in responses of plants from nutrient-rich and nutrient-poor systems to elevated CO_2 is expected based on the literature reviewed in this chapter. These differences in response may not always be related to the nutritional environment per se but may be a consequence of the properties of the species characteristic of these environments. In particular, the growth potential of plants under resource-rich conditions appears to be crucial, with the major impacts of elevated CO_2 evident in rapidly growing species under nutrient-rich conditions (Cure and Acock, 1986; Drake and Leadley, 1991; Mooney et al., 1991). Plants with slow growth rates respond poorly to nutrient additions, and from the limited number of studies published on natural ecosystems, a similar lack of response to elevated CO_2 has been noted (Norby et al., 1986; Tissue and Oechel, 1987).

The five Mediterranean ecosystems of the world are ideal areas in which to study plant and ecosystem responses to elevated CO_2. They have similar climatic conditions but contain a range of nutritional environments that contain both fast- and slow-growing species. The areas have been intensively studied at many levels, and although they are limited in size and are not major components of the global carbon budget, they offer a valuable opportunity to develop current understanding of ecosystem functioning under elevated CO_2 conditions. In particular, the links between leaf-level and ecosystem responses to elevated CO_2 can be addressed so as to overcome the lack of understanding of feedback effects. Against a background of existing information detailing the quantities and rates of carbon and nutrients being cycled, it will be possible to determine whether elevated CO_2 is going to affect carbon storage within ecosystems or lead to greater rates of carbon turnover.

There are many potential long-term effects of elevated CO_2, and these may include changes in species dominance through to a complete change in vegetation type. To attempt predictions of these long-term effects without knowledge of how species behave in competitive mixtures is impossible, and scaling up from single-species studies is inappropriate. Thus, small-scale ecosystem experiments are essential if we are to improve the accuracy of current predictions concerning the long-term impacts of elevated CO_2 on system functioning and stability.

Acknowledgments. GFM gratefully acknowledges the Syfrets-administered Rowland and Leta Hill Trust and the Southern African Nature Foundation for financial support to develop experimental greenhouse facilities.

References

Allison FE, Klein CJ. 1961. Comparative rates of decomposition in soil of wood and bark particles of several softwood species. Proc Soil Sci Soc Am 25:193–196.

Amthor JS. 1991. Respiration in a future, higher CO_2 world. Plant Cell Environ 14:13–20.

Arianoutsou M. 1989. Timing of litter production in a maquis ecosystem of north-eastern Greece. Acta Oecol 10:371–378.

Arp WJ. 1991. Effects of source-sink relations on photosynthetic acclimation to elevated CO_2. Plant Cell Environ 14:869–875.

Ausubel JH, Grübler A, Nakicenovic N. 1988. Carbon dioxide emissions in a methane economy. Climate Change 12:245–263.

Baas WJ. 1989. Secondary plant compounds, their ecological significance and consequences for the carbon budget. In: Lambers H (ed) Causes and Consequences of Variation of Growth Rate and Productivity of Higher Plants. SPB Academic Publishing, The Hague, Netherlands.

Bazzaz FA. 1990. The response of natural ecosystems to the rising global CO_2 levels. Ann Rev Ecol Syst 21:167–196.

Berendse F, Berg B, Bosattà E. 1987. The effect of lignin and nitrogen on the decomposition of litter in nutrient-poor ecosystems. Can J Bot 65:1116–1120.

Berendse F, Elberse WTH, Geerts RHME. 1992. Competition and nitrogen loss from plants in grassland ecosystems. Ecology 73:46–53.

Bloom AJ, Chapin FS, Mooney HA. 1985. Resource limitation in plants—an economic analogy. Ann Rev Ecol Syst 16:363–392.

Boden TA, Kanciruk P, Farrell MP. 1990. Trends '90: A compendium of data on global change. ORNL/CDIAC-36. Carbon Dioxide Information Analysis Center, Oak Ridge National Laboratory, Oak Ridge, TN.

Bowes G. 1991. Growth at elevated CO_2: photosynthetic responses mediated through Rubisco. Plant Cell Environ 14:795–806.

Bryant JP, Chapin FS, Klein DR. 1983. Carbon/nutrient balance of boreal plants in relation to vertebrate herbivory. Oikos 40:357–368.
Chapin FS. 1980. The mineral nutrition of wild plants. Ann Rev Ecol Syst 11:233–260.
Chapin FS. 1988. Ecological aspects of plant mineral nutrition. Adv Mineral Nutr 3:161–191.
Chapin FS, Vitousek PM, van Cleve K. 1986. The nature of nutrient limitation in plant communities. Am Nat 27:48–58.
Cody ML, Mooney HA. 1978. Convergence versus nonconvergence in Mediterranean-climate ecosystems. Ann Rev Ecol Syst 9:265–321.
Coleman JS, Rochefort L, Bazzaz FA, Woodward FI. 1991. Atmospheric CO_2, plant nitrogen status and the susceptibility of plants to an acute increase in temperature. Plant Cell Environ 14:667–674.
Conroy JP. 1992. Influence of elevated atmospheric CO_2 concentrations on plant nutrition. Aust J Bot 40:445–456.
Conroy JP, Milham PJ, Mazur M, Barlow WR. 1990a. Growth, dry weight partitioning and wood properties of *Pinus radiata* D. Don after 2 years of CO_2 enrichment. Plant Cell Environ 13:329–337.
Conroy JP, Milham PJ, Reed ML, Barlow EW. 1990b. Increases in phosphorus requirements for CO_2-enriched *Pine* species. Plant Physiol 92:977–982.
Cowling RM, Campbell B. 1980. Convergence in vegetation structure in the Mediterranean communities of California, Chile and South Africa. Vegetatio 43:191–197.
Cure JD, Acock B. 1986. Crop response to carbon dioxide doubling: a literature survey. Agric For Met 38:127–145.
Cure JD, Rufty TW, Israel DW. 1991. Assimilate relations in source and sink leaves during acclimation to a CO_2-enriched atmosphere. Physiol Plant 83: 687–695.
Curtis PS, Drake BG, Whigham DF. 1989. Nitrogen and carbon dynamics in C_3 and C_4 estuarine marsh plants grown under elevated CO_2 in situ. Oecologia 78:297–301.
Drake BG, Leadley PW. 1991. Canopy photosynthesis of crops and native plant communities exposed to long-term elevated CO_2. Plant Cell Environ 14:853–860.
Fajer ED, Bowers MD, Bazzaz FA. 1992. The effect of nutrients and enriched CO_2 environments on production of carbon-based allelochemicals in Plantago: a test of the carbon/nutrient balance hypothesis. Am Nat 140:707–723.
Field C, Mooney HA. 1986. The photosynthesis-nitrogen relationship in wild plants. In: Givnish TJ (ed) The Economy of Plant Form and Function, pp. 25–56. Cambridge University Press, Cambridge, UK.
Fog K. 1988. The effect of added nitrogen on the rate of decomposition of organic matter. Biol Rev 63:433–462.
Friedli H, Lötscher H, Oeschger H, Siegenthaler U, Stauffer B. 1986. Ice core record of $^{13}C/^{12}C$ ratio of atmospheric CO_2 in the past two centuries. Nature 324:237–238.
Grime JP. 1977. Evidence for the existence of three primary strategies in plants and its relevance to ecological and evolutionary theory. Am Nat 111:1169–1194.
Heddle EM, Specht RL. 1975. Dark Island Heath (Ninety-mile Plain, South Australia). VIII The effect of fertilizers on composition and growth, 1950–1972. Aust J Bot 23:151–164.
Horner JD, Gosz JR, Cates RG. 1988. The role of carbon-based plant secondary metabolites in decomposition in terrestrial ecosystems. Am Nat 132:869–883.

Hunt R, Hand DW, Hannah MA, Neal AM. 1991. Response to CO_2 enrichment in 27 herbaceous species. Func Ecol 5:410–421.
Kimball BA. 1983. Carbon dioxide and agricultural yield: an assemblage and analysis of 430 prior observations. Agron J 75:779–788.
Korner C, Arnone JA III. 1992. Responses to elevated carbon dioxide in artificial tropical ecosystems. Science 257:1672–1675.
Kramer PJ, Kozlowski TT. 1979. Physiology of Woody Plants. Academic Press, New York.
Larigauderie A, Hilbert DW, Oechel WC. 1988. Effect of CO_2 enrichment and nitrogen availability on resource acquisition and resource allocation in a grass, *Bromus mollis*. Oecologia 77:544–549.
Maggs J, Pearson CJ. 1977. Litterfall and litter decay in coastal scrub at Sydney, Australia. Oecologia 31:239–250.
Meentemeyer V. 1978. Macroclimate and lignin control of litter decomposition rates. Ecology 59:465–472.
Melillo JM, Aber JD, Muratore JF. 1982. Nitrogen and lignin control of hardwood leaf litter decomposition dynamics. Ecology 63:621–626.
Mitchell DT, Coley PGF, Webb S, Allsopp N. 1986. Litterfall and decomposition processes in the Coastal Fynbos vegetation, South-western Cape, South Africa. J Ecol 74:977–993.
Mooney HA, Drake BG, Luxmoore RJ, Oechel WC, Pitelka LF. 1991. Predicting ecosystem responses to elevated CO_2 concentrations. Bio Science 41:96–104.
Morison JIL. 1990. Plant and ecosystem responses to increasing atmospheric CO_2. Trends Ecol Evol 5:69–70.
Neftel A, Moor E, Oeschger H, Stauffer B. 1985. Evidence from polar ice cores for the increase in atmospheric CO_2 in the past two centuries. Nature 315:45–47.
Norby RJ. 1987. Nodulation and nitrogenase activity in nitrogen-fixing woody plants stimulated by CO_2 enrichment of the atmosphere. Physiol Plant 71:77–82.
Norby RJ, O'Neill EG. 1989. Growth dynamics and water use of seedlings of *Quercus alba* L. in CO_2-enriched atmospheres. New Phytol 111:491–500.
Norby RJ, O'Neill EG. 1991. Leaf area compensation and nutrient interactions in CO_2-enriched seedlings of yellow-poplar (*Liriodendron tulipifera* L.). New Phytol 117:515–528.
Norby RJ, O'Neill EG, Luxmoore RJ. 1986. Effect of atmospheric CO_2 enrichment on the growth and mineral nutrition of *Quercus alba* seedling in nutrient-poor soil. Plant Physiol 82:83–89.
Oberbauer SF, Sionit N, Hastings SJ, Oechel WC. 1986. Effects of CO_2 enrichment and nutrition on growth, photosynthesis, and nutrient concentration of Alaskan tundra plant species. Can J Bot 64:2993–2998.
Olson JS. 1963. Energy stroage and the balance of producers and decomposers in ecological systems. Ecology 44:322–331.
O'Neill EG, Luxmoore RJ, Norby RJ. 1987. Elevated atmospheric CO_2 effects on seedling growth, nutrient uptake, and rhizosphere bacterial populations of *Liriodendron tulipifera* L. Plant Soil 104:3–11.
Overdieck D. 1990. Effects of elevated CO_2-concentration levels on nutrient contents of herbaceous and woody plants. In: Greenhouse Effect and Primary Productivity in European Agro-ecosystems, pp. 31–37.
Peñuelas J, Matamala R. 1990. Changes in N and S leaf content, stomatal density and specific leaf area of 14 plant species during the last three centuries of CO_2 increase. J Exp Bot 41:1119–1124.
Poorter H, Bergkotte M. 1992. Chemical composition of 24 wild species differing in relative growth rate. Plant Cell Environ 15:221–229.

Scaglia J, Lensi R, Chalamet A. 1985. Relationship between photosynthesis and denitrification in planted soil. Plant Soil 84:37–43.
Schlesinger WH, Hasey MM. 1981. Decomposition of chaparral foliage: losses of organic and inorganic constituents from deciduous and evergreen leaves. Ecology 62:762–774.
Sinclair TR. 1992. Mineral nutrition and plant growth response to climate change. J Exp Bot 43:1141–1146.
Specht R. 1963. Dark Island Heath (Ninety-Mile Plain, South Australia). VII. The effects of fertilizers on composition and growth, 1950–1960. Aust J Bot 11:67–94.
Stock WD, Allsopp N. 1992. Functional perspective of ecosystems. In: Cowling RM (ed) The Ecology of Fynbos: Nutrients, Fire and Diversity. Oxford University Press, Cape Town.
Taylor BR, Parkinson D, Parsons WFJ. 1989. Nitrogen and lignin content as predictors of litter decay rates: a microcosm test. Ecology 70:97–104.
Thomas RB, Strain BR. 1991. Root restriction as a factor in photosynthetic acclimation of cotton seedlings grown in elevated carbon dioxide. Plant Physiol 96:627–634.
Thomas RB, Richter DD, Ye H, Heine PR, Strain BR. 1991. Nitrogen dynamics and growth of seedlings of an N-fixing tree (*Gliricidia sepium* (Jacq.) Walp.) exposed to elevated atmospheric carbon dioxide. Oecologia 88:415–421.
Tissue DT, Oechel WC. 1987. Response of *Eriophorum vaginatum* to elevated CO_2 and temperature in the Alaskan Tussock Tundra. Ecology 68:401–410.
Williams WE, Garbutt K, Bazzaz FA. 1988. The response of plants to elevated CO_2—V. Performance of an assemblage of serpentine grassland herbs. Environ Exp Bot 28:123–130.
Witkamp M. 1966. Decomposition of leaf litter in relation to environment, microflora, and microbial respiration. Ecology 47:194–201.
Witkowski ETF, Mitchell DT, Stock WD. 1990. Response of a Cape fynbos ecosystem to nutrient additions: shoot growth and nutrient contents of a proteoid (*Leucospermum parile*) and an ericoid (*Phylica cephalantha*) evergreen shrub. Acta Oecol 11:311–326.
Wong SC. 1990. Elevated atmospheric partial pressure of CO_2 and plant growth. II. Non-structural carbohydrate content in cotton plants and its effect on growth parameters. Photosyn Res 23:171–180.

17. Volatile Organics in Mediterranean Shrubs and Their Potential Role in a Changing Environment

Günther Seufert, Dimitris Kotzias, Carmelina Spartà, and Bruno Versino

Biogenic emissions have been the driving force in forming the global atmosphere during earth history. The current concentration of the major gases in the atmosphere, N_2 and O_2, is the result of photosynthesis and microbial activity. Important characteristics of the atmosphere, such as the radiation and energy balance, the concentration of oxidants in the troposphere, and the absorption of UV radiation in the stratosphere, are controlled by biogenically emitted compounds like O_2, CO_2, H_2O, CH_4, and N_2O, in a complex, interconnected system that has been portrayed by Lovelock (1979) as the Gaia system. Today, we have become aware of the potential effects of anthropogenic emissions on the chemical and physical characteristics of the atmosphere and reciprocal effects on the biosphere. The above characteristics of the atmosphere are part of the so-called global change phenomenon, including such global concerns as greenhouse warming or the polar ozone hole. Understanding such symptoms of a changing global environment will require understanding of the underlying natural processes.

Although material exchange between biosphere and atmosphere is undisputably prominent among the dynamic features in biogeochemical cycles, the knowledge of processes and magnitude of fluxes is not at all well grounded (Andreae and Schimel, 1989). This is especially true for the reactive trace gases emitted from plants: because their concentrations are generally low and their atmospheric lifetimes are short, it is extremely difficult to determine the magnitude of sources and sinks for the multiplicity of compounds emitted from plant species. Only isoprene and selected monoterpenes have been extensively studied, mostly in North American and some Central European coniferous and deciduous trees, but nonterpenoids can also be a significant fraction of the total emissions from plants. In addition to coniferous forests and tropical rain forests as the globally dominating source biomes, the Mediterranean shrublands could constitute a significant regional source of nonmethane hydrocarbons (NMHC) to the atmosphere. In the Mediterranean region, which is now recognized as one of the major source areas of tropospheric ozone in Europe, the whole topic of biosphere–atmosphere interactions requires urgent investigation.

This chapter briefly reviews some chemical and ecological aspects of volatile organics in Mediterranean plants, mainly of terpenoids, with respect to their effect on the atmosphere and on the global change issue. It will be argued that this topic would be of particular relevance in Mediterranean regions, where (1) high emission rates can be assumed due to the disproportionate share of plants producing volatile, aromatic compounds; and (2) chemical transformation of the emitted compounds is promoted by high temperature and radiation as well as by the omnipresent anthropogenic pollutants, thus suggesting an important role of biogenic volatiles in atmospheric chemistry. In anticipated warmer and drier climate conditions of the future, the relevance of biogenic emissions to atmospheric chemistry could further increase. Based on such arguments and realizing that nothing has been done yet in this field, the CEC-Joint Research Centre recently initiated a project on Biogenic Emissions in the Mediterranean Area (BEMA) that will be presented at the end of the chapter with some preliminary results and recommendations for research needs in Mediterranean Ecosystems.

Types of Emissions from Plants

Plants emit a wide array of organic compounds, including ethylene, isoprene, mono- and sesquiterpenes, alkanes, alcohols, carbonyls, volatile aromatics, etc. In the terpenoid class of compounds, mainly isoprene (C_5), monoterpenes (C_{10}), and sesquiterpenes (C_{15}) are volatile enough to be emitted. Isoprene is emitted immediately upon production, whereas mono- and sesquiterpenes are often stored in specialized tissues of the plants.

The main aspects of isoprene synthesis and emission have been recently reviewed by Sanadze (1991), Sharkey et al. (1991), and Monson et al. (1991). Isoprene is synthesized inside chloroplasts, and production and emission occur only in the light and are closely related to photosynthetic activity. In addition to light, isoprene emission is stimulated by high temperature. At present, more than 200 mainly woody plants are reported to be isoprene emitters (Sanadze, 1991), while emissions from crops are often low or not detectable (Winer et al., 1992). Tropical forests could be a major contributor to the global emissions of isoprene (Rasmussen and Khalil, 1988). Based on the measurements of Zimmerman et al. (1988), performed in summer 1985 in the Amazonia region of Brazil, Müller (1992) computed yearly isoprene emissions from tropical ecosystems in the range of 208 Tg, and of 41 Tg for all other ecosystems. Data of Monson et al. (1991) indicate that 49% of the biomass in deciduous oak forests belongs to high-isoprene emitting species. In contrast, 75% and 90% of biomass in scrublands and coniferous forests, respectively, is from nonemitting species. It has been estimated that global isoprene emissions are of the same order of magnitude as the emission of methane or the sum of mono- and sesquiterpene emissions (Zimmermann et al., 1978; Mooney et al., 1987).

The few studies considering nonterpenoid volatile compounds clearly indicate that they comprise a significant fraction of total NMHC emissions from plants. Winer et al. (1992) determined hydrocarbon emission rates for more than 30 of the dominant agricultural and natural plant species of the Central Valley in California. Besides mono- and sesquiterpenes, oxygenated compounds have been observed, with cis-3-hexen-1-ol and cis-3-hexenylacetate being the most dominant (Arey et al., 1991). Isidorow et al. (1985) determined 3-hexenylacetate as a major emission from *Vaccinium myrtillus* and n-heptane together with n-undecane as a significant compound in volatile oils of conifers. Müller (1992) assummed that global emissions of nonterpenoid compounds are about half the terpene emissions. For the oxygenated compounds, the pathway and site of synthesis and the factors controlling emissions are not well known. According to a review by Tingey et al. (1991), the enzymes forming the 6-carbon-aldehydes and -alcohols are located in the chloroplasts of green leaves; the concentrations vary with enzyme activity in different seasons and are enhanced after mechanical injury. The emissions follow the diurnal pattern of temperature and light; n-hexanal emissions are increased under drought conditions.

Monoterpenes and sesquiterpenes are ubiquitous in higher plants as intermediates for phytohormones, phytosterols, and carotenoids (Banthorpe et al., 1972), but significant amounts accumulate only in some 50 families of higher plants with specialized secretory and storage structures for volatile oils (Gildemeister and Hoffmann, 1955–1960). With respect to the factors controlling emissions, it is important to know the site of synthesis and storage in the plant, which determines the physical pathway of emissions. The following incomplete compilation is taken from the general phytochemical and plant morphological literature (e.g., Kisser, 1958; Hegnauer, 1962–1992) and indicates different types of organs containing volatile oils in some of the families (and genera) common in Mediterranean regions.

Volatile oil secretory organs are internal as in the case of resin canals in needles and bark of conifers. Although resins are defined as non-volatile products of plants (Dell and McComb, 1978), they often, perhaps always, contain some volatile oils, so that the compounds contained in resin canals of Coniferae (*Araucaria*, *Pinus*, *Picea*, *Abies*, *Larix*, *Tsuga*, *Taxodium*, *Cupressus*, *Thuja*, *Chamaecyparis*, *Juniperus*) and Anacardiaceae (*Pistacia*, *Cotinus*, *Rhus*) are sometimes referred to as oleoresins. Cistaceae (*Cistus*, *Helianthemum*) contain so-called aromatic resins in external glandular hairs. Glandular trichomes as sources of volatile oils are common in Lamiaceae (with many aromatic genera of Mediterranean shrublands: *Salvia*, *Rosmarinus*, *Majorana*, *Satureja*, *Lavandula*, *Thymus*, *Melissa*, *Mentha*), Cannabaceae (*Humulus*, *Cannabis*), Solanaceae (*Lycopersicon*), Geraniaceae (*Geranium*, *Pelargonium*), Juglandaceae (*Juglans*), and Myricaceae (*Myrica*). Internal structures that accumulate volatile oils can be specialized cells or idioblasts as in *Laurus* and *Magnolia* leaves, in *Acorus* and *Zingiber* rhizomes, and in *Cinnamomum* bark. Common oil-accumulating organs are the oil cavities of lysigenous origin as in Rutaceae (leaves and fruits of *Citrus* sp.) and Myrtaceae (leaf and bark of *Myrtus*, *Eucalyptus*). Schizogenous oil ducts are observed in Umbelliferae (fruits, leaves, and roots of *Daucus*, *Carum*, *Coriandrum*, *Pastinaca*) and Hypericaceae (*Hypericum*), as well as in the large family of Asteraceae (*Artemisia*, *Chrysanthemum*, *Carthamus*, etc.). Flower petals, e.g., of Rosaceae and Violaceae, accumulate fragrant oil compounds in epidermal and mesophyllic cells or cell groups (Bergström, 1991); often (e.g., *Arum*) the fragrances are contained in special organs, the osmophores.

Factors Controlling the Emission of Volatile Terpenes

The pool size and composition of volatile terpenes in plants is under strict genetic control but is also influenced by environmental factors. The monoterpene composition has been used as a basis for chemotaxonomic studies, to separate, for example, different ecotypes among provenances of *Pinus halepensis* (Schiller and Grunwald, 1987) or *Pinus nigra* (Fineschi and Grossoni, 1981). The role of environmental factors in controlling terpene pools in plants will be considered in the rent section. The interactions between genetic and environmental factors can be illustrated using the *Mentha piperita* studies of Burbott and Loomis (1967): commercial producers of peppermint oil utilize *Mentha piperita* plants that have been vegetatively propagated from special varieties, emphasizing the importance of genetic factors. However, cloned material is not enough to produce peppermint oil of acceptable commercial quantity and quality containing menthone instead of less desirable pulegone or mentofuran. Such oil can be produced only in certain geographic areas; major variations in oil composition are caused by variations of temperature, light, daylength, nutrient and water supply, as well as date and time of harvest.

In the case of external glandular hairs, the physical emission pathway is short; the oil accumulates beneath the cuticle of the head cells and is released continuously through the cuticle or through cuticular pores. In some species the cuticle may expand with increasing oil amount to the point of rupture, as observed in different Lamiaceae species (Kisser, 1958). If the volatiles are stored internally as in Pinaceae, they exit through stomata, but stomatal conductance has little influence on emission rates (Tingey et al., 1991). The monoterpene emitted from a resin duct must diffuse through the epithelial cells and the fibers that surround the canals. This pathway would have extremely high resistances, because the epithelial cells are often suberized (as in *Pinus*), since the content of the cavity is under pressure and the cavity must be well protected from gaseous diffusion (Dell and McComb, 1978). The hypothesis of Banthorpe et al. (1972) that two distinct pools of terpenes may exist, one of which is susceptible to outside influences and the other of which is more inert, is still under discussion. The relative size of the pools could vary in different species and under different circumstances. Thus the controls on monoterpene production and pools could be different from the controls on emissions, because some volatile monoterpenes are synthesized in the mesophyll and diffuse directly into intercellular air spaces (Lerdau, 1991).

For a given pool size and diffusive pathway, the emission rate of volatile terpenes is controlled mainly by temperature-induced changes in the vapor pressure of the compound of interest. (Knöppel et al., 1981; Tingey et al., 1991). However, Tingey et al. (1991) analyzed various *Pinus* species and found that the increase in α-pinene emissions with increasing temperature was always 1.2 to 3.2 times greater than that predicted by changes in vapor pressure alone. Some of the discrepancy may arise from the fact that although vapor pressure of the compound is a function of leaf temperature, in many of the studies only air temperature was measured. In a poorly mixed enclosure, leaf temperature may rise several degrees above air temperature, leading to greater emissions than predicted. In any case, the diurnal and seasonal deviation from the vapor pressure control of emissions indicates that there must be other factors involved. For example, at constant air temperature, the emissions from *Pinus densiflora* showed a maximum near noon and a twofold increase in values during the day versus the night (Yokouchi and Ambe, 1984). In contrast, the emission rates of monoterpenes from *Salvia mellifera* were the same at constant temperature in both light and dark and were directly proportional to leaf temperature (Tyson et al., 1974). In a companion paper, the same authors reported that monoterpene emissions were proportional to humidity but not proportional to transpiration in both intact and excised branches of *S. mellifera* (Dement et al., 1974).

Humidity and surface moisture stimulated monoterpene emissions from *Mentha piperita* four- to fivefold under sprinkler irrigation and 70–80% in high-humidity environments (Croteau, 1977). The emission rates of α-pinene in a Douglas fir (*Pseudotsuga menziesii*) forest, measured using a gradient profile method, were closely related to relative humidity. Branch emission rates measured with the enclosure method were an order of magnitude higher for wet branches than for dry branches (Lamb et al., 1985). These effects of surface moisture and humidity on monoterpene volatilization fit well with the theory that cuticular transport is passive transport of compounds through cuticular pores formed by hydration of polar functional groups in the presence of water (Schönherr, 1982). It seems worth noting that this plausible influence of water on the diffusive pathway of monoterpenes has been observed both in Lamiaceae species with external glandular trichomes and in Coniferae with internal resin ducts.

An indirect influence of temperature and moisture on ecosystem-level emissions can be expected by the acceleration of litter mineralization. Taking the data of Schindler and Kotzias (1989) on monoterpene contents and emission rates in *Picea abies* and *Pseudotsuga menziesii*, one can calculate that less than 5% of the pool is emitted per year. For *Salvia mellifera*, Tyson et al. (1974) calculated the volatilization loss of carbon from the terpene pool as 0.06% per month. The rest, if not catabolized, will be volatilized during litter mineralization, directly or after chemical transformation, possibly together with other decomposing terpenoids like carotenes.

Considering the likely parameters controlling biogenic emissions from plants, it may deduced that this topic is not sufficiently understood to calculate regional emission inventories (Lamb et al., 1987; Lübkert and Schöpp, 1989; Anastasi et al., 1991) with algorithms based on temperature alone as developed in phytotron studies with *Pinus elliottii* (Tingey et al., 1980). In a detailed study to obtain a gridded emission inventory of biogenic compounds for California's Central Valley, Winer et al. (1992) estimated that the uncertainty of the mean emission rates of the different plant species investigated in their study was as large as a factor of five. However, the main error arises from extrapolating such figures over large areas of heterogeneous terrain.

A general problem in assessing the likely errors in determining emission factors used for inventories is linked to the fact that ecologically or experimentally relevant parameters like mechanical stress due to wind, or due to enclosure mounting of the branch, insect infestation, or surface moisture, may easily alter emission rates by orders of magnitude, but are difficult to quantify in space and time. The same problem arises if other relevant compartments and phenological dynamics of terrestrial ecosystems such as emissions of volatile organics from barks, roots, blooms, soil fungi, and microorganisms are included. A large but unknown amount of hydrocarbons should be emitted during pruning, harvesting, especially whole-tree chipping of conifers or mowing of grasses, and burning. Another unknown source of monoterpenes is the emission from bees, ants, and particularly termites, which use these compounds for defense or as solvents for more toxic defense compounds (Francke, 1991). In any case, improving our understanding of the factors controlling biogenic emissions at the leaf level, the canopy level, and the ecosystem level (Monson et al., 1991) is a prerequisite to providing realistic figures on biogenic emissions of organics at the regional or global scale, as needed for atmospheric chemistry models.

Role of Volatile Organics in Mediterranean-Type Ecosystems

A specific role of volatile organics in Mediterranean-type ecosystems (MTEs) can be deduced from the observation that essential oils and resins with their associated volatiles are far more prominent in plants of semiarid regions than in those of other climates. Ross and Sombrero (1991) calculated that 49% of the plant species growing in the MTEs of the world produce aromatic volatile oils. About 90 of the 153 plant genera in 50 families accumulating volatile oils have a geographical distribution in Mediterranean ecosystems. In these regions, essential oils and resins are typical of perennial, evergreen, xeromorphic shrubs, but nof of the drought-avoiding deciduous species and annuals. Obviously there must be climatic/edaphic parameters in MTEs favoring perennial xerophytes and the disproportionate share of volatile oil producers. In extreme conditions, between 5% and 40% of fixed carbon may be allocated to the biosynthesis of volatile oils (Ross and Sombrero, 1991), and beneficial aspects of this significant investment are documented in the extensive literature. The few examples given below focus on likely interactions of volatile organics with some stress conditions typical of Mediterranean regions: hot and dry summers; high leaf temperature; high radiation fluxes; nutrient/nitrogen-poor, degraded soils; high fire frequency; and herbivore pressure.

Biotic Interactions

Herbivory by sheep, deer, and goats provides a significant selective pressure on Mediterranean plants. Morphological and chemical adaptations against such herbivory are common (Mooney and Gulmon, 1982), which again may be coevolutonary counteradapted by the animals (Harborne, 1988). In ruminants it was observed that monoterpene alcohols like linalol and α-terpineol inhibited digestive processes (Arnold and Hill, 1972).

The literature dealing with mono- and sesquiterpenes as insect control agents, including different roles as repellents and attractants, has recently been reviewed by Pickett (1991). The role of conifer oleoresins in the prevention of insect attack is clearly established (Hanover, 1972); however, the volatile components of the same oleoresin may serve as an attractant for insects. During flowering, plants emit complex blends of mono- and sesquiterpenes, benzenoids, and volatile fatty acid derivatives to attract pollinators, probably the ecologically most significant role of volatile organics (Bergström, 1991). The amount of fragrance emitted in the luxuriant blooming Mediterranean macchia/garrigue is high enough to be easily detected with human olfaction.

Allelopathy, defined by Molisch (1937) as biochemical interactions and chemical competition between plants, is caused primarily by volatile terpenes and phenolic compounds. As reviewed by Rice (1984), Harborne (1988, 1991), and Fischer (1991), the allellopathic actions between plant species and within species (autotoxicity) are legion. Many examples include competition between a shrub or tree (e.g., *Citrus*, *Eucalyptus*, *Juglans*, *Pinus*) and understory plants like herbs and grasses. From Mediterranean regions, there are several examples where volatile terpenes play an important role in the fire cycle. Bare zones of 1–2 m are observed around the common shrubs of the Californian chaparral, *Salvia leucophylla*, and *Artemisia californica*. Of several monoterpenes, 1,8-cineole and camphor are the most effective in supressing seed germination of the adjacent annuals (Muller, 1970). In the sandhill shrublands of Florida, several monoterpenes emitted from the shrubs *Ceratiola ericoides* (Empetraceae), *Calamintha ashei*, and *Conradina canescens* (Lamiaceae) strongly inhibited germination and radicle growth of the surrounding grasses. Because the shrub communities are nearly devoid of herbaceous or graminoid covers, surface fires sweeping through the sandhills are often extinguished by this abrupt ecotone with low or no fuel supply (Fischer, 1991). Other reports on allelopathic effects of volatile terpenes in Mediterranean shrubs are reported for *Artemisia absinthum* (Asteraceae), *Sassafras albidum* (Lauraceae; Harborne, 1988), *Coridothymus capitatus* and *Mentha spicata* (Lamiaceae; Fischer, 1991); the effects are often enhanced by interactions of terpenoids with soil fungi and microorganisms.

Interactions with the Abiotic Environment

The hypothesis that volatile organics are part of an adaptive strategy to Mediterranean climate stress is indirectly supported by the observed seasonal fluctuations of terpene pool sizes often peaking during summer. This has been observed in several Mediterranean aromatic shrubs (*Thymus capitatus*, *Satureja thymbra*, *Teucrium polium*, *Rosmarinus officinalis*) grown in Greece (Vokou and Margaris, 1986). However, it is not clear if the observed increases are caused by enhanced biosynthesis of terpenes or indirectly by reduced growth. In *Satureja douglasii*, the monoterpene content per leaf dry weight increased from 17 to 21 mg/g in parallel to decrease in leaf water potential from −6 bar to −9 bar, but the content per leaf did not change significantly because the growth of the water-stressed plants was reduced (Gershenzon et al., 1978). For *Mentha spicata* (Gershenzon et al., 1989), *Salvia officinalis* (Croteau et al., 1981) and *Thymus vulgaris* (Yamaura et al., 1989), it has been shown that monoterpenes are almost exclusively contained in the glandular trichomes which are formed during the first stage of leaf expansion. Thus, a reduction in leaf size at a certain quantity of trichomes may result in an indirect increase in oil concentration.

During the Mediterranean summer, plants must avoid both excessive surface temperatures and excessive water loss. In conjunction with other xeromorphic features, hairs and trichomes at the leaf surface meet both requirements by lowering the surface temperature and by increasing the resistance of the laminar boundary layer, thus reducing transpirational water loss. Volatile organics contained in surface hairs may contribute with some cooling during vaporization, or by initiating stomatal closure. The latter effect was observed with highly concentrated vapors of oxygenated sesquiterpenes like farnesol, phaseic acid, and xanthoxin (Harborne, 1988), which are closely related in their biosynthesis and structure to the phytohormone abscisic acid (ABA). ABA is the phytohormone that regulates stomatal closure, thus enabling water-stressed plants to conserve water via stomatal regulation. Unfortunately, the effects of the possible ABA substitutes have not yet been investigated in plants producing these compounds; according to Sandermann (1968), farnesol is produced by Gramineae, Leguminosae, and Rutaceae.

The secretion of resins from surface glands has been identified as another adaptation of plants to Mediterranean summer conditions (Dell and McComb, 1978). As in the case of essential oil glands, the resin glands are formed and start secreting during early stages of leaf formation; they protect the young bud and leaf with a hydrophobic coating during the period of low water supply and high potential evapotranspiration. In addition, resin surfaces are often shiny, thus reducing leaf temperature markedly. The leaves of the Australian shrub *Beyeria viscosa* (Euphorbiaceae) are matt-like in normal conditions; however, at 44°C the leaf resin coalesces within 12 min, giving the leaf a mirrorlike appearance and reducing leaf temperature (Dell and McComb, 1978).

Volatile Organics Are Part of an Adaptive Strategy in Mediterranean Habitats

Behind the specific roles of volatile organics in helping plants to compete and survive in their environment, there may be part of a general strategy, because the environmental conditions promoting the shrub life-form coincide partly with those favoring the accumulation of essential oils and other secondary plant products.

Shrubs represent a life-form of low-growing woody plants with multiple stems deriving from multiple stress interactions. The evolutionary convergence of evergreen shrubs in the five MTEs of the world has been attributed mainly to direct and indirect interactions of water and nutrient stresses (Rundel, 1991). Evergreen, sclerophyllous leaves are drought tolerant and can amortize their cost of production over a long period (Mooney, 1983); their resprouting capability matches well with the unpredictability of sometimes damaging stress factors like drought, fire, and herbivory; their flexible root architecture allows maximum use of below-ground available water and nutrients as well as storage of the carbohydrates needed for regrowth after damage of above-ground parts (Mooney and Dunn, 1970; Miller, 1983; Rundel, 1991). The evergreen shrub seems to be the morphological life-form best adapted to Mediterranean stress conditions with seasonally restricted growth. The hypothesis that accumulations of volatile oils and other related plant products could represent the corresponding stress defense system at the biochemical level is supported by the following observations:

1. In conditions of moderate nutrient (Chapin, 1980) or water (Bradford and Hsiao, 1982) limitations, plant growth generally declines more sharply than does photosynthesis; thus, a carbon surplus accumulates above the actual requirements of plant growth, as is the case before bud break in the spring. In these periods, woody plants generally accumulate either starch or carbon-based defense compounds which contain no nitrogen or phosphorus (Tuomi et al., 1988). If herbivory is important, starch accumulation in above-ground parts would only increase herbivory pressure, whereas investing in defensive compounds would help to preserve the nutrients the plant already has.
2. In Mediterranean climates, the periods with restricted growth are normally coincident with high light and temperature levels. Because the risk of photodamage is enhanced at high temperatures, and because an evergreen leaf cannot easily adapt its light interception capacity, the risk must be reduced by utilizing the captive energy for carbon fixation. Thus, the accumulation of volatile oils may serve as an excess energy dissipation system during periods of restricted growth (Ross and Sombrero, 1991).
3. During stress-induced dormancy, the biosynthesis of essential oils could maintain the appropriate enzyme systems in an active state to keep the metabolism prepared for the unpredictable situation that provides favorable conditions for rapid regrowth after dormancy (Banthorpe et al., 1972).
4. In addition, the volatile oil could provide ATP and a pool for the secondary biosynthesis of primary metabolites like amino acids or

sugars when other sources have been depleted. It is now well established that monoterpenes are by no means inert waste products but play a dynamic role in plant metabolism and are catabolized in a highly specific and ordered fashion (Croteau, 1987). Metabolic turnover has been divided into short-term effects, which can be observed by diurnal fluctuations or by rapid $^{14}CO_2$ incorporation patterns within a few hours, and long-term effects, which have been reported mainly during senescence of those species containing oils in external glands, like the Lamiaceae, where monoterpene losses exceeded 50% of the oil content within a few weeks (Croteau, 1987). In this context, it seems worth noting that the half-life of ^{14}C-monoterpenes in *Pinus silvestris* with internal resin canals was up to 6 months, in contrast to a few hours in Lamiaceae with external organs (Banthorpe et al., 1972).

In spite of all the examples underlining the ecochemical significance of volatile organics in oils and resins, it should be stated that there is no obvious essential function of these compounds for plant life; on the contrary, the compounds seem to be by-products of the formation of essential compounds like phytohormones, phytosterols, or carotenoids. During evolutionary selection these compounds became part of adaptive strategies in several plant species and families which had to evolve ways of dealing with growth-restricting environmental stress factors.

In spite of lacking essential functions of volatile organics in biological systems, there are clear indications that terpenoid volatiles are particularly abundant in Mediterranean regions in comparison, for example, with temperate regions. Measurements to support this assumption are still scarse. Future measurements must concentrate on relevant species and ecosystems to develop the emission factors and algorithms needed to calculate realistic regional emission fluxes for the Mediterranean area. This requires consideration of the role and ecological function of volatile organics and of the process of biosynthesis and emission, as outlined above. Without a realistic assessment of emission fluxes, we may not be able to understand the chemical processes in the atmosphere and, for example, the formation of tropospheric ozone.

Potential Role of Volatile Organics in a Changing Environment

Considering the potential role of volatile organics in Mediterranean shrubs with respect to a changing environment, the likely effects of these compounds on the changing atmosphere and climate should be included (see Fate of Biogenic NMHC in the Atmosphere), as well as the feedback effects of a changing environment on volatile organics and the plants producing them (see Effects of a Changing Mediterranean Environment on Biogenic NMHC). Compiling such arguments from existing literature seems to be a huge task, but it is not, because of the paucity of existing information. The interest in understanding the amounts and the atmospheric fate of biogenic emissions started some years ago, but was enlarged only recently as it became evident that monoterpenes and isoprene could play an important role in the global change issue by regulating the oxidation potential of the troposphere (Went, 1960; Crutzen, 1974; Altshuller, 1991) and by contributing to ozone and aerosol formation.

Fate of Biogenic NMHC in the Atmosphere

Perhaps the first observation of biogenic emissions affecting the atmosphere was made, according to Went (1960), by Leonardo da Vinci, who described in his notebooks the distant mountains of Tuscany as being enveloped in a blue haze. Went observed similar hazes above pine forests, Australian eucalypt forests, aromatic shrublands of the Australian interior, and creosote-covered Californian semidesert. He assumed that haze development was due to submicron-size particles formed by light- or ozone-induced oxidation of biogenically emitted terpenes, providing cloud condensation nuclei, and thus playing an important role in the heat balance of the earth.

Due to their chemical nature, mostly unsaturated hydrocarbons, monoterpenes react very fast with atmospheric constituents, e.g., O_3 and OH radicals, and may form various oxidation products partly as aerosols; it is assumed that the Criegee intermediates are the main oxidizing species in the terpene/O_3 reaction (Gäb et al., 1985; Jay and Stieglitz, 1989; Kotzias et al., 1990). Reviewing the aerosol-forming potential of natural volatile organics, Altshuller (1983) concluded that α-pinene is a significant aerosol producer, whereas isoprene is not. However, at rural and urban sites in the United States, acid sulfate species of anthropogenic origin are by far the dominant sources of fine aerosols and cloud condensation nuclei (Altshuller, 1983). Conducting flow reaction experiments with highly concentrated ozone–monoterpene mixtures, Yokouchi and Ambe (1985) were tentatively able to identify pinonaldehyde as an aerosol product. Moreover, the gas-phase reaction of selected monoterpenes (α-, β-pinene, limonene) in systems with [O_3, SO_2] and [O_3, SO_2, NO_2] yielded H_2SO_4, organic nitrates as well as nopinone, pinonaldehyde, and other products (Kotzias et al., 1990).

A second role of biogenic NMHC was attributed to their possible contribution to the formation of gaseous pollutants like CO, O_3, PAN, carbonyls, and organic acids. Chameides et al. (1988) estimated that the failure of Atlanta, GA to meet the air quality standard for ozone was due more to biogenic than to anthropogenic NMHC emissions. Two other southern cities and possibly other US cities also emit enough biogenic NMHC to cause them to exceed the air quality standards. In urban plumes of Rome (Ciccioli et al., 1987) or London (MacKenzie et al., 1991) there was no indication for enhanced ozone formation due to biogenic emissions, but it has been shown very clearly that anthropogenic NO_x and volatile organics in urban plumes can adversely affect the air quality of the surrounding rural areas (Altshuller, 1988). In contrast to the situation in urban areas, for a rural site in the eastern United States, Trainer et al. (1987) demonstrated that isoprene can have a significant impact on ozone formation in stagnant high-pressure conditions and in the presence of NO_x transported in from source areas. In the central Amazon forest, the soil represents a strong source of NO and the forest canopies are a source of isoprene and some monoterpenes. The model of Jacob and Wofsy (1988) indicated that under these conditions, biogenic emissions do not enhance the formation of ozone, but lead to the formation of organic acids, as was confirmed by the measurements of Andreae et al. (1988).

Recent reviews of the topic (Altshuller, 1991; Brasseur and Chatfield, 1991) indicate that the role of isoprene and monoterpenes in atmospheric chemistry is not well understood, but there is agreement that these compounds follow the general reaction scheme of methane, with oxidation by the hydroxyl radical to CO and CO_2 via complex processes. The time constant of the chemical destruction by the OH radical was given as 0.1 and 0.3 days for isoprene and α-pinene, respectively, in contrast to 10 years for the very stable methane (Brasseur and Chatfield, 1991). The OH radical is the major sink for both; this is why methane is so abundant in the atmosphere. Carbon monoxide, the final product of hydrocarbon oxidation by OH, is increasing globally at rates of 1–2% per year; increasing CO can deplete the OH radical if NO_x concentrations are at or below 1 ppb (Altshuller, 1991); in NO-rich air, the OH radical can be reformed by oxidizing NO to NO_2; NO_2 is photolyzed to yield ozone (Brasseur and Chatfield, 1991). The importance of ozonolysis in the NO_x photochemical system in initiating the ultimate fate of monoterpenes and in producing potentially important intermediate products has been often demonstrated and can be regarded as a feature in studies of the atmospheric reactions of biogenic hydrocarbons (Gäb et al., 1985; Jay and Stieglitz, 1989). It is generally accepted that oxidation of terpenoids can produce significant amounts of CO, comparable to the oxidation of methane. An inventory calculated for the United States gives the anthropogenic contribution to CO production as 135 Tg/year and the total contribution of biogenic sources as 35 Tg/year (Altshuller, 1991).

To summarize the potential effects of biogenic NMHC emissions on the global environment, the effects are largely indirect, such as:

- Depleting OH radicals by producing CO may increase the lifetime of radiatively active gases such as CH_4, thus increasing greenhouse warming;

- Formation of aerosols and cloud condensation nuclei may enhance cloud albedo and reduce greenhouse warming;
- Formation of organic acids may enhance acidic deposition in remote areas;
- Tropospheric ozone may be produced or consumed, depending on the NO concentration; ozone is a radiatively active gas, a UV-B absorber, and the most common phytotoxic air pollutant.

Effects of a Changing Mediterranean Environment on Biogenic NMHC

With respect to the concept of global change, we take the view that the topic should not be limited to the "greenhouse effect" but should include the following aspects, listed in the order in which they became a subject of scientific and public concern: (1) enhanced and further increasing concentration of tropospheric ozone; (2) acid deposition, the enhanced deposition and altered biogeochemical cycling of nitrogen and sulfur on a regional and global scale; (3) depletion of beneficial stratospheric ozone and resulting increase in tropospheric UV-B; (4) increasing emission of greenhouse gases, enhanced CO_2 concentration, and resulting global warming. Today, the most serious concern is the potential effects of global warming, but this topic is the most difficult to assess, among other reasons due to the unknown magnitude of the cloud feedback effect (Ramanathan et al., 1989). Although these phenomena are usually examined in isolation, they are tightly coupled and in need of simultaneous analysis, linked by the common cause of anthropogenic emissions perturbing the chemistry of the atmosphere and the natural biogeochemical cycles (Schlesinger, 1991).

Calculating the effects of greenhouse warming on biogenic emissions in the Mediterranean area might be attempted by simply combining the temperature-based algorithms as used for biogenic emission inventories in the United States (Lamb et al., 1987) or in Europe (Lübkert and Schopp, 1989) with the isolines of the anticipated temperature increase, as simulated with general circulation models (GCMs) for the Mediterranean region (Jung and Bach, 1986). However, given the present state of the art, this procedure is unlikely to yield realistic predictions. As indicated in Factors Controlling the Emission of Volatile Terpenes, temperature cannot be used as the only driving parameter to model biogenic emissions from vegetation, least of all in the case of Mediterranean shrublands. Examining past climate variations in the Mediterranean Basin (Berger, 1986) or future scenarios of CO_2 increase (Jung and Bach, 1986) by running different GCM simulations, the main conclusion was that methodology has to be improved before the GCMs can be used to predict adequately on a regional scale.

However, the Mediterranean climatic region is transitional between the suboceanic climate of the cool-temperature woodland zone and the dry climates of the high-pressure subtropical belt and should be particularly vulnerable to changes in temperature and precipitation. On the other hand, Mediterranean systems are particularly sensitive to anthropogenic disturbance. Intensive deforestation since ancient times has caused changes in the water balance and surface runoff and resulted in widespread soil erosion. Even today desertification is increasing in parts of Spain, Greece, Sicily, and Sardinia due to mechanized agriculture on formerly terraced slopes, man-made fires, and impacts of air pollution (Fantechi and Margaris, 1986). The Mediterranean shrubs referred to in this chapter became dominant in the macchia/garrigue shrublands in the Mediterranean region of today mainly as degradation stages of the naturally prevalent oak forests (CORINE, 1987). It is quite reasonable that these shrublands result largely from human land use rather than from the Mediterranean climate per se, possibly in contrast to the situation in other MTEs of the world. It seems difficult to imagine how potential effects of gradual future climate change could be separated from the ubiquituous land use changes.

In any case, in considering potential effects of climate change on contents and emissions of volatile organics in Mediterranean shrubs, we need to consider the specific type of emissions from the different species, the role of volatile oils in adaptive strategies to meet Mediterranean stress factors, and the biochemical mechanisms controlling emissions, as has been discussed earlier. Future increases in CO_2 concentrations and temperature would modify biogenic emissions in different ways. Isoprene emissions are linked to photosynthetic CO_2 assimilation and are stimulated by increasing temperatures. In some cases, the emissions are inhibited by higher CO_2 concentrations. Monoterpene emissions will certainly increase with temperature, but variations in volatile oil contents are likely to be important too. The main influence of a warmer, drier climate on biogenic emissions should be mediated through shifts in species composition, effects on leaf area or growth dynamics, changes of phenological patterns, or altered stomatal conductance, as described in other chapters of this volume. The arguments discussed in Role of Volatile Organics in Mediterranean-Type Ecosystems allow the general assumption that the role of xeromorphic plants which are rich in aromatic oils would be more important in conditions of the anticipated future climate.

In considering effects of other aspects of the changing global environment, it can first be noted that all the MTEs of the world (Hidy et al., 1978; maps shown in Krupa and Kickert, 1989), and the Mediterranean basin in particular (Millán, 1992), are highly susceptibile to photochemical smog. In addition, they should be sensitive to enhanced nitrogen deposition, due to their general nitrogen paucity (Kruger et al., 1983), but this topic is discussed elsewhere in this volume.

There are a few studies investigating effects of air pollutants on pools or emissions of volatile organics. Renwick and Potter (1981) observed quantitative changes of terpene emissions from Balsam fir after fumigation with 2 ppm of SO_2. In SO_2-fumigated *Pinus silvestris*, significant changes of monoterpene content and emission appeared only after necrosis (Bucher, 1984). Needles of *Picea abies* growing in the Black Forest in SW Germany with symptoms of montane yellowing differed significantly from green needles in terpene content and composition (Jüttner, 1988). No significant effects on monoterpene content or emission appeared in *Picea abies* clones after 5 years of continuous treatment with low concentrations of SO_2, O_3, or acid fog, alone or in combination (Bufler et al., 1990).

Thus, volatile organics in conifers do not seem to be significantly affected by ambient pollutant concentrations; however, pollutant effects on volatiles in Mediterranean shrubs as well as on the shrubs themselves have scarcely been investigated. Krupa and Kickert (1989) reviewed hundreds of experimental studies that screened hundreds of plants for effects of increased ozone, UV-B, CO_2, and temperature, but they were not able to include any study performed in Mediterranean countries. This is surprising because the phytotoxic oxidants were first shown to be responsible for widespread damage to grapes, citrus, tobacco, and pines in MTEs of southern Californiia (Richards and Taylor, 1961). Westman (1979) observed ozone damage in the Californian coastal sage shrub, dominated by *Salvia*, *Eriogonum*, and *Encelia*. The only study reporting ozone damage in the Mediterranean Basin is from Israel, where Naveh et al. (1980) ascribed declines of *Pinus halepensis* and the disappearance of the lichen *Xanthoria parietina* from maquis shrubs to effects of ozone.

The present state of knowledge does not allow any assessment of air pollutant effects on plant species and ecosystems in the Mediterranean basin. However, experience gained in other regions gives circumstantial evidence that they are likely to occur, at least with regard to photosmog exposure. Obviously, there is an urgent need for integrated studies on effects of a changing global environment in Mediterranean ecosystems, including the anticipated as well as the already existing effects. Investigating anthropogenic and biogenic emissions on the regional scale would promote understanding the chemistry of the atmosphere and the natural biogeochemical cycles as the common cause of global change phenomena.

The BEMA Project: Objectives and First Results

BEMA: Presentation

The Biogenic Emissions in the Mediterranean Area (BEMA) project is a pan-European program with the primary aim of quantifying the emission rates of selected reactive trace gases from the vegetation to the atmosphere in the Mediterranean Basin in order to contribute to the understanding of the chemical composition of the atmosphere in that area and, hence, its potential contribution to ozone formation and to global changes. The core of the study is a series of intensive field campaigns to be conducted at representative sites in Italy, France, and Spain. Both emission fluxes and the ambient concentrations of the primary compounds and their reaction products will be determined together with concentrations and deposition fluxes of the main anthropogenic compounds. Longer-term measurements and less intensive campaigns are also carried out at these sites to characterize seasonal and annual variability of emission rates and of selected environmental parameters controlling emission.

The first measuring campaign is scheduled for June 1993 at the Castelporziano test site near Rome, Italy; it will be joined by 13 scientific laboratories from different European countries. Measurement of trace gas fluxes and concentrations will be performed in a *Pinus pinea*/*Quercus ilex* forest with understory formed by typical Mediterranean shrubs, and in a seminatural grassland. The gases considered will include the biogenic VOC emissions, the deposition of some anthropogenic pollutants, and the gas exchange of the plants via photosynthesis and transpiration. Flux measurements will be methodically based on a balanced combination of enclosure methods with micrometeorological, tracer, and modelling techniques.

The field campaigns are complemented by a series of other activities, including (1) the identification of the major vegetation types in the Mediterranean Basin; (2) intercomparison exercises to ensure the compatibility of the sampling, analytical, and enclosure methods; (3) laboratory experiments in chemical reaction chambers to identify hydrocarbon reaction products; and (4) controlled greenhouse experiments with selected Mediterranean plants to identify emitted species and to search for the factors controlling emissions.

First Greenhouse Measurements

Within the last-named activity, experiments were performed with Mediterranean plant species cultivated in the greenhouse to identify the type and amount of volatile compounds emitted from and stored in the plants. The data presented at this stage are mostly qualitative; nevertheless they could be of some interest because the chemical identification of the compounds contained and emitted by Mediterranean shrubs is not trivial. Data are presented for the Lamiaceae shrubs *Rosmarinus officinalis*, *Thymus vulgaris*, *Salvia officinalis*, *Lavandula latifolia*, *Spartium junceum* (Leguminosae); *Laurus nobilis* (Lauraceae); and *Quercus ilex* (Fagaceae).

Emissions were measured using dynamic cuvettes with volumes between 100 and 400 L, covered with Teflon film and suited to investigate the mass

Table 17-1. Volatile Organics in the Leaf Pools Compared to Emissions of Selected Mediterranean Shrub Species. Compounds are Listed in Order of Their Relative Abundance

Species	Content	Emission
Thymus vulgaris, Jul 92	Thymol Cymene Δ3-Carene Thymoquinon Carvacrol Linalool Borneol Eugenol Caryophyllene	Cymene Thymol-methylether Dehydronaphthalene? Camphene Caryophyllene
Rosmarinus officinalis, Jun 92	Myrcene Cineol Camphor β-Caryophyllene Δ3-Carene α-Pinene Camphene α-Terpinen β-Pinene	α-Pinene Camphene Cineol β-Pinene Camphor
Salvia officinalis, Apr 92 (with flowers)	Manool? Cineol Thujon Caryophyllene Camphor Borneol Camphene α-Pinene β-Pinene Bornylacetate	Copaene Thujon Cariophyllene Camphene Cineol α-Pinene β-Pinene Borneol Bornylacetete
Salvia officinalis, Jun 92 (no flowers)	Manool? Borneol Camphor Cineol Thujon Caryophyllene	Thujon Camphor

balance of trace gas exchange. The cuvettes were continuously stirred and partly temperature controlled; air was provided from a pure air generator via mass flow controllers. The volatile organics emitted from the aromatic shrubs were trapped from cuvette air on charcoal-filled adsorption tubes and analyzed by gas chromatographic/mass spectrometric (GC/MS) methods. Search of mass spectra was performed by combining our own library with the data given by Adams (1989). The volatile oil content of leaf and bark was analyzed by GC/MS from 0.5–1.0 g of crushed plant material extracted with 2 mL Hexan and then Dichlormethan, both for 24 hours.

Table 17-1. *Continued*

Species	Content	Emission
	Camphene Bornylacetate Copaene α-, β-Pinene	
Lavandula latifolia, Mar 92	Coumarin Borneol Cineol + Limonene Camphor	Cineol Limonene Camphor Camphene α-Pinene Borneol
Laurus nobilis, Mar 92	Methyleugenol Cineol Terpenylacetate Linalool Eugenol α-, β-Pinene Terpineol	Alkanes Camphene α-Pinene Limonene Cineol Decanal β-Pinene
Spartium junceum, Apr 92 (with flowers)	Butylated Hydroxytoluene Compound not identified 2-Hepten-ol	Formamid (2-methylphenyl?) Benzaldehyde (2.2.5 trimethyl) Pyridine (4-methyl) Limonene Phenyl-ethylacetate Elemicin
Quercus ilex, Sep 92		α-Pinene β-Pinene Sabinene Ocimene? Myrcene Limonene

The results summarized in Table 17-1 show the specific pools and emissions of volatile organics for each species; in the case of *Quercus*, only emitted compounds are presented, because essential oil volatiles have not been detected in leaf or bark. Figures 17-1, 17-2, and 17-3 show the chromatograms of volatile oil pools and emission patterns of *Thymus vulgaris*, a Lamiacae with external glandular trichomes; of *Laurus nobilis*, with internal specialized cells (idioblasts) as oil containers; and of *Quercus ilex*, without any essential oil content.

Comparing pools and emissions of the aromatic shrubs, it becomes evident that knowing the relative composition of the pools is not enough to allow prediction of emission rates. In *Thymus* it was observed that thymolmethylether was emitted instead of the parent compound thymol, indicating the possibility that the plant can convert polar compounds into

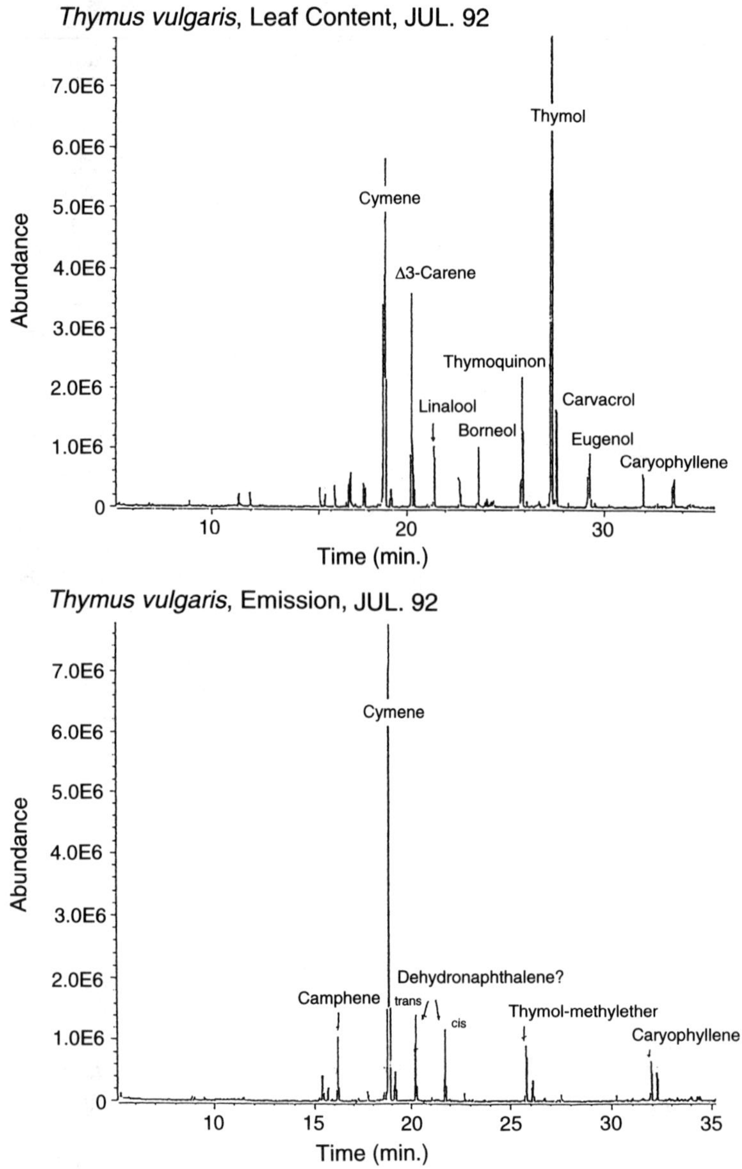

Figure 17-1. GC-MS separation of compounds contained in the leaves and in the emissions of *Thymus vulgaris* (Lamiaceae).

more volatile ones by methylation. The Lamiaceae shrubs tested emit mainly terpenoids and some phenolics, but volatile oxygenated compounds seem to be more frequent than in conifers. *Laurus* emissions are composed of different alcane species and some monoterpenes. During

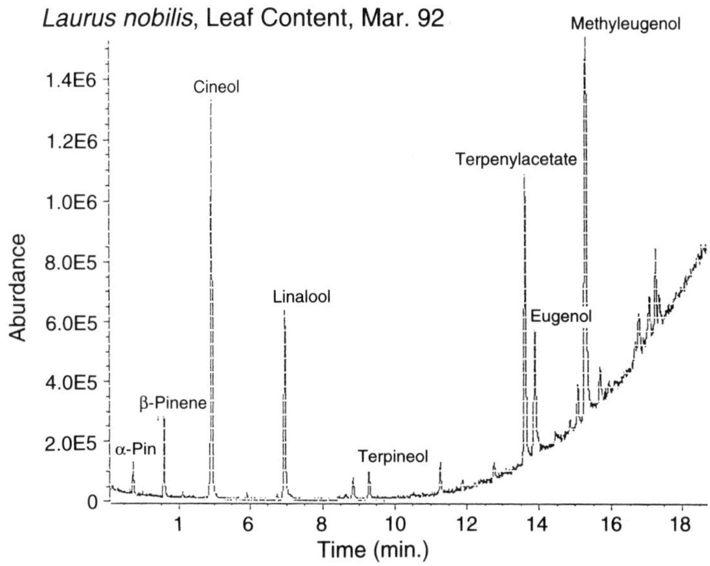

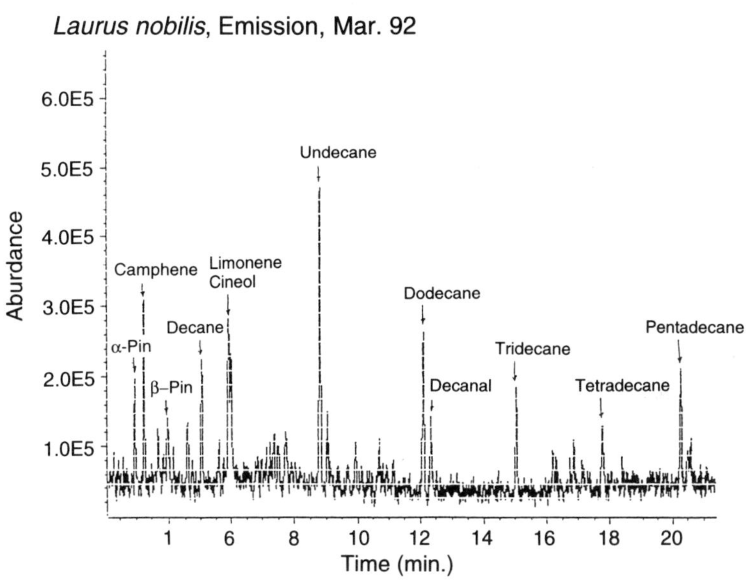

Figure 17-2. GC-MS separation of compounds contained in the leaves and in the emissions of *Laurus nobilis* (Lauraceae).

flowering, additional compounds are emitted in great quantities, as can be seen with *Salvia*, where the sesquiterpenes copaene and cariophyllene appear as main extra fragrance components. The fragrance which is emitted from *Spartium junceum* is composed of some limonene and

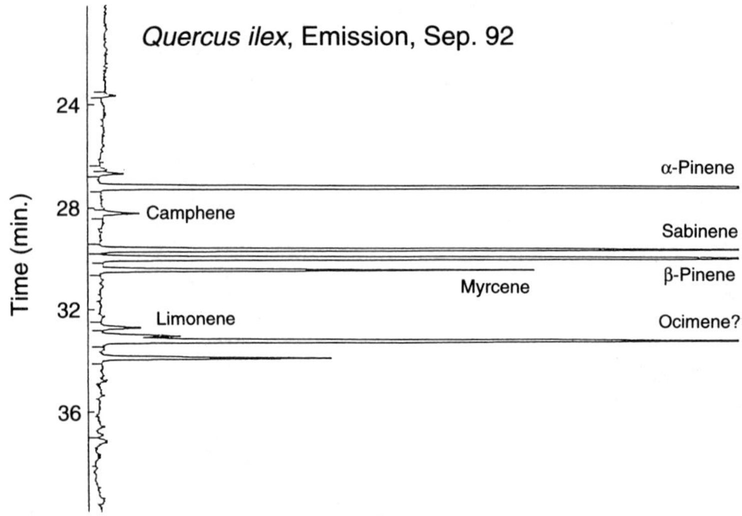

Figure 17-3. GC-MS separation of compounds emitted from *Quercus ilex* (Fagaceae) (relative abundance of compounds in the 2–40 min part of the run).

mainly of phenolics including the N-containing compounds formamid and pyridine.

Of particular interest are the results observed with *Quercus ilex*. Instead of emitting isoprene, as most other oak species do, this evergreen, sclerophyllous oak emits monoterpenes. The monoterpene emission rates observed during daytime seem to be higher than any other values reported in the literature for other plant species. Because holm oak has no monoterpenes accumulated in leaves or bark, the emitted compounds should derive from an instant, unknown metabolic pathway. Work is under way to study this topic in controlled conditions in the lab and in the field.

Conclusions and Research Recommendations

The *Quercus ilex* results presented above may exemplify the questionability of the procedure discussed in Factors Controlling the Emission of Volatile Terpenes, to calculate, at the present state of knowledge, inventories of biogenic emissions at regional to global scales. The results can further demonstrate that without process-level research on ecological functions and on biosynthesis and emissions of volatile organics, as outlined in Role of Volatile Organics in Mediterranean-Type Ecosystems, we may easily fail in developing the emission factors and algorithms needed for a realistic assessment of the role of biogenic emissions in atmospheric chemistry and ozone formation. The poor understanding of the potential role of volatile organics in the (changing) environment, as compiled from literature in Potential Role of Volatile Organics in a Changing Environment, could partly be ascribed to the hitherto existing omission of process-level research and of cooperation between ecological and atmospheric scientists.

Future research in MTEs on biogenic emissions should not be limited to this aspect, but should be designed from the very beginning to include all the relevant features that are needed to understand the whole topic of biosphere–atmosphere interactions and of atmospheric chemistry. In addition, such research requires that results be extrapolated in space and time from individual sites to regions and over relatively long periods. Thus, some general design criteria for MTE research on biosphere–atmosphere interactions could be to (1) establish a network of test sites representing common ecosystems in a region, (2) install long-term measurements of basic parameters, and (3) organize intensive measurements based on joint protocols for the individual sites.

Measuring the fluxes of trace gas exchange could link the different compartments of the ecosystems as well as the respective scientific disciplines. Once the basic instrumentation is properly installed at a site to observe trace gas exchange at the leaf, branch, canopy, and ecosystem level, it makes no sense to measure only, e.g., biogenic emission. Instrumentation can be easily extended to include pollutant deposition and photosynthetic gas exchange. To yield the maximum profit from these measurements, ecological and effects research should be included as well, which in turn will stimulate the understanding of flux measurements.

References

Adams RP. 1989. Identification of Essential Oils by Ion Trap Mass Spectroscopy. Academic Press, London.

Altshuller AP. 1983. Review: natural volatile organic substances and their effect on air quality in the United States. Atmos Environ 17:2131–2165.

Altshuller AP. 1988. Some characteristics of ozone formation in the urban plume of St. Louis, MO. Atmos Environ 22:499–510.

Altshuller AP. 1991. The production of carbon monoxide by the homogenous NO_x-induced photooxidation of volatile organic compounds in the troposphere. J Atmosph Chem 13:155–182.

Anastasi C, Hopkinson L, Simpson VJ. 1991. Natural hydrocarbon emissions in the United Kingdom. Atmosph Environ 25A:1403–1408.

Andreae MO, Schimel DS (eds). 1989. Exchange of trace gases between terrestrial ecosystems and the atmosphere. Dahlem workshop report 47. John Wiley & Sons, Chichester, UK.

Andreae MO, Talbot R, Andreae T, Harriss R. 1988. Formic and acetic acid over the central Amazon region, Brazil. 1. Dry season. J Geophys Res 93D: 1616–1624.

Arey J, Winer AM, Atkinson R, Aschmann SM, Long WD, Morrison CL. 1991. The emission of (Z)-3-hexen-1-ol, (Z)-3-hexenylacetate and other oxygenated hydrocarbons from agricultural plant species. Atmosph Environ 25A: 1063–1075.

Arnold GW, Hill JL. 1972. Chemical factors affecting selection of food plants by ruminants. In: Harborne JB (ed) Phytochemical Ecology. Academic Press, London, pp. 71–101.

Banthorpe DV, Charlwood BV, Francis MJO. 1972. The biosynthesis of monoterpenes. Chem Rev 72:115–155.

Berger A. 1986. Desertification in a changing climate with a particular attention to the Mediterranean countries. In: Fantechi R, Margaris NS (eds) Desertification in Europe. D. Reidel Publ., Dordrecht, pp. 15–34.

Bergström G. 1991. Chemical ecology of terpenoid and other fragrances of angiosperm flowers. In: Harborne JB, Tomas-Barberan FA (eds) Ecological Chemistry and Biochemistry of Plant Terpenoids. Clarendon Press, Oxford, pp. 287–296.

Bradford D, Hsiao T. 1982. Physiological responses to moderate water stress. In: Lange O, Nobel P, Osmond C, Ziegler H (eds) Encyclopedia of Plant Physiology. Springer-Verlag, Berlin, Vol. 8, pp. 263–324.

Brasseur GP, Chatfield RB. 1991. The fate of biogenic trace gases in the atmosphere. In: Sharkey TD, Holland EA, Mooney HA (eds) Trace Gas Emissions by Plants. Academic Press, San Diego, CA, pp. 1–28.

Bucher JB. 1984. Emissions of volatiles from plants under air pollution stress. In: Koziol MJ, Whatley FR (eds) Gaseous Air Pollutants and Plant Metabolism. Butterworth, London, pp. 329–412.

Bufler U, Seufert G, Jüttner F. 1990. Monoterpene patterns of different tissues and plant parts of Norway spruce (*Picea abies* L. Karst.). In: Special issue on the Hohenheim long-term experiment. Environ Poll 68:367–375.

Burbott AJ, Loomis WD. 1967. Effects of light and temperature on the monoterpenes of peppermint. Plant Physiol 42:20–28.

Chameides WL, Lindsay RW, Richardson JM, Kiang CS. 1988. The role of biogenic hydrocarbons in urban photochemical smog. Science 241:1473–1475.

Chapin FS. 1980. The mineral nutrition of wild plants. Ann Rev Ecol Syst 11:261–285.

Ciccioli P, Brancaleoni E, DiPalo L, Brachetti A, Cecinato A. 1987. Daily trends of photochemical oxidants and their precursors in a suburban forested area. A useful approach for evaluating the relative contributions of natural and anthropogenic hydrocarbons to the photochemical smog formation in rural areas in Italy. In: Angelletti G, Restelli G (eds) Physico-Chemical Behaviour of Atmospheric Pollutants. Reidel, Dordrecht, pp. 551–559.

CORINE. 1987. Map of natural vegetation of the member states of the European Community and of the Council of Europe. Off. for Official Publ. Europ. Comm., Luxembourg.

Croteau R. 1977. Effect of irrigation method on essential oil yield and rate of oil evaporation in mint grown under controlled conditions. Hort Sci 12:563–565.

Croteau R. 1987. Biosynthesis and catabolism of monoterpenoids. Chem Rev 87:929–954.

Croteau R, Felton M, Karp F, Kjonaas R. 1981. Relationship of camphor biosynthesis to leaf development in sage (*Salvia officinalis*). Plant Physiol 67:820–824.

Crutzen PJ. 1974. Photochemical reactions initiated by and influencing ozone in unpolluted tropospheric air. Tellus 26:47–57.

Dell B, McComb AJ. 1978. Plant resins – their formation, secretion and possible functions. Adv Bot Res 6:277–316.

Dement WA, Tyson BJ, Mooney HA. 1974. Mechanism of monoterpene volatilization in *Salvia mellifera*. Phytochemistry 14:2555–2557.

Fantechi R, Margaris NS (eds). 1986. Desertification in Europe. Proc Symp EEC Progr Climatology, Mytilene, Greece, 15–18 Apr 1984.

Fineschi S, Grossoni P. 1981. Contenuto in monoterpeni di oleoresine xilematiche in provenience diverse di pino laricio. Italia Forestale Montagna 36:232–239.

Fischer NH. 1991. Plant terpenoids as allelopathic agents. In: Harborne JB, Tomas-Barberan FA (eds) Ecological Chemistry and Biochemistry of Plant Terpenoids. Clarendon Press, Oxford, pp. 377–398.

Francke W. 1991. Semiochemicals: Mevalogenins in systems of chemical communication. In: Müller PM, Lamparsky D (eds) Perfumes: Art, Science, and Technology. Elsevier, London, pp. 61–99.

Gäb S, Hellpointner E, Turner WV, Korte F. 1985. Hydroxymethyl hydroperoxide and bis(hydroxymethyl)peroxide from gas-phase ozonolysis of naturally occurring alkenes. Nature 316:535–536.

Gershenzon J, Lincoln DE, Langenheim JH. 1978. The effect of moisture stress on monoterpenoid yield and composition in *Satureja douglasii*. Biochem System Ecol 6:33–43.

Gershenzon J, Maffei M, Croteau R. 1989. Biochemical and histochemical localization of monoterpene biosynthesis in the glandular trichomes of spearmint (*Mentha spicata*). Plant Physiol 89:1351–1357.

Gildemeister E, Hoffman F. 1955–1960. Die ätherischen Öle. Akademie Verlag, Berlin, 8 vol.

Hanover JW. 1972. Factors affecting the release of volatile chemicals by forest trees. Mitt. Forstl. Bundesversuchsanstalt, Wien 97:625–644.

Harborne JB. 1988. Introduction to Ecological Biochemistry. Academic Press, London.

Harborne JB. 1991. Recent advances in the ecological chemistry of plant terpenoids. In: Harborne JB, Tomas-Barberan FA (eds) Ecological Chemistry and Biochemistry of Plant Terpenoids. Clarendon Press, Oxford, pp. 399–425.

Hegnauer R. 1962–1992. Chemotaxonomie der Pflanzen. 8 vol. Birkhäuser Verlag, Basel.

Hidy GM, Mahoney JR, Goldsmith BJ. 1978. International aspects of the long range transport of air pollutants. Final Report US Department of State, Washington, DC.

Isidorow VA, Zenkevich IG, Ioffe BV. 1985. Volatile organic compounds in the atmosphere of forests. Atmos Environ 19:1–8.

Jacob DJ, Wofsy SC. 1988. Photochemistry of biogenic emissions over the Amazon forest. J Geophys Res 93D:1477–1486.

Jay K, Stieglitz L. 1989. Identifizierung chemischer-photochemischer Umsetzungsprodukte von biogenen Kohlenwasserstoffen mit anthropogenen Luftschadstoffen (O_3, NO_2, SO_2). Kernforschungszentrum-PEF, Karlsruhe, ISSN 0931-2749, Berichte Bd. 53, 123 pp.

Jung HJ, Bach W. 1986. The effects of model-generated climatic changes due to a CO_2 doubling on desertification processes in the Mediterranean area. In: Fantechi R, Margaris NS (eds) Desertification in Europe. D Reidel Publ, Dordrecht, pp. 35–48.

Jüttner F. 1988. Changes of monoterpene concentrations in needles of pollution-injured *Picea abies* exhibiting montane yellowing. Physiol Plant 72:48–56.

Kisser JG. 1958. Die Ausscheidung von ätherischen Ölen und Harzen. In: Ruhland W (ed) Encyclopedia of Plant Physiology, Vol 10: The Metabolism of Secondary Plant products. Springer-Verlag, Berlin, pp. 91–131.

Knöppel H, Versino B, Peil A, Schauenburg H, Vissers H. 1981. Quantitative determination of terpenes emitted by conifers. In: Versino B, Ott H (eds) Proc 2nd Eur Symp on Phys Chem Behav Atmos Pollutants, Varese, Italy. Reidel Publ, Dordrecht, pp. 89–98.

Kotzias D, Fytianos K, Geiss F. 1990. Reaction of monoterpenes with ozone, sulphur dioxide and nitrogen dioxide-gas phase oxidation of SO_2 and formation of sulphuric acid. Atmos Environ 24A:2127–2132.

Kruger FJ, Mitchell DT, Jarvis JUM. 1983. Mediterranean-type ecosystems. The role of nutrients. Springer-Verlag, Berlin.

Krupa SV, Kickert RN. 1989. The greenhouse effect: Impacts of ultraviolet-B (UV-B) radiation, carbon dioxide (CO_2), and ozone (O_3) on vegetation. Environ Poll 61:263–393.

Lamb B, Westberg H, Allwine G. 1985. Biogenic hydrocarbon emissions from deciduous and coniferous trees in the United States. J Geophys Res 90: 2380–2390.

Lamb B, Guenther A, Gay D, Westberg H. 1987. A national inventory of biogenic hydrocarbon emissions. Atmos Environ 21:1695–1705.

Lerdau MT. 1991. Plant function and biogenic emission. In: Sharkey TD, Holland EA, Mooney HA (eds) Trace Gas Emissions by Plants. Academic Press, San Diego, CA, pp. 121–134.

Lovelock JE. 1979. Gaia. A New Look at Life on Earth. Oxford University Press, London.

Lübkert B, Schöpp W. 1989. A model to calculate natural VOC emissions from forests in Europe. Working paper WP-89-082. International Institute for Applied Systems Analysis, Laxenburg, Austria.

MacKenzie AR, Harrison RM, Colbeck I, Hewitt CN. 1991. The role of biogenic hydrocarbons in the production of ozone in urban plumes in southeast England. Atmos Environ 25A:351–359.

Millán MM. 1991. Mesometeorological cycles of air pollution in the Iberian peninsula. CEC-Air Pollution Research Report Series No 44, Guyot, Brussels, 219 pp.

Miller PC. 1983. Canopy structure of Mediterranean-type shrubs in relation to heat and moisture. In: Kruger FJ, Mitchell DT, Jarvis JUM (eds) Mediterranean-Type Ecosystems. The Role of Nutrients. Springer-Verlag, Berlin, pp. 133–166.

Molisch H. 1937. Der Einfluß einer Pflanze auf die andere – Allelopathie. Fischer Verlag, Jena.

Monson RK, Guenther AB, Fall R. 1991. Physiological reality in relation to ecosystem- and global-level estimates of isoprene emission. In: Sharkey TD, Holland EA, Mooney HA (eds) Trace Gas Emissions by Plants. Academic Press, San Deigo, CA, pp. 185–208.

Mooney HA. 1983. Carbon-gaining capacity and allocation patterns of Mediterranean-climate plants. In: Kruger FJ, Mitchell DT, Jarvis JUM (eds) Mediterranean-Type Ecosystems. The Role of Nutients. Springer-Verlag, Berlin, pp. 103–119.

Mooney HA, Dunn EL. 1970. Convergent evolution of Mediterranean-climate evergreen sclerophyll shrubs. Evolution 24:292–303.

Mooney HA, Gulmon SL. 1982. Constraints on leaf structure and function in reference to herbivory. Bioscience 32:189–206.

Mooney HA, Vitousek PM, Matson PA. 1987. Exchange of materials between terrestrial ecosystems and the atmosphere. Science 238:926–932.

Muller CH. 1970. Phytotoxins as plant habitat variables. Rec Adv Phytochem 3:106–121.

Müller JF. 1992. Geographical distribution and seasonal variation of surface emissions and deposition velocities of atmospheric trace gases. J Geophys Res 97:3787–3804.

Naveh Z, Steinberger EH, Chaim S, Rotmann A. 1980. Photochemical air-pollutants—a threat to Mediterranean coniferous forests and upland ecosystems. Environ Conserv 7:301–309.

Pickett JA. 1991. Lower terpenoids as natural insect control agents. In: Harborne JB, Tomas-Barberan FA (eds) Ecological Chemistry and Biochemistry of Plant Terpenoids. Clarendon Press, Oxford, pp. 297–313.

Ramanathan V, Cess RD, Harrison EF, Minnis P, Barkstrom BR, Ahmad E, Cohen M. 1989. Cloud-radiative forcing and climate: results from the Earth radiation budget experiment. Science 243:57–63.

Rasmussen RA, Khalil MAK. 1988. Isoprene over the Amazon Basin. J Geophys Res 93:1417–1421.

Renwick JAA, Potter J. 1981. Effects of sulfur dioxide on volatile terpene emission from Balsam Fir. J Air Poll Contr Assoc 31:65–66.

Richards BL, Taylor OC. 1961. Status and redirection of research on the atmospheric pollutants toxic to field grown crops in Southern California. J Air Poll Contr Assoc 11:125–128.

Rice EL. 1984. Allelopathy. Academic Press, New York.

Ross JD, Sombrero C. 1991. Environmental control of essential oil production in Mediterranean plants. In: Harborne JB, Tomas-Barberan FA (eds) Ecological Chemistry and Biochemistry of Plant Terpenoids. Clarendon Press, Oxford, pp. 83–94.

Rundel PW. 1991. Shrub life-forms. In: Mooney HA, Winner WE, Pell EJ (eds) Response of Plants to Multiple Stresses. Academic Press, San Diego, CA, pp. 345–370.

Sanadze GA. 1991. Isoprene effect-light dependent emission of isoprene by green plants. In: Sharkey TD, Holland EA, Mooney HA (eds) Trace Gas Emissions by Plants. Academic Press, San Diego, CA, pp. 136–152.

Sandermann W. 1968. Terpenoids: structure and distribution. In: Florkin M, Mason HS (eds) Comparative Biochemistry. Academic Press, New York, pp. 503–630.

Schiller G, Grunwald C. 1987. Resin monoterpenes in range-wide provenanced trials of *Pinus halepensis* Mill. in Israel. Silvae Geneticae 36:109–114.

Schindler T, Kotzias D. 1989. Comparison of monoterpene volatilization and leaf-oil composition of conifers. Naturwissenschaften 76:475–476.

Schlesinger WH. 1991. Biogeochemistry. An Analysis of Global Change. Academic Press, San Diego, CA.

Schönherr J. 1982. Resistance of plant surfaces to water loss: Transport properties of cutin, suberin and associated lipids. In: Lange O, Nobel P, Osmond C, Ziegler H (eds) Encyclopedia of Plant Physiology. Springer-Verlag, Berlin, Vol 12B, pp. 153–179.

Sharkey TD, Loreto F, Delwiche CF. 1991. The biochemistry of isoprene emissions from leave during photosynthesis. In: Sharkey TD, Holland EA, Mooney HA (eds) Trace Gas Emissions by Plants. Academic Press, San Diego, CA, pp. 153–184.

Tingey TD, Turner DP, Weber JA. 1991. Factors controlling the emissions of monoterpenes and other volatile compounds. In: Sharkey TD, Holland EA, Mooney HA (eds) Trace Gas Emissions by Plants. Academic Press, San Diego, CA, pp. 93–120.

Tingey TD, Manning M, Grothaus LC, Burns WF. 1980. Influence of light and temperature on monoterpene emission rates from Slash Pine. Plant Physiol 65:797–801.

Trainer M, Williams EJ, Parrish DD, Buhr MP, Allwine EJ, Westberg HH, Fehsenfeld FC, Liu SC. 1987. Models and observations of the impact of natural hydrocarbons on rural ozone. Nature 329:705–707.

Tuomi J, Niemelä P, Chapin FS, Bryant JP, Sirèn S. 1988. Defensive responses to trees in relation to their carbon/nutrient balance. In: Mattson W, Levieux J, Bernard-Dagan C (eds) Mechanisms of Woody Plant Defense Against Insects. Springer-Verlag, Berlin, pp. 57–72.

Tyson BJ, Dement WA, Mooney HA. 1974. Volatilisation of terpenes from *Salvia mellifera*. Nature 252:119–120.

Vokou D, Margaris NS. 1986. Variation of volatile oil concentration of Mediterranean aromatic shrubs *Thymus capitatus* Hoffmag et Link, *Satureja thumbra* L., *Teucrium polium* L. and *Rosmarinus officinalis*. Int J Biometeor 30: 147–155.

Went FW. 1960. Blue hazes in the atmosphere. Nature 187:641–643.

Westman WE. 1979. Oxidant effects on Californian coastal sage scrub. Science 205:1001–1003.

Winer AM, Arey J, Atkinson R, Aschmann SM, Long WD, Morrison CL, Olszyk DM. 1992. Emission rates of organics from vegetation in California's Central Valley. Atmos Environ 26A:2647–2659.

Yamaura T, Tanaka S, Tabata M. 1989. Light-dependent formation of glandular trichomes and monoterpenes in thyme seedlings. Phytochemistry 28:741–744.

Yokouchi Y, Ambe Y. 1984. Factors affecting the emission of monoterpenes from red pine (*Pinus densiflora*). Plant Physiol 75:1009–1012.

Yokouchi Y, Ambe Y. 1985. Aerosols formed from the chemical reaction of monoterpenes and ozone. Atmos Environ 19:1271–1276.

Zimmermann PR, Greenberg JP, Westberg CE. 1988. Measurements of atmospheric hydrocarbons and biogenic emission fluxes in the Amazon boundary layer. J Geophys Res 93:1407–1416.

Zimmerman PR, Chatfield RB, Fishman J, Crutzen PJ, Hanst PL. 1978. Estimation of the production of CO_2 and H_2 from the oxidation of hydrocarbon emission from vegetation. Geophys Res Lett 5:679–682.

18. Using Catchments of Contrasting Hydrological Conditions to Explore Climate Change Effects on Water and Nutrient Flows in Mediterranean Forests

Josep Piñol, Jaume Terradas, Anna Avila, and Ferran Rodà

Increased CO_2 concentrations can affect ecosystems directly through effects on the plant photosynthetic system, and indirectly through the induced climatic change. Plants respond to a CO_2-enriched atmosphere by enhancing photosynthesis, reducing stomatal conductance and transpiration, and consequently increasing their water use efficiency (Eamus and Jarvis, 1989; Bazzaz, 1990). Nevertheless, very few experiments have been carried out in natural ecosystems under field conditions: only four experiments were initiated by February 1992, all with low-height vegetation (Strain and Thomas, 1992).

The indirect effect of CO_2 comes from its greenhouse properties, which will probably increase the mean temperature of the Earth. Ecological systems will be affected not only by higher temperatures, but also by a higher potential evapotranspiration. The hydrological effects of a CO_2-enriched atmosphere have been explored using global circulation models (GCMs), with the majority predicting a drier climate for most regions of the Earth (Manabe and Wetherald, 1986; Gleick, 1987; Rind et al., 1990, 1992).

In this contribution we will explore the possible effects of climate change using a different approach from that of plant physiologists and global climate modelers. In particular, we will compare the hydrological and hydrochemical results obtained for two different forested areas of northeast Spain experiencing different precipitation amounts and evaporative demands. The interest of this comparison regarding the effects of global change lies in the fact that: (1) both experimental areas are forested with the same species, have a similar bedrock, and have a similar Mediterranean-type climate; and (2) both sites have a different water availability: one (Prades) receives less precipitation (P) and has a higher potential evapotranspiration (E_T) than the other (Montseny). This similarity in the biological and physical characteristics of the catchments, and the difference in their aridity index (P/E_T), allows us to make inferences about the hydrological and hydrochemical changes that would follow a hypothetical change in the aridity conditions of the western Mediterranean region. The basic hypothesis is that the more humid Montseny catchment would change its behavior in the direction of that currently observed at Prades.

Study Areas

L'Avic and la Teula are two adjacent catchments located in the Prades massif (41°13′N, 0°55′E). TM9 is a small catchment in the Montseny mountains (41°46′N, 2°21′E). Linear distance between the Prades and Montseny sites is 120 km. The areas of the catchments are 51.6 ha for l'Avic, 38.5 ha for la Teula, and 5.9 ha for TM9. TM9 area was estimated from the chloride balance for an 8-year period. Unmeasured dry deposition of chloride was assumed to be 25% of bulk deposition. As the catchment is very small and its water divides are poorly defined, the area obtained with this method is presumably more accurate than the previous reported value of 4.3 ha obtained from 1:10.000 topographic maps (Avila and Rodà, 1988, 1990; Piñol and Avila, 1992). The altitude range is approximately 700–1,050 m for each of the three catchments. A more detailed description of the physiography of the catchments is given in Piñol and Avila (1992).

All catchments are covered by holm oak (*Quercus ilex* L.) forests that were coppiced for charcoal production until some 40 years ago. Now they are mainly composed of stump resprouts and are accreting biomass. Forest cover is nearly 100%. All catchments are underlain by phyllites. Soils are dystric or typic xerochrepts of very variable depth (15–150 cm). Texture is clay-loam at both locations.

The climate is Mediterranean, with moderate precipitation, summer drought, and relatively warm temperatures. Prades received less precipitation (mean 547 mm/year during the period 1981–1988) than Montseny (mean 884 mm/year during the period 1984–1992) and had a more pronounced summer drought. This usually causes the interruption of flow during the summer months at the Prades catchments, whereas the Montseny catchment is perennial.

Methods

Methods for measurement of rainfall and runoff at the catchments and for chemical analysis of streamwater are fully documented elsewhere (Rodà et al., 1990; Piñol et al., 1991; Piñol and Avila, 1992).

Annual E_T was estimated by the method of Samani-Hargreaves (Samani and Hargreaves, 1985, quoted in McKenney and Rosenberg, 1993). It was chosen because it gives reasonably estimates of E_T (compared to the Penman-Monteith method) both in the absolute sense and in response to simulated changes in climate (McKenney and Rosenberg, 1993). This method calculates daily E_T (mm) as a function of the daily extraterrestrial radiation (R_a, in mm H_2O equivalent), the mean daily temperature (tF, in degrees Fahrenheit), and the difference between minimum and maximum daily temperature (tD, in degrees Farenheit):

$$E_T = 0.00094 * R_a * t\mathrm{F} * t\mathrm{D}^{1/2}$$

where R_a was computed by the method described by Allen et al. (1989).

Maximum and minimum daily temperatures were measured at two different topographic positions within the catchment of l'Avic during 1989. Using these two temperature series and the estimated R_a, two values of annual E_T were calculated. The value taken as representative of l'Avic was their average.

At TM9, E_T was calculated from temperatures measured at the outlet of the catchment during 1993. As the available temperature data set was shorter than a year, the values needed to complete and entire year were estimated by least squares linear regression from temperatures measured at two meteorological stations located in the Montseny mountains.

Comparative Results of Prades and Montseny Catchments

Water Balance

The distribution of annual precipitation (P) between actual evapotranspiration (E_A) and runoff (Q) was very different at Prades and at Montseny (Table 18-1): the water yields of l'Avic and TM9 catchments were approximately 1/10 and 1/3 of annual precipitation, respectively.

The interannual variation of Q and E_A is also very different at both sites. At Prades, E_A is highly correlated with P, whereas Q is not. The opposite holds for the Montseny catchment (Figure 18-1). This noticeable difference in hydrological behavior can be easily understood by the relative magnitude of P vs. the potential evapotranspiration (E_T). When P is lower than E_T, as in Prades, then the water supply in the wetter years is still not enough to satisfy the transpiration and evaporation possible in

Table 18-1. Annual Precipitation (P), Runoff (Q), Actual Evapotranspiration (E_a, Computed as $E_A = P - Q$), and Potential Evapotranspiration (E_T) at the Catchments of l'Avic and TM9

Catchment	Location	Period	P	Q	E_A	E_T
L'Avic Mean	Prades	81–88	548	45	502	1,025
TM9	Montseny	84–85	873	295	578	
TM9	Montseny	85–86	719	239	480	
TM9	Montseny	86–87	808	231	577	
TM9	Montseny	87–88	1,049	529	519	
TM9	Montseny	88–89	641	130	511	
TM9	Montseny	89–90	689	94	594	
TM9	Montseny	90–91	1,179	501	677	
TM9	Montseny	91–92	1,111	423	688	
TM9 Mean	Montseny	84–92	884	305	578	876

All figures in mm.
Prades data taken from Piñol et al. (1991).

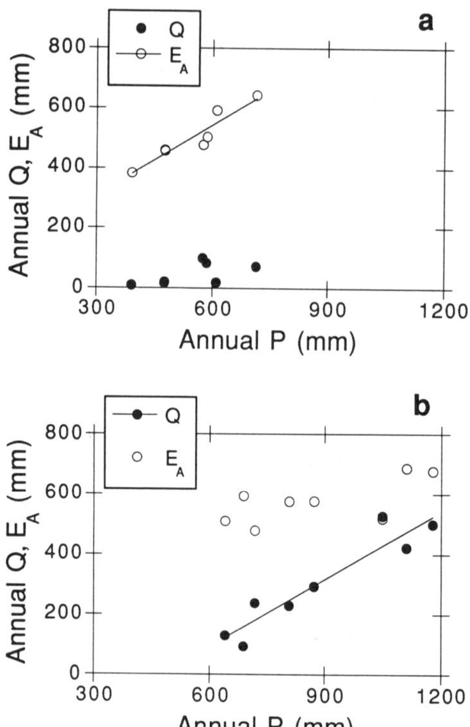

Figure 18-1. Relationships of Q and E_A vs. P for l'Avic (a) and TM9 (b) catchments. Significant linear regressions of E_A vs. P at l'Avic ($E_A = 77 + 0.78P$; $r^2 = 0.89$, $n = 7$, $p = 0.0014$) and of Q vs. P at TM9 ($Q = -358 + 0.75P$; $r^2 = 0.89$, $n = 8$, $p = 0.0005$) are also shown. The other two linear regressions (E_A vs. P at TM9 and of Q vs. P at l'Avic) were not statistically significant ($p > 0.05$).

this climate. Consequently, annual E_A increases with annual P, whereas annual Q depends more on the temporal distribution of precipitation events than on the annual P. The Prades catchments behave in a similar way to those of other dry regions of the world, like California, USA (Lewis, 1968) and Queensland, Australia (Prebble and Stirk, 1988). In contrast, when P is greater than E_T, as is the case in many catchments of colder and humid regions of the world (Likens et al., 1977; Hudson, 1988), the evaporative demand can always be totally satisfied and the remaining water becomes runoff. Annual P and E_T at the Montseny catchment are very similar. The relationships between annual E_A and Q vs. annual P are intermediate between those of Prades and of colder and humid regions: both annual Q and E_A increase with annual P, but only the linear regression of Q vs. P was statistically significant ($p < 0.05$).

Nevertheless it is worth noticing that not all the differences in the water balance between different catchments are simply a consequence of their climate, even for catchments that are "very similar" in their physical characteristics. An example of this is the difference in the water balance of the two adjacent Prades catchments, l'Avic and la Teula. During two years of simultaneous hydrological record, la Teula drained almost twice as much water as l'Avic (81 and 44 mm/year, respectively; Piñol et al., 1991). Such a difference cannot be attributed to different P or E_T at l'Avic and la Teula. In this case the difference was attributed to a higher water retention capacity at l'Avic than at la Teula (Piñol et al., 1991).

Then, if we are interested in the comparison of Prades and Montseny because they have different climatic characteristics, we have an important question to solve: which part of the differences in the water balance of Prades and Montseny is due to climatic differences and which is due to physiographic characteristics of the catchments? In Piñol et al. (1991) an expression was proposed that can be used to answer this question. The equation is defined in terms of the main components of the annual water balance (P, Q, E_A) divided by E_T (all in mm):

$$p = P/E_T; \quad e_A = E_A/E_T; \quad q = Q/E_T$$

The relations, based on a generalization of those proposed by Hsuen-Chen (1988), are the following:

$$e_A = [p^k/(1 + p^k)]^{(1/k)} \tag{18-1}$$

and, given that $p = q + e_A$

$$q = p - [p^k/(1 + p^k)]^{(1/k)} \tag{18-2}$$

Hence, actual evapotranspiration and runoff are a function of the quotient between P and E_T (the effect of climate), and of a parameter k

specific for each catchment. The value of k depends on the catchment evaporative capacity, which is the result of some factors related to catchment physiographic characteristics, of vegetation cover, and of the seasonal rainfall distribution. As l'Avic and TM9 are very similar in their vegetation cover and in the seasonal pattern of precipitation, we assume that eventual differences in their k values will mainly arise from their different physiographic characteristics.

Extreme values of k represent also the extremes of water partition between Q and E_A.

When k tends toward infinite:

$$\text{if} \quad p < 1 \ (\Rightarrow P < E_T) \quad \text{then} \quad e_A = p \ (\Rightarrow E_A = P)$$
$$\text{if} \quad p > 1 \ (\Rightarrow P > E_T) \quad \text{then} \quad e_A = 1 \ (\Rightarrow E_A = E_T)$$

This hypothetical catchment behaves like a huge reservoir where no runoff is produced until E_T is satisfied. If $P < E_T$, then no runoff is produced.

$$\text{When } k \text{ tends to } 0, \text{ then} \quad e_A = 0$$

This other extreme represents a catchment with no reservoir. All the precipitation is converted to runoff, and no water is available for evapotranspiration.

Actual catchments are located between these two extremes. Catchments with a high water retention capacity and/or high vegetation cover will have high k values, and catchment with low retention capacity will have low k values.

The parameter k has been fitted to $e_A - p$ values for l'Avic and to $q - p$ values for TM9, because these two relationships were statistically significant, whereas the other two ($e_A - p$ at TM9, and $q - p$ at l'Avic) were not. The fitted value of k is higher for l'Avic ($k = 2.48$; S.E. = 0.29) than for TM9 ($k = 1.70$; S.E. = 0.15), indicating a higher water retention capacity for the former catchment, and consequently, a higher opportunity for l'Avic to evapotranspire water. Observed e_A and q vs. p values, and Equations 18-1 and 18-2 with the corresponding fitted k values, are shown in Figure 18-2.

The k value is used here to discriminate which part of the differences between the water balance of the Prades and Montseny catchments are due to climate and which part to their physiographic characteristics. If we calculate the water balance of TM9 ($k = 1.70$) under a hypothetical climatic scenario like that currently observed at l'Avic ($P = 548$ mm, $E_T = 1{,}025$ mm), the water balance would not be the same as that currently observed at l'Avic ($E_A = 502$ mm and $Q = 45$ mm), but $E_A = 460$ mm and $Q = 88$ mm. Under the current Prades climatic scenario, annual Q of

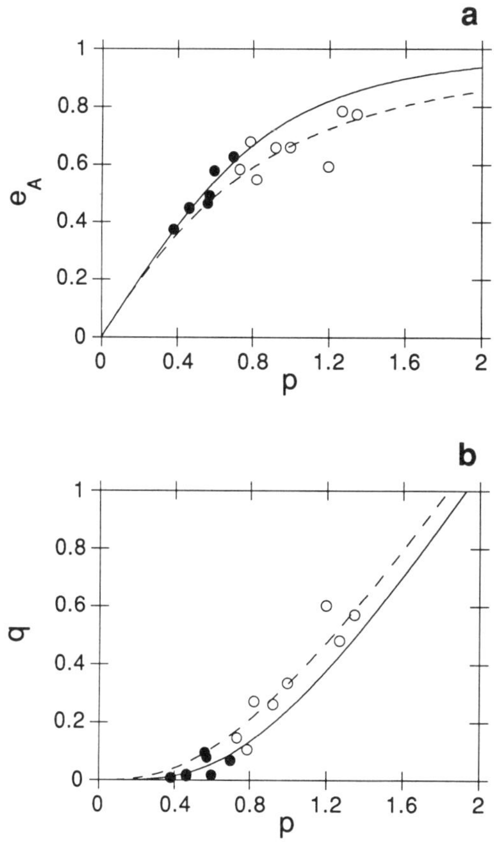

Figure 18-2. Relationships between (a) e_A ($e_A = E_A/E_T$) and p ($p = P/E_T$) and between (b) q ($q = Q/E_T$) and p for l'Avic (•) and TM9 (O) catchments. Equation 18-1 (a) and equation 18-2 (b) with the fitted k values of 2.48 (l'Avic, solid line) and 1.70 (TM9, dashed line) are also shown.

TM9 would be twice as that currently observed at l'Avic because of the physical characteristics of TM9 catchment.

Using Equations 18-1 and 18-2 with the corresponding fitted values of k, we have calculated the water balance components of l'Avic and TM9 catchments with current values, −10% and +10% of P, and current E_T for two scenarios of mean annual temperature increases: +2°C and +4°C. The water balance of these hypothetical scenarios is shown in Table 18-2. The change in the mean temperature of +2°C and +4°C would produce an increase of annual E_T of 6–7% and 12–13%, respectively. In the driest scenario (−10% P and +12% E_T) the proportion of evaporative demand which is satisfied (E_A/E_T) in the Montseny catchment would be slightly reduced compared to the current value, but it would be still

Table 18-2. Annual Actual Evapotranspiration (E_A), Runoff (Q), and Some Quotients of the Water Balance Variables Under Different Climatic Scenarios of Precipitation (P) and Potential Evapotranspiration (E_T) at l'Avic and TM9

AVIC

P	E_T	E_A	Q	P/E_T	E_A/E_T	P/Q
548 (0)	1025 (0)	507 (0)	41 (0)	0.53 (0)	0.49 (0)	13.4 (0)
548 (0)	1086 (+6)	512 (+1)	36 (−12)	0.50 (−6)	0.47 (−4)	15.2 (+13)
548 (0)	1147 (+12)	516 (+2)	32 (−22)	0.48 (−9)	0.45 (−8)	17.2 (+28)
493 (−10)	1025 (0)	464 (−8)	29 (−29)	0.48 (−9)	0.45 (−8)	17.0 (+26)
493 (−10)	1086 (+6)	467 (−8)	26 (−37)	0.45 (−15)	0.43 (−12)	19.3 (+44)
493 (−10)	1147 (+12)	470 (−7)	23 (−44)	0.43 (−19)	0.41 (−16)	22.9 (+63)
603 (+10)	1025 (0)	548 (+8)	55 (+34)	0.59 (+11)	0.53 (+8)	10.9 (−19)
603 (+10)	1086 (+6)	554 (+9)	49 (+20)	0.56 (+6)	0.51 (+4)	12.4 (−8)
603 (+10)	1147 (+12)	560 (+10)	43 (+5)	0.53 (0)	0.49 (0)	13.9 (+4)

TM9

P	E_T	E_A	Q	P/E_T	W_A/E_T	P/Q
884 (0)	876 (0)	585 (0)	299 (0)	1.01 (0)	0.67 (0)	2.96 (0)
884 (0)	933 (+7)	604 (+3)	280 (−6)	0.95 (−6)	0.65 (−3)	3.15 (+6)
884 (0)	990 (+13)	621 (+6)	263 (−12)	0.89 (−12)	0.63 (−6)	3.36 (+14)
796 (−10)	876 (0)	554 (−5)	242 (−19)	0.91 (−10)	0.63 (−6)	3.29 (+11)
796 (−10)	933 (+7)	570 (−3)	226 (−24)	0.85 (−16)	0.61 (−9)	3.52 (+19)
796 (−10)	990 (+13)	585 (0)	211 (−29)	0.80 (−21)	0.59 (−12)	3.76 (+27)
972 (+10)	876 (0)	612 (+5)	360 (+20)	1.11 (+10)	0.70 (+4)	2.70 (−9)
972 (+10)	933 (+7)	633 (+8)	339 (+13)	1.04 (+3)	0.68 (+1)	2.87 (−3)
972 (+10)	990 (+13)	652 (+11)	320 (+7)	0.98 (−3)	0.66 (−1)	3.04 (+3)

P, E_T, E_A, and Q in mm.
Hypothetical E_T was calculated using current temperatures, +2°C and +4°C, by the method of Samani-Hargreaves. The first row corresponds to the current annual P and E_T. E_A and Q have been calculated with equations 1 and 2, and with the fitted values of $k = 2.48$ for l'Avic and $k = 1.70$ for TM9. Percentages of difference with respect to current figures (first row) are in parentheses.

higher than that currently observed at Prades. In consequence, a reduction in forest cover is not expected at Montseny even in the driest scenario. At TM9 water yield would be importantly diminished in absolute terms (−88 mm/year); this reduction in water yield would negatively affect the regional water supply. At Prades, this same scenario would mean a very dry situation ($E_A/E_T = 0.41$), probably too dry to maintain the present 100% forest cover. In fact, in an especially dry summer (1989), some holm-oak trees of l'Avic suffered an important defoliation which took some time to recover an adequate water status (Sala, 1992).

The results of this forecasting exercise should be considered very carefully, for several reasons. The most important one is that it has not taken into account the interaction between a warmer climate and the higher water use efficiency of plants in a CO_2-enriched atmosphere. Depending on the importance of the decrease in transpiration in a CO_2- enriched atmosphere, it is possible that plants would maintain a healthy water status even in a warmer world (Rosenberg et al., 1990). Some authors have concluded that the antitranspiratory effect of high concentrations of CO_2 will more than compensate for the climatic warming, and thus have predicted an increase in runoff (Aston, 1984; Idso and Brazel, 1984). That effect could be offset or reduced by a parallel increase in plant growth and leaf area (Eamus and Jarvis, 1989).

Another important question that would affect the water balance, and that we have not considered, is the seasonal distribution of precipitation. If it changed in the future, then the water balance would probably change accordingly, even with the same annual precipitation.

Stream Water Chemistry

Stream water at Prades catchments is much more mineralized than that of Montseny (Table 18-3). The differences in stream water chemistry between Prades and Montseny are related to the ratio precipitation/runoff, which is much higher at Prades than Montseny. Stream water is more mineralized at Prades due to the higher concentrating effect of evapotranspiration and because soil water and ground water become enriched in rock weathering products as water residence time increases (Piñol and Avila, 1992). Also, the weathering rates are probably enhanced at Prades, compared to Montseny, due to a higher pCO_2 in the soil atmosphere (Piñol et al., in the press).

The temporal variation in concentrations of certain solutes in the stream waters of Prades and Montseny is also very different. The first stream water drained at Prades after the summer dry period of no runoff flushes out a large quantity of salts accumulated during the dry period, especially chloride and sulfate (Figure 18-3). At Montseny, as the dry period is much shorter and there is no flow interruption at TM9, this flushing effect is much less noticeable (Piñol et al., 1992).

Table 18-3. Flow-Weighted Mean Solute Concentrations ($\mu eq\,L^{-1}$) in Weekly Stream Water Samples at the Prades and Montseny Catchments

	Prades		Montseny
	L'Avic	La Teula	TM9
pH	7.49	7.75	7.46
HCO_3^-	2,868	1,797	346
Ca^{2+}	2,822	1,741	248
Mg^{2+}	595	467	155
Na^+	608	470	276
K^+	36	34	11
Cl^-	349	281	112
SO_4^{2-}	778	558	208

Number of samples: L'Avic, 61; La Teula, 62; and TM9, 192. Data set was from August 1984 to July 1988 at TM9, and from July 1986 to June 1988 at l'Avic and la Teula. Prades data taken from Piñol et al. (1992).

Very few studies have dealt with the effects of climate change on stream water chemistry or water quality (Jacoby, 1990). Therefore let us consider what would happen with the stream water chemistry at Prades and at Montseny if the climate became drier. Solutes in soil water and in ground water are concentrated by the effect of evapotranspiration. For solutes that are mainly derived from atmospheric deposition and that do not interact closely with the soils and plants of the catchments, such as chloride and, to a lesser extent, sulfate, the expected mean concentration in stream water is the product of their concentration in rain water and the quotient P/Q. The computed values are not necessarily equal to the observed ones because the dry deposition of these elements should also be taken into account. Nevertheless, we do not need to relate future stream water concentrations to rain water concentration; we can relate them to the current mean stream water concentrations:

$$C_f = C_a * [(P_f/Q_f)/(P_a/Q_a)] \qquad (18\text{-}3)$$

where C is the solute concentration in stream water, subscript a refers to the present values, and subscript f to the future forecasted values.

In the driest scenario of Table 18-2 (-10% P, $+12\%$ E_T), the quotient between forecasted and current enrichment factors $[(P_f/Q_f)/(P_a/Q_a)]$ is 1.71 for l'Avic and 1.27 for TM9. According to Equation 3, using sulfate concentration as an example, its mean concentration would increase from 778 to 1,330 $\mu eq\,L^{-1}$ at l'Avic, and from 208 to 264 $\mu eq\,L^{-1}$ at TM9 in such a scenario. In general, a decrease in the water yield of the catchment will produce an increase in the concentration of all the solutes in stream water. Obviously, this will also apply to water pollutants discharged into

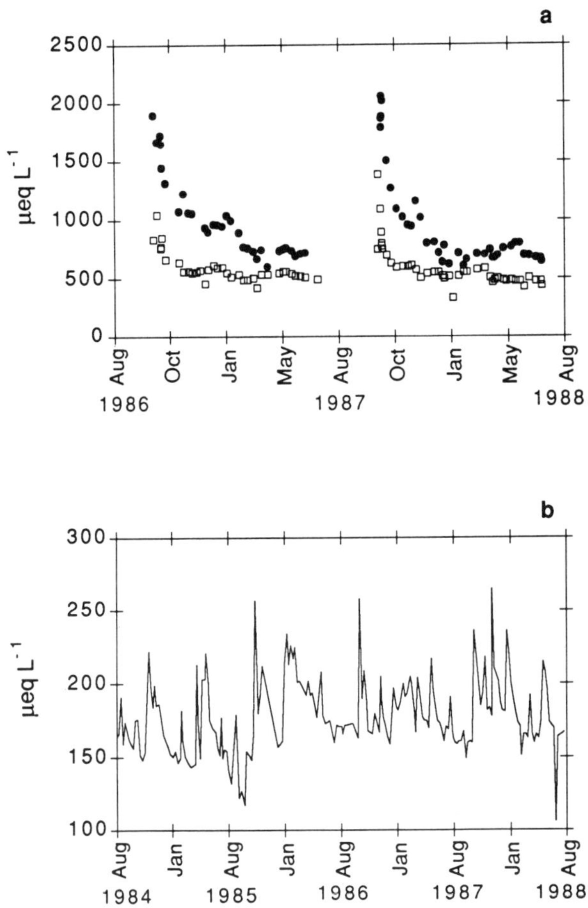

Figure 18-3. Temporal variation of sulfate concentration in stream water at l'Avic and la Teula (a) and at TM9 (b). Data set was from August 1984 to July 1988 at TM9, and from July 1986 to June 1988 at l'Avic and la Teula. (Taken from Piñol et al. (1992).)

streams, and consequently, river pollution is expected to increase in a drier world.

Eventually, as the stream water solute concentration increases, the solubility product of some salts will be reached. Then, the concentration of these solutes in the stream water will no longer increase but they will start to accumulate in a precipitated form within the catchment. This process seems to be currently operating at the Prades catchments. Stream water of l'Avic and la Teula is often oversaturated with respect to calcite. Consequently, a decrease in the water yield of the Prades catchments would produce an accumulation of calcite in the soil and in the stream channel. Perhaps this process has already started, as indicated by the relationship between pH and alkalinity in streamwater (Figure 18-4). At TM9 the relationship is the same as that usually observed in more humid catchments (Kleissen et al., 1990; Neal et al., 1990): pH and the decimal logarithm of alkalinity are positively correlated. At Prades the relationship is the opposite and the data points are very close to the calcite saturation condition, suggesting that these waters are equilibrated with a solid phase of calcite (Piñol and Avila, 1992). This result is certainly surprising, since both the Prades and the Montseny catchments are underlined by silicate bedrock, with no evidence of carbonate content. In fact, some preliminary analyses of soil carbonate content indicate that calcite is present at low concentrations (<0.5%) in some soil samples of l'Avic, especially in the surface horizons. A future drier climate would produce an accumulation of calcite in soils even in silicate areas, a process that is fully documented in soils of arid and semiarid regions (Schlesinger, 1982).

Conclusions

In this chapter we have compared the hydrological and hydrochemical characteristics of catchments located in two different areas of northeast Spain. These two areas, Prades and Montseny, are very similar in their ecology and geology but have a marked difference in water availability.

The water balance and stream water chemistry of both sites are very different: at the drier Prades catchments very little runoff is generated, stream water is very mineralized, and it is highly enriched in solutes after summer as a consequence of the washing out of salts accumulated in the dry period. At Montseny, one-third of precipitation is converted into runoff, stream water is much less mineralized, and the washing-out effect after the summer drought is much less noticeable. With the main difference between Prades and Montseny being their relative aridity index, we have allowed ourselves to speculate about what would happen if the climate became drier.

We have speculated that in a warmer world, the water balance of Montseny would move toward that of Prades, i.e., higher evapotranspiration and reduced runoff. In this warmer scenario, the Prades catchments would hardly be able to maintain a forest such as the present one. Dense sclerophyllous forests at Prades require almost all the precipitation water available, and only a small fraction drains. In the western Mediterranean, under present climate, dense evergreen-oak forests are usually found only where mean annual precipitation is higher than 400–450 mm. An increase in potential evapotranspiration can raise this limit and the forest limit can shift northwards.

Obviously, we are aware of the high uncertainty of our forecasts, mainly because there are many possible relevant feedbacks. For instance,

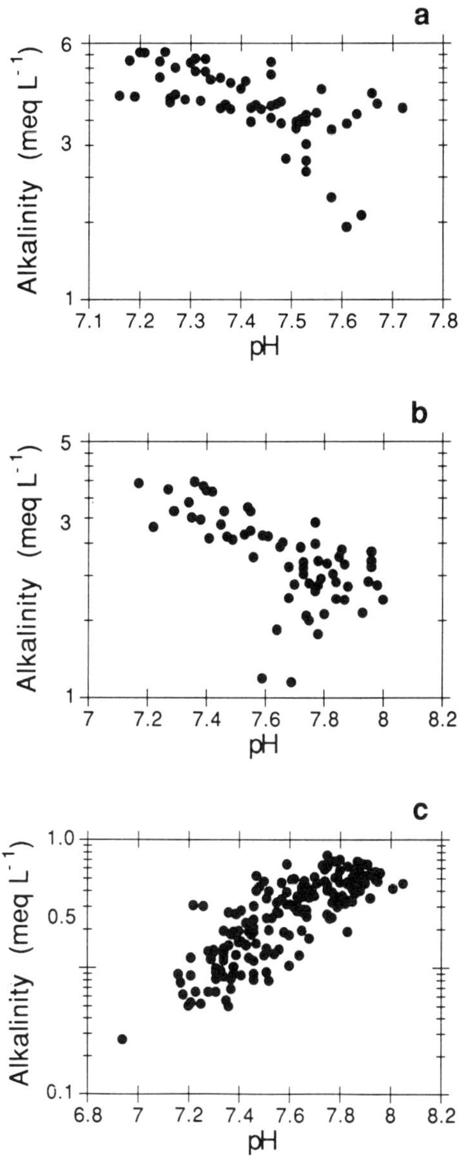

Figure 18-4. Relationship between alkalinity (log scale) and pH in stream water of l'Avic (a), la Teula (b), and TM9 (c). Data set was from August 1984 to July 1988 at TM9, and from January 1987 to June 1988 at l'Avic and la Teula. (Taken from Piñol and Avila (1992).)

we have not considered the antitranspiratory effect that would be exerted by a CO_2-enriched atmosphere.

In a warmer and drier world, stream water would become more mineralized mainly as a consequence of the concentrating effect of evapotranspiration on ground water and soil water solutes. This could have important effects on water quality. Stream water solute concentrations can increase until the equilibrium product of a salt is reached. This is most likely to occur with salts of low solubility, like calcite. If the calcite saturation was reached, as seems to be happening at Prades, then the catchment would begin to accumulate calcite in its soils. The accumulation of inorganic carbon in soils would represent a negative feedback on the increase of CO_2 in the atmosphere. A major and little-explored uncertainty for predicting the effects of climate change on stream water chemistry lies in how mineral weathering rates will be affected.

Acknowledgments. We thank Keith Beven, Paul Quinn, and an anonymous reviewer for their useful comments on our manuscript and Hanna Green for English corrections. Daily temperature data for the Prades catchment were kindly supplied by Anna Sala. Data used in this paper were obtained in research projects funded by the EC (EV4V-0109-E and ENCORE) and CAICYT (2129/83). Logistic aid from the Departament d'Agricultura, Ramaderia i Pesca de la Generalitat de Catalunya has been essential.

References

Allen RG, Jensen ME, Wright JL, Burman RD. 1989. Operational estimates of reference evapotrans piration. Agron J 81:650–662.

Aston AR. 1984. The effect of doubling atmospheric CO_2 on streamflow: a simulation. J Hydrol 67:273–280.

Avila A, Rodà F. 1988. Export of dissolved elements in an evergreen-oak forested watershed in the Montseny mountains. Catena supplement 12:1–11.

Avila A, Rodà F. 1990. Water budget of a broadleaved sclerophyllous forested catchment. In: Hoogart JC, Posthumus CWS, Warmerdam PMM (eds) Hydrological Research Basins and the Environment. The Netherlands Organization for Applied Scientific Research, The Hague, pp. 29–40.

Bazzaz FA. 1990. The response of natural ecosystems to the rising global CO_2 levels. Annu Rev Ecol Syst 21:167–196.

Eamus D, Jarvis PG. 1989. The direct effects of increase in the global atmospheric CO_2 concentration on natural and commercial temperate trees and forests. Adv Ecol Res 19:1–55.

Gleick PH. 1987. Regional hydrologic consequences of increases in atmospheric CO_2 and other trace gases. Climatic Change 10:137–161.

Hsuen-Chun Y. 1988. A composite method for estimating annual actual evapotranspiration. Hydrol Sci J 33:345–356.

Hudson JA. 1988. The contribution of soil moisture storage to the water balances of upland forested and grassland catchments. Hydrol Sci J 33:289–309.

Idso SB, Brazel AJ. 1984. Rising atmospheric carbon dioxide concentrations may increase streamflow. Nature 312:51–53.

Jacoby HD. 1990. Water quality. In: Waggoner PE (ed) Climate Change and U.S. Water Resources. John Wiley & Sons, New York, 496 pp.
Kleissen FM, Wether HS, Beck MB, Harriman R. 1990. Conservative mixing of water sources: analysis of the behaviour of the Allt A'Mharcaidh catchment. J Hydrol 116:365–374.
Lewis DC. 1968. Annual hydrologic response to watershed conversion from oak woodland to annual grassland. Water Resour Res 4:59–72.
Likens GE, Bormann FH, Pierce RS, Eaton JS, Johnson NM. 1977. Biogeochemistry of a Forested Ecosystem. Springer-Verlag, New York, 146 pp.
Manabe S, Wetherald RT. 1986. Reduction in summer soil wetness induced by an increase in atmospheric carbon dioxide. Science 232:626–628.
McKenney MS, Rosenberg NJ. 1993. Sensitivity of some potential evapotranspiration estimation methods to climate change. Agric For Meteorol 64:81–110.
Neal C, Robson A, Smith CJ. 1990. Acid neutralization capacity variations for the Hafren forest stream, Mid-Wales: inferences for hydrological processes. J Hydrol 121:85–101.
Piñol J, Avila A. 1992. Streamwater pH, alkalinity, pCO_2 and discharge relationships in some forested Mediterranean catchments. J Hydrol 131:205–225.
Piñol J, Alcañiz JM, Rodà F (in press). Carbon dioxide efflux and pCO_2 in soils of three *Quercus ilex* montane forests. Biogeochemistry.
Piñol J, Avila A, Rodà F. 1992. The seasonal variation of streamwater chemistry in three forested Mediterranean catchments. J Hydrol 140:119–141.
Piñol J, Lledó MJ, Escarré A. 1991. Hydrological balance of two Mediterranean forested catchments (Prades, northeast Spain). Hydrol Sci J 36:95–107.
Prebble RE, Stirk GB. 1988. Hydrological effects of land use change in small catchments at the Narayen Research Station, Queensland. Aust J Soil Res 26:231–242.
Rind D, Rosenzweig C, Goldberg R. 1992. Modelling the hydrological cycle in assessments of climatic change. Nature 358:119–122.
Rind D, Goldberg R, Hansen J, Rosenzweig C, Ruedy R. 1990. Potential evapotranspiration and the likelihood of future drought. J Geophys Res 95: 9983–10004.
Rodà F, Avila A, Bonilla D. 1990. Precipitation, throughfall, soil solution and streamwater chemistry in a holm oak (*Quercus ilex*) forest. J Hydrol 116:167–183.
Rosenberg NJ, Kimball BA, Martin P, Cooper CF. 1990. From climate and CO_2 enrichment to evapotranspiration. In: Waggoner PE (ed) Climate Change and U.S. Water Resources. John Wiley & Sons, New York, 496 pp.
Sala A. 1992. Water relations, canopy structure and canopy gas exchange in a *Quercus ilex* forest: variation in time and space. PhD dissertation, Universitat de Barcelona, Barcelona, Spain.
Samani ZA, Hargreaves GH. 1985. Water requirements, drought and extreme rainfall manual for the United States. International Irrigation Center, Utah State University, Logan, UT, 105 pp.
Schlesinger WH. 1982. Carbon storage in the caliche arid soils: a case study from Arizona. Soil Sci 133:247–255.
Strain BR, Thomas RB. 1992. Field measurement of CO_2 enhancement and climate change in natural vegetation. Water Air Soil Pollut, 64(1–2):45–60.

19. Water Balance of Mediterranean Ecosystems Under a Changing Climate

S. Rambal and G. Debussche

As Walter Oechel and others present in this book, increasing atmospheric carbon dioxide will generally induce a positive effect on net primary production for the plants. Other undetermined effects will likely result from a changing climate. Negative effects are predictable if the amount of rainfall decreases or if, the mean amount being fixed, the frequency of rare events (drought periods, . . .) increases. To anticipate the effect of global change, it is necessary to determine the intersect of these two contradictory effects.

This chapter focuses on the effects of a changing climate on Mediterranean ecosystems and mainly on the effects of precipitation changes. The discussion is limited to some evergreen oak ecosystems.

Climate Variability: Near Past, Future, and Present

Does the Climate Change? A Centuries-Old Concern

The question "Does the climate change?" has undoubtedly preceded the beginning of daily measurements of both air temperature and rainfall. The development of ideas on this topic is illustrated by some examples taken in works on the climatology of Montpellier under the Mediterranean-type climate in southern France.

The current measurement of air temperature was first performed by Bon in 1702 with an unshielded, north-exposed Amonton's gas thermometer. These data were first converted into Réaumur degrees and finally, into centigrade degrees. Bon observed 38.1°C on 30 July 1705 and −16.1°C on 11 January 1709, and affirmed "to have caught in a short time interval the two extremes of the climate." The assumption of stationarity of the thermal climate, implicitly formulated by Bon, will be true for a long time. The coldest temperature was equalled on 31 December 1870 (Martins, 1871) and exceeded on 12 February 1956, −16.9°C (Commission Météorologique de L'Hérault, 1956). Bon's highest value was first exceeded on 3 July 1847 with 39°C (Rodrigues, 1855). The absolute maximum, 42.9°C, was reached on 19 July 1904 (Chassant, 1906).

In the nineteenth century, two contradictory studies followed a virulent debate on the stability of the climate which set the physicist Arago (1835) against Dupin (1836, 1837). These two studies were done to test the stationarity of the thermal climate. Rodrigues (1855) compared three temperature series, 1756–1792, 1794–1806, and 1806–1818, and his own measurements for the 1840–1852 period. He observed a progressive decline of the mean annual temperature and asked, "Are these facts purely accidental and the capricious results of the combination of various causes which act on the highest manifestation of the strengths of Nature, or must we recognize the expression of a great natural law? Is the temperature of the Earth going to decrease in a progressive irregular way? Are there upward and downward phases with more or less duration?" Roche (1882) also compared the 1756–1792 series and his own data for the 1857–1867 period. The seasonal decreases were 0.5, 0.4, 1.0, and 0.3°C for winter, spring, summer and autumn, respectively. He assumed the 0.5°C mean annual cooling to be of instrumental origins. The three above-mentioned examples shown the difficulties of concluding, on short climatological series and with a biased estimation of the variance, on a true shift of the mean annual temperature or on its stationarity.

Concerning the precipitation, Poitevin (1803) analyzed the annual rainfall of Montpellier from 1767 to 1787. For this period, the mean reached 762 mm. This value was not significantly different from that published by Angot (1918) for the 1851–1900 period or calculated from today's data. On the other hand, Poitevin analyzed his data considering three 7-year periods separately. For these periods, the means were 800, 754, and 731 mm. The cause of this increasing dryness, suggested by Poitevin, is interesting to recall, even it is based on a weak statistical analysis. He wrote "The cause of such a phenomenon, if it is only local, must be attributed with some basis to abuses in forest clearing and tree destroying in the neighborhood of Montpellier." So he anticipated the fundamental role of the vegetation cover in the exchanges of matter and energy at the Earth's surface and consequently, in the origin of climates, and laid the basis of the interactions which control the functioning of general circulation models (GCMs).

A Short Review of Scenarios from GCMs

The three-dimensional character of the climate may be simulated using GCMs. Nineteen predictions of global warming and change in precipitation that follow a doubling of the CO_2 content of the atmosphere are given in Cubasch and Cess (1990). Values group around +3.8°C (standard deviation SD = 0.9°C) and +8.5% (SD = 3.3%), respectively, giving anticipated average increases of about 0.75°C and 1.7% per decade over the next 50 years.

Mitchell et al. (1990) in the IPCC Report averaged the regional results of three GCMs. For southern Europe (35°–50°N, 10°W–45°E), the warming is about 2°C in winter and varies from 2° to 3°C in summer. There is some indication of increased precipitation in winter, but summer precipitation decreases by 5% to 15%. The Commission of European Communities (Warrick et al., 1990) published "state-of-the-art" composite scenarios by averaging the regional results from four GCMs (Figure 19-1). In these scenarios, most of the southern region winters are 4°C warmer. For the summer, average temperatures are 4°–4.5°C warmer

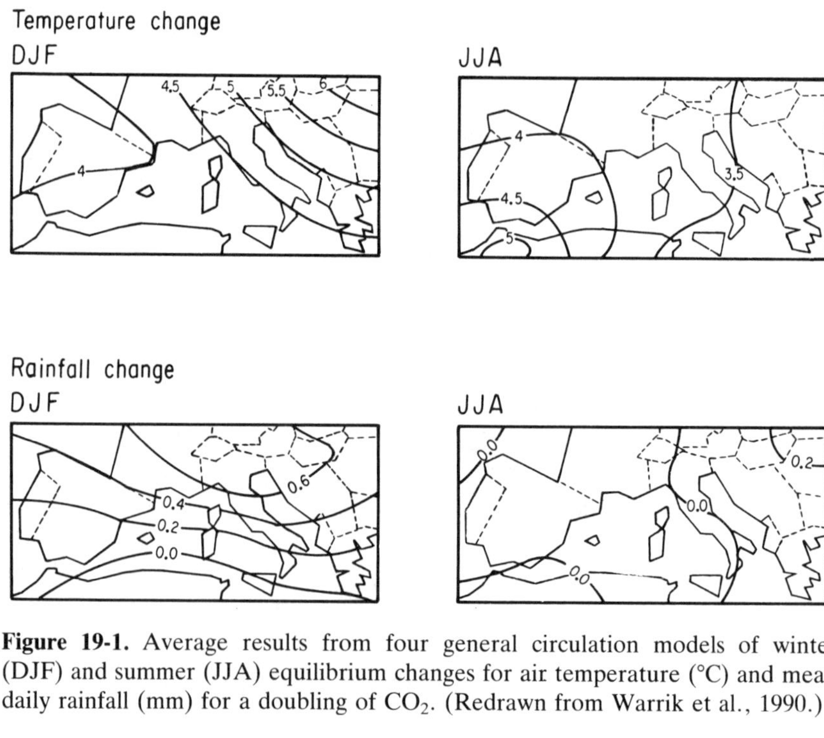

Figure 19-1. Average results from four general circulation models of winter (DJF) and summer (JJA) equilibrium changes for air temperature (°C) and mean daily rainfall (mm) for a doubling of CO_2. (Redrawn from Warrik et al., 1990.)

throughout all Spain, Portugal, and southern France. The scenario for winter precipitation shows an increase of 0.2 to 0.4 mm/day, or 6 to 12 mm/month, in the Mediterranean area. In contrast with the winter scenario, the results for summer show little change in precipitation. Several GCMs have reported that with enhanced greenhouse conditions, their models indicate an increase in the frequency of intense convective rainfall and a possible decrease in less intense nonconvective rain. That is an average increase in total rain, but a decrease in the area covered by precipitation at any one time. Using the CSIRO4 GCM, Pittock et al. (1991) and Gordon et al. (1992) showed that the regionally averaged total annual rainfall increases at all latitude. However, in middle latitude (40°N), the average increase is small (+8.1%) and the number of rain days decreases (−3.5%). But in these middle latitudes, the decrease in the number of rain days is associated with a decrease in the frequency of low rainfall days. This means that there will be an increased frequency of dry spells, despite an increase in the number of heavy rain events. Histograms of percentage changes in frequency of occurrence of daily rainfall in the indicated classes are shown in Figure 19-2. The frequency distributions generated by the CSIRO4 GCM indicate that there is a systematic shift toward more days of heavy rain and less of light rain. These are based on 10 years of control and 10 years of doubling atmospheric CO_2, and are for grid points that likely damp extreme values.

Since the errors associated with the prediction of today's climate are of the same order as the prediction associated with CO_2 doubling, great care must be taken in the use of the predictions. However, because of con-

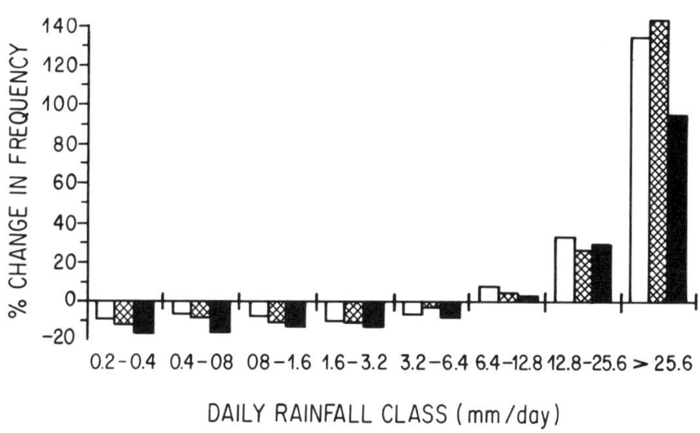

Figure 19-2. Percentage changes of probabilities of occurrence of daily rainfall, in indicate classes, between the enhanced greenhouse and the control conditions. Results were simulated with the CSIRO4 general circulation model and averaged over all grid points at 40°N (blank), 40°S (crosshatch), and over Austrialia. (Redrawn after Pittock et al., 1991.)

siderable interest, there is pressure to use GCM predictions for regional studies and to develop realistic projections at mesoscale (Giorgi and Mearns, 1991). Nevertheless, Tyson (1990) observed that "the stage has not been reached where this can be done with any confidence owing to incertainties in the models and to the poor resolution of many of many of them. In presenting some regional scenarios, these facts must be remembered and the prognoses must not be overinterpreted beyond providing the bases for what-if scenarios." The what-if scenarios used in this chapter concern (1) the change of both mean and variability of the annual amount of precipitation and (2) the changes in daily rainfall type and intensity.

Can Statistics Help to Define Rainfall Scenarios Better?

Climate modelers have usually focused only on the average performance of GCMs, as opposed to their ability to produce variability and consequently to estimate the relative frequency of extreme events. "Climate variability is typically regarded as noise" (Katz, 1992). An extreme event is here defined as an event where a climate variable had a low relative frequency of occurrence. Indeed, much of the recent need for the consideration of how variability and the frequency of extreme events might change with a change in average climate has come from water resource management, agronomy, or plant ecology. This concern arises naturally, since the impacts of climate are realized largely through the incidence of variation about normal conditions or extreme events (Wigley, 1985). For instance, severe impacts would come from droughts, which are evident in the frequency distribution of precipitation rather than its mean alone. Because extreme events are inherently rare and because of the limited sample sizes produced by GCMs, it may not possible to make statistical inferences about the frequency of extreme events directly from GCM outputs.

Waggoner (1989) suggested relating the mean and some parameters of the frequency distribution of precipitation and using these relationships to infer the new frequency distribution if climate changes alter the mean. Examining annual precipitation at 360 stations scattered over the earth, Conrad (1941) long ago found a relationship between interannual variability and mean precipitation. Conrad's equation was originally based on average variability and not on variance or standard deviation. The average variability is defined as the absolute average variability (the average absolute deviation from the mean) expressed in percent of the mean. Using the approximation that the average variability is 4/5 the coefficient of variation, Waggoner (1989) modified Conrad's equation. After conversion of inches to millimeters, the following equation is obtained:

$$\mathrm{SD} = 0.163M + (44.96/M + 59.94) \tag{19-1}$$

Waggoner (1989) used data from 55 US stations. The regression for this observation between mean and standard deviation is (n = 55, coefficient of correlation $r = 0.91$, $P < 0.01$):

$$SD = 2.65M^{0.625} \tag{19-2}$$

He also proposed a curvilinear relationship derived from monthly data ($n = 660$, $r = 0.95$, $P < 0.01$):

$$SD = 2.76M^{0.655} \tag{19-3}$$

For 73 stations located in a 100 km × 100 km area around Montpellier, a linear relationship ($n = 73$, $r = 0.87$, $P < 0.01$) between mean and standard deviation is calculated (Figure 19-2):

$$SD = 0.255M + 36.2 \tag{19-4}$$

The annual means and standard deviations for the precipitations at these stations are plotted in Figure 19-3 with the corresponding regression line (Equation 19-4, solid line). Waggoner's equation derived from monthly data (Equation 19-3, long dashed line) overestimates SD for small means (smaller than 700 mm) and underestimates SD for large

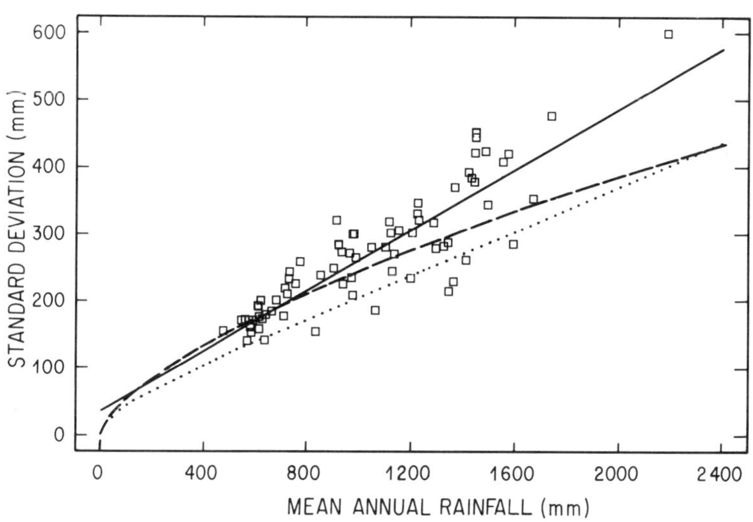

Figure 19-3. Standard deviations and means of annual precipitation from 73 stations located around Montpellier, sourthern France. Solid line represents the regression line. Lower curves represent respectively Waggoner's equation derived from monthly data (long dashed line) and Conrad's equation (dotted line).

Table 19-1. Some Estimations of the Coefficient of Variation for Mean Annual Rainfall Amounts of 1,000 mm and for Various Locations. *n* Is the Number of Stations

References	Source	*n*	CV
Conrad (1941)	World	360	0.21
Waggoner (1989)	USA	55	0.20 0.25[a]
Le Houérou (1992)	North Africa, Near East	407	0.18
	Sahel, Sudan	228	0.17
	East Africa	300	0.16
Our results	Languedoc (France)	73	0.29

[a] Based on monthly data.

means. Conrad's equation (Equation 19-1, dotted line) underestimates SD for any mean annual amounts. Table 19-1 presents some estimations of the coefficient of variation CV for a 1,000 mm mean. The values range between 0.16 and 0.29. The higher value is obtained applying Equation 19-4. The lowest values (0.16–0.18) are proposed by Le Houérou (1992) for a large set of locations in Africa and the Near East.

Conrad (1941) discussed positive and negative "anomalies of variability." Negative anomalies occurred when the variability is less than predicted by Equation 19-1 and particularly in temperate climates. He concluded "the continental climate and its effects are regulators of the precipitation." He also identified variability more than the rule or positive anomalies in some dry areas. As shown in his paper, positive anomalies of 3%, 7%, and 10% were observed in Marseille (southern France), San Diego (California, USA), and Alicante (southern Spain), respectively. These findings agreed with those of Waggoner who observed that "frequency distributions with much larger variance than expected are in the Mediterranean climates" of California.

Ecosystem Water Balance and Rainfall Amounts

For the following two paragraphs, data on the water balance of a dense continuous canopy of shrubs, dominated by the evergreen oak, *Quercus coccifera* L., and located 10 km north of Montpellier at Saint Gély du Fesc, were used. The yearly average of leaf area index for this canopy was about 2.4, and the rooting depth reached 4.7 m (see a more detailed description in Rambal, 1984).

Relating Annual Balance to Annual Rainfall

The annual water balance during seven consecutive years was measured (Rambal, 1984). From these data, a linear regression equation (Figure

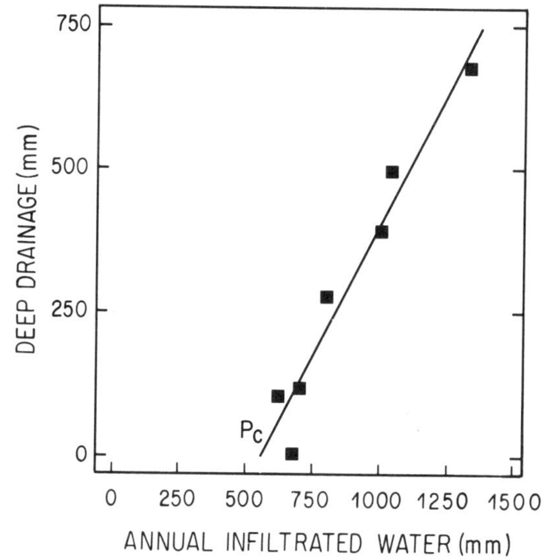

Figure 19-4. Relationship between the deep drainage and the infiltrated water, expressed on a yearly basis, for a *Quercus coccifera* ecosystem.

19-4) between drainage below the root zone (D), or water yield, and annual infiltrated rainfall (P) was fitted:

$$D = 0.907(P - 578) \qquad (19\text{-}5)$$

This result indicated that when annual rainfall was less than P_c = 578 mm, deep drainage loss was negligible and almost all precipitation infiltrating the soil was lost by actual evapotranspiration. Above this threshold, the canopy consumed a small amount ($1 - 0.907 = 0.093$ or 9.3%) of the additional precipitation. For instance, with an annual amount of 868 mm (i.e., 300 mm above the threshold value), the actual evapotranspiration was only $578 + 0.093 \times 300 = 596$ mm. The water tapping system of *Q. coccifera* is rigid and adjusted to a low annual rainfall. Poole and Miller (1981) observed that the boundary of chaparral and coastal sage was defined when annual transpiration per unit leaf area index reached about 200 mm. Miller and Poole (1979) gave similar annual transpiration rates in mixed and chamise chaparral. At this value, Miller (1981), from field measurements and model simulations of the carbon budget, found the annual carbon fixed in photosynthesis to be completely used to maintain the biomass. For *Q. coccifera*, using a transpiration/actual evapotanspiration ratio of 0.8 (Rambal, 1993), a transpiration of $0.8 \times 578 = 462$ mm was obtained, and a transpiration per unit of leaf area index of 193 mm was deduced. This value is similar to the one proposed above. Hence, 200 mm is a threshold value necessary for this plant to maintain a positive carbon balance.

Frequency distributions of precipitation amounts vary from nearly normal to highly skewed. For the 43 annual amounts observed at Saint Gély du Fesc (1949–1991 period), the mean is 901 mm and the standard deviation is 324 mm. The distribution fits well with the gamma distribution function (Figure 19-5) with its dimensionless shape parameter, $G = 8.95$ (G = mean^2/variance), and scale parameter, $B = 9.92\ 10^{-3}$ (B = mean/variance).

The climate change is considered to involve a combination of two statistical operations at the same time: (1) the distribution function is shifted, inducing a change in the mean, and (2) the distribution function is rescaled, inducing a change in the standard deviation. Under a given climate, an extreme event may be assumed to occur when a climate variable, here the annual amount of rainfall, falls below a critical threshold P_c value (for instance, 578 mm). The probability of such an event is given by the area under the tail of the distribution, where P is less than P_c. How much this probability does change depends on the change in either the mean or the standard deviation. The simple probabilistic aspects of this problem are easy to demonstrate (Katz, 1992; Katz and Brown, 1992; see also Rambal, 1991). Figure 19-6 compares the probability of occurrence, i.e., the tail area for three different distributions: (1) the reference case defined by the mean M_1 and the standard deviation SD_1 (Figure 19-6a); (2) the case in which the mean is decreased to a new value M_2,

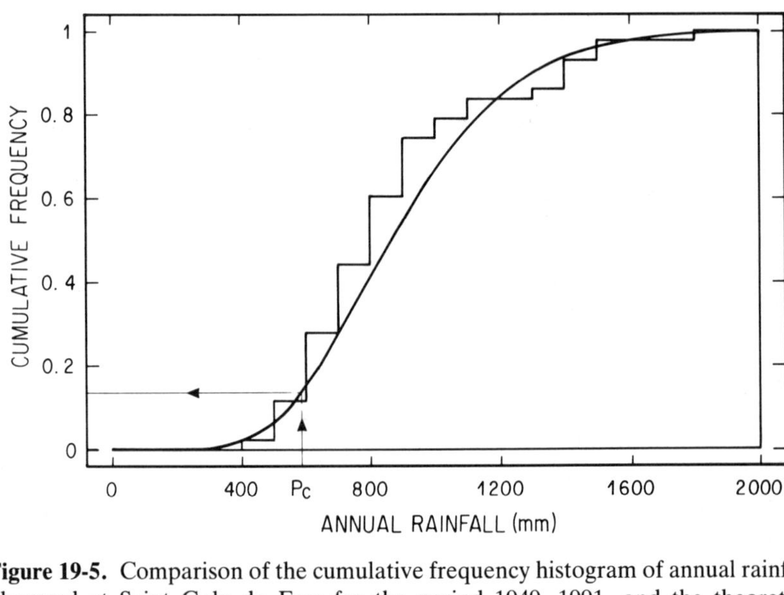

Figure 19-5. Comparison of the cumulative frequency histogram of annual rainfall, observed at Saint Gely du Fesc for the period 1949–1991, and the theoretical cumulative gamma distribution function.

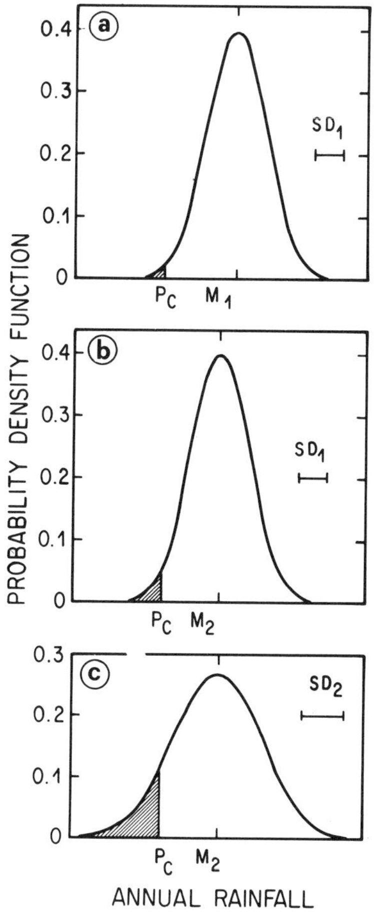

Figure 19-6. Probability of occurrence of an extreme event or tail area for three distributions: (a) the reference case characterized by the mean M_1 and the standard deviation SD_1; (b) the case in which the mean decreases to M_2 keeping SD_1 fixed; (c) the case in which both the mean and the standard deviation are changed.

keeping SD_1 fixed (Figure 19-6b); (3) the case in which both mean and standard deviation are changed to new values M_2 and SD_2 (Figure 19-6c).

The probability that annual rainfall will be 578 mm is 0.12, that is, a return interval of 1/0.12 = 8.3 years (Figure 19-6a, see also Figure 19-5). If we now shift the mean of the distribution towards P_c, this area will increase in a highly nonlinear way, as shown in Figure 19-6b. For example, if the annual mean precipitation of 901 mm over the study site fell by 10%, a drought that was expected one year in a hundred (RP1 = 100) would become roughly 3 times more frequently expected (RP1/RP2 = 31/100; see Table 19-2). If we changed both mean and variability using the results

Table 19-2. Anticipated Return Periods (Years) for Events Occurring Once Each 100, 20, and 10 Years in the Case in Which Both Mean and Standard Deviation Are Changed

Percent Decrease Today Return Period RP1	Anticipated Return Period RP2			
	5	10	15	20
100	58 (55)	36 (31)	22 (20)	15 (12)
20	14 (15)	9.6 (9.0)	7.0 (6.5)	5.3 (4.8)
10	7.4 (7.2)	5.5 (5.3)	4.3 (4.1)	3.4 (3.3)

The numbers in parentheses represent return periods in the case in which the mean alone is changed.

of the above paragraph (Equation 19-4), that is an SD decrease of about 2.5% $\Delta SD/\Delta M = 25\%$). The results are slightly different; the ratio RP1/RP2 = 36/100. In Table 19-2, we present the new return periods, expressed in years, for events occurring at the study site once each 10, 20, or 100 years in the case in which both mean and standard deviation are changed. The mean was decreased by 5%, 10%, 15%, and 20%, and the standard deviation followed Equation 19-4. The numbers in parentheses represent the return periods in the case in which the mean is decreased by the previous values keeping the standard deviation fixed.

The return period of the extreme event P_c is more modified by the decreases in the mean than by the changes in both mean and standard deviation. The anticipated return periods are, respectively, 6.1, 4.6, 3.6, and 2.9 years as a consequence of 5%, 10%, 15%, and 20% decreases. The return period in the reference case is 8.3 years. The corresponding values in the case in which we modified both the mean and the standard deviation are slightly different: 6.2, 4.8, 3.8, and 3.1 years. The larger decrease of the mean annual rainfall associated with global changes, the more frequent will be failure of the ecosystem. Consequently, an adjustment of the transpirating surface will be expected. This reduction of the leaf area index will maintain the plant water stress at the same level as in the reference case and will lead to a decrease of P_c toward a new P_c value with the same probability of occurrence.

Simulating the Effects of Expected Changes in Rainfall Intensity

In order to determine the effect of a changing climate on the water balance and the plant water status, a scenario involving change in daily rainfall distribution was used to drive a canopy simulation model (see detailed description in Rambal, 1993) on the same ecosystem as in the annual water balance analysis. This biophysical model simulates on a daily basis the water balance components (transpiration, evaporation, and deep drainage or water yield) and the water status of the rooted soil layer and of the plant canopy (predawn and minimum leaf water potentials). The evaporation component includes the evaporation from bare soil under the plant canopy and the evaporation of intercepted rain water. The significant aspect of the CSIRO4 scenarios we selected as what-if scenarios is an increase, of the order of 100%, in the frequency of occurrence of the largest rainfall class. For modeling purpose, the data set for Saint Gély du Fesc (1949–1991 period; 3,270 wet days) was used.

Daily rainfall sequences are generated by two separated processes, the rainfall occurrence generally denoted as wet-dry process and another process for the rainfall amounts on wet days. The frequency distribution of rainfall amounts on wet days is usually J-shaped. When all days with a positive rainfall amount are called wet, there may be some problems, since small values can also be caused by dew. To clear up this difficulty, days with a rainfall amount smaller than 1 mm will also considered as dry days. Since a wet day was defined as a day with a rainfall amount above a threshold, the lower limit of the distribution is 1 mm. The last value was subtracted from be original amounts and then the distribution functions were fitted. The experimental distribution of the 2,531 remaining wet days fits well with the gamma distribution function with its dimensionless shape parameter, $G = 0.68$, and scale parameter, $B = 0.50\ 10^{-2}$, or with the log-normal distribution ($M = 16.2$, SD $= 43.8$).

It was also possible to calculate the experimental 24-hour rainfalls for given return periods from 1 month to 5 years. These are shown by squares in Figure 19-7. The relationship between return periods, on logarithmic scale, and daily amounts can be fitted by a straight line. This means that the distribution of daily rainfall amounts greater than about 5 mm and less than 80 mm can be approximated by an exponential density function. Hence, we estimated daily rainfall of 16.8, 39, and 64.5 mm for return periods of 3 months, one year, and 5 years, respectively. This approximation cannot be applied for rare daily events (here greater than 80 mm). Also shown in Figure 19-7, by a gray area, are the hypothetical corresponding relationships in the enhanced greenhouse conditions.

Two consecutive years receiving 636 and 673 mm of rainfall were chosen. These values produced very close actual evapotranspiration, 576 and 573 mm, of which evaporation from the soil was 116 and 115 mm and yielded very contrasting drainages below the root zone, 161 and 19 mm. The corresponding transpirations were 460 and 462 mm, respectively. The lowest simulated values for predawn and minimum leaf water potentials were −0.92 and −4.4 MPa on 15 July for the first year, and −0.82 and −3.7 MPa on 17 September for the second one.

The enhanced daily amounts were calculated from the observed amounts, assuming that the wet–dry sequences are unchanged. The present amounts were reported in Figure 19-7 and the corresponding enhanced amounts (gray area) were assumed to have the same return

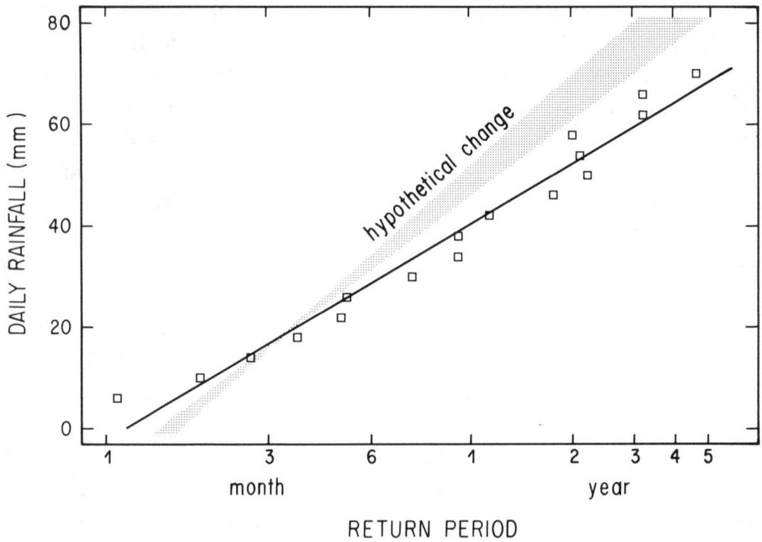

Figure 19-7. Relationship between daily rainfall and return periods for Saint Gely du Fesc (period 1949–1991). The gray area represents a hypothetical relationship under enhanced greenhouse conditions which takes into account a signifcant shortening of the return period for daily rainfall greater than 20 mm.

periods. Changing the behavior of rainfall amounts on wet days induced changes in monthly (Figure 19-8) and annual amounts, and consequently in all the terms of the water balance components. For the monthly amounts, the main negative differences (present amount larger than anticipated amount) occurred in winter. The larger difference was observed in January of the first simulation year, 73.5 vs. 40 mm. This is a 46% decrease. The main positive differences were observed the second year in February, 87.2 vs. 123.9 mm (+42%), and in June, 109 vs. 172.9 mm (+59%). This last value highlights the role of convective rains. The annual amounts reached 513 and 684 mm, respectively, i.e., percentage changes of −19% and 2%. These correspond to an average decrease of 9% in the same order as those in some climate scenarios.

As a result, actual evapotranspiration reached 546 and 570 mm, respectively, i.e., percentage descreases of 5% and 1% compared to the reference values. The soil evaporation (100 and 115 mm) was very close to the reference values (116 and 115 mm). The reduction of both transpiration and soil evaporation did not favor an increase in water yields. They reached 102 and 6 mm for the 2-year simulation, i.e., percentage decreases of 37% and 66% (Figure 19-9). During the first year, the plant undergoes a greater water stress consequent to the 19% decrease in precipitation. The lowest value of predawn potential was −1.6 MPa. The minimum leaf

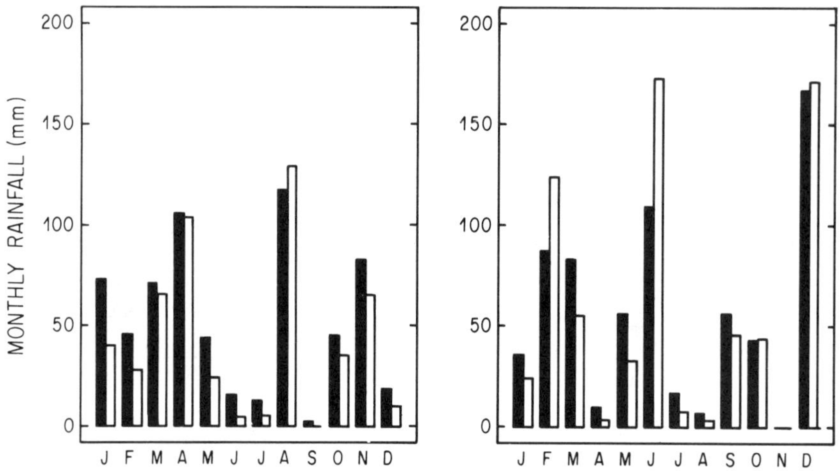

Figure 19-8. Comparison of observed monthly rainfall (solid bars) and those simulated under enhanced greenhouse conditions assuming changes in the return periods of daily rainfall (open bars) for 2 consecutive years.

water potentials were lower than −4.2 MPa from 7 to 14 August. We will discuss this threshold value further. In the second year, predawn and minimum potentials reached −0.89 and −4.0 MPa the same day as in the reference year.

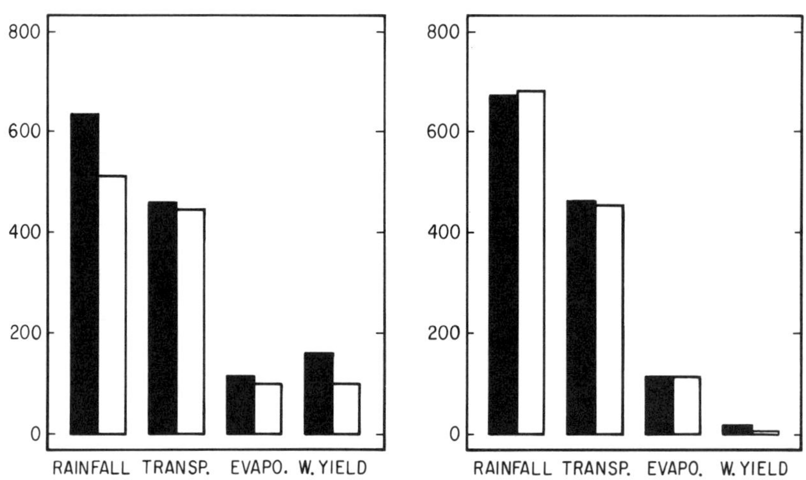

Figure 19-9. Comparison of annual rainfall and the water balance terms (transpiration, soil evaporation, and deep drainage or water yield) under reference conditions (solid bars) and under enhanced greenhouse conditions (open bars) assuming changes in the return periods of daily rainfall.

GCMs tend to give annual or seasonal rainfall changes between enhanced greenhouse and normal conditions which differ between modes, even in sign, at a regional level (Mitchell et al., 1991). The predicted changes in rainfall intensity are more statistically significant and more spatially and seasonally uniform in sign than changes in total rainfall amounts (Pittock et al., 1991). As shown in this paragraph, the decrease in frequency of low rainfall means that there will be an increased frequency of occurrence of dry spells despite an increase in the number of heavy rain events and, for the second simulation year, a slight increase in the annual rainfall amount. This leads to an increased frequency of periods with low leaf water potential for the plant.

A Working Hypothesis to Anticipate the Plant's Response to a Changing Climate

For this purpose, data for leaf water potential measured on Mediterranean evergreen oak species in (1) six stands of *Quercus ilex*, three across a soil water gradient in southern France (Puéchabon, Camp Redon and Mont Ventoux), two in Spain and the last in Italy; (2) two stands of *Quercus coccifera* in southern France and in Portugal; and (3) various locations of *Quercus dumosa*, *turbinella*, and *agrifolia* in California and Arizona were used. The data of Valette (1981) were pooled into one data set. The same procedure was applied to the *Quercus dumosa* and *turbinella* data of Poole and Miller (1981) and of Blake-Jacobson (1987).

Regardless of the study site and the amount of rainfall during a given period, minimum and predawn leaf water potentials were always higher than −3.8 and −4.4 MPa (Table 19-3), the values reached by *Q. ilex* in a location in northeast Spain and by *Q. coccifera* near Montpellier. The mean values are −3.3 MPa (n = 10, SD = 0.3 MPa) and −4.1 MPa (n = 12, SD = 0.2 MPa) for minimum and predawn leaf water potentials, respectively.

Duhme (1974) measured minimum potential of −4.4 MPa for *Q. coccifera*, a similar value to that in Table 19-3. In the low *Q. coccifera* garrigue of Saint Gély du Fesc, a −3.4 MPa predawn potential (see previous description) was observed. This measurement was made on 25 October 1985, the last day of the summer drought. The amount of rainfall in 1985 was the lowest since 1948, when records began, reaching only 418 mm during a period when the average value is 950 mm. For *Q. ilex* in the mesic location at Camp Redon, predawn leaf potential reached at the end of the 1986 summer drought was −3.2 MPa, a potential slightly lower than that of the xeric site at Puéchabon the same year, −3.1 MPa. In this case the separation between mesic and xeric sites is not based on the amount of rainfall but rather on rough estimation of water availability in the soil (Rambal, 1992). Other examples well illustrated this assumption

Table 19-3. Some Literature Values of the Lowest Leaf Water Potentials (Predawn and Minimum) Measured in Mediterranean Evergreen *Quercus* Species

Species	Sites	Leaf Water Potential (MPa)		References
		Predawn	Minimum	
Quercus ilex	Puechabon–southern France	−3.1	−3.9	Rambal (1992)
	Camp Redon–southern France	−3.2	−4.1	Rambal (1992)
	Mont Ventoux–southern France	−3.2	−3.8	Valette (1981)
	Near Rome–Italy	−3.6	−4.1	De Lillis and Fontanella (1992)
	Near Barcelona–northeast Spain	−3.8	−4.3	Castell (1992)
	Prades Mountains–northeast Spain	−3.6	−4.0	Sala (1992)
Quercus coccifera	Saint-Gély-du-Fesc–southern France	−3.4	—	Rambal (1992)
	Near Montpellier–southern France	—	−4.4	Duhme (1974)
	Quinta Sao Pedro–Portugal	−2.9	−3.7	Tenhunen et al. (1985)
Quercus agrifolia	Central California	−2.5 to −3.1	—	Griffin (1973)
Quercus dumosa	Southern California (5 locations)	—	−4.2	Poole and Miller (1981)
	Laguna Mountains–southern California	—	−4.2	Hasting et al. (1989)
	Echo Valley–southern California	—	−4.2	Poole et al. (1981)
Quercus turbinella	Southern California and Arizona (3 locations)	−3.4	−4.2	Blake-Jacobson (1987)

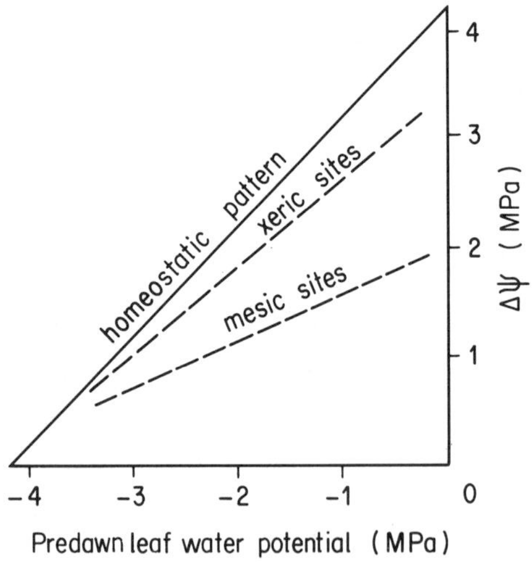

Figure 19-10. Relationship between the daily amplitude of the leaf water potential, Δ_ψ, that is, the difference between the predawn and the minimal potential, and the predawn leaf water potential. The solid line represents the homeostatic pattern. The long dashed lines show the observed patterns in mesic and xeric conditions.

of a lower boundary of water stress even during a very dry year. Griffin (1973) measured leaf water potential on three woodland oaks of central California. He observe, at the end of the driest year in a 32-year period, (only 300 mm during a period when the mean stood at 510 mm), predawn potentials of evergreen black oak *Quercus agrifolia*, in the more xeric location, of between −2.5 and −3.1 MPa. In an arid to semiarid Mediterranean-climate transect (350 to 600 mm annual precipitation), Poole and Miller (1981) studied the distribution of plant water stress and vegetation characteristics. In all the five study sites where *Quercus dumosa* was present, they observed that it maintained higher potentials (above −4.2 MPa) than did the other species. The months preceding the measurement period were unusually dry. San Diego received only 85 mm from November to April, which is approximately half the average amount.

Nevertheless, the trajectories followed by minimum and predawn leaf water potentials to reach these limits were very different according to location and species. In Rambal (1992), we analyzed the relationship between the daily amplitude of the leaf water potential Δ_ψ, that is the difference between the predawn potential and the minimum potential. In the five studied cases, pointed out by asterisks in Table 19-3, Δ_ψ decreased to zero when the predawn potential decreased or water stress increased. This pattern of evolution was first observed by Waring and Cleary (1967) on Douglas fir, but was not initially considered as being a general phenomenon. Indeed, Hickman (1970), based on measurements of 44 species, concluded that the pattern in which Δ_ψ increases with the water stress is the most common pattern, corresponding to species that can be characterized as "conformers" (Aussenac and Granier, 1978). Species with the same pattern as our observations were named "regulators." Hickman (1970) suggested that "this pattern is probably typical of most plant species in areas with modified (sic) Mediterranean climates." This has also been described by Oechel et al. (1972) for *Larrea divaricata* and by Aussenac and Valette (1982) for Mediterranean forest species. As proposed by Ritchie and Hinckley (1975), "It is tempting to compare species based upon these curves using both the slope and intercept as indicators of species differences."

A simple application of Ohm's law relating flow conductance L_p to water potential and transpiration rate has been widely empolyed to estimate total flow conductance from soil to leaf expressed on a leaf area basis. Several simplifications, inherent to the underlying model, were made in the calculations: (1) the capacitance effect in the plant is low; (2) steady-state transpiration conditions exist; and (3) predawn potential is a good estimator of the soil water potential. A hierarchy of L_p from the highest soil-to-leaf conductances at the more mesic site to the lowest at the more xeric site was established. More generally, *Q. turbinella* showed the steepest decrease of Δ_ψ and the most negative slope, and *Q. ilex* at Camp Redon the slowest decrease. The hierarchy of slopes is the same as the hierarchy of L_p (Figure 19-10). The presence and magnitude of the difference in flow conductance suggests that this attribute may be an important component of the hydraulic architecture of woody plants.

Changes in plant conductances under water stress are attributable to effects on the roots and xylem. As the soil dries, decreased permeability by root suberization and/or increased fine root mortality can reduce the balance between water extraction capacity and transpiring leaf area. Xylem loss of vascular transport by cavitation also causes an increase in plant resistance. There is evidence of embolism formation in temperate deciduous *Quercus* species, one of the genera which most commonly have ring-porous xylem anatomy (Cochard and Tyree, 1990; Cochard et al., 1992). These authors found a 50% loss of hydraulic conductivity in current-year shoots at a xylem water potential between −2.5 and −3 MPa. At −4 MPa, a close value to the lower boundary of Table 19-3, the conductivity was only 5% of its maximal value. These results agree well with those of Lo Gullo and Salleo (1993). For 1-year-old twig segments of *Q. ilex*, the loss of conductivity began at −1.8 MPa and linearly increased to reach a total xylem cavitation at −4.35 MPa. The loss of conductivity is approximately 50% at the turgor loss point. Generally, minor branches are more affected by embolism than major branches (Cochard et al., 1992). This result supports Zimmerman's (1983) plant segmentation hypothesis. He suggested that plants are designed hydraulically to sacrifice highly vulnerable minor branches. When catastrophic xylem dysfunction occurs, Tyree and

Sperry (1988) showed that minor branches begin to die, leading to a loss of leaf area and a reduction in the water flow, which causes an improved water balance in the remaining living stems.

The key role of the leaf area index in the water balance of ecosystems was previously highlighted. The control of the carbon budget (Rambal, 1993) was also suggested. Evaluating the consequences of climate changes or global changes on the water balance should include an evaluation of the new leaf area index. It is now suggested that woody plants growing in water-limited situations and, particularly, Mediterranean evergreen oaks adjust the ratio of total hydraulic conductance to transpiring leaf area to maintain water potential relatively stable (Winkel and Rambal, 1993). Consequently, the leaf area index is under the control of water relations, rather than under the control of the carbon budget.

Acknowledgments. This work, as part of the Fortissimo project, was partly funded by the French program IGBP Ecosystems. I thank in advance any colleagues who may be able to supply published and/or unpublished data on leaf water potential in deciduous or evergreen Mediterranean *Quercus* species to extend the analysis of the last paragraph.

References

Angot A. 1918. Etudes sur le climat de la France. Régime des pluies. Deuxième partie, régions du sud-ouest et du sud. Annales du Bureau Central Météorologique pour l'année 1912. I. Mémoires: 101–214.

Arago F. 1835. Changements de climat en France. Bulletin de la Société d'Agriculture du département de l'Hérault, 22ème année: 110–118.

Aussenac G, Granier A. 1978. Quelques résultats de cinétique journalière du potentiel de sève chez les arbres forestiers. Ann Sci For 35:19–32.

Aussenac G, Valette JC. 1982. Comportement hydrique estival de *Cedrus atlantica* Manetti, *Quercus ilex* L. et *Quercus pubescens* Willd. et de divers pins dans le Mont Ventoux. Ann Sci For 39:41–62.

Blake-Jacobson ME. 1987. Stomatal conductance and water relations of shrubs growing at the chaparal-desert ecotone in California and Arizona. In: Tenhunen JD, Catarino FM, Lange OL, and Oechel WD, (eds). Plant Response to Stress. Functional Analysis in Mediterranean Ecosystems. NATO, ASI series, Vol. G15. Springer-Verlag, Berlin, pp 223–245.

Bon FX. 1705–1709. Premières observations météorologiques faites à Montpellier par M. Bon depuis 1705 jusqu'en 1709. Archives de la Société Royale des Sciences de Montpellier. Manuscrit convervé aux Archives Départementales de l'Hérault, Série D 138, fol. 1–30.

Castell Puig C. 1992. Ecofisiologia d'individus adults i rebrots de dues espècies esclerofilles mediterrànies: *Arbutus unedo* i *Quercus ilex*. Memoria de Doctor en Ciències, Seccio Biologiques. Universitat Autonoma de Barcelona, 280 p.

Chassant M. 1906. La plus haute température observée en France. Bulletin Météorologique du département de l'Hérault pour l'année 1905:74–78.

Cochard H, Tyree MT. 1990. Xylem dysfunction in *Quercus*: vessel sizes, tyloses, cavitation and seasonal changes in embolism. Tree Physiol 6:393–407.

Cochard H, Breda N, Granier A, Aussenac G. 1992. Vulnerability to air embolism of three European oak species (*Quercus petrea* (Matt) Liebl., *Q. pubescens* Willd., *Q. robur* L.). Ann Sci For 49:225–233.

Commission Météorologique de L'Hérault. 1956. Observations de 1956. Bulletin annuel, nouvelle série, 6.

Conrad V. 1941. The variability of precipitation. Mon Weath Rev 69:5–11.

Cubasch U, Cess RD. 1990. Processes and modelling. In: Houghton JT, Lenkins GJ, and Ephraums JJ, (eds). Climate Change: The IPCC Scientific Assessment. Cambridge University Press, pp. 74–91.

Duhme F. 1974. Die kennzeichneung des ökologischen konstitution von gehölzen im hinblick auf den wasserhaushalt. Dissertationes Botanicae 28:1–143.

Dupin P. 1836. Examen de cette question: Le climat a-t-il changé en France? Les étés sont-ils moins chauds qu'autrefois, par l'effet des défrichements? Bulletin de la Société d'Agriculture du département de l'Hérault, 23ème année:97–113.

Dupin P. 1837. Mémoire sur le changement et la détérioration du climat (suivi de «Corrections et additions» au Mémoire sur le changement du climat). Bulletin de la Société d'Agriculture du département de l'Hérault, 24ème année:297–328, 403–405.

Giorgi F, Mearns LO. 1991. Approaches to the simulation of regional climate change: a review. Rev Geophy 29:191–216.

Gordon HB, Whetton PH, Pittock AB, Fowler AM, Haylock MR. 1992. Simulated changes in daily rainfall intensity due to the enhanced greenhouse effect: implications for extreme rainfall events. Climate Dynam 8:83–102.

Griffin JR. 1973. Xylem sap tension in three woodland oaks of central California. Ecology 54:152–159.

Hastings SJ, Oechel WC, Sionit N. 1989. Water relations and photosynthesis of chaparral resprouts and seedlings following fire and hand clearing. In: Keeley SC (ed). The California Chaparral: Paradigms Reexamined. Science Series 34, Natural History Museum of Los Angeles County:108–113.

Hickman JC. 1970. Seasonal course of xylem sap tension. Ecology 51:1052–1056.

Katz RW. 1992. Role of statistics in the validation of general circulation models. Climate Res 2:35–45.

Katz RW, Brown BB. 1992. Extreme events in a changing climate: variability is more important than averages. Climatic Change 21:289–302.

Le Houérou HN. 1992. Relations entre la variabilité des précipitations et celle des productions primaire et secondaire en zone aride. In: Le floc'h E. et al., (eds). L'aridité: une contrainte au developpement. ORSTOM:197–220.

de Lillis M, Fontanella A. 1992. Comparative phenology and growth in different species of the Mediterranean maquis of central Italy. Vegetatio 99–100:83–96.

Lo Gullo MA, Salleo S. 1993. Different vulnerabilities of *Quercus ilex* L. to freeze- and summer drought-induced xylem embolism: an ecological interpretation. Plant Cell Environ 16:511–519.

Martins C. 1871. L'hiver de 1870–1871 dans le Jardin des Plantes de Montpellier et aux environs de cette ville. Bulletin de la Société Centrale d'Agriculture et des Comices agricoles du département de l'Hérault, 58ème année, 191–203.

Miller PC. 1981. Similarities and limitations of resource utilization in Mediterranean-type ecosymstems. In: Miller PC, (ed). Resource use by chaparral and matorral. A comparison of vegetation function in two Mediterranean-type ecosystems. Ecol Stud 39:369–407. Springer-Verlag, Berlin.

Miller PC, Poole DK 1979. Patterns of water use by shrubs in southern California. For Sci 25:84–98.

Mitchell JFB, Manabe S, Meleshko V, Tokioka T. 1990. Equilibrium change and its implications for the future. In: Houghton JT, Jenkins GJ, and Ephraums JJ, (eds). Climate Change: The IPCC Scientific Assessment. Cambridge University Press, pp. 137–172.

Oechel WC, Strain BR, Odening WR. 1972. Photosynthesis rates of a desert shrub, *Larrea divaricata* Cav., under field conditions. Photosynthetica 6: 183–188.

Pittock AB, Fowler MA, Whetton PH. 1991. Probable changes in rainfall regimes due to the enhanced greenhouse effect. Proceedings of International Hydrology and Water Resource Symposium "Challenges for sustainable development," 2–4 October 1991, Perth, Western Australia:182–186.

Poitevin J. 1803. Essai sur le climat de Montpellier. Montpellier, Paris, 164 p.

Poole DK, Miller PC. 1981. The distribution of plant water stress and vegetation charateristics in Southern California chaparral. Am Midl Nat 105:32–43.

Poole DK, Roberts SW, Miller PC. 1981. Water utilization. In: Miller PC (ed). Resource use by chaparral and matorral. A comparison of vegetation function in two Mediterranean-type ecosystems. Ecol Stud 39:123–149. Springer-Verlag, Berlin.

Rambal S. 1984. Water balance and pattern of root water uptake by a *Quercus coccifera* L. evergreen scrub. Oecologia 62:18–25.

Rambal S. 1991. 2030: conjectures sur la ressource en eau en Languedoc karstique. La météorologie 38:2–7.

Rambal S. 1992. *Quercus ilex* facing water stress: a functional equilibrium hypothesis. Vegetatio 99–100:147–153.

Rambal S. 1993. The differential role of mechanisms for drought resistance in a Mediterranean evergreen shrub: a simulation approach. Plant Cell Environ 16:35–44.

Ritchie GA, Hinckley TM. 1975. The pressure chamber as an instrument for ecological research. Adv Ecol Res 9:165–254.

Roche E. 1882. Le climat actuel de Montpellier comparé aux observations du siècle dernier. Académie des Sciences et Lettres de Montpellier. Mémoires de la Section des Sciences. Tome X, 1880–1884, IIème fascicule, 329–384.

Rodrigues H. 1855. Clinique médicale de Montpellier. In: Dumas JA, (ed). Montpellier: 422 p.

Sala i, Serra A. 1992. Relacions hiriques, estructura de la capçada i intercanvi de gasos en un bosc de *Quercus ilex*: variacions en el temps i a l'espai. Memoria de Doctor en Ciències, Seccio Biologiques. Universitat Autonoma de Barcelona, Barcelona, Spain. 171 p.

Tenhunen JD, Lange OL, Harley PC, Beyschlag W, Meyer A. 1985. Limitations due to water stress on leaf net photosynthesis of *Quercus coccifera* in the Portuguese evergreen scrub. Oecologia 67:23–30.

Tyree MT, Sperry JS. 1988. Do woody plants operate near the point of catastrophic xylem dysfunction caused by dynamic water stress? Plant Physio 88:318–330.

Tyson PD. 1990. Modelling climatic change in southern Africa: a review of available methods. So Afr J Sci 86:318–330.

Valette JC. 1981. Comportements hydriques du Cèdre, des Chênes et des Pins méditerranéens. Institut National de la Recherche Agronomique. Station de sylviculture méditerranéenne, Avignon, document interne 81–1:43 p.

Waggoner PE. 1989. Anticipating the frequency distribution of precipitation if the climate change alters its mean. Agric Forest Meteorol 47:321–337.

Waring RH, Cleary BD. 1967. Plant moisture stress: evaluation by pressure bomb. Science 155:1248, 1953–1954.

Warrick RA, Barrow EM, Wigley TML. 1990. The greenhouse effect and its implications for the European Community. Commission of the European Communities. Report EUR 122707 EN, 30 p.

Wigley TML. 1985. Impact of extreme events. Nature 316:106–107.

Winkel T, Rambal S. 1993. Influence of water stress on grapevines growing in the field: from leaf to whole-plant response. Aust J Plant Physiol 20:143–157.

Zimmerman MH. 1983. Xylem structure and the ascent of sap. Springer-Verlag, Berlin, Heidelberg, New York.

20. Patterns of Fire Occurrence Across a Climatic Gradient and Its Relationship to Meteorological Variables in Spain

Antonio Vázquez and José M. Moreno

The relationships between climate and fire are manifold. Climate controls primary productivity and vegetation type (Lieth and Whittaker, 1975; Woodward, 1986), and, through that, fuel quality, distribution, and accumulation rate (Mutch, 1970; Philpot, 1977; Olson, 1981; Rundel, 1981). The prevailing climate of a given area is characterized by specific values of temperature, precipitation, moisture deficit and wind, among other characteristics, which, in turn, determine the probability of fire ignition and spread (Rothermel, 1983). Therefore, changes in climate are likely to induce variations in the most critical biotic and abiotic factors that control the fire regime.

Climate changes in the past have produced variations of fire frequency and intensity, and produced different landscapes (Clark, 1988; Baisan and Swetnam, 1990; Jonhson and Larsen, 1991). Meyer et al. (1992) showed how climate change, by altering fire size and intensity, can produce very significant changes in the structure and functioning of an alluvial system. Even changes in climate associated with El Niño have been shown to affect the fire regime (Swetnam and Betancourt, 1990; Brenner, 1991).

The changes in temperature and precipitation expected to occur as a result of releasing greenhouse gases are likely to induce increased fire risk in various fire prone-regions (Beer et al., 1988; Flannigan and Van Wagner, 1991; Torn and Fried, 1992). Nevertheless, to predict whether the number of fires and actual area burned will increase is more difficult because fire incidence depends not only on weather conditions but also on ignition sources. Predicting this is probably no simple task. This is particularly so in areas where fire is mostly dominated by human influence. For instance, increasing temperatures may increase the fire season, but may also modify land use and/or recreational use patterns. These changes may interact with the number of ignitions from various sources and affect total surface burned or the time when this occurs in an unpredictable manner.

In order to better understand how climate change may alter fire incidence, we may first learn how the temporal and size fire patterns (season of fire occurrence and fire size structure) are responsive to today's climate and weather, and also how these patterns vary in relation to the source of ignition. This is particularly important in areas where man is the main source of ignition. Most studies available on past climate changes are based on natural fires, but these play a minimal role in many fire prone-areas. Extrapolating from the present to future scenarios with realized climate change based on past patterns may not be realistic if we only take into account the weather as the main driver in fire occurrence. Fire incidence may be strongly influenced by changes in the patterns of human activities, either in preventing fire or in causing it, intentionally or unintentionally.

In this paper we have explored how areas that have different climates differ in the seasonal occurrence of fire and in their fire size patterns, in particular in fire size inequalities. Additionally, the study across a climate gradient has been complemented with studies of the yearly variability of such fire patterns and its relationship to the weather. This has been done for different types of fires separated according to their ignition source. In an earlier paper (Vázquez and Moreno, 1993), we showed that fire occurrence (total number of fires and total area burned) responded very different to mean yearly meteorological variables in different climate areas. Specific relationships were detected depending on the climate of the area of study as well as on the ignition source. The time of the year for a fire to occur as well as its size are important from the management point of view and also from the ecological point of view. Fire impacts may be sensitive to changes of these components of the fire regime.

Objectives and Hypothesis

The objectives of this work were first, to determine how fire occurrence (number of fires and surface burned) developed through the year in areas that differ in their climate and for fires ignited by different sources. We wanted also to study this through the years in order to determine interannual variability and its relationship to the extant meteorological conditions. The second objective was to determine how fire size distribution and inequalities, i.e., the variability in fire size, changed in areas with a different climate and for fires ignited by different sources. We equally wanted to assess the interannual variability of fire size patterns and its relationship to the extant meteorological conditions. The main hypothesis were as follows:

Fire Temporal Patterns

(1) In general, the period of fire occurrence during the year would be smaller in cooler and wetter areas (i.e., areas less susceptible to fire) than in warmer and drier areas. (2) The time required to burn a certain percentage of the total area burned in the year should be smaller than the time required to have the same percentage of the total number of fires, since the probability of a fire to spread would not be equal through the year. (3) The fire season for fires originated by negligence (see later for definition) or from natural causes should be longer than for fires caused intentionally (i.e., fires occurring more by chance vs. fires made specifically to propagate, hence made under a more restricted set of conditions). (4) Fires started by pasture burning (i.e., started when they escaped from such activity) would have a longer fire season during the year than fires started by other causes (pasture burning by farmers may occur whenever conditions are suitable for burning but that are not too dangerous. These conditions may be present at any time during the year). (5) The variability of the period for fire occurrence should be greater in warmer and drier areas than in rainier and cooler areas due to the greater variability in weather events, particularly precipitation, in the first type of areas. (6) The relationship between period of fire occurrence and meteorological variables should be greater in cooler and wetter areas than in warmer and drier areas due also to the greater variability of the weather in the second type of areas.

Fire Size Patterns

(1) Fire size inequalities should be greater in warmer and drier areas than in cooler and wetter areas due to the greater likelihood of extreme events conducive to very large fires. (2) The variability from year to year in fire size inequalities should be greater in warmer and drier areas than in cooler and wetter areas because drier climates are more variable, particularly for precipitation. (3) Fire size inequalities should be greater in fires likely to start under the most extreme conditions (i.e., by negligence or by lightning) than in fires caused under a more restricted set of conditions (pasture burning, intentional fires). (4) The relationship between meteorological variables and fire size inequalities should be stronger in cooler and wetter areas than in warmer and drier areas where

the probability of extreme events may be larger and mean weather values may be less meaningful.

Materials and Methods

Study Areas

We selected three areas within Spain (Figure 20-1). (1) The Northern area comprehends most of the Spanish territory within the Eurosiberian phytogeographical region (Rivas-Martínez, 1981, 1987; Moreno et al., 1990). This is characterized by high precipitation throughout the year and mild temperatures (Figure 20-2A). (2) The Central area comprehends most uplands around and including the Central System, which are included mainly in the Carpetano-Ibérico-Leonesa phytogeographical province, within the Mediterranean region (Rivas-Martínez, 1987). This area is more continental, with less rainfall and a clear dry summer season (Figure 20-2B). (3) The Levant area comprehends a strip along the Mediterranean coast of Spain, which includes the biogeographical Valenciano-Tarraconense sector of the Catalano-Valenciano-Provenzal-Balear phytogeographical province (Rivas-Martínez, 1987). This area is characterized by a warmer and drier climate than the other two areas, with a prolonged summer drought (Figure 20-2C). Additionally, the precipitation patterns of the North and Central areas are more related to meteorological situations coming from the northwest to the southwest, whereas the Levant is more related to situations developing from the east (Villa et al., 1985).

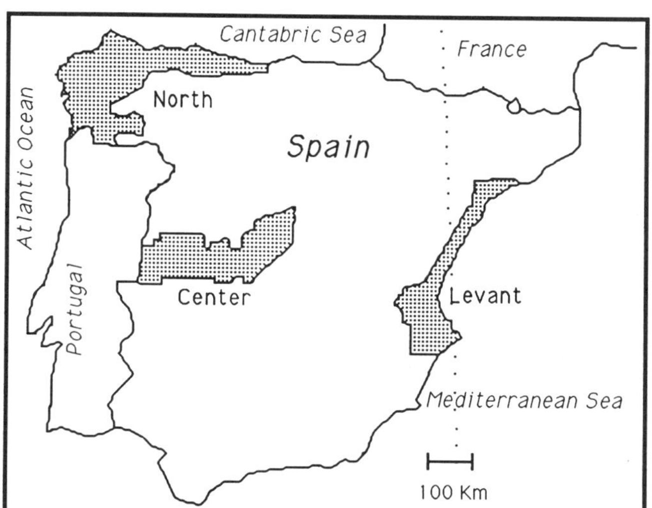

Figure 20-1. Location of the study areas within Spain.

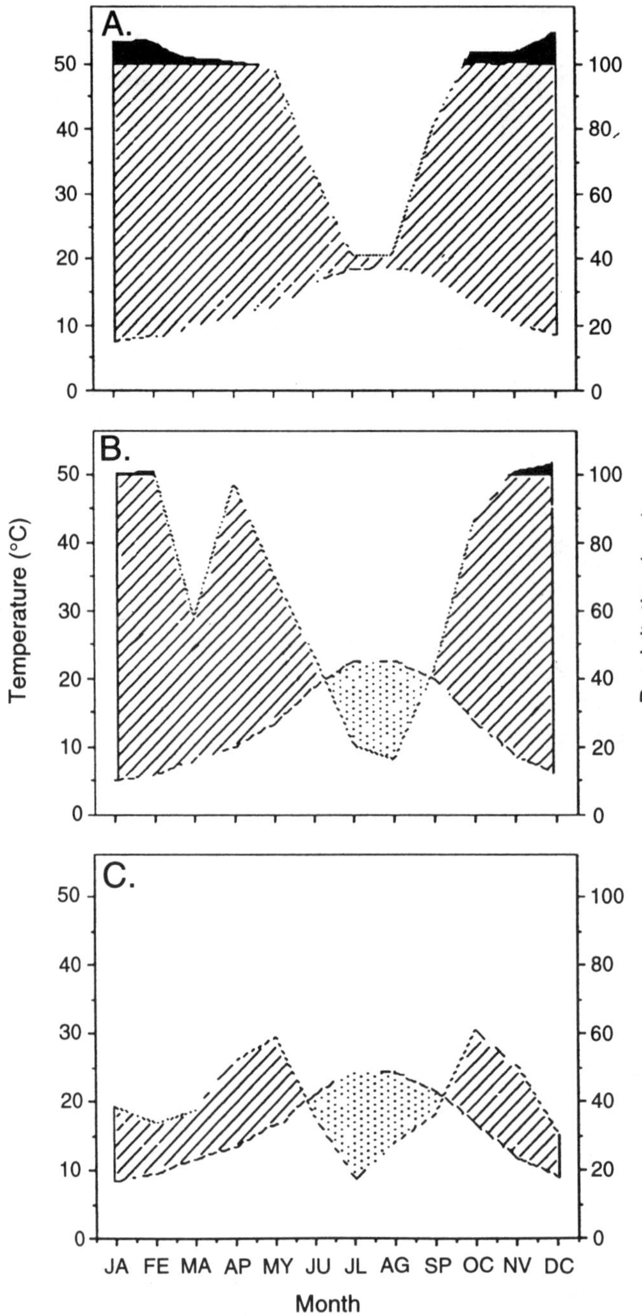

Figure 20-2. Climate diagrams for each of the three study areas based on 19 stations from each area and on the 15 years of observations of the study period: (A) North; (B) Center; (C) Levant. Hatched areas represent relatively wet periods of the year. Dotted areas represent relatively dry periods of the year. Solid areas correspond to 1/10 reduction in the scale for precipitation exceeding 100 mm.

Data

We used all available fire records from the data bank of the Spanish National Institute for the Conservation of Nature (ICONA), from 1974 to 1988. In the Levant area the records were incomplete for 1979 and 1988. The 15 years chosen were the only available years with geographical coordinates in the fire record when this study was started. This information was needed to ascribe the fire to one of the three study areas. Based on the information supplied in the fire records, we have grouped the causes of fires into 5 categories: fires caused by pasture burning (this is, fires that escaped when this activity was being carried out, becoming a wildfire), by lightning, by arsonists (i.e., intentional fires), by unknown sources, and fires caused by negligence or accidentally. In this last category we have included a miscellaneous set of fire causes: sparks from trains, cigarette stubs, fires set by people with mental disorders, fires related to forestry or nonforestry operations with fire, fires from military maneuvers, etc. From here on, they will be referred to as negligence fires.

The yearly weather of each area was characterized by means of all available meteorological records from 19 of the best meteorological stations located within each of the three study areas. Missing data were interpolated by using multiple regression from all available data for all stations and cases. Among the meteorological variables available we selected the following: *Ptot*, total year precipitation; *Dp1*, total number of days during the year with precipitation equal to or greater than 1 mm; *Dp10*, total number of days during the year with precipitation equal to or greater than 10 mm; *Dp30*, total number of days during the year with precipitation equal to or greater than 30 mm; *Tm*, mean monthly temperature during the year; *Tmx*, mean maximum monthly temperature during the year; *Dm20*, total number of days during the year with minimum temperature equal to or greater than 20°C; *DM25*, total number of days during the year with maximum temperature equal to or greater than 25°C; *DM30*, total number of days during the year with maximum temperature equal to or greater than 30°C.

Procedures

Based on the date of the commencement of the fire supplied in the fire records, we calculated the mean Julian day of the fires during the year and its standard deviation. Additionally, we grouped all fires by the fortnight of occurrence. All months were given two fortnight periods of equal duration (15 days). We constructed cumulative frequencies of fire occurrence (number and surface burned) by fortnights. Based on this we calculated the period needed to have 50% of fire occurrence (PFO_{50}) either of number of fires or of surface burned, this being the time elapsed between cumulative frequencies at 25% and 75%. This was done for each year and for all fires combined, regardless of their ignition source, for fires started by negligence, and for fires caused intentionally. Due to the small number of fires, lightning fires and pasture burning fires were not included in this analysis. Fires caused by unknown sources were similarly not included in the analysis. The differences between the three study areas in the length of the fire season, as measured by the PFO_{50}, either for the number of fires or area burned was tested by means of a Wilcoxon matched-pairs signed-ranks test between pairs of study areas. These comparisons were considered as planned comparisons, and two areas were considered statistically different when $P \leqslant 0.05$. The relationship between mean Julian day, its standard deviation and the PFO_{50} and each of the meteorological variables (see above) was tested by product-moment correlation. As we were mainly interested in correlation patterns, no correction was made for testing the significance of the correlations performed.

Differences in the cumulative fire size frequency distribution of the same type of fire between pairs of study areas, or for different pairs of types of fires within the same area, was tested by the two-sample Kolmogorov-Smirnov test. This was done for all fires recorded in the data bank and ≥0.1 ha, or for fires ≥5 ha, or for fires ≥50 ha. Fire size inequalities between the three study areas and for fires ignited by various sources were compared by plotting the extreme proportion function "$EP(p)$" against "p" (a given proportion of the largest fires). $EP(p)$ specifies the proportion of the total area burned that is caused by the largest $100p\%$ of the fires (Strauss et al., 1989). The extreme proportion function was calculated for fires caused by all sources combined or by each of the five causes considered for fires occurred during the whole study period. Furthermore, this same function was calculated for each individual year for all fires combined, for fires caused by negligence, and for fires caused intentionally for those years when the number of fires of each individual ignition source exceeded 50. This procedure was not possible for fires caused by pasture burning or lightning, due to the small number of fires available in some years and/or at some of the study areas. Fires from unknown causes were also excluded from this analysis. Additionally, we fitted various functions to the $EP(p)$ vs. p plots. The truncated Pareto distribution yielded the best fit and was used for further analysis. We used the nonlinear regression program of the BMDP statistical software to fit the real data to the truncated Pareto distribution. This function takes the form

$$Ep(p) = ((pb^{\alpha} + q)^{1-1/\alpha} - 1)/b^{\alpha-1} - 1$$

where p is a given proportion, $q = (1 - p)$, α is the shape parameter, and b is the ratio of largest to smallest possible fires. Based on the fitted function we calculated for each area and for all fires, for fires caused by negligence and for fires caused intentionally the percentage of largest fires needed to affect a given percentages of the total surface burned (P_i; i

75%, 80%, . . . , 90%, 95%). The patterns obtained for the various areas were similar, and only the rather extreme P_{90} will be shown. P_{90} values were compared across areas by means of paired t-tests among pairs of study areas. The level of significance for these planned comparisons was set to $P \leqslant 0.05$. Additionally, similar analysis were done within each study zone comparing fires caused by negligence or intentionally, in which case the level of significance was set to $P \leqslant 0.01$.

Results

There were large numbers of fires and a large total area burned per year in each of the three study areas during each year of the study period (Figure 20-3). The three areas selected comprised ≈47% of all fires and ≈39% of the total area burned in Spain during the period of study. The proportion of fires caused by each different source varied among themselves and between areas (Figure 20-4). Fires caused by pasture burning or by lightning were a small fraction of the total number of fires and represented also a small fraction of the total area burned. The proportion of lightning-caused fires increases from the North to the Levant area, in which area this kind of fires represented ≈10% of the total area burned. Most fires were related to human activities. The differences in the proportion of fires caused by arsonists between the three areas are noteworthy, with the highest percentage in the North (Figure 20-4). Additionally, the large percentage of fires caused by unknown sources is also noteworthy.

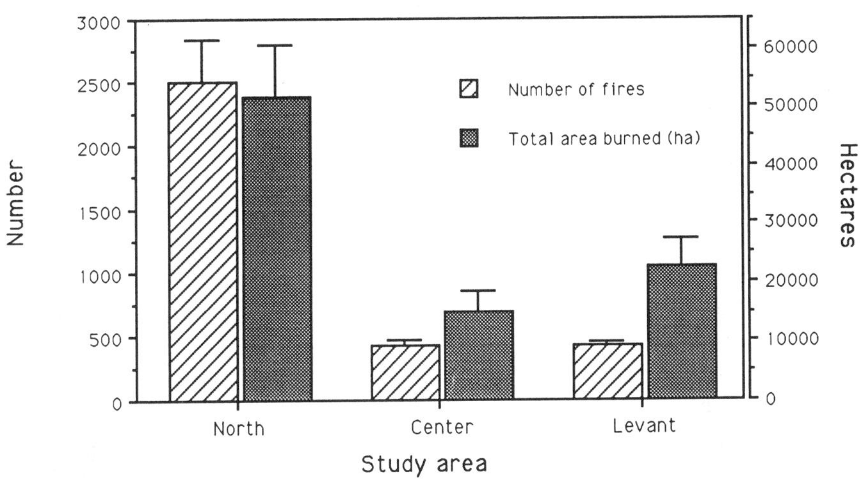

Figure 20-3. Mean (±1 SE) number of fires per year and mean total area burned per year in each of the three study areas from 1974 to 1988, except for the Levant, where no complete data were available for 1979 and 1988.

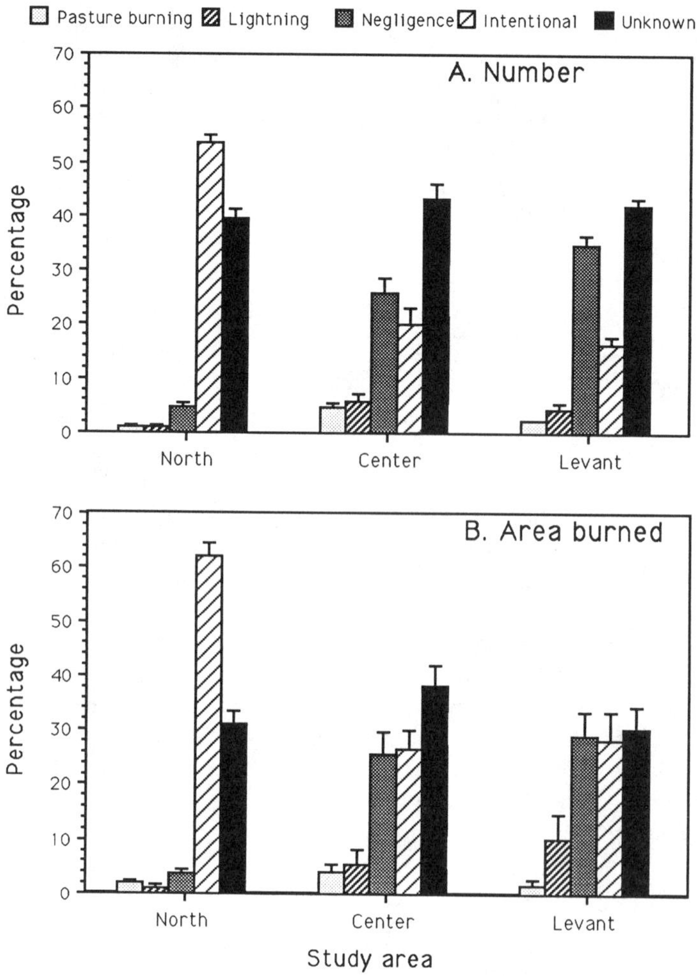

Figure 20-4. Proportion of fires caused by different sources of ignition for each of the three study areas: (A) total number of fires per year; (B) total area burned per year.

Fires were less widespread during the year in number and surface burned in the Center and more widespread, in general, in the Levant, although the North was between both areas, with a large area burned in spring (Figure 20-5A,B). The largest number of fires and greatest surface burned took place in just a few fortnights in the summer months. Fires caused by pasture burning were widespread throughout the year in number and area burned, particularly in the North. The Levant and Center areas showed greater fire occurrence during the summer months, more so in the Central area (Figure 20-5C,D). Fires caused by lightning occurred exclusively during the summer months, burning earlier more surface in the Levant than in the Center, with the North somewhere in between and more widespread in time (Figure 20-5E,F). Fires caused by negligence were more widespread in time in the Levant than in the other areas, and so too for surface burned, although in this case the North also had a large proportion of area burned in spring. The Central area presented a pattern of very high concentration of surface burned during the summer months (Figure 20-5G,H). Fires caused by arsonists were more widespread in the Levant and less widespread in the Central area, less so for surface burned, which was very much concentrated during the summer months, than for number of fires in both the Center and Levant. The North was between the two other areas (Figure 20-5I,J). Fires caused by unknown sources were widespread in the Levant and less so in the Center. The North was also between the other two areas (Figure 20-5K,L). In general, fires started by pasture burning were the most widespread in time, followed by fires started by negligence. On the other hand, fires caused intentionally were the most restricted in time, followed by fires caused by lightning. Fires caused by unknown sources were close to intentional fires in the North, to fires caused by negligence or intentional fires in the Center, and to pasture burning and fires started by negligence in the Levant.

The period for fire occurrence of 50% (PFO_{50}) of the total number of fires or of the total area burned by fires started by all sources (Figure 20–6A,B) was significantly longer and more variable in the Levant than in the Center. Additionally, there was a pattern in both areas for values of the PFO_{50} of area burned to be smaller (shorter duration) than that for the number of fires. On the other hand, the North had similar values for the PFO_{50} for number of fires than the Center, and also similar values for PFO_{50} for area burned than the Levant. Additionally, the greatest variability in PFO_{50} was observed in the North for both number of fires and surface burned. Furthermore, PFO_{50} for area burned was more variable and of similar magnitude to number of fires. These patterns were similar when fires were separated by the ignition source (negligence or intentional fires) (Figure 20-6), although in the case of PFO_{50} for area burned the significant differences between the Levant and Center were not maintained for both, negligence or intentional fires. In other cases, it was the North that did not have statistically significant different values in PFO_{50} for number of fires (intentional fires) or surface burned (negligence and intentional fires) than the Levant.

The relationship between the period of fire occurrence and the mean yearly meteorological variables was very poor. No significant relationships were obtained for PFO_{50} for number of fires or surface burned and precipitation variables in any of the study areas (not shown). In the case of the temperature variables, significant relationships appeared only in the North area (Table 20-1). The number of days with maximum tem-

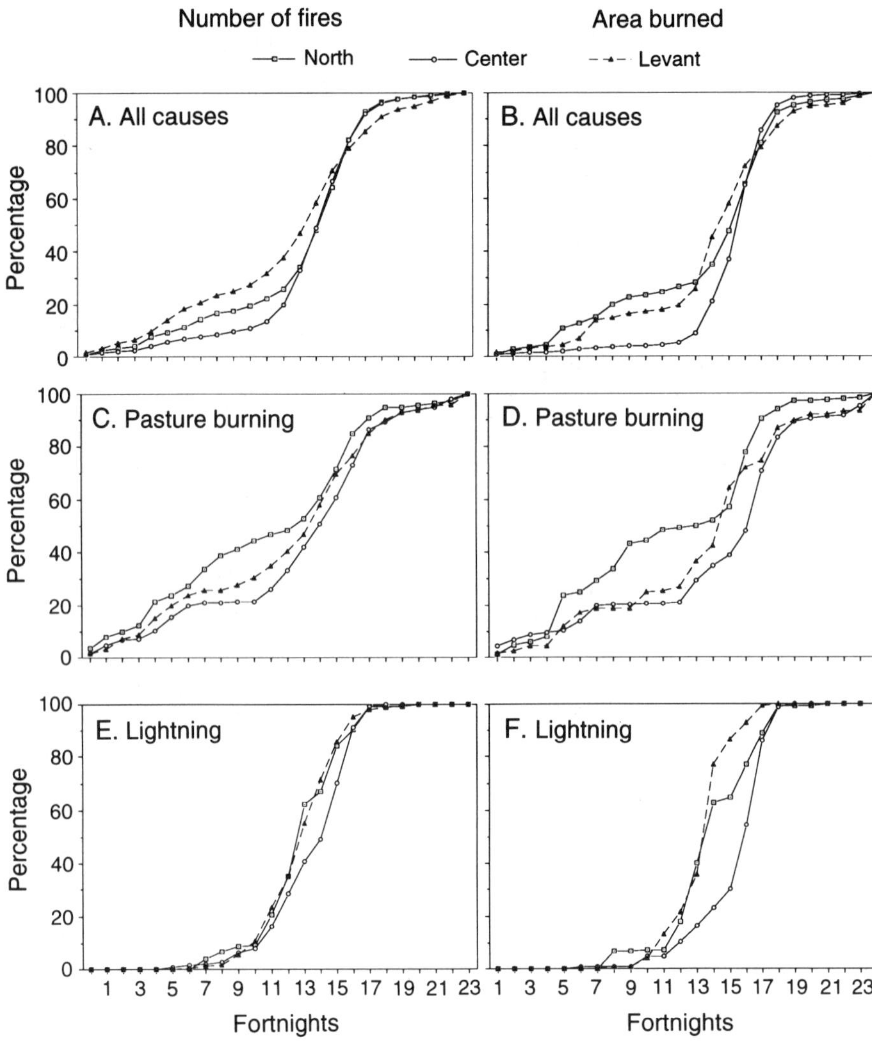

Figure 20-5. Cumulative frequency of number of fires (left panels) and area burned (right panels) through the year for fires caused by: (A, B) all sources; (C, D) pasture burning; (E, F) lightning;

peratures exceeding 25° or 30°C were negatively related to PFO_{50} for either number of fires or area burned for all fires combined, for intentional fires, and less so for negligence fires. Thus, the more warm days there were, the more concentrated in time were the fires. In the Center, PFO_{50} for number of fires and mean maximum monthly temperatures was close to the significance level in the case of all fires combined, but in this case the relationship was positive, as would have been expected (r =

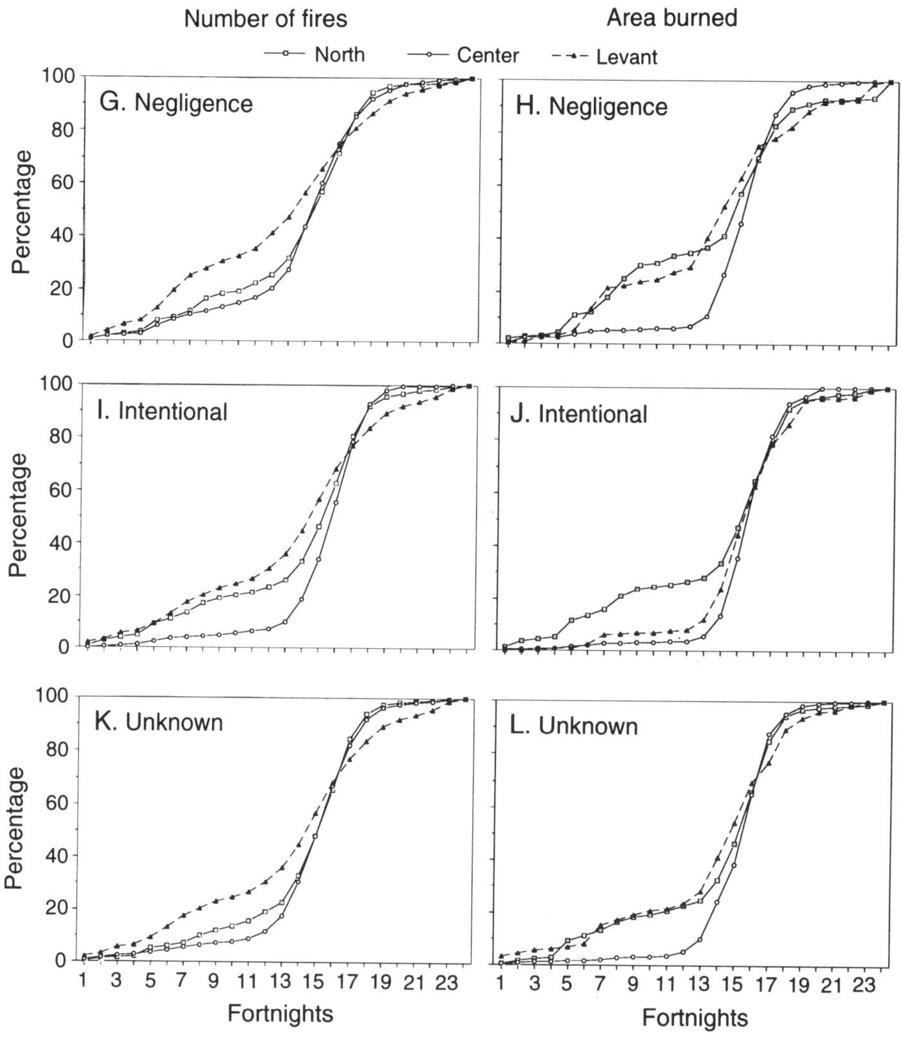

Figure 20-5. *Continued* (G, H) negligence; (I, J) intentional; (K, L) unknown sources.

+0.51; $P > 0.05$). Additionally, the mean Julian day of the year when a fire was started was not significantly related to the meteorological variables in any of the areas tested (not shown). The standard deviation of the mean Julian day of the year when the fire was started for all sources of ignition was significantly negatively correlated with days above 30°C in the North ($r = -0.59$; $P < 0.05$), and was not in the other areas or for all the other variables.

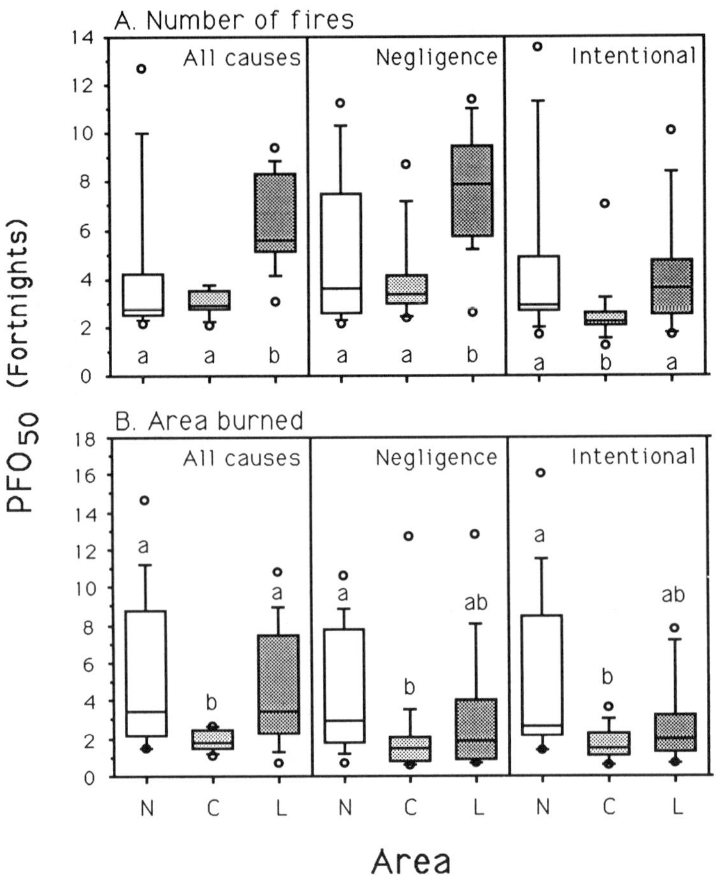

Figure 20-6. Box-plots of the period (in fortnights) of fire occurrence for 50% (PFO_{50}) of (A) total number of fires and (B) total area burned during the year for fires caused by all sources, by negligence, and intentionally in each of the three study areas. Plots show the median, the quartiles, the 10% and 90% range and extreme values. For a given type of fire, PFO_{50} of areas sharing the same letter are not significantly different ($P > 0.05$) according to a Wilcoxon matched-pairs signed-ranks test. (N = North; C = Center; L = Levant.)

The mean size of fires (±SE) was 20.3 (±0.35), 35.2 (±2.38) and 53.1 (±5.36) ha for the North, Center and Levant, respectively. However, the distributions were highly skewed towards smaller size fires. Actually, the median size of fires was 4, 2 and 2 ha for the North, Center and Levant, respectively (Figure 20-7). The differences in the cumulative size frequency distribution between the three study areas was highly statistically significant (Figure 20-7; Table 20-2). These differences between the three study areas were maintained when fires were separated by the source of

Table 20-1. Coefficient of Correlation (*r*) Between PFO_{50} for Number of Fires and Area Burned and Yearly Temperature Variables in the North Study Area for Fires Started by All Sources Combined, by Negligence or Intentionally

	All Fires		Negligence		Intentional	
Variable	Number of Fires	Area Burned	Number of Fires	Area Burned	Number of Fires	Area Burned
Tm	+0.14	+0.14	−0.01	−0.36	+0.01	−0.01
Tmx	−0.01	−0.01	−0.26	−0.37	−0.01	−0.20
Dm20	−0.36	−0.26	−0.50	+0.17	−0.35	−0.22
DM25	−0.50	−0.50	−0.51	−0.36	−0.51	−0.50
DM30	−0.51	−0.60*	−0.49	−0.14	−0.53*	−0.54*

* = Significant correlations ($P \leqslant 0.05$). For names of meteorological variables see Material and Methods ($n = 15$).

Table 20-2. Significance (*P*) of the Two-Sample Kolmogorov-Smirnov Test for the Cumulative Size Frequency Distribution for Fires ⩾0.1 ha, ⩾5 ha and ⩾50 ha Between Pairs of Study Areas for Fires Caused by All Causes, by Pasture Burning, by Lightning, by Negligence, Intentionally, and by Unknown Sources (N = North; C = Center; L = Levant)

		Size of Fire		
Type of Fire	Comparison	⩾0.1 ha	⩾5 ha	⩾50 ha
All causes	N vs. C	<0.001	<0.001	<0.001
	N vs. L	<0.001	<0.001	<0.001
	C vs. L	<0.001	0.112	0.130
Pasture burning	N vs. C	<0.001	0.044	0.755
	N vs. L	<0.001	0.085	0.057
	C vs. L	0.002	0.028	0.311
Lightning	N vs. C	<0.001	0.645	0.111
	N vs. L	0.059	0.515	0.060
	C vs. L	0.005	0.943	0.996
Negligence	N vs. C	<0.001	0.001	0.020
	N vs. L	<0.001	0.003	0.032
	C vs. L	<0.001	0.969	0.950
Intentional	N vs. C	<0.001	<0.001	<0.001
	N vs. L	<0.001	<0.001	<0.001
	C vs. L	<0.001	<0.001	0.045
Unknown	N vs. C	<0.001	<0.001	<0.001
	N vs. L	<0.001	<0.001	<0.001
	C vs. L	<0.001	0.554	0.785

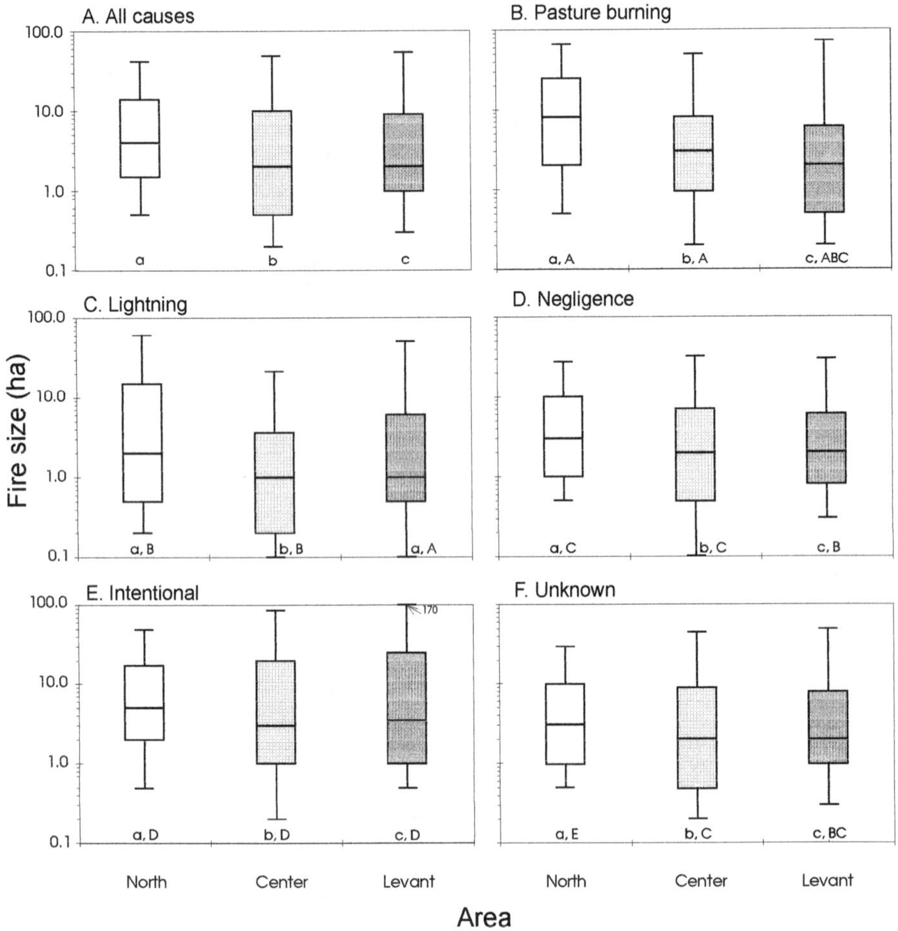

Figure 20-7. Box-plots of the size of fires of the three study areas: (A) all fires; (B) fires caused by pasture burning; (C) fires caused by lightning; (D) fires caused by negligence; (E) Fires caused intentionally; (F) Fires caused by unknown sources. Plots show the median, the quartiles and the 10% and 90% range. For a given type of fire, areas sharing the same letter (lower case letter) are not significantly different ($P > 0.05$) according to a two-sample Kolmororov-Smirnov test. For a given area, types of fires sharing the same letter (upper case letter) were not significantly different ($P > 0.01$) according to a similar statistical test.

Table 20-3. Significance (*P*) of the Two-Sample Kolmogorov-Smirnov Test for the Cumulative Size Frequency Distribution for Fires ≥0.1 ha (Normal Type), and ≥5 ha (Bold Type) Between Pairs of Types of Fires (Caused by Pasture Burning, by Lightning, by Negligence, Intentionally, and by Unknown Sources) Within Each Study Area

	Type of Fire				
	Pasture Burning	Lightning	Negligence	Intentional	Unknown
North					
Pasture burning	—	**0.277**	**0.009**	**0.200**	**0.001**
Lightning	<0.001	—	**0.001**	**0.024**	**<0.001**
Negligence	<0.001	<0.001	—	**<0.001**	**0.931**
Intentional	<0.001	<0.001	<0.001	—	**<0.001**
Unknown	<0.001	<0.001	0.004	<0.001	—
Center					
Pasture burning	—	**0.680**	**0.518**	**0.002**	**0.034**
Lightning	<0.001	—	**0.744**	**0.289**	**0.671**
Negligence	0.001	<0.001	—	**<0.001**	**0.186**
Intentional	0.006	<0.001	<0.001	—	**0.003**
Unknown	0.007	<0.001	0.064	<0.001	—
Levant					
Pasture burning	—	**0.725**	**0.096**	**0.867**	**0.212**
Lightning	0.402	—	**0.204**	**0.406**	**0.239**
Negligence	0.318	0.005	—	**<0.001**	**0.087**
Intentional	<0.001	<0.001	<0.001	—	**<0.001**
Unknown	0.029	<0.001	0.014	<0.001	—

ignition, except in the lightning ignited fires, in which case the North and Levant had similar size frequency distributions (Figure 20-7; Table 20-2). Within a given area, the size frequency distribution of fires originated by the various causes were all different from each other in the North. In the Center, all fires except those caused by negligence and unknown causes were also different from each other. On the contrary, in the Levant only the size frequency distribution of the intentional fires was significantly different from all the other causes, whereas the other four causes had various degrees of non-significant differences among each other, with lightning fires being on one extreme, having the lowest median size, and the intentional fires on the other extreme (Figure 20-7; Table 20-3).

When the comparison between the three areas are made using only fires ≥5 ha or ≥50 ha, the patterns of differences changed (Table 20-2). For all fires, the difference between the North and the two Mediterranean zones was maintained, not so the differences between the Center and Levant which at the ≥5 ha size were already similar. When specific types of fires were considered, in general, differences in the size distribution between the North and the two Mediterranean areas were maintained for fires caused intentionally, by unknown sources and, to a lesser extent, by negligence. For fires caused by lightning and by pasture burning only the differences between the North and Levant tended to prevail, not so the differences between the North and the Center. In general, differences between the Center and Levant tended no to be significantly different except in the case of intentional fires.

When similar analysis were performed comparing the various types of fires within a given area the results also changed (Table 20-3). For fires ≥5 ha, out of ten comparisons made, in six occasions differences were significant in the North, in four occasions in the Center and in two occasions in the Levant. At the ≥50 ha size all comparisons rendered non-significant differences at $P \leq 0.01$ (not shown).

Fire size inequalities, as determined by the proportion of area burned by a given proportion of the largest fires, was smaller in the North and larger in the Center and Levant, the last two areas being closer together (Figure 20-8). For the fifteen year study period, the percentage of total area burned "$EP(p)$" for a given proportion "p" of the largest fires was smallest in the North and largest in the Levant. Hence, for all the fires occurred during the 15 year study period, 10% of the largest fires affected ca. 70% of the total area burned in the North, nearly 90% in the Center and still a little more in the Levant. Additionally, there was a considerable degree of variation from year to year for the three study areas, as can be seen for the maximum (more convex) and minimum (closer to a straight line) curves obtained (Figure 20-9). The North exhibited a smaller range of variation than the Center and the Levant.

The pattern of fire size inequalities caused by different types of fires was similar to the general one (Figure 20-8B–F). However, within the same area there were differences among the various types of fires. Thus, fires caused by lightning tended to be the most extreme, with a small percentage of the largest fires affecting a larger percentage of the surface burned. It is noteworthy that in the Center and more even so in the Levant, 10% of the largest lightning-ignited fires accounted for more than 95% of the total area burned by fires originating from this source. On the other hand, fires caused by pasture burning or by arsonists were at the other extreme, producing a pattern of lesser size inequalities, i.e., greater fire size homogeneity. Fires caused by negligence were closest to those caused by lightning, and fires caused by unknown sources were closer to intentional or pasture burning fires. In general, differences among the various types of fires were greater in the Levant and Center areas, whereas in the North, with the exception of the lightning ignited fires, all other fires had similar fire size inequalities, with all curves overlapping over most of the range of values.

Among the various models tested, the truncated Pareto distribution gave the best fit to the observed data (Figure 20-10). As expected from

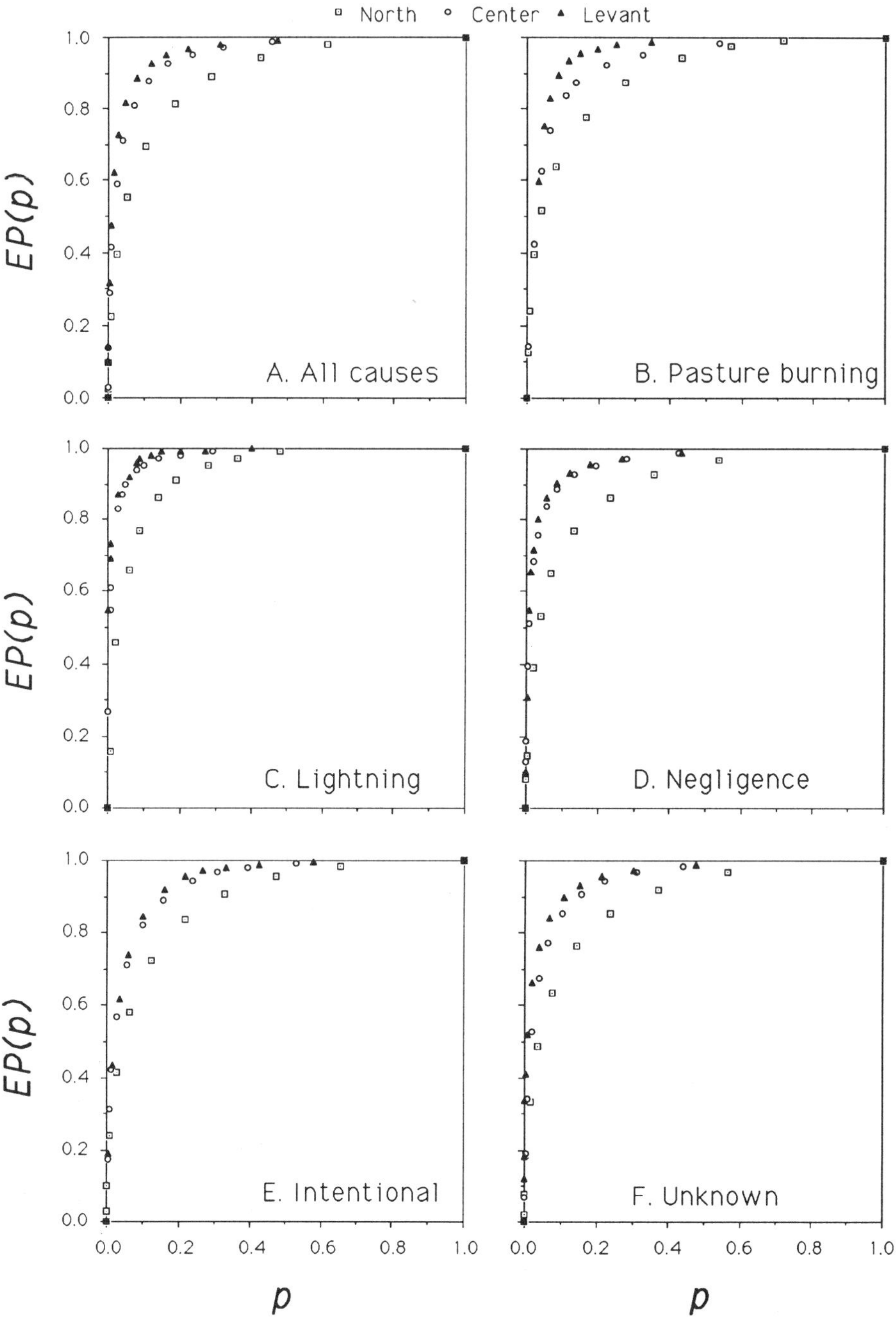

Figure 20-8. Plot of the extreme proportion $[Ep(p)]$ against p (a given proportion of the largest fires) for fires during 1974–1988 in three study areas (North, Center and Levant) in Spain. (A) all fires; (B) fires caused by pasture burning; (C) fires caused by lightning; (D) fires caused by negligence; (E) fires caused intentionally; (F) fires caused by unknown sources.

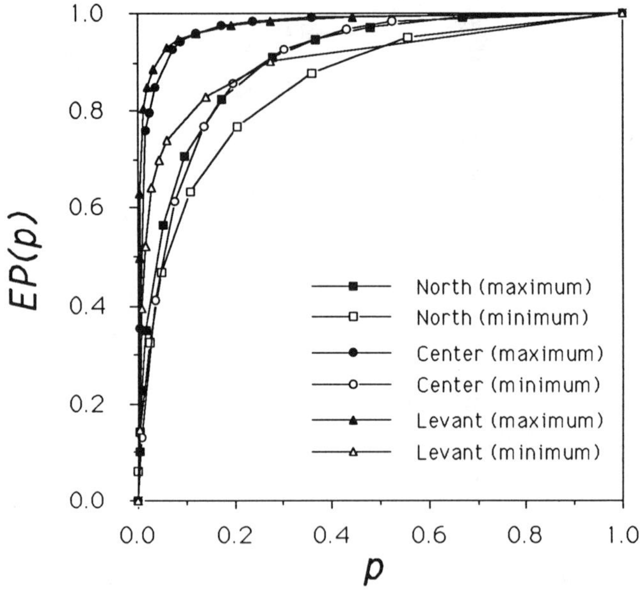

Figure 20-9. Plot of the maximum (more convex) and minimum (more straight) extreme proportion [*Ep* (*p*)] against *p* (a given proportion of the largest fires) for fires during 1974–1988 in three study areas (North, Center and Levant) for fires caused by all sources.

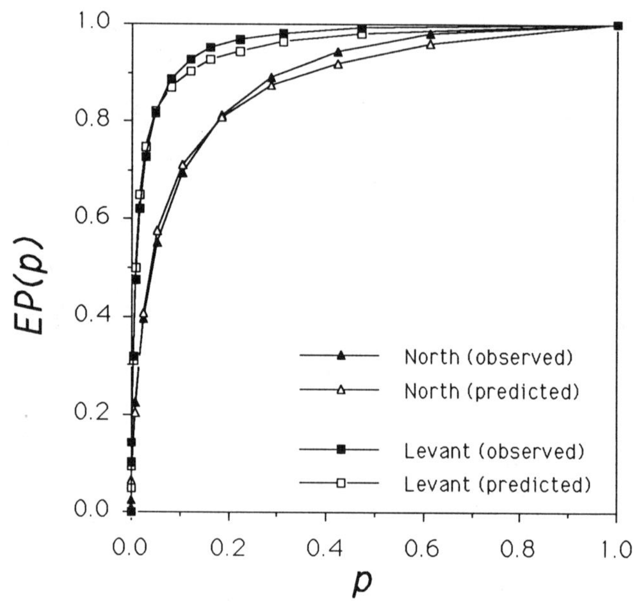

Figure 20-10. Plot of the extreme proportion [*Ep*(*p*)] against *p* for observed and predicted values obtained by fitting the truncated Pareto function to fires caused by all sources in the North and Levant study areas. The values obtained for the α parameter were 0.69, 0.61 and 0.63 and for the b parameter 450, 3076 and 6754 for the North, Center and Levant, respectively.

the curves described earlier, the mean value of the α parameter for fires caused by all sources decreased from the North to the Mediterraneann areas, the Center having the lowest value. On the other hand, parameter *b* increased from the North to the Levant.

The proportion of largest fires necessary to burn 90% of the area (P_{90}), as obtained from the functions generated by fitting the truncated Pareto function to the observed yearly data, shows that all three areas had significantly different P_{90} values for all fires combined, the North having the largest value, the Levant the smallest one (Figure 20-11). The differences between the Levant and Center were small. The general pattern was maintained for fires caused by negligence or intentionally, in the first case the two Mediterranean areas had non-significantly different P_{90} values. It is noteworthy that P_{90} tended to be larger for intentional fires than for fires caused by negligence although differences between the two types of fires within each area were never significant ($P \gg 0.01$).

The correlation between the various values of P_i (from P_{75} to P_{95}, in increments of 5%) and the temperature variables showed that no significant relationships were obtained for the Levant and any of the temperature variables used. In the Center, P_i and the number of days above 30°C (*DM30*) tended to reflect a pattern of significant relationships at P_{75} or lower for fires of all causes together (Figure 20-12). No other significant relationship was obtained for the other types of fires and variables.

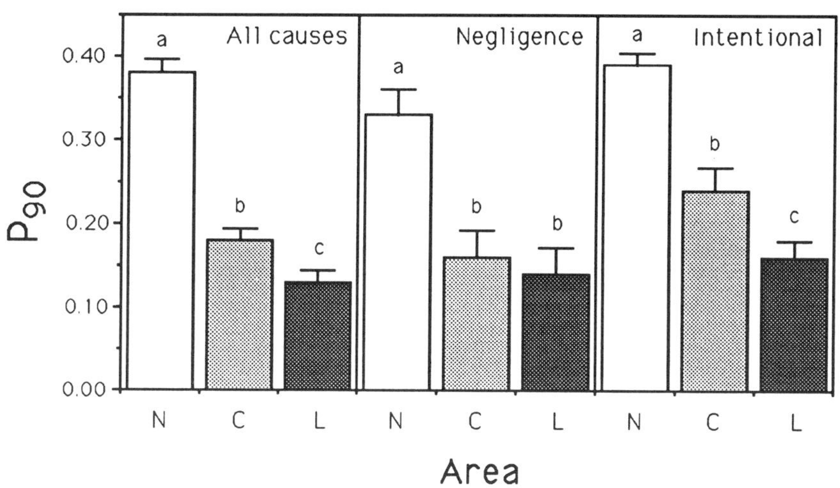

Figure 20-11. Mean (±1 SE) percentage of the largest fires needed to affect 90% of all the yearly surface burned (P_{90}) in the three study areas and for fires caused by all causes, by negligence and intentionally. For a given type of fire, areas sharing the same letter are not significantly different ($P > 0.05$) according to a t-test for paired samples. (N = North; C = Center; L = Levant.)

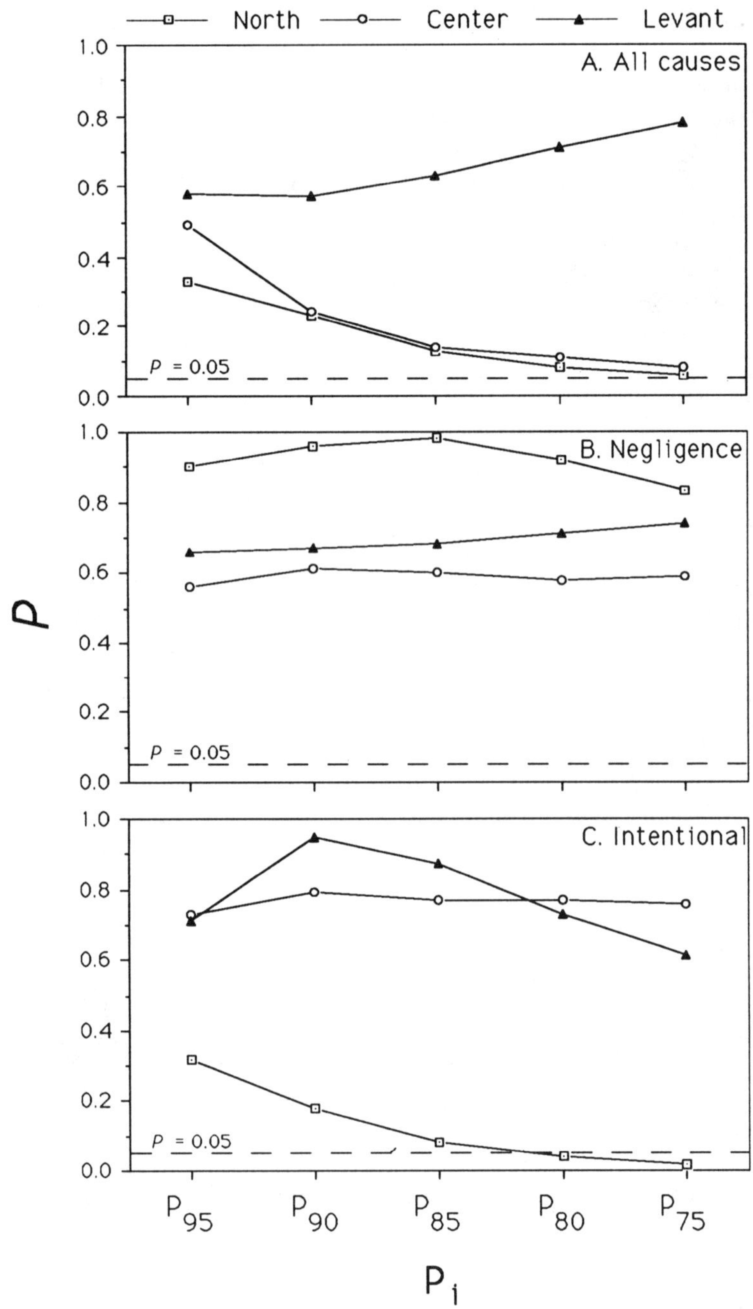

Figure 20-12. Significance (P value) of the correlation between various values of Pi (from P_{75} to P_{95}) and the number of days above 30°C for the three study areas for (A) fires caused by all sources; (B) fires caused by negligence; (C) fires caused intentionally. Points at or below the dashed line (P = 0.05) would render a significant correlation.

Similar patterns between P_i and *DM30* were obtained for the North, for both, all fires and intentional fires (Figure 20-12). In all these cases, a negative relationship was obtained, indicating that the greater the number of days above 30°C, the smaller the Pi value, this is, a smaller fraction of the largest fires was needed to account for similar percentage of the total area burned. The coefficients of determination were small (the maximum r^2 value obtained was 0.36 for intentional fires and P_{75}). Also in the North, a similar pattern, although less powerful, was obtained for the mean maximum temperature of the month and both, all fires ($P = 0.096$ for P_{75}) and intentional fires ($P = 0.033$ for P_{75}) (not shown).

On the other hand, also in the North a pattern of increasing correlations between P_i and mean temperature of the month (*Tm*) or the number of days with minimum temperatures ≥20°C (*Dm20*) from P_{75} to P_{95} was obtained ($P = 0.05$; $r = +0.58$ for *Tm* and P_{90}; and $P = 0.05$, $r = +0.58$; for *Dm20* and P_{90}) (not shown). This relationships indicate that the warmer it was the mean temperature of the month during the year, or the higher the number of days with minimum temperatures ≥20°C, the largest the percentage of the largest fires that it was required to have a given (90%) of the area burned from fires ignited by negligence.

No significant correlations ($P > 0.05$), or clear correlations patterns were obtained for precipitation variables and any type of fires in any of the three areas studied (not shown).

Discussion

The fire period, either for number of fires or surface burned, conformed well to our predictions in the Levant and Center areas, but not for the North area. Hence, the period of peak fire activity was longer and more variable from year to year in the Levant than in the Center. The area burned was more concentrated in time than the total number of fires. Intentional fires were less widespread in time than natural fires or fires caused by negligence. Fires caused by pasture burning were very widespread in time, although in the Levant not much more than fires started by negligence or from unknown sources. These trends were entirely different in the North. The period of fire was long, nearly as long as in the Levant; the area burned was less concentrated in time than the number of fires; intentional fires were more widespread in time than natural fires and similar to fires started by negligence. The only prediction met was that fires from pasture burning were the most widespread in time, with fires virtually all year round.

The differences among areas and particularly the sharp contrast between the North and the other two areas, in an entirely unpredictable manner, indicates the strong influence that man may have upon the patterns of fire occurrence during the year. Not surprisingly, the area with greatest incidence of fires caused intentionally was the most unpredictable. Even fires caused by negligence responded differently than expected, which suggests that negligence and accidental fires may not be the same in all areas.

These findings point to an important caveat for predicting the effects of global change upon fire occurrence, namely, that the role of man should not be forgotten. How man may respond to the predicted change may strongly influence overall fire occurrence. Changes in land use may affect his behavior and lead to most important changes in fire. For instance, in the Levant and Center, fires caused intentionally occur already at the period of higher fire risk, but not in the North, where intentional fires and fires from unknown causes follow more closely the pattern of fires caused by pasture burning. Should these be related, as the actual patterns suggest, then changes in land use may have important consequences. The pattern of sizes of fires caused by lightning indicates that fires could be more catastrophic if they occurred at different times of the year and/or in different places. Therefore, a change in the behavior of arsonists could very well alter the overall pattern of fire incidence. How much global change may be needed to modify man's behavior is purely speculative. Our earlier work indicated that arsonists in the North and Center were very sensitive to warm years, increasing the number of fires and surface burned when maximum temperatures were higher (Vázquez and Moreno, 1993).

The significant negative relationship detected between PFO_{50} for number of fires and surface burned and days with high maximum temperatures in the North indicates that in such warm years most burning occurred in fewer days. Furthermore, in such warm years fire activity is very much increased, as shown by our earlier work (Vázquez and Moreno, 1993). Therefore, the fire season was sensitive not so much to warming in general (mean temperatures) but to how it occurred (warm days) through the year. This points to a complex interaction between climate change and fire activity. Whether this relationship will prevail is impossible to predict, but fire services should be alert to changes not only in overall activity of arsonists but also to changes in their time of activity and further sensitivity to periods of warm weather. This could give an indication of future trends and risks.

Our results show that the three study areas were very different in the pattern of fire sizes. This was reflected both in the cumulative size frequency distribution and in the extreme proportion function. As predicted, fire size inequalities were larger in the warmer and drier Levant than in the Center, and in these two, in turn, larger than in the North. Additionally, size inequalities varied importantly through the years in any of the three areas. The variability tended to be greater where size inequalities were also greater, i.e., in the warmer and drier Levant. Assuming that the differences in fire size inequalities and in interannual variability among the three study areas reflect responses to the climatological gradient analyzed, these results would point to important consequences for the predicted climate change. The warmer and drier and area will become, the greater the fire size inequality, i.e., the greater the overall weight of fires of large size, undoubtedly the most damaging. Since fire area may also be positively related with fire intensity, changes in fire size may also lead to other critical changes in many ecosystem dynamics (Moreno and Oechel, 1994). Nevertheless, one should be cautious because fire size inequalities may also be sensitive to differences in the landscape related to intrinsic structural differences in the territory (i.e., topography), land use, fire fighting capability, management, etc. (Minnich, 1983, 1989; Turner and Romme, 1994). Because these would be most difficult to quantify for any of the three areas, we cannot exclude that the trend detected may also in part be due to these factors.

The pattern of relationships between P_i and the meteorological variables confirmed our prediction in the sense that, if found, significant relationships would be obtained in the North, rather than in the Levant. The overall relationships were not very powerful, which indicates that many other factors come into play. Yet it is the pattern that it is indicative of a closer relationships between mean weather variables and fire size patterns. One point of interest is that it was the first few largest fires (affecting about 75% of all surface burned) which were best related to the temperature variables. Precipitation related variables did not seem to be related at all to the fire size patterns even though they had a significant relationships with overall fire occurrence in the Levant (Vázquez and Moreno, 1993). Our results confirm that areas of mild weather (North) may be sensitive to small variations in weather from year to year as far as fire activity is concerned. Paradoxically, such less fire-prone areas may be better indicators than more fire-prone areas for detecting changes in the impact of climate change. On the other hand, the lack of significant relationships between meteorological variables and fire size inequalities in the Levant indicates that variations in mean temperature may not be readily reflected in changes in the size structure of fires, particularly in warmer and drier areas. This could be relevant to understanding the impact of climate change on fire. The structure of fire sizes is very sensitive to extreme events, i.e., to large fires. These may occur under very specific meteorological conditions, which may be sporadic and not necessarily be reflected in mean values. Hence, it seems likely that increases in such extreme conditions may be more meaningful than just changes in mean overall values. Therefore, predictions of future climate change based on mean values may not reflect the range of potential impacts. It has been shown that future climate may be characterized by an increased frequency of extreme meteorological conditions, such as drought periods (Rind et al., 1990) or periods of high temperature (Mearns et al., 1984). Given the strong weight of extreme events (large fires), which is greater the warmer and drier a given area is, it indicates that small changes in the frequency of such events may have a most pronounced effect. A final conclusion might be that mild areas, which as shown from our data may be more predictable in terms of the interactions of weather and fire, as they warm up they may become less predictable in terms of fire size structure. This definitely will have important consequences at all socioeconomic and ecological levels.

One further conclusion from our study of the fire size distributions is that the warmer and drier and area the less important are the various agents in producing different fire size patterns. Our data show that it was in the cooler and wetter North where fires of different type tended to produce fire size frequency distributions more differentiated from each other, even when only large fires (≥50 ha) were considered. It would appear that under more extreme conditions, these conditions override the importance of the small modifications that may be brought about by the source of ignition through changes in the time, place, etc., of the fire, as it would be expected from fires originated by different causes.

On the other hand, if the type of fire is important for determining the overall fire size structure, particularly in cooler and wetter areas, its significance with respect to the overall fire size inequalities generated varies also depending on the area. Hence, fire size inequalities were usually different depending on the source of ignition, except in the North where all fires produced a similar fire size extreme proportion function, excluding the lightning caused fires. The fact that the greatest fire size inequalities were obtained for lightning-caused fires and then by fires caused by negligence indicates that the potential for a greater role of extreme events is higher than it is today. As we had originally hypothesized, arsonist-caused fires would produce lesser fire size inequalities than fires originated more by chance, indicating that arsonists act within restricted sets of conditions or at sites that may not include the most extreme conditions. A similar finding was obtained in southern California by Mees (1991) when trying to determine whether arsonist activity was related to the most severe conditions.

Acknowledgments. We thank the Instituto Nacional para la Conservación de la Naturaleza (ICONA) for supporting this research.

References

Baisan CH, Swetnam TW. 1990. Fire history on a desert mountain range: Rincon Wilderness, Arizona, USA. Can J For Res 20:1559–1569.

Beer T, Gill AM, Moore PHR. 1988. Australian bushfire danger under changing climate regimes. In: Greenhouse 87, CSIRO, Victoria, Australia, pp. 421–427.

Brenner J. 1991. Southern oscillation anomalies and their relationship to wildfire activity in Florida. Int J Wildland Fire 1:73–78.

Clark JS. 1988. Effect of climate change on fire regimes in northwestern Minnesota. Nature 334:233–235.

Flannigan MD, Van Wagner CE. 1991. Climate change and wildfire in Canada. Can J For Res 21:66–72.

Johnson EA, Larsen CPS. 1991. Climatically induced change in fire frequency in the southern Canadian Rockies. Ecology 72:194–201.
Lieth H, Whittaker RH. 1975. Primary Productivity of the Biosphere. Ecological Studies 14. Springer-Verlag, Berlin, p. 339.
Mearns LO, Katz RW, Schneider SH. 1984. Extreme high-temperature events: changes in their probabilities with changes in mean temperature. J Clim Appl Meteorol 23:1601–1613.
Mees R. 1991. Is arson associated with severe fire weather in Southern California? Int J Wildland Fire 1:97–100.
Meyer GA, Wells SG, Balling RC, Timothy Jull AJ. 1992. Response of alluvial systems to fire and climate change in Yellowstone National Park. Nature 357:147–150.
Minnich RA. 1983. Fire measures in Southern California and Northern California. Science 219:1287–1294.
Minnich RA. 1989. Chaparral fire history in San Diego County and adjacent northern Baja California: An evaluation of natural fire regimes and the effects of suppression management. In: Keeley SC (ed) The California Chaparral. Paradigms Reexamined. Natural History Museum of Los Angeles County, Science Series No. 34:37–47.
Moreno JM, Oechel WC. 1994. Fire intensity as a determinant factor of postfire plant recovery in southern California chaparral, pp. 26–45. In: Moreno JM, WC Oechel (eds) The Role of Fire in Mediterranean-Type Ecosystems. Springer-Verlag, Ecological Studies 107, New York.
Moreno JM, Pineda FD, Rivas-Martínez S. 1990. Climate and vegetation at the Eurosiberian-Mediterranean boundary in the Iberian Peninsula. J Veg Sci 1:233–244.
Mutch RW. 1970. Wildland fires and ecosystems: a hypothesis. Ecology 51: 1046–1052.
Olson JS. 1981. Carbon balance in relation to fire regimes. Proceedings of the Conference on Fire Regimes and Ecosystem Properties. USDA For Serv Gen Tech Rep WO-26:327–378.
Philpot CW. 1977. Vegetative features as determinants of fire frequency and intensity. In: Mooney HA, Conrad CE (eds) Symposium on the Environmental Consequences of Fire and Fuel Management in Mediteranean Ecosystems. USDA For Serv Gen Tech Rep WO-3:12–16.
Rind D, Goldberg R, Hansen J, Rosenzweiz C, Ruedy R. 1990. Potential evapotranspiration and the likelihood of future drought. J Geophys Res 95:9983–10004.
Rivas-Martínez S. 1981. Les étages bioclimatiques de la végétation de la Peninsule Ibérique. Anales Jard Bot 37:251–268.
Rivas-Martínez S. 1987. Mapa de las Series de Vegetación de España. Ministerio de Agricultura, Pesca y Alimentación, Madrid.
Rothermel RC. 1983. How to predict the spread and intensity of forest and range fires. USDA For Serv Gen Tech Rep INT-143.
Rundel PW. 1981. Structural and chemical components of flammability. In: Mooney HA, Bonnicksen TM, Christensen NL, Lotan JE, Reiners WA (tech coord) Fire Regimes and Ecosystem Properties. Proceedings of the Conference. USDA For Serv Gen Tech Rep WO-26:183–207.
Strauss D, Bednar L, Mees R. 1989. Do one percent of forest fires cause ninety-nine percent of the damage? Forest Sci 35:319–328.
Swetnam TW, Betancourt JL. 1990. Fire-southern oscillation relations in the southewestern United States. Science 249:1017–1120.
Torn MS, Fried JS. 1992. Predicting the impacts of global warming on wildland fire. Clim Change 21:257–274.

Turner MG, Romme WH. 1994. Landscape dynamics in crown fire ecosystems. Landscape Ecol 9:59–77.

Vázquez A, Moreno JM. 1993. Sensitivity of fire occurrence to meteorological variables in Mediterranean and Atlantic areas of Spain. Landscape Urban Plan 24:129–142.

Villa SD, Guerra SJ, Corres AR. 1985. Análisis estadístico de la pluviometría en la Península Ibérica. Instituto Nacional de Meteorología, A-132, Madrid.

Woodward FI. 1986. Climate and Plant Distribution. Cambridge University Press, Cambridge, UK.

21. Sensitivity of Fire Regime in Chaparral Ecosystems to Climate Change

Frank W. Davis and Joel Michaelsen

The fire regime of an area is characterized by the type, intensity, size, return interval, and spatial pattern of fires that have occurred there over recent history. Fire regime is an emergent property of landscapes, expressing the interaction of climate, vegetation, topography, and land management. Thus the potential impact of climatic change on chaparral fire regime must be analyzed within a specific regional context. Ideally, the analysis should also account for the spatial arrangement of topography, vegetation types, ignition sources, and land cover types that serve as fire breaks, because fire regime and fire hazard may vary widely under the same climate, depending on how fire spread is regulated by these and other geographic factors.

In this chapter we analyze recent fire history and climate data from central coastal California in order to better understand the relationship between wildfire and climate in this region. We then employ a spatially explicit fire regime simulation model to study the sensitivity of one particular landscape in central coastal California to possible changes in seasonal temperature and precipitation associated with a doubling of atmospheric CO_2.

Several studies have analyzed potential impacts of global climate change on fire regime in Mediterranean ecosystems. Malanson and Westman (1991), and Westman and Malanson (1992), simulated growth of five dominant shrubs of the Venturan coastal sage scrub association under changing moisture gradients, fuel loads, and associated fire intensity (measured as total energy released). They also modeled interactions between climate change and air pollution effects. General circulation model (GCM) predictions of increased rainfall over southern California produced increased fuel loads and fire intensity. The higher fire intensity resulted in significant changes in community composition because it favored shrubs regenerating by seed over those regenerating primarily by resprouting. Taken together, ozone stress, increased rainfall, and increased fire intensity led to the elimination of all species under short fire return interval (10 years), and to significant changes in composition under 20- and 30-year fire return intervals.

Torn and Fried (1992) simulated the effect of global climate change on wildfire hazard in northern California by applying a mathematical model of fire intensity and spread to historical and predicted daily weather data. Model outputs were coupled to a fire-fighting module that predicted the kinds and cost of fire suppression activities that would be required to control the burns under different fuel conditions and population densities. Use of daily weather data rather than monthly or seasonal averages allowed them to capture the extreme meteorological conditions of high winds and low humidities under which most burning occurred in that area. Although results varied considerably among different GCM scenarios and fuel types, fire danger (potential rate of spread) was generally predicted to increase under a $2CO_2$ climate due to increased wind speeds and decreased relative humidities. The increase in spread rate, intensity, and area burned was large for grassland fuels, small for short and tall brush fuels, and negligible for redwood fuels.

We have studied the sensitivity of chaparral ecosystems in southern California to global climate change using REFIRES (Burrows, 1988; Davis and Burrows, 1993), a spatial simulation model that is designed to analyze the relationships between fire history and vegetation pattern in heterogeneous landscapes. Like the FINICS model used by Westman and Malanson (1992), REFIRES tracks fuel development and fire intensity over many fire cycles. Unlike FINICS, REFIRES lacks an autecological framework linking fire intensity to community composition. REFIRES is similar to the CCFMS model of Torn and Fried (1992) in that fire ignition and spread conditions are calculated from daily meteorological data using mathematical models of fire behavior. REFIRES has the added feature that fire behavior depends on the spatial distribution of fuels over the landscape, which are themselves a function of fire history. We believe that this is especially useful in understanding the consequences of climate change on fire-prone ecosystems over more than a single fire cycle.

Like the previous modeling efforts cited above, ours builds from the Rothermel fire model. This model has many limitations for predicting chaparral fire behavior. The assumption of a uniformly arranged fuel bed is rarely met in the field. The model predicts little to no spread in fuel beds consisting of only live vegetation, when in fact fire will burn through such fuels under high winds (P. Riggan, personal communication). Perhaps most importantly, the model was not formulated to predict fire spread for large fires burning under conditions of high winds and extreme atmospheric turbulence that characterize the modern chaparral fire regime.

Recognizing these limitations, not to mention the uncertainty associated with projecting future climate under increasing atmospheric CO_2 (see below), we have designed our modeling experiment to test how sensitive the fire regime in a heterogeneous California landscape could be to climatic change. Accordingly, we have parameterized REFIRES to span a reasonable range of future climate and fuel conditions. We will show that both actual and simulated fire histories are very sensitive to climatic variation, but that they are sensitive for somewhat different reasons. By combining empirical and theoretical analyses, we hope to demonstrate the potentially large consequences of global climate change on fire regime in chaparral ecosystems, but also to highlight some major sources of uncertainty currently limiting our predictive capabilities.

Existing Fire Regime

We analyzed fire history and climate data from the Los Padres National Forest (LPNF), California (Figure 21-1). The LPNF consists of 709,725 ha of the central Coast Ranges and the Transverse Ranges, stretching from Monterey (36.5°N, 122°W) to Castaic (34.5°N, 118.5°W). It is divided into two segments, the main division and the Monterey division. The terrain is rugged and largely inaccessible or accessible only via unpaved service roads. Elevations range from near sea level to over 1,750 m, with most LPNF lands situated between 200 and 1,200 m. Annual precipitation in the main division ranges from 250 mm to 1,000 mm or more at higher elevations, with 95% falling between November and March. The Monterey division is slightly wetter, receiving 500–1,500 cm of annual rainfall. Roughly 68% of the LPNF is shrublands, predominantly chaparral, but with coastal sage scrub on more xeric sites. Wildfires have been actively suppressed in LPNF since the early 1900s (Minnich, 1983), even in designated Wilderness Areas. However, suppression efforts became much more effective after 1950, when large air tankers were first employed routinely (Fritts Cahill, personal communication).

Fire history data were compiled by Fritts Cahill of the Los Padres National Forest from field notes, maps, and air photos. His fire history maps and database cover the period 1911–1992 and provide information on fire location and perimeter, ignition source, date of ignition, and area burned inside and outside the LPNF.

Based on LPNF fire history data, the modern fire regime of the LPNF can be summarized as follows:

Figure 21-1. Location map of the Los Padres National Forest.

Most Fires Are Started by Humans

Since 1950, 62 of 70 (89%) large (>400 ha) fires that occurred in LPNF were human ignited. Lightning is not as frequent here as in other chaparral regions in California (Keeley, 1982), but lightning ignition is still significant. Of the total burned area, 24.5% was burned in lightning fires, and the largest burn during the period was a 72,010 ha lightning event in 1977 in the Monterey division.

The comparison of ignitions from humans versus lightning is not straightforward, because the two differ systematically in their location and weather conditions. Human ignitions are concentrated along roads and campgrounds, and are often timed by arsonists to coincide with extreme weather conditions, whereas lightning ignitions cluster on peaks and along ridges and usually accompany milder weather in which wildfire is more easily contained. In any case, unless climate change includes a very large increase in lightning events in the region, the ignition regime is likely to remain dominated by humans.

Most Burning Occurs in Infrequent, Large Wildfires

Based on fossil charcoal records, the prehistoric fire regime of this region was characterized by relatively long fire return intervals of 60–100 years. Although fire size cannot be reconstructed with certainty from the

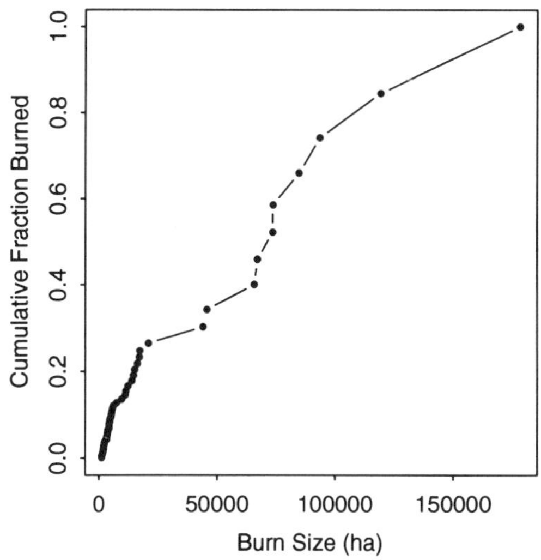

Figure 21-2. Cumulative fraction of total area burned in the Los Padres National Forest between 1950 and 1990 as a function of fire size, for burns greater than 400 hectares (unpublished data from F. Cahill, Los Padres National Forest).

charcoal records, very large fires were apparently not uncommon (Byrne et al., 1977).

Minnich (1983, 1989) asserts that in southern California fire suppression has reduced fire frequency, but that this has led to increased fire size. Certainly, low fire frequency allows older chaparral stands to develop extensive, high-mass fuels. An estimated 48% of chaparral in the Los Padres is greater than 50 years old, while only 24% is younger than 10 years. Between 1950 and 1991, fires in or adjacent to the LPNF consumed 465,898 ha. Of this area, 95% was in chaparral, and 96.7% of the total area burned in events exceeding 400 ha (1,000 acres). Perhaps more significantly, six burns accounted for nearly half of the total burned area during the period (Figure 21-2).

Nearly All Significant Fires Occur Between June and October During Extreme Summer Heat Waves and/or Under Strong Santa Ana Conditions

Large burns were nearly always initiated during extreme summer heat waves or during warm, very windy Santa Ana conditions that are most frequent between September and December. A similar relationship of weather to burning pattern occurs in other chaparral landscapes to the south (e.g., Radtke et al., 1977; Minnich, 1989). Under these extreme conditions, the probability of fuel ignition and the rate of fire spread both increase dramatically, defying efforts at fire suppression. The extreme

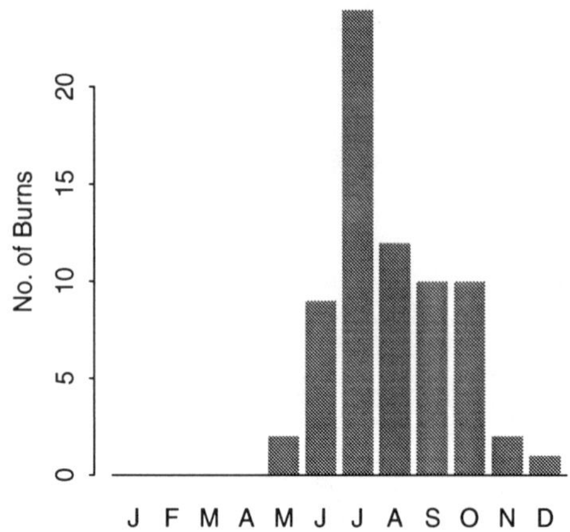

Figure 21-3. Monthly totals for area burned in the Los Padres National Forest, 1950–1991.

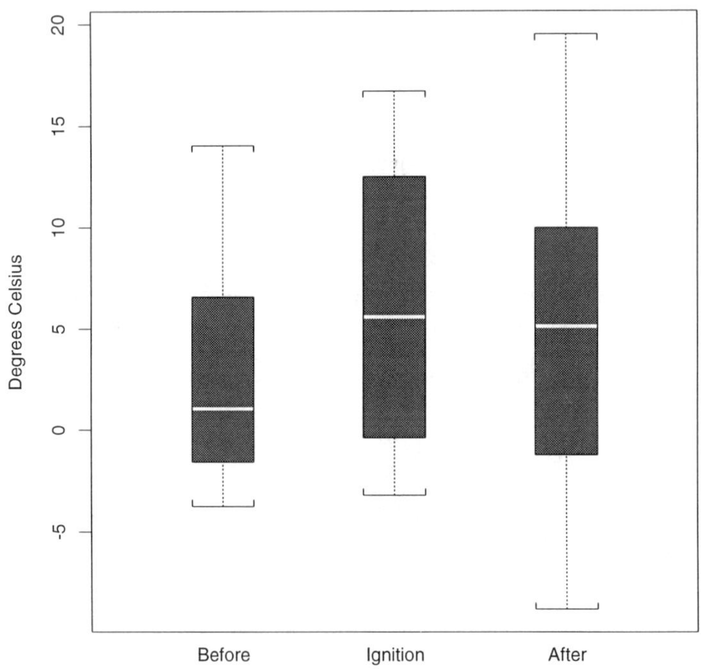

Figure 21-4. Boxplots summarizing the statistical distribution of daily maximum temperatures recorded at Santa Barbara, California on days when large (>400 ha) fires ignited in the Los Padres National Forest versus the previous or following days. Boxplots show the median, quartiles, and range of the data.

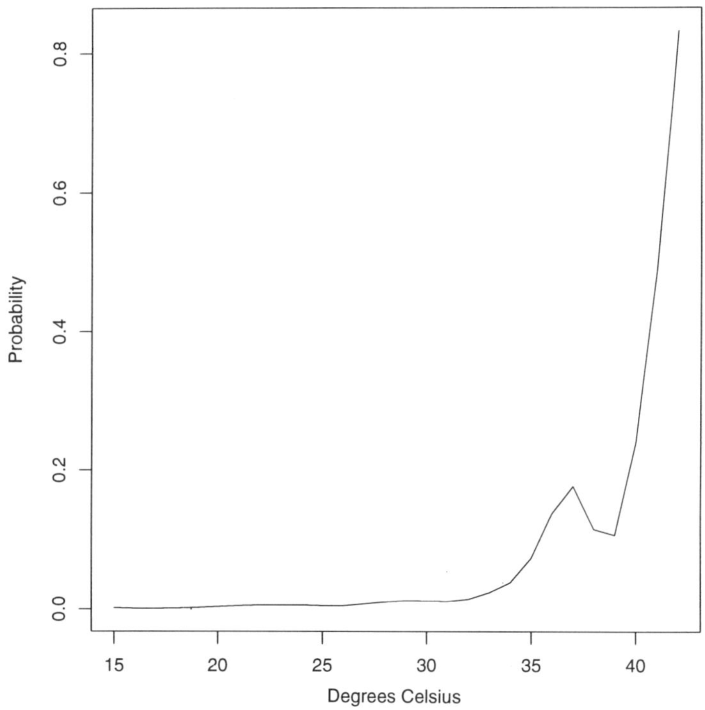

Figure 21-5. Probability of a large fire in the Los Padres National Forest as a function of daily maximum temperature at Santa Barbara, California.

conditions do not have to persist for the duration of a fire event, only long enough for the fire to grow too large for suppression forces to contain, at which point the fire may spread at lower rates under more moderate weather conditions (P. Riggan, personal communication).

Practically all large wildfires occurred between June and October (Figure 21-3). These events show a strong relationship to daily temperature data from the Santa Barbara airport, which, despite its coastal location, is strongly correlated with more interior localities during Santa Ana conditions when warm, dry air flows predominantly from the northeast. In the LPNF, 45 of 70 (64%) of large fires occurred during unusually warm periods (daily maximum temperatures at the airport greater than 30°C). Thirty-two of 70 (46%) occurred during Santa Anna conditions. Temperatures on the days when fires ignited were generally 3–5°C warmer than the monthly mean, and these conditions developed very suddenly (Figure 21-4). The frequency of such days is not correlated with average temperatures during the fire season. The probability of a large wildfire occurring somewhere in the main division of LPNF is close to zero on days in which maximum daily temperatures recorded at Santa Barbara are less than 25°C, and increases dramatically with increasing temperature for days exceeding 30°C (Figure 21-5).

Large Fires Are Much More Likely to Occur in Years with Very Low Spring Precipitation

Literature on chaparral combustibility has tended to focus on chamise (*Adenostoma fasciculatum*), which is readily ignited and can propagate fire under more moderate weather conditions than most chaparral species because of its relatively high proportion of fine twigs, high fuel surface-to-volume ratios, large dead fuel fraction, low live fuel moisture content, and large fraction of volatile chemicals in the foliage (e.g., Muller et al., 1968; Rothermel and Philpot, 1973; Hanes, 1971). However, many chaparral dominants of the central coast, including *Ceanothus megacarpus*, *C. spinosus*, *Quercus berberidifolia*, and *Arctostaphylos* spp., are coarser fuels with a low dead fuel fraction and lower potential velocities of combustion, and thus are much less likely to ignite and carry a fire than chamise under normal conditions (e.g., Riggan et al., 1988). Under dry, windy conditions, however, the same species yield very intense burns (Riggan et al., 1988). This relationship may explain in part the extreme sensitivity of wildfire hazard in the region to late winter and spring rainfall (Figure 21-6). Between 1915 and 1991, spring (March–May) rainfall at Santa Barbara averaged 114.8 mm, only 26% of the annual total. However, 82% of total burned area was contributed by years when spring rainfall was below 100 mm (52% of years), and 48% of the total burned following spring rain of less than 50 mm (21% of years). The correlation(r) between spring and total precipitation is only 0.62.

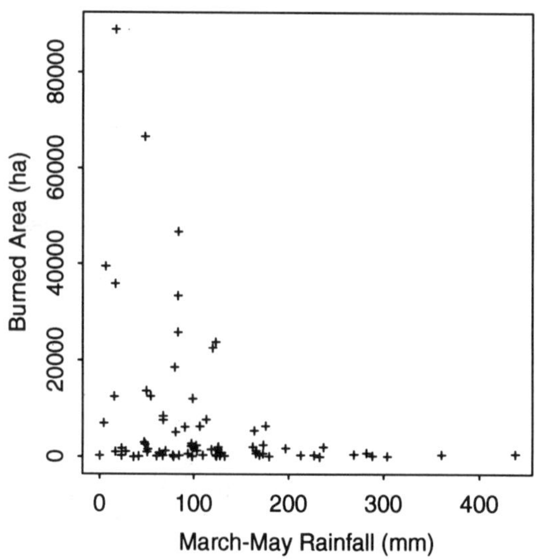

Figure 21-6. Total area burned in the southern division of the Los Padres National Forest as a function of total spring precipitation.

The Fire Interval Is Generally Much Longer Than the Time Required for Recovery of the Shrub Canopy

A rough estimate of fire recurrence interval may be obtained by dividing the total chaparral area of LPNF by the average hectares burned. Based on an estimated chaparral area of 482,607 ha and the assumption that 85% of the total area burned each year is chaparral, the fire recurrence interval for the period 1915–1991 is 58 years, and is 66 years for the period 1950–1991. For the Monterey Division, the return intervals are 55 and 42 years for the total record and modern period, respectively. While admittedly crude, the estimated recurrence interval is three to four times longer than the time required for reestablishment of a closed chaparral canopy. Whether fire can ignite and spread through the canopy is a complex function of dead fuel amount and size distribution, canopy structure and chemistry, and weather conditions. However, it appears that wildfire is generally not fuel limited at present, but is determined instead by the interaction between weather conditions and the effectiveness of suppression efforts.

Modeling the Sensitivity of Chaparral Fire Regime to Climatic Change

Modeling Fire Regime Using REFIRES

The REFIRES model was written by Burrows (1988), who provides a detailed description of the model as well as the source code. A summary of the model's formulation can also be found in Davis and Burrows (1993).

REFIRES operates on a spatial database of hexagonal cells, using maps of topography and potential vegetation together with daily weather and fuel moisture data from the fire season. The model simulates fuel development, fire start, and fire spread for multiple fire events over many fire seasons to develop a simulated fire history. Each fire event involves a possible ignition, an actual ignition, and fire spread. The number of possible ignitions per year is modeled as a Poisson process with the mean number per year set by the user. A possible ignition occurs when an ignition source is randomly placed in the region under weather conditions sampled from the daily weather records. An actual ignition can occur at that location depending probabilistically on weather and fuel conditions (Bradshaw et al., 1983). If actual ignition occurs, fire spread is simulated using the fire spread equations of Rothermel (1972) and Albini (1976) based on local conditions of topography, stand age, and meteorological conditions. One weather day is used per event, but temperature and humidity conditions alternate between daytime and nighttime values for that day to simulate fires that burn over one or more diurnal cycles.

The user defines a rate of spread below which fires are no longer assumed to spread. If the calculated rate of spread is less than this threshold, fires spread with probability equal to the calculated spread rate divided by the threshold spread rate. This threshold can be set low to simulate fire spread in unsuppressed conditions or set high to simulate fire spread under active suppression.

If fire spreads through the cell, the cell's age is set to zero and the fire then continues to spread to neighboring cells with probabilities and rates calculated on a cell-by-cell basis as determined by local fuel conditions, topography, and wind. Based on calculated fire spread rates and distances, the program keeps track of the time elapsed between ignition and when the burn reaches a cell. Daily maximum temperatures and minimum humidities are applied to cells that burn during daylight hours, and minimum temperatures and humidities are applied during nighttime hours. Wind speed and velocity are kept constant over each period.

Fires spread only by contagious diffusion, and not by spotting. Fires cease to spread when they reach the edge of the grid or encounter conditions for which the predicted spread rate falls below the threshold spread rate and spread is stopped probabilistically. Fuel recovery in burned cells is a function of the age and the potential vegetation of the cell and is independent of fire intensity.

After the model is run for a specified number of years, the model outputs fire regime statistics and fire history maps. Statistics include average ignitions per year, number of possible ignitions, number of actual ignitions, number of fires (by cell), fire size distribution, stand age frequency distribution, patch size distribution, and fire recurrence interval distribution.

In previous applications, the model was applied to test the relationships among fire size and recurrence interval, ignition frequency, and vegetation fragmentation for a small area of chamise and mixed chaparral near Lompoc, California. Climate is Mediterranean with a strong maritime influence and prevailing northwesterly winds. The topography is gently rolling uplands interrupted by wide stream valleys with short, steep slopes. Modeling was conducted here as part of longer-term studies on the fire history and ecology of the region (Davis et al., 1988). We operated REFIRES over a 1,865-ha region represented by 20,720 0.09-ha cells. Although relatively small compared to the extensive chaparral areas of the nearby LPNF, this test area is uniform in climate and physiography and is still large enough to study the interactions among ignition frequency, weather, and landscape heterogeneity.

For the present study, we used the same 1,865-ha study region, but modeled it as continuous chamise or mixed chaparral, unbroken by roads or development. Thus stand age and topography were the only sources of spatial variation. Topography was modeled using 30-m digital elevation data from the U.S. Geological Survey. Daily temperature and humidity data from nearby Santa Maria, California for the period 1975–1985 were obtained from the National Oceanic and Atmospheric Administration. A frequency distribution of hourly wind speed and direction was based on 12 years of data from Vandenberg Air Force Base, as summarized by Ogden (1975). Green fuel moisture readings for the period 1977–1990 were obtained from the Los Padres National Forest (Fritts Cahill, unpublished data).

Based on previous experience with the model, the frequency of possible ignitions was fixed at 1.0/yr. To simulate the effect of fire suppression, the threshold rate of spread was set at a relatively high level of 5 m/min. This threshold provides a reasonable approximation to the modern fire regime, based on model runs using a range of values from 1 to 10 m/min.

Modeling Future Climates

Available GCM climate scenarios for California under doubled atmospheric CO_2 provide a very weak basis for projecting future chaparral fire regimes, for several reasons. In general, there is a severe mismatch in spatial scales that makes any regional climate scenarios highly suspect. For example, the Goddard Institute for Space Science (GISS) GCM cell encompassing the study area is roughly equal proportions of ocean and land, and extends from northern Baja California to Sacramento, enclosing a wide range of climates. Very different energy budgets for ocean and land surfaces are a major driving force in current regional climates, particularly during the summer fire season. There is no straightforward way to infer this subgrid scale information from coarse resolution GCM outputs.

Looking more specifically at simulating changing fire regimes, it is apparent that GCMs are not particularly good at modeling the most important factors. Based on the relationships between climate and the existing fire regime discussed above, changes in spring precipitation and frequency of Santa Ana conditions during late summer and fall are the most important variables for predicting future fire regimes. The hydrologic cycles in GCMs are quite crude, and, as a result, they generally do a poor job of modeling soil moisture and precipitation (e.g., Dickinson, 1989; Milly, 1992). Perhaps most importantly, the wind conditions associated with Santa Anas are not likely to be modeled at all well. First, GCMs do not model surface wind patterns with enough accuracy and resolution to relate to local and regional fire hazard (Torn and Fried, 1992). (Consider that there is only a single roughness parameter for the entire GCM cell over California.) Second, the particular wind conditions that dominate the modern fire regime are short-term meteorological extremes that are not strongly related to monthly or seasonal climate variations over larger scales, i.e., the variables which GCMs do predict with some skill.

With these caveats in mind, it is possible to make some statements about the most likely changes in the aspects of climate important for the fire regime of central coastal California. It is unlikely that fire season temperatures will decrease. A temperature increase is reasonably likely, but the magnitude is quite uncertain, with a range of 1° to 4°C spanning the predictions for fire season months from the NASA Goddard Institute for Space Science (GISS), Princeton Geophysical Fluid Dynamics Laboratory (GFDL), and United Kingdom Meteorological Organization (UKMO) models. These increases were added to the daily maximum and minimum temperatures used as input to REFIRES. We adjusted the associated relative humidities to the new temperatures based on a linear regression of humidity versus temperature from local station data ($r^2 = 0.53$). As mentioned above, there is no relationship between average summer temperatures and the frequency of extreme Santa Ana conditions. Thus, we did not change the frequency of temperature outliers in our modeling.

As noted above, changes in precipitation are even more uncertain, but the various GCM results are in general agreement, showing that winter and spring precipitation will increase. For example, applying the GISS predictions to local station data yields a 36% increase in precipitation for the months of November through March, and a 59% increase in precipitation for the total in March and April. The GFDL model predicts a 21% increase in rainfall between November and March, and a 17% increase in the March/April total. The UKMO model predicts increases of 21% and 27%, respectively. These results are not sufficiently certain to rule out precipitation decreases, but we have chosen to assume that it will either remain constant or increase. Any decrease in precipitation will presumably exacerbate changes in fire regime that result from an increase in temperature. For the purposes of this sensitivity analysis, we considered an increase of 30% in spring precipitation, with an associated 5.25% increase in live fuel moisture during the fire season (based on regression of fuel moisture levels versus spring precipitation at Santa Barbara, Figure 21-7).

To study the sensitivity of fire regime to changes in vegetation structure and composition, two different fuel models were used to represent stand level fuel properties as a function of time since fire. The first model, which will be referred to as the Low Biomass model, is based on equations for chamise (*Adenostoma fasciculatum*) chaparral from Rothermel and Philpot (1973) (Figure 21-8a). The High Biomass model describes mixed chaparral, such as might occur on more mesic sites, with roughly twice the biomass of the chamise model (Figure 21-8b).

Both fuel models were run under present conditions and under higher fire season temperatures, with and without increased spring precipitation, to study the interactions among climate factors and site productivity in affecting fire regime. We should note that both fuel models predict a large increase in dead fuel fraction with increasing age. Recent studies

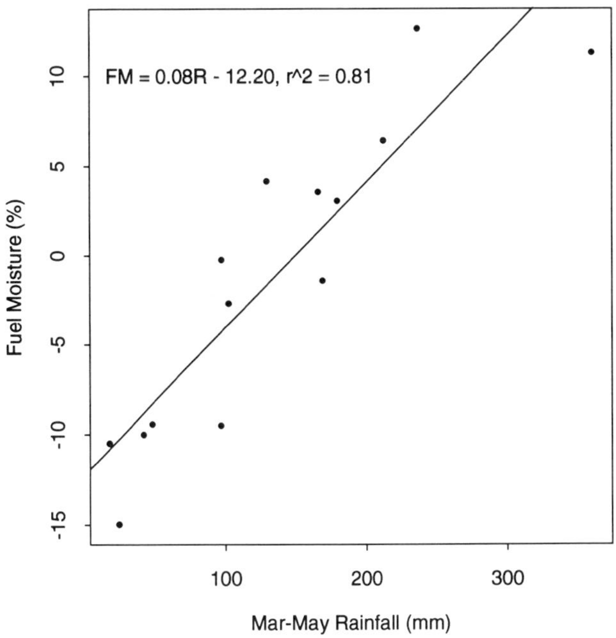

Figure 21-7. Fire season fuel moisture (chamise) for a station near Santa Barbara as a function of spring precipitation at the Santa Barbara airport, 1977–1990. Fuel moisture is expressed as the average departure from monthly means for June–October.

suggest that the actual amount of dead fuel is highly variable among chamise stands (Paysen and Cohen, 1990) and is also generally lower for other chaparral species (Riggan et al., 1988). Thus, the fuel models used here suggest a stronger relationship between age and combustibility than is actually observed under field conditions, and also depict fuels with higher combustibility than observed under many field conditions.

In summary, we have modeled fire regime under the following climate change scenarios:

1. Increased fire season temperatures and associated lower relative humidities, with no change in winds or live fuel moisture status (1° and 4°C increases);
2. 1° or 4°C increases in fire season temperatures, accompanied by a 5.25% increase in live fuel moisture during the fire season;
3. Changes in chaparral productivity under present climate conditions and under scenarios 1 and 2 above.

We have not included GCM predictions of changes in surface wind speed because of the modeling uncertainties mentioned above. Instead, modeling has been conducted using probability distributions for wind speed and

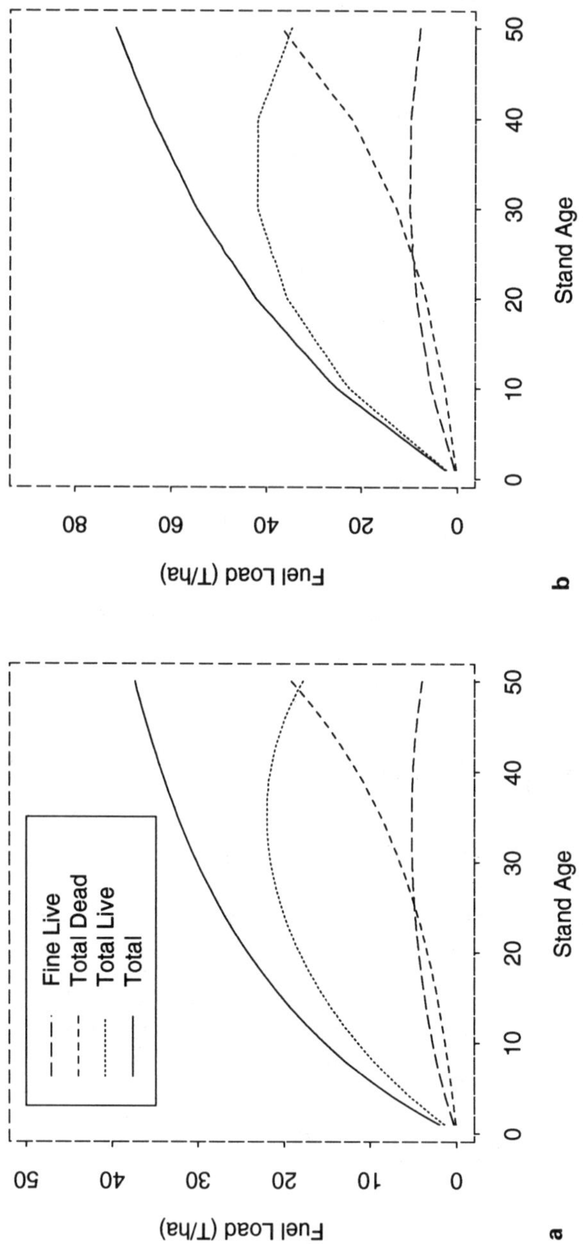

Figure 21-8. Fuel recovery models used in (a) the Low Biomass fire regime simulations, and (b) the High Biomass fire regime simulations.

direction derived from daily local station data. In summary, the model has been run under 10 different scenarios. Model runs of 1,000 years were run 15 times for each combination of climate conditions and fuel model.

Model Results

The modeled fire regime based on current climate data is similar to the actual regime in having many ignitions (229/1,000 years), but in being dominated by relatively infrequent, large burns. The average recurrence interval for the Low Biomass case was 66 years, and the largest 8% of the burns accounted for 50% of total burned acreage. For the High Biomass model, the recurrence interval was significantly shorter (48 years). Once again, the largest 8.3% of burns accounted for half of the total burned area.

The number of possible ignitions was relatively constant among model runs (due to constant frequency of possible ignition), but the ratio of actual ignitions to possible ignition, which depends largely on the amount and moisture content of fine dead fuels, was very sensitive to increased fire season temperatures (Figure 21-9). Compared to the value for current

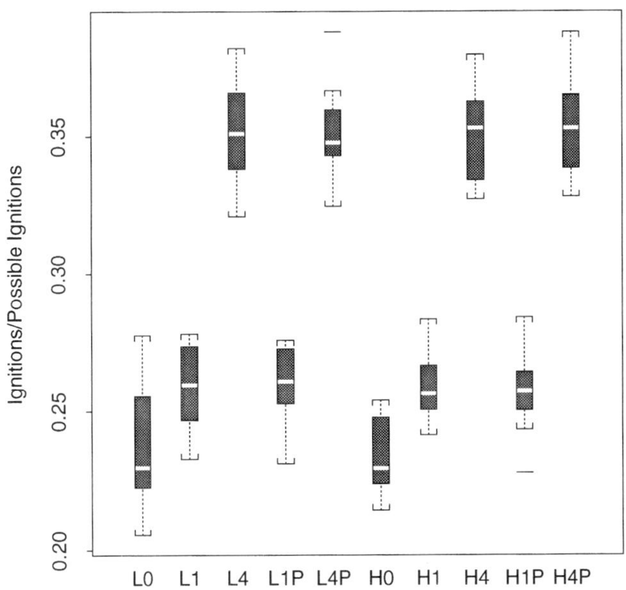

Figure 21-9. Boxplots of the ratio of actual to possible ignitions for 10 different fuel/climate scenarios based on Low (L) or High (H) biomass fuel models, current fire season temperatures (0), or increases of 1° or 4°C, and with current or 30% increase (P) in spring precipitation. For example, L1P is the low-biomass fuel model using climate data with a 1°C temperature increase and increased precipitation. Boxplots show the median, quartiles, and range of 15 1,000-year runs for each scenario.

Table 21-1. Statistical Summary of Relative Changes (%) in Simulated Fire Regime Under Various Future Climate Scenarios Relative to Current Climate Using the Low Biomass Model (L0)

	L0	L1	L4	L1P	L4P	H0	H1	H4	H1P	H4P
Actual/possible ignitions	0	9	49	10	49	−1	10	48	9	51
Fires/ignitions	0	−2	−5	−2	−4	−12	−12	−17	−14	−17
Fires/1,000 yrs	0	7	41	10	42	−11	−2	24	−6	27
Burned cells/ 1,000 yrs	0	9	31	0	19	36	42	58	29	51
Recurrence interval (yrs)	66	61	51	66	55	48	47	42	51	44

The mean fire recurrence interval (yrs) is also shown. Modeled conditions include Low (L) or High (H) biomass, current climate (0), 1° or 4°C increase in mean fire season temperatures with associated lower humidities (1 or 4), and a 30% increase in spring precipitation (P). Percentages are the means of 15 1,000-year model runs.

climate data using the Low Biomass model, this ratio increased by 9% and 49% for 1° and 4°C increases, respectively (Table 21-1). The ratio is not related to the fuel model, nor to increased live fuel moisture levels associated with increased spring precipitation.

For an ignition to evolve into a fire requires a combination of fuel, topography, and weather conditions sufficient to generate relatively high rates of spread. The number of fires per ignition actually decreases with increasing temperatures and increasing site productivity, less so with increasing precipitation (Figure 21-10, Table 21-1). This may seem contrary to the result that both the number of fires and the total area burned increase with increasing temperature (Figures 21-11 and 21-12). However, the decreasing fire recurrence interval under warmer or more productive conditions leads to a decrease in overall landscape combustibility. Thus, although ignitions increase, a smaller fraction of them actually occur on sites with sufficient fuels to promote a fire. This effect is most pronounced for the High Biomass model under a 4°C temperature increase, where the recurrence interval is only 42 years compared to 66 years for the Low Biomass model under present climate, and where fires/ignition decreases by 17% while the area burned increases by 42%.

Fire frequency is equally sensitive and total area burned is more sensitive to increased temperatures for Low Biomass versus High Biomass fuel models. For the less productive chaparral, fire frequency increases by 7% and 41% for 1°C and 4°C temperature increases, compared to 9% and 40% for the more productive case (Figure 21-11). However, total area burned using the Low Biomass model increases by 9% and 31% for 1° and 4°C increases, compared to only 4% and 16% for the High Biomass model (Figure 21-12). This is because increased temperatures have a larger proportional effect on spread rates through low fuel loads than

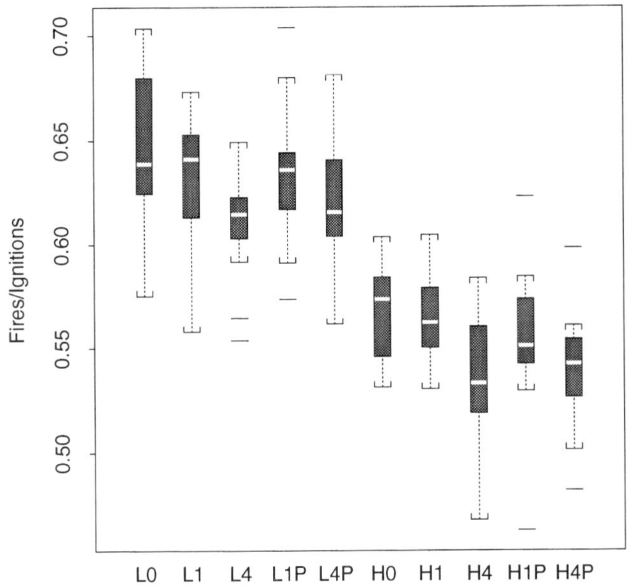

Figure 21-10. Boxplots of the ratio of fires to actual ignitions for the different fuel/climate scenarios. See Figure 21-9 for the abbreviation code and explanation of boxplots.

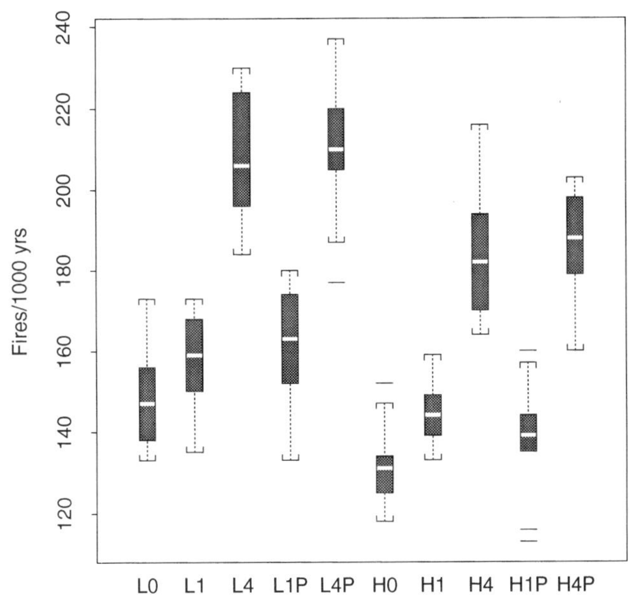

Figure 21-11. Boxplots of the number of fires per 1,000 years under the various fuel/climate scenarios, as explained in Figure 21-9.

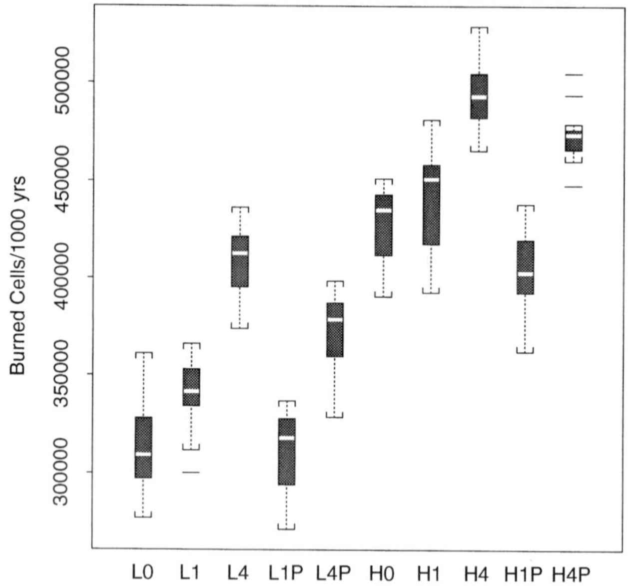

Figure 21-12. Boxplots of the total number of cells burned per 1,000 years, as explained in Figure 21-11.

through high fuel loads, where spread rates are more likely to exceed the suppression threshold even under current conditions.

Calculated rates of spread decrease with increased spring precipitation, which for the Low Biomass model effectively offsets a temperature increase of 1°C in terms of total area burned and fire recurrence interval (Table 21-1). The effect is even more pronounced for the High Biomass model, where the total area burned is predicted to decrease for a 1°C increase and wetter spring compared to the High Biomass model under the current climate. The effect of increased spring precipitation is similar in direction but relatively less important with a 4°C warming.

Conclusions

There are myriad direct physical and indirect biological relationships between regional climate and fire regime in chaparral. Obvious direct controls include lightning frequency and the effects of air temperature, relative humidity, and wind speed on dead fuel moisture, ignition, and fire spread. An example of indirect control is the relationship between seasonal soil water availability and plant growth and water status, which affect ignition probability, spread rate, and fire intensity. Chaparral species differ widely in their combustion characteristics, and thus climate-induced changes in stand composition can also affect fire regime variables such as fire recurrence interval, size, and severity. Predicting future fire regimes is especially complicated because increasing CO_2 may not only affect climatic variables, but may also alter plant physiology, water use efficiency, canopy structure, interspecific interactions, and community development after fire.

In this study we have only considered the direct consequences of climatic change on fire ignition and spread in chaparral, and have not treated the interactions among climate, fire frequency, and intensity, and potentially large changes in chaparral composition (e.g., Malanson and Westman, 1991). Incorporating a submodel into REFIRES to account for compositional dynamics would be relatively straightforward. Unfortunately there is not much information available to parameterize such a model.

The retrospective analysis of wildfire history in the Los Padres National Forest reveals a modern fire regime dominated by infrequent large, fires. These fires occur between June and October during short-lived meteorological events characterized by extreme high temperatures and winds. The timing and frequency of these events is not obviously related to seasonal or annual mean temperature conditions. Fire hazard also depends strongly on live fuel moisture levels, which are a function of late winter and spring precipitation. Fire hazard is greatly reduced in years when total spring rainfall at Santa Barbara exceeds 200 mm, and is extremely high in years where spring rain totals less than 50 mm. On average, fire spread does not appear to be limited by fuel availability, since most stands are mature, closed-canopy chaparral.

GCM predictions for future climate in southern California are very unreliable, particularly given the importance of extreme events in the fire climatology of chaparral. Another large source of uncertainty is the fire spread and fuel models used to simulate fire behavior. These models are particularly undependable under the high wind conditions typical of modern fire events. Nonetheless, the models give us an indication of how ignition, climate, vegetation, terrain, and fire management could interact to affect regional fire regime under a changed climate.

Using current weather data, fire regimes simulated by REFIRES resemble the actual regional regime in terms of fire recurrence interval and relative fire size distribution. The simulated regime is very sensitive to predicted increases in fire season temperatures, and is equally as sensitive to large changes in stand productivity. Total area burned is predicted to increase and fire recurrence interval to decrease significantly, even for a modest 1°C temperature increase. The effect of increased temperature is more pronounced for a less productive chaparral, where rates of spread are more frequently limited by both fuels and weather conditions. This effect of a 1°C increase in temperature is more than offset by an increase of 5.25% in summer fuel moisture levels. Increased fuel moisture does not affect the frequency of initial ignitions; rather, it operates to reduce spread rates by decreasing fire intensity and increasing the amount of heat required to bring neighboring fuels up to ignition temperature. Modeled fire regime is not as sensitive to spring precipitation as the observed fire regime, a result that is probably due to the use of a chamise fuel model rather than a fuel model closer to the mix of coarse, predominantly live fuels that actually characterize the region.

These results are based on present wind conditions, and could change dramatically depending on changes in that parameter. Most GCM models predict an increase in summer wind speeds under a $2CO_2$ atmosphere, which would certainly increase fire hazard as predicted by fire behavior models. However, wind is unquestionably the most difficult variable in any analysis of future fire regimes. Existing GCM models and fire behavior models are both very weak in their treatment of this variable. Some of the main difficulties in modeling wind effects on fire behavior include adequate characterization of spatial and temporal variability, especially in rugged terrain, accounting for local effects of the fire itself, and use of a single velocity to characterize a dynamic vertical profile.

The predicted increase in wildfire hazard under global warming is greater than that predicted by Torn and Fried (1992) for northern California, despite their specification of a future increase in wind velocities. This is partly due to regional differences in fire season climate, but is also due to differences in fuel models and model formulation. Because it includes the spatial and temporal dynamics of fuel loading on the landscape, REFIRES is probably more responsive to changes in ignition frequency and calculated rates of spread than the model used by Torn and Fried (1992), which treated chaparral as a single, uniform fuel.

We are currently exploring other interactions between climate and fire regime in this region using a much larger grid and more rugged topography. We also intend to analyze different wind scenarios, as well as sensitivity of the chaparral fire regime to variance in climate parameters. Results to date appear reasonable and certainly indicate the potential impact of global warming on fire in chaparral ecosystems. However, REFIRES itself requires many additional improvements, including a subroutine for fire spotting and a shorter time step for meteorological conditions (Davis and Burrows, 1993).

Given the importance of fire regime in chaparral ecosystems and in Mediterranean-climate ecosystems generally, and the apparent sensitivity of fire regime to climate and vegetation productivity and composition, we believe that research into the possible consequences of global changes in atmospheric CO_2 and climate on regional fire regimes must be given very high priority. This research should integrate both empirical, quantitative comparisons of fire regimes in areas under different climates and fire management systems, as well as simulation studies incorporating fire behavior models and ecological models of community development after fire. In particular, we need more information on fuel characteristics of different chaparral species and communities as a function of age and site conditions, and information on interactions among fire intensity, climate, and postfire vegetation development.

Acknowledgments. We would like to thank Mark Borchert and Fritts Cahill of the Los Padres National Forest for providing fire history, climate, and fuel moisture data for the forest. We also gratefully acknowledge David Burrows for assisting in parameterization and operation of REFIRES. Philip Riggan and a second reviewer provided very constructive reviews of the draft manuscript. Computing support for this research was provided by the Digital Equipment Corporation Sequoia 2000 project.

References

Albini FA. 1976. Estimating wildfire behavior and effects. USDA Forest Service General Technical Report INT-30. Inter-Mountain Forest and Range Experiment Station, Ogden, UT.

Bradshaw LS, Deeming JE, Burgan RE, Cohen JD. 1983. The 1978 National Fire Danger Rating System: Technical Documentation. USDA Forest Service General Technical Report INT-169, Inter-Mountain Forest and Range Experiment Station, Ogden, UT.

Burrows DA. 1988. The REFIRES (Regional FIre REgime Simulation) model: a C program for regional fire regime simulation. MA Thesis, University of California, Santa Barbara, CA.

Byrne R, Michaelsen J, Soutar A. 1977. Fossil charcoal as a measure of wildfire frequency in southern California: a preliminary analysis. Pages 361–367 in HA Mooney and CE Conrad, eds, Proceedings of the Symposium on the consequences of fire and fuel management in Mediterranean Ecosystems, Palo Alto, California. USDA Forest Service General Technical Report WO-3. Washington, DC.

Davis FW, Burrows DA. 1993. Modeling fire regime in Mediterranean landscapes. Pages 247–259 in S Levin, T Powell, and J Steele, eds, Patch Dynamics. Springer-Verlag Lecture Notes in Biomathematics, New York.

Davis FW, Hickson DE, Odion DC. 1988. Composition of maritime chaparral related to fire history and soil, Burton Mesa, Santa Barbara County, CA. Madrono 35:169–195.

Dickinson RE. 1989. Uncertainties in estimates of climatic change: a review. Climatic Change 15:5–13.

Hanes TL. 1971. Succession after fire in the chaparral of southern California. Ecol Monog 41:27–52.

Keeley J. 1982. Distribution of lightning and man-caused wildfires in California. In: CE Conrad and WC Oechel, eds, Dynamics and Management of Mediterranean-type Ecosystems. USDA Forest Service General Technical Report PSW-58, pp. 431–437.

Malanson GP, Westman WE. 1991. Modeling interactive effects of climate change, air pollution and fire on a California shrubland. Climatic Change 18:363–376.

Milly PCD. 1992. Potential evaporation and soil moisture in General Circulation Models. J Climate 5:209–226.

Minnich RA. 1983. Fire mosaics in southern California and northern Baja California. Science 219:1287–1294.

Minnich RA. 1989. Chaparral fire history in San Diego County and adjacent northern Baja California. In: SC Keeley, ed, The California Chaparral: Paradigms Re-examined. Natural History Museum of Los Angeles County Science Series 34:37–48.

Muller CH, Hanawalt RB, McPherson JK. 1968. Allelopathic control of herb growth in the fire cycle of California chaparral. Bull Torrey Bot Club 95:225–231.

Ogden GL. 1975. Differential response of two oak species to far inland advection of sea-salt spray aerosol. PhD Dissertation, University of California, Santa Barbara, CA.

Paysen TE, Cohen JD. 1990. Chamise chaparral dead fuel fraction is not reliably predicted by age. West J Appl Forestry 5:127–131.

Radtke KW-H, Arndt A, Wakimoto RH. 1977. Fire history of the santa monica Mountains. In: CE Conrad and WC Oechel, eds, Dynamics and Management of Mediterranean-type Ecosystems. USDA Forest Service General Technical Report PSW-58, pp. 438–443.

Riggan PJ, Goode S, Jacks P, Lockwood RN. 1988. Interaction of fire and community development in chaparral of southern California. Ecol Monog 58:155–176.

Rothermel RC. 1972. A mathematical model for predicting fire spread in wildland fuels. USDA Forest Service Research Paper INT-115. Inter-Mountain Forest and Range Experiment Station, Ogden, UT.

Rothermel RC, Philpot CW. 1973. Predicting changes in chaparral flammability. J Forestry 71:640–643.

Torn MS, Fried JS. 1992. Predicting the impacts of global warming on wildland fire. Climatic Change 21:257–274.

Westman WE, Malanson GP. 1992. Effects of climate change on Mediterranean-type ecosystems in California and Baja California. In: RL Peters and TE Lovejoy, eds, Global Warming and Biological Diversity. Yale University Press, New Haven, CT, pp. 258–256.

22. Global Environmental Change to and the Future of Mediterranean Forest Avifauna

Tomás Santos and José Luis Tellería

The antiquity and intensity of human colonization and exploitation of Mediterranean countries (Naveh and Dan, 1973; Le Houérou, 1981) has caused profound changes in the distribution and composition of their forests and associated fauna (Blondel, 1986; Purroy et al., 1990; Tellería, 1992). Human use (Le Houérou, 1981) and the physiographic complexity of this region (Bradbury, 1981) have led to highly diversified and heterogeneous landscapes, in which the reduction and fragmentation of the original forests is the predominant note. Land-use changes produced by human activity are a fundamental component of global change (Meyer and Turner, 1992; Vitousek, 1992). Land-use alterations take two main forms: land conversion and land modification (Meyer and Turner, 1992). In the Mediterranean basin, the duration and intensity of human land use has led to widespread processes of habitat modification and conversion, to the point where land use changes may perhaps be considered as the fundamental component of the recent past environmental change in the region.

Biodiversity and endemicity are very high in the Mediterranean basin (Blondel, 1986). The current patterns of species richness, however, are extremely variable among taxonomic groups, as some taxa are highly diverse and rich in rare species (Blondel, 1990a), while others, such as forest avifauna, are impoverished versions of their European counterparts (Herrera, 1980; Blondel, 1984). These differences are partly due to the distinct biogeographic histories of each group (Moreau, 1954; Chaline and Mein, 1979; Sutcliffe, 1985), but they may well also reflect their differing sensitivity to the long historical processes of environmental changes in the Mediterranean area. Land-use change will probably be the most important component of global environmental change in coming decades (Vitousek, 1992). The study of the consequences of these processes may therefore be the key to understanding the species composition and the current distribution patterns of many Mediterranean taxonomic groups, and also to predicting their future in the context of environmental change in the region.

The avifauna of the European Mediterranean forests can be defined as impoverished palaearctic (Herrera, 1980; Blondel, 1984). The scarcity of palaearctic forest birds in Mediterranean forests may be due to the peripheral location of the Mediterranean basin (Blondel, 1986), with specific climatic conditions to which the pool of palaearctic forest species are not adapted. Moreover, there is a remarkable lack of endemic Mediterranean forest species, with the exception of some that have evolved under isolated island conditions (Harrison, 1982; Blondel, 1986, 1990a).

The present paper postulates that the alteration and fragmentation of Iberian Mediterranean forests is an important cause of their current paucity in forest bird species. The effects of these components of global environmental change on the avifauna are approached on a dual spatial scale. First, we discuss their influence on a broad geographic gradient (Iberian Peninsula) along which some climatic parameters, to which the palaearctic forest birds are sensitive, change. Second, on a local scale we analyze the selective effects of fragmentation on these bird species, and compare their consequences in Mediterranean and Eurosiberian localities.

Changes on a Regional Scale

The palaearctic forest avifauna appears to reach its optimum in deciduous forests of central Europe, from which its abundance and diversity diminish southwards (Herrera, 1980; Blondel, 1986). Even quite ubiquitous, euryoecious species such as the small foliage gleaners tend to diminish towards the southwest of the Iberian Peninsula (Tellería and Santos, 1993; Figure 22-1), suggesting a deterioration of the optimum conditions for forest birds in extreme southwestern Europe. This pattern could be interpreted using purely biogeographic criteria, such as a simple peripheral effect of the distribution areas (Brown, 1984), accentuated in the Iberian area by its peninsular isolation (Blondel, 1986).

On the Iberian Peninsula gradient, precipitation is associated significantly with the abundance of 40% of the forest passerine species. This

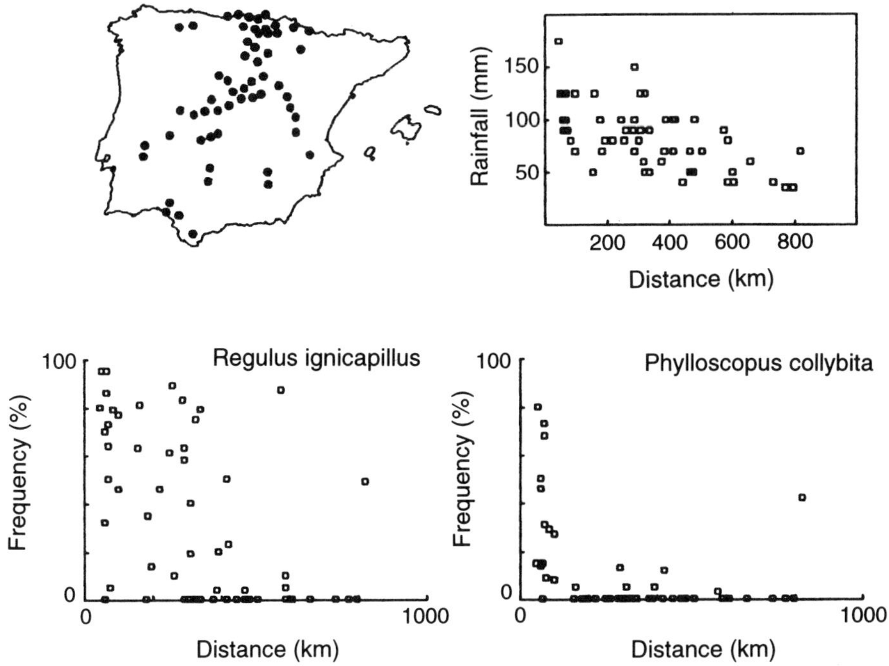

Figure 22-1. Frequency of occurrence of two insectivorous passerine species in 58 Iberian forests according to distance from the French border (distance 0). May precipitation in the nearest locality to each forest is also indicated. See Tellería and Santos (1993).

relationship varies according to the geographic origin of the species (Voous, 1960), so that the northern and palaearctic species tend to associate positively with precipitation, while the Mediterranean and south palaearctic species are associated negatively (Tellería and Santos, 1994; Figure 22-2). This close link between precipitation and bird distribution has been revealed in several regional studies along gradients of xericity in both forest (Smith, 1977; Kendeigh and Fawyer, 1981) and pasture and steppe habitats (Serventy, 1971; Maurer, 1985; Tellería et al., 1988). A common explanation is that precipitation, through its control over productivity, drastically restricts the distribution of many mesic birds in xeric environments. Given that precipitation declines in a southward direction on the Iberian Peninsula (Figure 22-1), this hypothesis fits the abundance patterns of the northern and palaearctic birds in Iberia.

This biogeographic explanation runs the risk, however, of being incomplete, as human land use has been accentuating the intensity of the Eurosiberian-Mediterranean climatic gradient for the last several thousand years. The structure and composition of Iberian Mediterranean

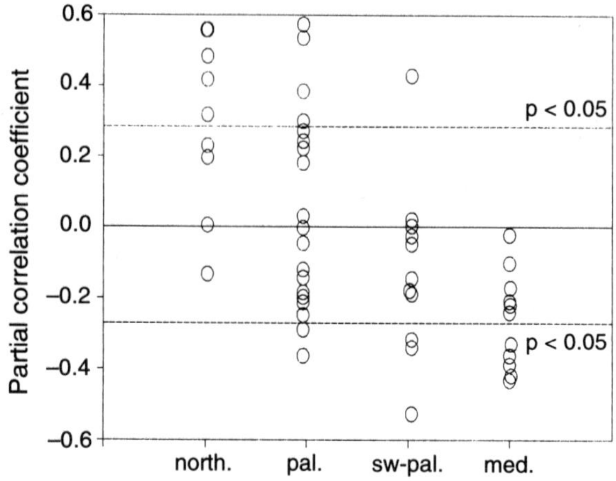

Figure 22-2. Partial correlation between abundance of 50 passerines censused in 58 Iberian forests (see Figure 22-1) and annual average rainfall (excluding effects of temperature and taxonomic structure and composition of tree mass). North: northern species; pal: palaearctic species; sw-pal: sw-palaearctic species; med: mediterranean species. See Tellería and Santos (1994).

forests has been changed profoundly by human activity since the Holocene (Costa et al., 1990; Ruíz de la Torre, 1990). Most mature forests have been replaced by shrub vegetation or cleared to favor pasture growth, giving rise to the "dehesas" (open savannah-like woodlands) of holm oak (*Quercus rotundifolia*) and cork oak (*Q. suber*) that predomiate in the southwestern corner of the Iberian Peninsula. This structural modification, generally produced with the help of fire, has deeply changed the original microclimatic conditions, favoring the propagation of heliophitic (*Q. rotundifolia*) and pyrophitic (*Q. suber*) tree species (Le Houérou, 1981; Costa et al., 1990). In fact, available palynological data reveal that the Holocene Mediterranean vegetation was composed of mixed fagaceous forests, in large areas predominantly deciduous mesic trees such as *Q. pyrenaica* and *Q. faginea* (Costa et al., 1990). Fire, logging, grazing, and other age-old land uses have caused the gradual replacement of these species by others that are adapted to more xeric conditions, producing a dramatic "mediterraneanization" of the forest vegetation in large areas of the Iberian range (Le Houérou, 1981; Trabaud, 1981; Costa et al., 1990).

The features of the bird communities studied in mixed mesic forests (equivalent to the original Mediterranean forests: forests 2 and 3 in Figure 22-3) and altered monospecific forests (forests 4 and 5) suggest that the original Mediterranean forest avifauna was quite similar to that now occupying the moist European forests (forest 6). This idea is also corroborated by the composition of the bird communities in typically

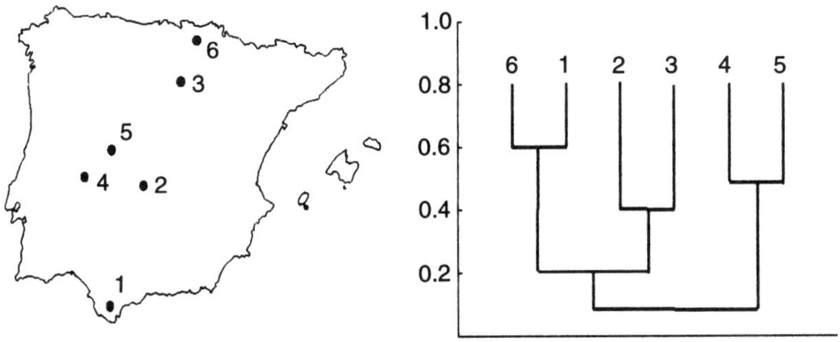

Figure 22-3. Percentage similarity dendrogram constructed by the UPGMA method among six breeding communities of forest passeriforms censused in Spain. (1) Cork oak *Quercus suber* (Cádiz); (2) cork oak *Q. suber* (Toledo Mountains); (3) mixed forest of holm *Q. ilex* and Lusitanian oaks *Q. faginea* (Lerma); (4) holm oak dehesa (Trujillo); (5) holm oak dehesa (Monteagudo); (6) deciduous forest of *Q. robur* and beech *Fagus sylvatica*. All communities were censused by menas of the line-transect method (Järvinen and Väisänen, 1975) using a 50-m main belt. Features of the study areas in Santos and Alvarez (1990), Tellería and Galarza (1990), Tellería et al. (1991, 1992).

Mediterranean forests located in moist southern enclaves, e.g., the mixed cork oak forests of Cádiz (forest 1). The loss of forest species in the southern French forests can be interpreted in a similar anthropogenic context (Blondel, 1984, 1986, 1990a).

These results support the hypothesis that forest bird distribution and abundance on the Iberian Peninsula has been heavily influenced by human action. Consequently, the human impact will have to be added to the loss of species along the north-south climatic gradient, in parallel with the propagation of xeric microclimates linked to the deterioration of the forests.

Changes on a Local Scale

The fragmentation of the original habitats is the inevitable result of any change in land use (Forman and Godron, 1986). Land uses are associated with the division of previous habitats, so that fragmentation is one of the most common, ubiquitous conservation problems (Harris, 1984; Wilcox and Murphy, 1985; Wilcove et al., 1986; Opdam, 1990). The gradual decline in the size of habitat fragments increases their permeability to the surrounding habitats, because the perimeter/area ratio increases as the fragment size falls. This atomization effect in forest habitats has been found to cause an increase in the negative impact by various external agents, which ultimately changes the essential traits of the habitat and produces a loss of forest species (Janzen, 1983). Alterations associated with the increase in the perimeter/area ratio, known as the edge effect, can be classified into two broad groups according to whether they affect physical variables (abiotic factors) or species that interact with the forest organisms (biotic factors).

Abiotic Factors

Land-use changes not only affect organisms directly, due to the reduction of their original habitats (Wilson and Peters, 1988), but also indirectly via their interactions with other components of global change such as climate (Davis, 1990; Shukla et al. 1990; Dickinson, 1991; Vitousek, 1992). The influence of landscape structure on local, regional, and even global climate is by now commonly recognized (Mintz, 1984; Pielke and Avissar, 1990; Meyer and Turner, 1992; Vitousek, 1992).

On a microclimatic scale, some ecological studies have revealed that the decline in the size of forest fragments reduces their climatic homoeostasis (Janzen, 1986), increasing insolation and wind exposure, altering the moisture and temperature regimes, and accentuating daily and seasonal oscillations of these two parameters (Jordan, 1986; Lovejoy et al., 1986). Lovejoy et al. (1986) have revealed that many animal populations, especially ectothermic species (e.g., butterflies), are extremely sensitive to these microclimatic variations, although endothermic species can also be affected, either secondarily, via changes in the interactions with other species (Janzen, 1986; Lovejoy et al., 1986; Wiens, 1989), or directly, due to physiological restrictions on their adaptive ability (Wiens, 1989; Blondel, 1990b; Walsberg, 1993, for birds).

The microclimatic effects described may be accentuated in the peripheral distribution areas, where the species will be subjected to high physiological stress due to deviations from optimal climatic conditions (Walsberg, 1993). One may thus expect considerable local variation in the response to fragmentation according to the geographic location of the fragments. In the Mediterranean basin, the duration and intensity of human land use has led to widespread processes of habitat fragmentation. In the case of the Iberian geographic gradient (Figure 22-1), the deviation of the climatic conditions to which genuine forest birds are adapted accentuates in a southerly direction due to the progressive increase in xericity. These species are therefore more sensitive to fragmentation the further south the fragments, a hypothesis that permits one to infer two predictions: 1) northerly fragments will have a larger number of forest birds than southern fragments of an equal size; 2) the loss of forest species will occur earlier in the southerly fragmentation processes (i.e., the loss will commence in larger fragments than in more northerly localities). Figure 22-4 illustrates this prediction with a paradigmatic example.

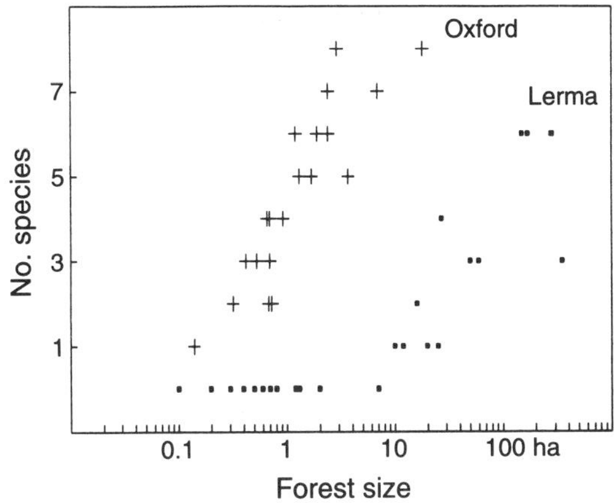

Figure 22-4. Patterns of richness of pariforms and related species (genus *Parus*, *Sitta*, *Certhia*, *Aegithalos*, *Regulus*, and *Phylloscopus*) in relation to fragment size in 20 fragments of deciduous forest in Oxford (Great Britain; see Ford, 1987) and in 31 fragments of Mediterranean forest in Lerma (central Spain; sampling methods as in Figure 22-3; see study area in Tellería et al., 1991).

Biotic Factors

In addition to the deterioration of the habitat's microclimatic conditions, forest species that settle in fragments are subjected to biological interferences. The predominance of the edge areas favors ecotonal and generalist species from the surrounding habitats, which may invade the fragments and act as competitors, predators, or nest parasites on the forest species (Janzen, 1983, 1986; Yahner, 1988). Predation appears to be one of the main causes of loss of forest species. With regard to birds, it has been proven that predation can act via two different mechanisms.

On the one hand, direct predation on eggs and nestlings has been observed in many ecotonal species (e.g., crows, carnivores), and the predation rate has been shown to rise as forest size falls (Wilcove, 1985; Tellería and Santos, 1992; Figure 22-5). In forest birds, nest predation tends to diminish with an increase in vegetation cover (Salathé, 1987; Martin and Roper, 1988), so that in deciduous forests, where the leaf phenology produces temporal changes in cover, the probability of nest predation falls as the reproductive season advances (Slagsvold, 1982). The advantages of the increase in cover can collapse in Mediterranean fragmented forests due to a dramatic increase in predation (Santos and Tellería, 1991, 1992). Thus, for example, in a study in forest fragments in Lerma (central Spain), where artificial nests were placed in deciduous (*Quercus faginea*) and evergreen (*Q. rotundifolia*) species of tree, no relationship was found between leaf cover and the predation rate: 86.25%

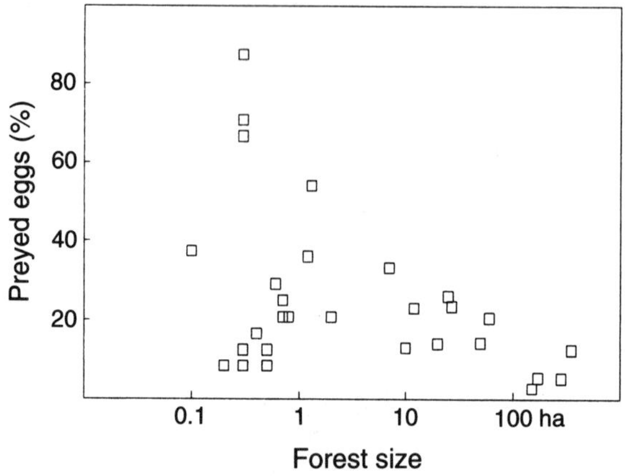

Figure 22-5. Relationship between percentage of eggs taken by predators and forest area (0.1 to 350 ha) in 31 forest fragments studied in Lerma (central Spain; see Tellería and Santos, 1992).

of the nests placed in the holm oaks were subject to predation, while the Lusitanean oak rate ranged from 81.25% before leafing to 90% after leafing (Santos and Tellería, 1991).

Clearing or degradation of the vegetation structure in the fragment cores can have similar effects to those described if the disturbed areas are colonized by generalist predator species. This "internal edge" effect can thus be added to the direct and indirect effects originating from the habitats surrounding the fragments, amplifying the intensity of predation (Wilcove, 1990). In fact, this "double edge effect" is probably typical in the advanced processes of fragmentation, as the large amount of external edge facilitates the human exploitation of the fragments. In Lerma, the interiors of the small forest fragments are extremely degraded. The consequence is that the availability of an authentic forest habitat in many fragments is much lower than their area would indicate.

The secondary biological interactions caused by ecotonic species are another type of mechanism acting on forest species (Janzen, 1983, 1986; Lovejoy et al., 1986). Some ecotonic animals feed upon the fruits and seeds of forest plants, endangering the regeneration capacity of the fragments and the survival of the animal species linked to them. This is the case with the heavy predation exercised by certain ecotonic rodents on the fruits and seeds of some forest plants in fragmented forests (Wästljung, 1989).

In Lerma, for example, the wintering populations of mice gather in untilled fields (forest fragments or shrublands), where they find a refuge from low temperatures (Alcántara and Tellería, 1991). In mixed forests of juniper (*Juniperus thurifera*) and holm oak (*Q. rotundifolia*), field mice

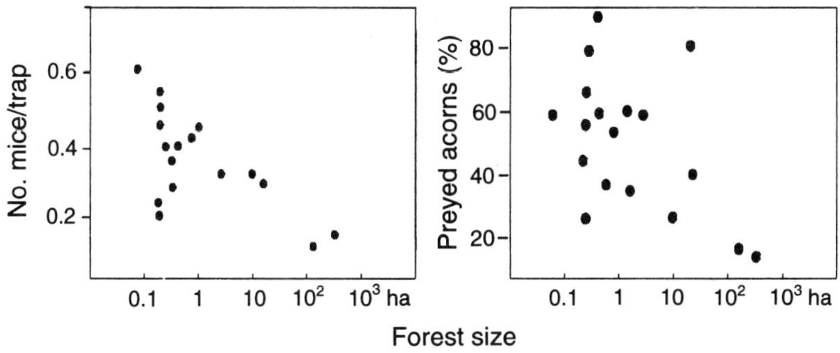

Figure 22-6. Left: relationship between abundance of mice (*Apodemus sylvaticus*) and forest area (0.1 to 270 ha) in 17 forest fragments in Lerma. Right: relationship between fragment area and percentage of acorns consumed by mice in the same fragments. See Tellería et al. (1991).

(*Apodemus sylvaticus*) are nine times more abundant in fragments between 0.1 and 16 ha than in forests over 150 ha (Tellería et al., 1991; Figure 22-6). This leads to an extremely heavy predation on the holm oak and juniper fruits in the small fragments which might affect zoochorous dispersers. This is the case with thrushes (Gen. *Turdus*), the main juniper dispersers in the Lerma study forests, whose density falls drastically in the small fragments (Santos and Tellería, 1994; Figure 22-7).

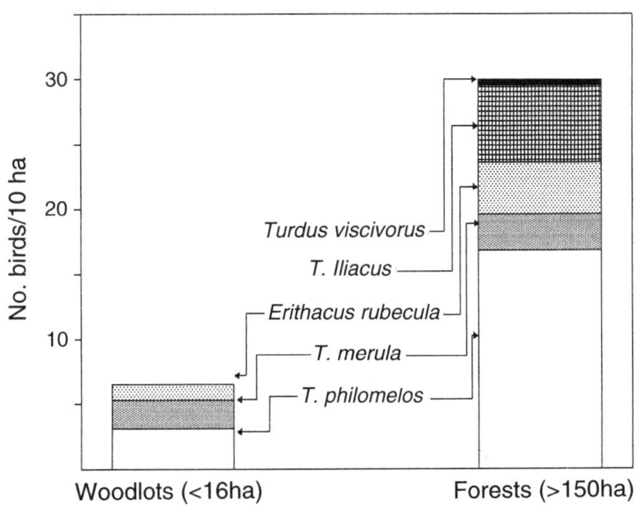

Figure 22-7. Mean abundance of thrushes and blackbirds in eight small forest fragments (0.2 to 16 ha) and two large forests (150 and 270 ha) studied in Lerma in a mast year of juniper (*Juniperus thurifera*) cones. See Santos and Tellería (1994).

Conclusions

Regional and local variations in forest bird communities suggest the existence of a heavy deterioration of the optimum conditions for the Iberian forest avifauna as the climatic xericity intensifies and the size of forest fragments falls. The fragmentation of a habitat reduces the population sizes of the species associated with the resulting habitat fragments (Wilcox, 1980). In the past, this might have led to an atomization of the populations of Iberian forest birds, and the subsequent process of extinction would have increased with the fall in the forest size (Wilcove et al., 1986; Opdam, 1990) and the increase in the distance from the central distribution area of the species (Brown, 1984).

The larger species, and those that are more demanding in requirements or that depend on scarce resources, are those most exposed to numerical reduction, soon reaching dangerous demographic levels (Harris, 1984). The heavy environmental deterioration of the southernmost forest enclaves on the Iberian Peninsula suggests that the bird species which are most sensitive to fragmentation might have become extinct in this region in the distant past. Species such as *Tetrao urogallus*, *Scolopax rusticola*, *Dryocopus martius*, *Dendrocopos medius*, *D. leucotos*, *Aegolius funereus*, *Certhia familiaris*, and *Parus palustris* fit this scenario (Bernis, 1954). Their southernmost breeding areas in the western Palaearctic are now found in certain forest areas of the northern half of the Iberian Peninsula (Harrison, 1982).

Given the long history of human settlement in the Mediterranean region, the deterioration of the vegetation and the associated microclimatic conditions are likely to have taken place in the last 10,000–12,000 years, coinciding with the development of the Mediterranean climatic conditions. In the same way as human interaction has impeded the development of an authentic climax forest in many Mediterranean sectors (Naveh and Dan, 1973; Le Houérou, 1981), it also may have impeded the settlement of taxonomically diversified forest bird communities, particularly the species with the strictest demand for forest conditions. In fact, there is general agreement that the Mediterranean climate has not changed in the last 3,000 years (Le Houérou, 1981), the most intense period of human land use, which further reinforces the hypothesis that human action has played a fundamental role in determining the Mediterranean forest avifauna.

For forest birds, and undoubtedly many other forest organisms in the Mediterranean region, land use change is a very old process to which they have been exposed and whose impact intensifies in a southerly gradient towards the driest sectors in the extreme South. In these sectors, human impact rapidly modifies the microclimatic conditions required for the maintenance of the most demanding forest species. The impending global change, characterized climatologically by an increase in xericity in the

Mediterranean area (Houghton et al., 1990), will only intensify this age-old process of species loss, especially if forest fragmentation continues. The complex interactions among the components of global change, along with the uncertainty about the future course of many of them and some of their consequences (Vitousek, 1992), makes prediction an exercise of little applied value in many scenarios of global change in the medium and long term. The results set out in the present chapter may have the value of a verbal predictive model, the main utility of which would be in indicating some of the most productive lines of research to decipher the future alternatives to preserve the avifauna and perhaps other Mediterranean forest organisms.

Acknowledgments. We thank the Spanish Ministry of Education and Science for reporting this project (DGICYT No. PB92-20238).

References

Alcántara M, Tellería JL. 1991. Habitat selection of the Wood Mouse (*Apodemus sylvaticus*) in cereal steppes of Central Spain. Z Saugetierk 56:347–351.

Bernis F. 1954. An ecological view of Spanish avifauna with reference to the Nordic and Alpine birds. Actas XI Orn Int Congr 1954, pp. 417–423.

Blondel J. 1984. Avifaunes forestieres méditerranéennes; histoire des peuplements. Aves 21:209–226.

Blondel J. 1986. Biogéographie Evolutive. Masson, Paris.

Blondel J. 1990a. Biogeography and history of forest bird faunas in the Mediterranean zone. In: Keast A (ed) Biogeography and Ecology of Forest Bird Communities. SPB Academic Publishing bv, The Hague, pp. 95–107.

Blondel J. 1990b. Long term studies on bird communities and populations in mainland and island Mediterranean forests. In: Keast A (ed) Biogeography and Ecology of Forest Bird Communities. SPB Academic Publishing bv, The Hague, pp. 167–182.

Bradbury DE. 1981. The physical geography of the Mediterranean lands. In: di Castri F, Goodall DW, Specht RL (eds) Ecosystems of the World 11. Mediterranean-type Shrublands. Elsevier, Amsterdam, pp. 53–62.

Brown JH. 1984. On the relationship between abundance and distribution of species. Am Nat 124:255–279.

Chaline J, Mein P. 1979. Les rongeurs et l'évolution. Doin, Paris.

Costa M, García M, Morla C, Sainz H. 1990. La evolución de los bosques de la Península Ibérica: una interpretación basada en datos paleobiogeográficos. Ecología, Fuera de Serie 1:31–58.

Davis MB. 1990. Biology and paleobiology of global climate change: Introduction. TREE 5:269–270.

Dickinson RE. 1991. Global change and terrestrial hydrology: a review. Tellus 43AB:176–181.

Ford HA. 1987. Bird communities on habitat islands in England. Bird Study 34:205–18.

Forman RTT, Godron M. 1986. Landscape ecology. John Wiley & Sons, New York.

Harris LD. 1984. The fragmented forest. University of Chicago Press, Chicago, IL.

Harrison H. 1982. An Atlas of the Birds of the Western Palaearctic. Collins, London.

Herrera CM. 1980. Composición y estructura de dos comunidades mediterráneas de Passeriformes. Doñana Acta Vertebrata 7:1–340.
Houghton JT, Jenkins GJ, Ephrauns JJ (eds). 1990. Climate change: the IPCC scientific assessment. Cambridge University Press, Cambridge, UK.
Janzen DH. 1983. No park is an island: increase of outside as park size increases. Oikos 41:402–410.
Janzen DH. 1986, The eternal external threat. In: Soulé ME (ed) Conservation Biology. The Science of Scarcity and Diversity. Sinauer Associates, Sunderland, MA, pp. 286–303.
Järvinen O, Väisänen RA. 1975. Estimating relative densities of breeding birds by line transect methods. Oikos 26:316–322.
Jordan CF. 1986. Local effects of tropical deforestation. In: Soulé ME, Wilcox B (eds) Conservation Biology. The science of scarcity and diversity. Sinauer Associates, Sunderland, MA, pp. 410–426.
Kendeigh SC, Fawyer BJ. 1981. Breeding bird populations in the Great Smoky Mountains, Tennessee and North Carolina. Wilson Bull 93:218–242.
Le Houérou HN. 1981. Impact of man and his animals on Mediterranean vegetation. In: di Castri F, Goodall DW, Specht RL (eds) Ecosystems of the World 11. Mediterranean-Type Shrublands. Elsevier, Amsterdam, pp. 479–521.
Lovejoy DH, Bierregard RO Jr, Rylands AB, Malcolm JR, Quintela CE, Harper LH, Brown KS Jr, Powell AH, Powell GVN, Schubart HOR, Hays MB. 1986. Edge and Other effects of isolation on Amazon fragments. In: Soulé ME (ed) Conservation Biology. The Science of Scarcity and Diversity. Sinauer Associates, Sunderland, MA, pp. 257–285.
Martin TE, Roper JJ. 1988. Nest predation and nest-site selection of a western population of the Hermit Thrush. Condor 90:51–57.
Maurer BA. 1985. Avian community dynamics in desert grasslands: observational scale and hierarchical structure. Ecol Monogr 55:295–312.
Meyer WB, Turner BL II. 1992. Human population growth and global land-use/cover change. Annu Rev Ecol Syst 23:39–61.
Mintz Y. 1984. The sensitivity of numerically simulated climates to land-surfaces conditions. In: Houghton JT (ed) The Global Climate. Cambridge University Press, Cambridge, UK, pp. 830–842.
Moreau RE. 1954. The main vicissitudes of the European avifauna since the Pliocene. Ibis 96:411–431.
Naveh Z, Dan J. 1973. The human degradation of Mediterranean landscapes. In: di Castri F, Mooney HA (eds) Mediterranean Type Ecosystems. Origin and Structure. Springer-Verlag, Berlin, pp. 373–390.
Opdam P. 1990. Dispersal in fragmented populations: the key to survival. In: Bunce RGU (ed) Species Dispersal in Agricultural Habitats. Belhaven Press, pp. 3–17.
Pielke RA, Avissar R. 1990. Influence of landscape structure on local and regional climate. Landscape Ecol 4:133–155.
Purroy FJ, Alvarez A, Clevenger AP. 1990. Bosque y fauna de vertebrados en España. Ecología, Fuera de Serie 1:349–363.
Ruiz de la Torre J. 1990. Distribución y características de las masas forestales españolas. Ecología, Fuera de Serie 1:11–30.
Salathé T. 1987. Crow predation on coot eggs: effects of investigator disturbance, nest cover and predator learning. Ardea 75:221–229.
Santos T, Alvarez G. 1990. Efectos de las repoblaciones con eucaliptos sobre las comunidades de aves forestales en un maquis mediterráneo (Montes de Toledo). Ardeola 37:319–324.
Santos T, Tellería JL. 1991. Effects of leafing and position on nest predation in a Mediterranean fragmented forest. Wilson Bull 103:676–682.

Santos T, Tellería JL. 1992. Edge effects on nest predation in Mediterranean fragmented forests. Biol Conserv 60:1–5.
Santos T, Tellería JL. 1994. Influence of forest fragmentation on seed consumption and dispersal of Spanish juniper *Juniperus thurifera*. Biol Conserv 70: 129–134.
Serventy DL. 1971. Biology of desert birds. Avian Biol 1:287–339.
Shukla J, Nobre C, Sellers P. 1990. Amazonian deforestation and climate change. Science 247:1322–1324.
Slagsvold T. 1982. Clutch size variation in passerine birds: the nest predation hypothesis. Oecologia 54:159–169.
Smith KG. 1977. Distribution of summer birds along a forest moisture gradient in an Ozark watershed. Ecology 58:810–819.
Sutcliffe AJ. 1985. On the Track of Ice Age Mammals. British Museum (Natural History), London.
Tellería JL. 1992. Gestión forestal y conservación de las aves en España peninsular. Ardeola 39:99–114.
Tellería JL, Galarza A. 1990. Avifauna y paisaje en el norte de España: efecto de las repoblaciones con árboles exóticos. Ardeola 37:229–245.
Tellería JL, Santos T. 1992. Spatio-temporal patterns of egg predation in forest islands: an experimental approach. Biol Conserv 62:29–33.
Tellería JL, Santos T. 1993. Distributional patterns of insectivorous passerines in the Iberian forests: does abundance decrease near the border? J Biogeogr 20:235–240.
Tellería JL, Santos T. 1994. Factors involved in the distribution of forest birds in the Iberian Peninsula. Bird Study 41:161–169.
Tellería JL, Santos T, Alcántara M. 1991. Abundance and food searching intensity of wood mice (*Apodemus sylvaticus* L.) in fragmented forests. J Mammol 72:183–187.
Tellería JL, Suárez F, Santos T. 1988. Bird communities of the Iberian shrubsteppes. Holarctic Ecol 11:171–177.
Tellería JL, Alcántara M, Asensio B, Cantos FJ, Díaz JA, Díaz M, Sánchez A. 1992. Evaluación del impacto del embalse de Monteagudo (Avila-Toledo) sobre la fauna de vertebrados terrestres. Confederación Hidrográfica del Tajo, Unpublished Report.
Trabaud L. 1981. Man and fire: impacts on Mediterranean vegetation. In: di Castri F, Goodall DW, Specht RL (eds) Ecosystems of the World 11. Mediterranean-type Shrublands. Elsevier, Amsterdam, pp. 523–537.
Vitousek PM. 1992. Global environmental change: An introduction. Annu Rev Ecol Syst 23:1–14.
Voous KH. 1960. Atlas of European Birds. Nelson, Amsterdan.
Walsberg GE. 1993. Thermal consequences of diurnal microhabitat selection in a small bird. Ornis Scand 24:174–182.
Wästljung U. 1989. Effects of crop size and stand size on seed removal by vertebrates in hazel *Corylus avellana*. Oikos 54:178–184.
Wiens JA. 1989. The Ecology of Bird Communities. Cambridge University Press, Cambridge, UK.
Wilcove DS. 1985. Nest predation in forest tracts and the decline of migratory songbirds. Ecology 66:1211–1214.
Wilcove DS. 1990. Forest fragmentation as a wildlife management issue in the Eastern United States. General Technical Report NE-140:1–5.
Wilcove DS, Mclellan CH, Dobson AP. 1986. Habitat fragmentation in temperate zone. In: Soulé ME (ed) Conservation Biology: The Science of Scarcity and Diversity. Sinauer Associates, Sunderland, MA, pp. 237–256.

Wilcox BA. 1980. Insular ecology and conservation. In: Soulé ME, Wilcox BA (eds) Conservation Biology: An Evolutionary Approach. Sinauer Associates, Sunderland, MA, pp. 95–117.

Wilcox BA, Murphy DD. 1985. Conservation strategy: the effects of fragmentation on extinction. Am Nat 125:879–887.

Wilson EO, Peters FM (eds). 1988. Biodiversity. National Academy of Sciences, Washington, DC.

Yahner RH. 1988. Changes in wildlife communities near edges. Conserv Biol 2:333–347.

23. Monitoring Environmental Change Through Amphibian Populations

Rafael Márquez and Pere Alberch

There has been much research on the use of biological indicators for the detection of pollution and specific pollutants in the developed world (see review in Spellerberg, 1991). On the other hand, the identification and monitoring of natural populations of species which best reflect the overall condition of the environment (indicator species) has not been given sufficient attention. Changes in the populations of these indicator species may best reflect long-term, large-scale changes at a regional or even global level. Extinction-prone species should be identified so that systematic monitoring programs can be established to ensure early detection of the causes of species decline (Beiswenger, 1986). Such long-term monitoring programs are among the most important tasks that modern museums of natural history face today. Museums have in the past gathered precious biogeographical information that may be contrasted with current censuses of species. Museums should actively participate in the development of standardized monitoring techniques as well as in the identification of species that can be used as early warning systems (a global version of the concept of "the canary in a coal mine"). Here we present the case for amphibians, as an example of a taxonomic group that has the potential to include many useful biological indicators.

Some taxonomic groups are remarkable for the number of species they contain. Amphibians are not. The class Amphibia rivals class Mammalia for being the least speciose class of terrestrial vertebrates, each including about 4,000 described species (Wilson, 1988). According to Duellman and Trueb (1985), the class Amphibia is composed of 163 described species of caecilians (Gymnophyona), 352 described species of salamanders and newts (Caudata), and 3,438 described species of frogs and toads (Anura). This number of species is well below that available for other groups of animals such as reptiles (6,300 species), birds (9,040 species), and bony fishes (18,150 species), and the number of amphibian species pales when compared to the 751,000 species of insects described (all data from Wilson, 1988).

However, amphibians may play relevant roles in the terrestrial ecosystems they inhabit. Their role as biological indicators is derived from several features of their biology.

Amphibians are the only group of vertebrates that generally have two phases in their life cycle (aquatic and terrestrial). This characteristic renders them vulnerable to alterations in both media, the fresh water medium affecting the embryonic and larval stages, and the terrestrial environment affecting adults. While on land, virtually all amphibians are primary predators (insectivorous).

In the aquatic environment, there is a marked dichotomy in the feeding specializations in the Amphibia. Anurans usually have aquatic larvae which can be considered herbivorous (primary consumers) or detritivorous (decomposers). On the other hand, larvae of Caudata are generally predators of aquatic invertebrates.

Therefore, in spite of an apparent lower diversity compared to other animal groups, temperate amphibians may have representatives at several levels in the trophic chain as a consequence of their complex ontogeny. Their presence in a terrestrial ecosystem is therefore directly conditioned by the correct functioning of both the terrestrial and the aquatic components of the habitat. In spite of the apparent scarcity of amphibians, their representation can be substantial from a biomass point of view (Burton and Likens, 1975).

At the individual level, there are also features of the physiology of amphibians that render them vulnerable to changes in the environment. Amphibian skin is a delicate and complex organ. Its role in the balance of hydric and osmotic conditions is crucial. For some species of amphibians, cutaneous respiration is the dominant mode of oxygen interchange with the environment (both aquatic and terrestrial). Typically, amphibian skin is highly permeable, as it is used for absorption of water from the substratum (Duellman and Trueb, 1985). This permeability renders adults more vulnerable to dissolved substances. Furthermore, given their thin epidermis, it has been argued that amphibians could be particularly sensitive to UV radiation, and that the alteration of the ozone layer by chlorofluorocarbon emissions (CFCs) may be increasing such radiation and hence affecting the amphibians (but see Wyman, 1990). Therefore, amphibians could be particularly sensitive to environmental change

(Beiswenger, 1986). Not only the larvae will be affected (Freda and Dunson, 1985; Pierce and Montgomery, 1989), but also adults are likely to respond to impacts such as acid rain (Pierce, 1985; Wyman, 1988, 1990) or other types of chemical pollution (Kirk, 1988).

Recent Evidence of Decline of Amphibian Populations

In September 1989, the First World Congress of Herpetology was held at Canterbury (UK). This unprecedented gathering of researchers on different aspects of the biology of amphibians and reptiles facilitated the exchange of information about the situation of many populations around the globe. Many accounts of amphibian population declines and extinctions were reported there. As a result of this meeting, Professor David Wake of the University of California (Berkeley) obtained funds from the Biology Board of the National Research Council of the USA to hold a meeting to further look into the subject. A workshop was then organized by D. Wake and H. Morowitz in Irvine, California in February 1990. (For a summary of the meeting see Anonymous, 1991; Barinaga, 1990; and Blaustein and Wake, 1990.) The goals of this meeting were to determine (1) whether the declines and extinctions of amphibian populations were a general trend throughout the world, and (2) whether these declines were the consequence of a single global cause. Many accounts were presented at Irvine reporting well-documented declines of particular amphibian populations or species (e.g., Corn and Fogleman, 1984; Hayes and Jennings, 1986; Heyer et al., 1988; Bradford, 1991). Several of these accounts were unpublished before the meeting. Overall, there was general consensus about the answer to the first question: We are witnessing a worldwide decline in amphibian populations. Evidence of such declines has been increasing (Wake, 1991), although some critical voices have been raised indicating that evidence was lacking for some areas of the world (Beebe, 1992). However, the second question, whether the widespread declines could be attributed to a single global phenomenon, remains to be clarified.

Undoubtedly, one of the main general causes for the decline in populations of amphibians is, of course, habitat destruction or alteration. Deforestation is usually the most conspicuous form of habitat destruction. The disappearance of the tropical forests, of course, affects amphibians together with the rest of the tropical fauna. Although tropical deforestation is a topic of current concern for all conservationists, it is not the best example to determine the usefulness of amphibians as environmental monitors. Another important, though less conspicuous type of environmental change that may be reflected by fluctuations in amphibian populations is the alteration of hydrological characteristics. Here we could cite the Spanish example, where the use of subterranean waters for irrigation is lowering the level of the water table in large areas of the interior of the Iberian Peninsula. This lowering of the water table may affect the duration of temporary ponds and streams, which are precious resources for amphibians in the Mediterranean ecosystems and often have whole communities depending on them (e.g., Rodríguez-Jiménez, 1988). Possibly this trend can be extrapolated to other parts of the mediterranean ecosystems. Unfortunately, although there are some cases documented in the popular press of whole regions with receding underground water levels such as the National Parks of Doñana and Tablas de Daimiel, no long-term study of the effect of these changes on amphibian populations has yet been completed in Spain.

Apart from the declines of amphibian populations due to obvious and direct environmental changes, there have been some cases reported of local extinctions of amphibians in apparently pristine habitats, sometimes within protected areas. In some cases, these declines were attributed to global anthropogenic impact such as pollution from pesticides (Kirk, 1988) or acid rain (Wyman and Hasksley-Lescault, 1987; Wyman, 1988, 1990). In other cases, introduction of exotic species such as the cane toad (*Bufo marinus*) or the bullfrog (*Rana catesbeiana*) may have had severe effects on local populations of amphibians (Moyle, 1973; Hayes and Jennings, 1986; Tyler, 1991). In Mediterrannean ecosystems, the practice of stocking lakes and streams with predatory fish (often exotic) for the enjoyment of sportsmen should be carefully monitored and critically analyzed (Ibarra-Vidal, 1989). Their effect on local amphibian populations can be quite dramatic. In general, the evidence presented was based on punctual observations, and the problem of separating the effect of natural fluctuations of populations and declines due to human impact have only been addressed in a few long-term studies (e.g., Pechmann et al., 1991).

Case Studies of "Natural" Extinctions

Throughout the evolution of life on earth, extincitons have occurred at all times. Extinctions can fall along a continuum between those that are the consequence of a clearly identifiable event ("dramatic extinctions") and those that are constantly occuring in isolated patches of habitats or islands ("background extinctions") (Diamond, 1984). While the previously described examples adequately fit the characteristics of dramatic extinctions, there are some examples of recent extinctions of amphibians that cannot clearly be attributed to a direct cause. A number of sites in southeast Brazil have been surveyed by herpetologists over the last 35 years. Long-term changes in species assemblages have been recorded over this span of time, particularly at Boraceiá, a field station run by the Universidade de São Paulo. This protected site is in the Atlantic Forest Morphoclimatic Domain, among the wettest areas of

Brazil, covered with continuous cloud forest and containing up to 63 species of frogs (Heyer et al., 1990). Heyer et al. (1988) analyzed the evolution of the frog populations of Boraceiá and neighboring sites. They described drastic population reductions for seven species of anurans and the possible local extinction of five additional species. Most of these events were attributed to unusually heavy frosts that occurred in 1979, although an accurate estimation of the magnitude of the population reduction was not possible due to the lack of a standardized monitoring program for the frog populations or for any other indicator species.

A second case study was provided by the disappearance of the Australian gastric brooding frog (*Rheobatrachus silus*) from its only known range: the Blackall and Connondale Ranges (southeast Queensland). This frog's unique reproductive system rendered it extremely interesting: the eggs and larvae developed in the stomach of the female and fully developed froglets emerged from the mouth. It was under extensive investigation and had a high public focus (Tyler, 1983). This frog was discovered and described in 1973. A year later, its unusual reproductive system was described (Corben et al., 1974). In 1981 the species totally disappeared from the wild, and a colony could only be kept for two further years in captivity. Careful analysis of the available evidence did not come to a conclusive answer as to what was the main reason for the disappearance (Tyler, 1991).

The story of the golden toad, *Bufo periglenes*, is probably the best documented case. This toad is endemic to a small area of undisturbed cloud forest within the Monteverde Cloud Forest Reserve in Costa Rica. The marked sexual dimorphism in coloration, with bright red males, made them an emblematic species for the reserve and is in sharp contrast with the secretive behavior of the species. Toads were regularly observed in the park every year from 1978 to 1986, though never in large numbers. In 1987 unusually large assemblages of toads (adding up to more than 1,500 adults) were seen from April to May, although most eggs and larvae they produced died because most pools subsequently dried. However, only 10 adult toads were sighted in 1988, and the sighting of a single adult male was reported in 1989. Since then no additional adults have ever been found. In the last 5 years no eggs or tadpoles have been observed, indicating that reproduction did not occur, and an extensive search carried out in 1990 failed to reveal any golden toads (Crump et al., 1992). The decline or possible extinction of this peculiar species led to the realization that other species of anurans coexisting in the area have suffered drastic declines in their populations (Crump et al., 1992). However, there is no general agreement as to what may be the cause of these negative changes in biodiversity. Some scientists suggest that complex climatic changes due to large scale shifts in the warm oceanic current known as EI Niño (which has been blamed for many alterations in the climate of the Pacific coast, from California to Peru) may be responsible for subtle changes in the weather to which toads may be particularly sensitive (Pounds, 1990, 1991, 1992). However, careful analysis of the available data cannot clearly support this hypothesis, although a slight decrease in rainfall for some months was reported for the period in which toads were not observed in Monteverde.

Mediterranean Ecosystems: A Call for More Information

Mediterranean ecosystems constitue unique transitional zones between temperate and dry zones. The climatic conditions that characterize these ecosystems include winter rainfalls, a distinct summer drought of variable length, mild to warm or hot summers, cool to cold winters, and high variability of precipitation from year to year (Di Castri, 1981). All these features have particular relevance for the composition of the biota of these ecosystems, and their transitional nature is particularly well reflected by the characteristics of the amphibian communities inhabiting these zones.

The effect of climate on the onset and length of the reproductive season imposes strict limitations on the persistence and recolonization of individual amphibian populations in Mediterranean ecosystems (Rodríguez-Jiménez, 1988; Salvador and Carrascal, 1990; Reques and Tejedo, 1991; Márquez, 1992). These limitations can affect both the terrestrial (adult) stage and the aquatic stage of the amphibian life cycle. Weather conditions can impose direct constraints on the behavior of adult individuals. The ectothermic condition of amphibians implies that the margins of ambient temperature in which amphibian terrestrial activity can be sustained have a lower limit at freezing temperatures. There is no evidence of the existence of internal heat production mechanisms that may increase body temperatures of amphibians above that of the environment. Moreover, the nocturnal habits of most amphibians curtail the possibility of thermoregulation by basking, leaving four basic mechanisms for heat loss or gain with the atmosphere of limited efficiency: convective heat exchange, heat conduction to or from the substrate, thermal radiation to the environment, and evaporative heat loss.

Another limitation to the terrestrial activity of adult individual amphibians is the availability of high levels of ambient moisture to counteract the permeability of their body covering. Breeding migrations are therefore often limited to periods of rainfall, or to the nights immediately following rainfall. In the Mediterranean climate, the presence of rainfall, mainly in the colder winter months, implies that adult amphibians have to trade off two parameters which are often incompatible: the presence of water and the existence of benign nocturnal temperatures.

Not only is the activity of adults limited by climate, but the presence of bodies of water in which to reproduce is an additional limiting factor.

Most amphibians from mediterranen ecosystems have completely aquatic egg and larval development (Salthe and Mecham, 1974; Lamotte and Lescure, 1977), which implies that they are highly dependent on the regular availability of bodies of water for reproduction. Their prolonged adult life span allows for the persistence of amphibian populations in areas where water avalilability for reproduction only occurs sporadically. In some cases, adaptations to ephemeral bodies of water have evolved, with extremely fast larval development times and usually small sizes at metamorphosis (Werner, 1986). These populations occupying marginal habitats often show a peculiar population structure with complete absence of some age classes, corresponding to the cohorts where recruitment was nil (Tejedo, 1989). Of course, when the drought period in a given area exceeds the maximum life span of the individuals, the population goes extinct. These local extinctions can be a consequence of large-scale regional processes such as the lowering of the levels of the underground water tables, a generalized phenomenon in some Mediterranean areas subjected to processes of desertification, either naturally or as a consequence of human overexploitation.

If a classification was to be made of the breeding habits of amphibians in Mediterranean ecosystems, a gradient could be established based on the variables that determined the onset of the breeding seasons (Márquez, 1992). At one extreme of the gradient, we could have species with a rainfall-mediated system, which usually have highly explosive mating systems and which are probably derived from the amphibian fauna of the neighboring arid regions (savanna, steppes, desert shrublands) At the other extreme of the gradient, we would find species having a temperature-dependent mating season, usually occurring in spring or early summer, which is typical of temperate regions. These species would generally inhabit the moister zones in the Mediterranean domain (evergreen and coniferous forests).

This gradation of species may be invaluable for the monitoring of changes at a global scale. Periodic surveys of selected natural areas may reveal changes in the composition of amphibian communities. A knowledge of the breeding habits and phylogenetic origin of the taxa may allow for a categorization of each species, and this would reveal whether the changes are generalized towards more mesic or more xeric species. The limited number of species available, the limited mobility of amphibians in general, and the relative ease of sampling (usually sampling is made in and around water bodies), make amphibians a particularly interesting group of vertebrates for monitoring climatic changes at a regional scale in Mediterranean ecosystems. Unfortunately, an adequate historical record of the presence of amphibians is not available for comparison with current information on the distribution an conservation status of species in Mediterranean ecosystems. In 1992, an international project was submitted to the European Commission by the Western

European Group of the Declining Amphibian Population Task Force. The project involved the resampling of a dozen areas throughout western Europe for which data on the composition of the amphibian fauna had been gathered 10 to 15 years ago. Originally, areas in Spain, Portugal, France, and Italy were included in the project, representing a variety of Mediterranean habitats. Unfortunately, the European Commission declined to fund the project.

At the Spanish level, there has been some progress since the Spanish Institute for the Conservation of Nature (ICONA) funded a project in 1991 and 1992 for the survey of areas of particular herpetological importance. This survey, carried out by 32 teams of members of the Asociación Herpetológica Española (AHE), provided the first information at a regional scale on the distribution and abundance of amphibians and reptiles over a wide variety of habitats using a standardized methodology. The survey included a total of 84 areas sampled in 10×10 km quadrats covering most of the Spanish insular geography and a sizeable percentage of its continental territory. Although this effort was significant, such surveys ought to be completed periodically in order to establish temporal trends which may reflect large-scale changes.

Reported Trends Revealed by Recent Surveys

In the last four years, the IUCN/SSC Declining Amphibian Populations Task Force has centralized information on general trends of amphibian populations around the world. Recently a first document was released summarizing the data gathered so far (Vial and Saylor, 1993). Although this report is meant to be only a first step, it presents the results of a few ongoing monitoring projects of amphibian populations throughout the world. The most comprehensive surveys deal with the amphibian fauna of California (Fellers and Drost, 1993; Martin, 1993; Schaffer et al., 1993). The projects surveyed extensive areas of protected and unprotected habitats for which adequate data on species composition and abundance were available through museum collection records. The results showed that substantial decines in amphibian species diversity were evident throughout the whole state. Fellers and Drost (1993) surveyed nine US National Parks and reported precipitous declines for six different species of anurans. Comparing to surveys conducted 75 years ago, 42% of the original populations had disappeared, and two montane species (*Rana cascadae* and *R. boylei*) were no longer present in such a substantial area of their range (western Sierra Nevadas) that there is room for concern about the extinction of the species. The authors suggested that the decline could only be accounted for by a combination of local and regional factors, which included: (1) the stocking of mountain lakes with nonnative predatory fish, (2) the gradual loss of forest meadows and associated aquatic habitats as a consequence of fire suppression policies, and (3) the loss of breeding habitats due to extended drought. The surveys conducted in the lower areas of California, including the San Joaquin valley (Schaff et al., 1993) and the coastal ranges (Fellers and Droste, 1993), report important losses in species diversity in the amphibian communities, mainly in the grassland inland communities, which suffered dramatic transformations due to their extensive agricultural use. On the other hand, dramatic changes in amphibian species diversity in the protected areas of the coastal ranges were not detected.

Conclusions

The above-mentioned examples of natural extinctions can only be explained by invoking complex climatic mechanisms. If this is the case, these instances provide some support for the role of amphibians as early environmental indicators. The International Union for the Conservation of Nature Species Survival Commission (IUCU/SSC) recently created a task force to organize a global monitoring program of amphibian populations. Among the primary objectives of this new organization is the development of a standardized methodology to monitor changes in amphibian populations. A second one is to promote long-term research projects that will provide more information in order to determine whether the declines observed are part of normal population oscillations, or are a consequence of generalized global trends (Pechmann et al., 1991). We therefore suggest that more attention be devoted to monitoring amphibian populations and that this effort should be particularly intense in transitional zones such as Mediterranean ecosystems where, more than anywhere else, amphibian populations can be outstanding indicators of environmental changes. In the words of David Wake, "They were here when the dinosaurs were here, and survived the age of mammals. They are tough survivors. If they are checking out now, I think this is significant" (Barinaga, 1990).

Acknowledgments. The authors' research is supported by project CYCIT PB 92-0091 (PI: P. Alberch) Ministerio de Educación y Ciencia (Spain).

References

Anonymous. 1991. Declining amphibian populations—a global phenomenon? Findings and recommendations. Alytes 9:33–42.
Barinaga M. 1990. Where have all the froggies gone? Science 247:1033–1034.
Beebe TJC. 1992. Amphibian decline? Nature 355:120.
Beiswenger R. 1986. An endangered species, the Wyoming toad *Bufo hemiophrys baxteri*. The importance of an early warning system. Bio Conserv 37:59–71.
Blaustein AR, Wake DB. 1990. Declining amphibian populations: A global phenomenon? Trends Ecol Evol 5:203–204.

Bradford DF. 1991. Mass mortality and extinction in a high elevation population of *Rana muscosa*. J Herpetol 25:174–177.

Burton TM, Likens GE. 1975. Salamander populations and biomass in the Hubbard Brook Experimental Forest, New Hampshire. Copeia 1975:541–546.

di Castri F. 1981. Mediterranean-type shrublands of the world. In: F di Castri, DW Goodall, and RL Specht (eds) Mediterranean-type shrublands. Ecosystems of the World Vol 11:1–52. Elsevier, Amsterdam.

Corben CJ, Ingram GJ, Tyler MJ. 1974. Gastric brooding: unique form of parental care in Australian frog. Science 186:946–947.

Corn PS, Fogleman JC. 1984. Extinction of montane populations of the northern leopard frog (*Rana pipiens*) in Colorado. J Herpetol 18:147–152.

Crump ML, Hensley FR, Clarck KL. 1992. Apparent decline of the golden toad: underground or extinct? Copeia 1992:413–420.

Diamond JM. 1984. "Normal" extinctions of isolated populations. In: MH Nitteki (ed) Extinctions. University of Chicago Press, Chicago, IL, pp. 91–246.

Duellman WE, Trueb L. 1985. Biology of Amphibians. McGraw-Hill, New York.

Fellers GM, Drost CA. 1993. Decline of anurans in national parks in western United States. Proceedings Second World Congress of Herpetology, University fo Adelaide, Australia, Dec 29 1993–Jan 6 1994.

Freda J, Dunson WA. 1985. Field and laboratory studies of ion balance and growth rates of ranid tadpoles chronically exposed to low Ph. Copeia 1985: 415–423.

Hayes MP, Jennings MR. 1986. Decline of ranid frog species in western North America: Are bullfrogs (*Rana catesbeiana*) responsible? J Herpetol 20:490–509.

Heyer WR, Rand AS, Gonçalvez da Cruz CA, Peixoto OL. 1988. Decimations, extinctions and colonizations of frog populations in Southeast Brazil and their evolutionary implications. Biotropica 20:230–235.

Heyer WR, Rand AS, Gonçalvez da Cruz CA, Peixoto OL, Nelson CE. 1990. Frogs of Boraceiá. Arq Zool S Paulo 31:237–410.

Ibarra-Vidal H. 1989. Impacto de las actividades humanas sobre la herpetofauna en Chile. Comun Mus Reg Concepción 3:33–39.

Kirk JJ. 1988. Western spotted frog (*Rana pretiosa*) mortality following forest spraying of DDT. Herp Rev 19:51–53.

Lamotte M, Lescure J. 1977. Tendances adaptatives à l'affranchissement du milieu aquatique chez les amphibiens anoures. Terre Vie 31:225–331.

Márquez R. 1992. Terrestrial paternal care and short breeding seasons: reproductive phenology of the midwife toads *Alytes obstetricans* and *A. cisternasii*. Ecography 15:279–288.

Martin DL. 1993. Standardized survey of anurans in the Sierra Nevada. Proceedings Second World Congress of Herpetology, University of Adelaide, Australia, Dec 29 1993–Jan 6 1994.

Moyle PB. 1973. Effects of introduced bullfrogs, *Rana catesbeiana*, on the native frogs of the San Joaquin Valley, California. Copeia 1973:18–22.

Pechmann JHK, Scott DE, Semlitsch RD, Caldwell JP, Vitt LD, Gibbons JW. 1991. Declining amphibian populations: the problem of separating human impacts from natural fluctuation. Science 253:892–895.

Pierce BA. 1985. Acid tolerance in Amphibians. BioScience 35:239–243.

Pierce BA, Montgomery J. 1989. Effects of short-term acidification on growth rates of tadpoles. J Herpetol 23:97–102.

Pounds AJ. 1990. Disappearing gold. BBC Wildldife December: 813–817.

Pounds AJ. 1991. The secret Sahara. BBC Wildlife 9:381.

Pounds AJ. 1992. New cues in the case of disappearing amphibians, Wildlife Conserv 16–18.

Reques R, Tejedo M. 1991. Fenología y hábitats reproductivos de una comunidad de anfibios en la Sierra de Cabra (Córdoba). Rev Esp Herp 6:49–54.

Rodríguez-Jiménez AJ. 1988. Fenología de una comunidad dc anfibios asociada a cursos fluviales temporales. Doñana Acta Vertebrata 15:29–43.

Salthe SN, Mecham JS. 1974. Reproductive and courtship patterns. In: B Lofts (ed) Physiology of the Amphibia. Academic Press, New York, pp. 309–521.

Salvador A, Carrascal LM. 1990. Reproductive phenology and temporal patterns of mate access in Mediterranean anurans. J Herpetol 24:438–441.

Schaffer BH, Scott S, Fisher RF. 1993. The genetics of decline: a case study of the California tiger salamander *Ambystoma californiense*. Proceedings Second World Congress of Herpetology, University of Adelaide, Australia, Dec 29 1993–Jan 6 1994.

Spellerberg IF. 1991. Monitoring ecological change. Cambridge University Press, Cambridge, UK.

Tejedo M. 1989. Exito reproductor y selección sexual en el sapo corredor *Bufo calamita*. Tesis Doctoral Universidad de Cordoba Facultad de Ciencias Biologicas.

Tyler MJ. 1983. The gastric brooding frog. Croom Helm, London.

Tyler MJ. 1991. Declining amphibian populations—a global phenomenon? An Australian perspective. Alytes 9:43–50.

Vial JL, Saylor L. 1993. The status of Amphibian populations, a compilation and analysis. International Union for the Conservation of Nature, Species Survival Commission, Declining Amphibian Populations Task Force Working Document No 1.

Wake DB. 1991. Declining amphibian populations. Science 253:860.

Werner EE. 1986. Amphibian metamorphosis: growth rate, predation risk, and optimal size at transformation. Am Nat 128:319–348.

Wilson EO. 1988. The current state of biological diversity. In: EO Wilson (ed) Biodiversity National Academic Press, Washington DC, pp. 3–20.

Wyman RL. 1988. Soil acidity and the distribution of amphibians in five forests of southcentral New York. Copeia 1988:394–399.

Wyman RL. 1990. What's happening to the amphibians? Conserv Biol 4:350–352.

Wyman RL, Hawksley-Lescault DS. 1987. Soil acidity affects distribution, behavior, and physiology of the salamander *Plethodon cinereus*. Ecology 68:1819–1827.

24. Conservation, Restoration, and Research Priorities for Mediterranean Uplands Threatened by Global Climate Change

Zev Naveh

There seems to be an almost general consensus among climatologists that global warming will occur as a result of rising levels of CO_2 and other anthropogenic polyatomic gases. Atmospheric concentrations of these greenhouse gases expressed in CO_2 equivalence have grown from 280 ppm in preindustrial times to 365 ppm now. At the same time, the mean global surface layer temperature has increased by about 0.5°C, and in this decade the six warmest years have been recorded. In addition, ice-core studies from Antarctica and Greenland showed that large natural fluctuations in CO_2 and methane were highly correlated with temperature changes over the last 160,000 years. These trends of climate warming have been identified in a number of comprehensive reports by international groups of prominent scientists as one of the most serious threats to the future of life on earth, e.g., Bolin et al. (1986), WCED (1987), Schneider (1989), IPCC (1990), IGBP (1990), and Jaeger and Ferguson (1991). Most of these studies assumed that a doubling of CO_2 could occur until 2050 if no countermeasures are taken. They attempted to predict its impact on the global climatic pattern with the help of General Circulation Models (GCMs) and estimated that consequently the average global surface air temperature will increase by about 2–4°C. In the Northern Hemisphere the warming in higher latidutes is predicted to the about double that in low latitudes. These models also suggest a global increase in sea surface temperatures and an increase in the average global rate of evaporation and precipitation.

These findings paved the way for the International Convention on Climate Change at the UN Conference on Environment and Development (UNCED) at Rio de Janeiro in 1992. It was generally agreed that the threats of global climate change can only be prevented by the same modern industrialized society that caused them. However, at this Rio Earth Summit conference–as well as at many other occasions–it became apparent that one of the major excuses for not taking any immediate effective action to reduce the emissions of heat-trapping and polluting gases is the fact that there are still large uncertainties with regard to the magnitude and rate of climate change.

Facing the Uncertainties of Climate Changes

The main difficulties in determining how changes in concentrations of these greenhouse gases will further affect the climate have been discussed recently by Graber and Cohen (1993). Most of these are related to the fact that the atmospheric warming is followed by other processes that may act as positive feedbacks to further increase temperatures, while others may act as negative feedbacks and inhibit this increase and stabilize it.

One of the main criticisms raised against GCMs are the coarse spatial and temporal resolutions on which their calculations are based. Their grids are too large for the simulation and evaluation of regional climatic changes in specific habitats, and their small-scale topographic and orographic landscape heterogeneity and their time scales are unsuitable for predictions of changes in climatic patterns, including short-term variability and extreme meteorological events, not shown by monthly and annual averages. Outlining these and other limitations of regional scenarios in the European Conference on Landscape-Ecological Impact of Climatic Change (LICC), Hulme et al. (1990) concluded that presently it is impossible to predict its future climate. Models at their current state of development are incapable of producing reliable simulations at the regional scale, and empirical studies based on past instrumental data as analogues for the future cannot serve as predictions. They therefore regarded their scenarios as possible climate outcomes for Europe in a warmer world rather than as probable outcomes. For this reason, the "what if" approach was adopted in this conference, focusing on the climate sensitivity of key biotic and abiotic processes operating in terrestrial landscapes (Boer and de Groot, 1990).

However, even such an approach faces the limitations of uncertainty because of our present lack of knowledge of the interactions between these biotic and abiotic processes. According to Woodwell (1992), bio-spheric positive feedbacks may act to affect the rate of warming. Thus a 1°C increase in temperature is widely recognized as increasing the rate of respiration by 10% to 30% while having little effect on photosynthesis and thereby stimulating the release of both CO_2 and methane, adding to the greenhouse effect. Biotic influences can be large enought to change the amount of CO_2 in the atmosphere by several percent within a few weeks. The destruction of the biota globally is not only contributing directly to biotic impoverishment but is also speeding the atmospheric warming of the earth. Therefore Woodwell suggests that the safest assumption is that warming will proceed rapidly over the next decades, especially in the higher and middle latitudes. Along with heating will also come increase in the frequency of destructive events such as fires, hurricanes, and drought. These increasing frequencies of destructive events, affecting biotic processes and patterns may be more important than temperature change itself in their threats to life.

The same may be true also for synergistic effects, increasing the risk factor introduced by the uncertainty in the responses of natural and cultivated vegetation to the synergistic interactions with other human-caused disturbances, such as acid rain, photochemical oxidants and other pervasive pollutants, stratospheric ozone, and habitat destruction. Another example of synergism, mentioned by Myers (1992) in reviewing this problem, is the increase in susceptibility to disease of plants under high temperature or reduced water stress. Conversely, diseased or otherwise damaged or weakened plants are less able to cope with the onset of unduly elevated temperatures or reduction in moisture. They could also become subject to pandemic diseases and pathogen-carrying insects and invasion of aggressive competitors, such as might occur through ecological disruptions of a greenhouse-affected world. He concluded that in light of such synergistic interactions, of which we know almost nothing, we could well be facing a greater and more rapid mass extinction than is generally expected to date. There is therefore an urgent need for well-coordinated interdisciplinary research that could be regarded as a creative challenge for conservation biologists and ecologists. Harte et al. (1992), discussing the uncertainties created by the existence of a myriad of indirect linkages between climate change and biodiversity, concluded that "we hedge our bets and implement a wide variety of protective policies, for no single approach is likely to be successful." They maintain that although parks will and should continue to be the core of our efforts, they will not be a panacea for protecting biodiversity in a world of changing climate. If parks become habitat islands, extinctions are more likely than invasions, but the consequences of the species reshuffling cannot be predicted. Given such an uncertainty of future distributions of species, we should minimize species extinctions from existing threats and look for conservation tools in addition to parks. These include the conservation of biodiversity in unprotected seminatural rangeland, forests, and deserts by a multitude of management policies, such as sustainable timbering practices, regulating recreational uses, constraining locations of roads and human dwellings, and confining grazing activities to protect streams and riparian zones. They stressed that the challenges for formulating a viable species conservation policy are not less important than the challenge for forecasting effects. But first of all we will have to overcome counter-productive disciplinary barriers to such research.

Recent unexpected results of deep ice cores in Greenland from the previous interglacial period, lasting around 20 thousand years and ending 115,000 years ago, require a reassessment of climatic control mechanisms (GRIP, 1993). These findings show that at that time Greenland experienced a succession of short-term oscillations of 10–30 years, switching between phases about 2°C warmer that now and about 5°C cooler, and indicating a great variability in the duration of the different climatic modes. Although the longest uninterrupted warm period was around 2,000 years, there were several cool (and warm) events lasting for only a few hundred years or less. Apparently there was an inherent instability in regional (and global) ocean circulation and ocean transport processes. This introduces a further risk factor, because enhanced greenhouse effects, due to human activities, could trigger abrupt and highly disruptive climatic changes, rather than gradual global warming (Williamson and Oeschger, 1993).

There are some clear indications that in the eastern Mediterranean we have already entered such a period of disruptive changes and of increase in climatological fluctuations and unpredictability. Recent findings of scientists from Greece and Israel identifed a trend of decreasing sea surface temperatures in the eastern Mediterranean (Lascaratos et al., 1991; Kutiel and Bar-Tuv, 1992), decreasing daily temperatures (Katsoulis, 1987), decreasing daily temperature ranges (Jaffe, 1992), and decreasing potential evapotranspiration rates (Mandel, 1986). There also seems to be a clear indication of the destabilization of the atmospheric circulation patterns and pressure systems affecting seasonal rainfall patterns, which may have far-reaching ecological implications. Sea level pressure fields show a retardation of about 6 weeks in the building of the Siberian High, resulting in a similar retardation in the rainfall regime in Turkey (Wigley and Farmer, 1982) and in Israel. The delay of this anticyclone could be attributed to the warming in the higher latitudes, due to an increase of greenhouse gases in the atmosphere (Kutiel, 1992). This subtropical high pressure belt, in general cause the prolonged summer drought and its decrease will enable some of the tropical late spring and summer monsoon rains to reach the southern parts of Israel. This pattern apparently reinforces the model of a multimodal course of rainfall (Kutiel, 1985), which further increases rainfall variability and decreases its monthly predictability.

These disruptive trends of the climate patterns have become even more apparent in recent years in Israel. They are characterized chiefly by an increase in the frequency of extreme meteorological events and an even clearer shift in spatial rainfall distribution, with relatively larger amounts in the southern part of Israel and smaller amounts in the northern parts, resulting in a further increase in the coefficient of rainfall variability (M. Kutiel and H. Steinberger, personal communication).

Thus, as in the two preceding rain seasons, also in 1990–91, the deficit in total rainfall in northern Israel was larger than in the central part. On the other hand, in southern Israel the rainfall was even above the long-term mean, and here some unusually high amounts of precipitation were recorded during several days in January. Contrary to the normal pattern, rainfall was much higher in March–May than in the early season, which was extremely dry and had some unusual warm days with 3°–4.5°C above the normal measures and up to 45% higher evaporation rates in certain locations. A similar very dry and even warmer early rain season is occurring presently between November 1993 and January 1993–94. However, winter 1991–92 was one of the wettest and coldest winters ever recorded in Israel. It was declared a natural disaster season for agriculture because of the combination of unusually extreme meteorological events and severe conditions of heavy and prolonged rainfall and lack of sunshine, floods, cold waves, frost, snow, and strong winds, causing heavy damages. In that year even the absolute rainfall amounts in central Israel were higher than in the northern parts. They reached 1,500–1,700 mm, 200% to 220% of the long-term mean, as compared to 140% in northern parts (Gat and Rubin, 1993).* In conclusion, judging fom all this sometimes even contradictary evidence for the expected warming and drying tendencies, presently only a clear tendency for the decrease in the predictability of eastern Mediterranean climate patterns seems to prevail, and consequently an increase in climatic uncertainty.

Further warming at the higher latitudes would probably cause an even greater distortion of Mediterranean rainfall patterns. At the middle latitudes of the Mediterranean, this could lead to the alarming scenario of increasing aridity with potentially far-reaching and wide-ranging environmental effects and mostly disastrous socioeconomic impacts. At the

* At the time of the final editing of this paper, in March 1995, the same trends were observed of increased climatic unpredictability and of higher rainfall amounts in central Israel, reaching, f.i. in Tel Aviv 792 mm (the annual long term mean is 560 mm), as compared with the Galilee in northern Israel with only 676 mm (the long-term meaning is 925 mm). However the distribution was again a very abnormal one, and this year it was concentrated chiefly in the early rainfall season, whereas the usually highest rainfall and coldest months, January and February were much drier and also warmer.

Mediterranean Working Group of the above-mentioned LICC meeting, it was assumed that an increase of 3°–4°C in average temperatures in the Mediterranean Basin could lead to an increase in potential evapotranspiration of 400 mm/year. Together with a changing frequency and magnitude of rainfall events and soil degradation processes decreasing its water-holding capacity, this could lead to an increase in aridity, even if the annual amount of precipitation were unchanged (Boer and De Groot, 1990).

According to the findings of a special UNEP Mediterranean task force, an assumed temperature increase of 1.5°C by the year 2025 will cause an increase of potential evapotransporation throughout the Mediterranean, coupled with a possible decrease in precipitation in the South and an increase in the North. Hot, dry summers and exceptional events of drought or rainfall and floods, marine storms, tidal surges, and water stagnation and eutrophication could become more frequent. However, future adverse impacts of exponential population growth and urban-industrial development may far exceed the direct impacts of climate change. They will cause continuous increase in society's vulnerability to climatic stress, particularly in the South (Jeftic, 1993). In the semiarid Mediterranean region, the resulting desert encroachment is presently estimated at more than 2% per annum, and in combination with climatic changes it will have truly catastrophic consequences (Le Houerou, 1988; Sestini et al., 1989).

In view of this chaotic and nonlinear behavior of atmosphere–biosphere interactions and of our almost complete lack of knowledge from the Mediterranean on the response of organisms, populations, communities, ecosystems, and landscapes at different time and space scales to these uncertainties, it would be futile, at this stage, to attempt any robust models and scenarios.

On the other hand, we can anticipate with much greater certainty that any increase in climatic stresses would further aggravate the already existing, alarming processes of overall landscape degradation. Therefore, in order to face these challenges of climatic uncertainties and to prevent, as far as possible, their adverse repercussions on the open Mediterranean landscape, we have to act without delay on a broad front of research and conservation issues.

For this purpose, I will first review briefly some of the most relevant ecological and evolutionary features and processes on which these conservation and research priorities should be based, and then outline the holistic ecological strategies and tools by which these priorities could be realized. For most of these issues, I will use Israel as an example, but the same principles apply to the other much more extensive regions, especially in southern Europe and North Africa, located in the drier fringes of the Mediterranean region.

Past and Present Landscape Dynamics, Stability, and Diversity

Most of the Mediterranean uplands, too steep and rocky or with soils too poor and shallow for profitable intensive agriculture, make up great proportions of all these countries. Although of low economic value, their vegetation canopy has great ecological importance for environmental and watershed protection of the densely populated valleys and coastal regions and as a genetic source for future utilization. Together with the rapidly vanishing coastal dunes and undrained wetlands and marshes, they are the last refuges of the natural, spontaneously occurring and reproducing Mediterranean plants and animals.

As described elsewhere in more detail (Naveh and Whittaker, 1979; Naveh and Lieberman, 1993; Naveh and Kutiel, 1990), these landscapes are distinguished by their striking floristic, faunistic, and structural diversity and scenic attractiveness. These are the result of the unique combination of both natural and cultural factors, inducing variability in time and space: 1) the great ecological micro- and macrosite heterogeneity, 2) the mild winter rainfall climates with great long- and short-term fluctuations, and 3) millenia-lasting human perturbations and defoliation pressures, first by periodic, prescribed burning and later also by grazing, cutting, and coppicing.

Their final shaping took place from the Middle Pleistocene onward in a coevolutionary process with Mediterranean paleolithic food gatherer-hunters. There were gradually replaced in the early Holocene by agro-pastoralists in closely interwoven natural and cultural processes of agro-pastoral, multifactorial land use functions, which lasted until modern times. These processes created complex and dynamic vegetation regeneration and degradation patterns in which the different woody and herbaceous strata were maintained in a dynamic flow equilibrium. In such a so-called homeorhesis (Greek: maintaining the flow), the system is going on to change along the same perturbation trajectory as in the past, as long as similar perturbations are driving these changes. It is thereby returning to its natural and cultural closely interwoven dynamics but not to an artificial, undisturbed, and stationary state of homeostasis, as assumed in the classical succession-to-climax theory. These perturbation-dependent systems and their landscapes have thereby acquired long-term adaptive resilience and metastability, and their homeorhetic flow equilibrium can be maintained and restored only by the continuation of optimum defoliation pressures. From a thermodynamic point of view, they can be regarded as "non-equilibrium, dissipative systems." According to Prigogine (1976), they create "order through fluctuation." Such dissipative structures are maintained and stabilized only by permanent energy/matter and entropy exchange processes. Driven by positive feedbacks of environmental and internal fluctuations, they move to new regimes that generate the conditions of renewal of higher entropy production. In the case of Mediterranean upland ecosystems, such new regimes are apparently created by the periodic perturbations of fire, grazing, and cutting (Naveh, 1987, 1991).

The outcome of these interactions between nature and culture is still present in a great number of not yet irreversibly despoiled seminatural and agro-pastoral landscapes. These form a matrix of fine-grained, heterogeneous mosaics of forests, woodlands, shrublands and grasslands, interspersed with cultivated field, vineyards, olive groves, and orchards. However, recent dramatic changes in land use have severely distorted the human-maintained flow equilibrium, with mostly adverse and far-reaching undesirable consequences for overall environmental quality and stability. This results from the unfortunate combination of exponential growth of populations and their growing need for land, coupled with deviation-amplifying (positive) "run-away" feedback loops with land-use pressures of even faster-growing mass tourism, together with the combined and synergistic processes of intensification of traditional and modern agricultural land uses and uncontrolled urban-industrial expansion.

As discussed recently in detail in a conference on the future of Mediterranean landscape (Farina and Naveh, 1993), in this process of "neo-technological landscape despoilation," nontillable and marginal uplands are either left to their fate of abandonment and neglect, or are exposed to agro-pastoral over-exploitation and uncontrolled urban industrial encroachment. Its major impact is the rapid loss and fragmentation of unspoiled open landscapes, their soil, plant, and animal resources, and their "life-supporting" production, protection, carrier, and regulatory functions. This process is aggravated by he gradual functional, structural, and visual degradation of the remaining open landscape with concomittant biological impoverishment and ecological disruption by accelerated erosion and soil, water, and air pollution from urban-industrial and agricultural sources.

Other, mostly adverse, impacts on the mountainous regions are the creation of monotonous and highly flammable conifer plantations and the spread of scrub thickets from abandoned terraces and pastures. Because of the accumulation of dry and dead branches and undecomposed litter and the increase in uncontrolled human and recreational activities, there is an alarming increase in more and more devastating and hotter wildfires. In the absence of efficient fire and fuel management policies, including prescribed burning, and in spite of the costly but futile efforts for total fire suppression, wildfire damage and sometimes even human losses are becoming greater from year to year.

In addition to the loss of open landscapes and their fragmentation by urban-industrial developments, the floristic and structural diversity of the natural vegetation, as well as the faunistic species richness and abundance, is constantly being reduced. This impoverishment is on the one hand the result of heavy human and livestock pressures in densely populated hill and mountain areas, chiefly in the Levant and in North Africa, and on the other, that of complete cessation of human interference in depopulated areas, as well as in protected nature reserves and parks. In many of these, moreover, heavy pressures of mass recreation are threatening floristic and structural diversity.

In general, after the release of human and livestock pressures, there is an initially vigorous vegetative regeneration of the woody sclerophyll vegetation from stunted shoots and encroachment of the shrub canopy. But lack of light and space leads gradually to the almost total suppression of the herbaceous understory in undisturbed and/or protected maquis and forests. These dense and monotonous thickets are dominated by a few species, mostly oak and/or pine, which become stagnant and senescent and more and more prone to hot and devastating wildfires. Thus, contrary to preconceived climax theories, noninterference is turning from a blessing to a curse.

The possibilities of the above-mentioned global and regional climatic changes in rainfall and temperature regimes may further aggravate these threats. In the more humid western Mediterranean biomes, these could further enhance the encroachment of brush in abandoned and neglected forests and maquis and their floristic impoverishment. Their increased fuel loads will cause even hotter and more destructive wildfires. However, in the subhumid biomes, such as those of northern Israel, increasing climate stresses, and especially the anticipated rise in summer temperatures and in evaporation rates, will have severe repercussions on the most mesic and richest plant communities, their diversity and stability. Most prone would be the more demanding eumediterranean herbaceous and woody plants in sheltered sites and more mesic western slopes on altitudes above 500–600 m, including evergreen and deciduous forests and maquis such as *Quercus calliprinos-Pistacia palestina galilea*, rich in arboreal species and distinguished by a wealth of rare and endemic species. In Israel, these plants enjoyed the most favorable climatic conditions with the most predictable and highest total rainfall of 800 mm and more. However, as mentioned above, there are already alarming indications of a destabilization of the favorable rainfall pattern and a rise in its variability.

In the drier semiarid ecotones of the Mediterranean, this process of dessication, in combination with heavy human and livestock pressures, may enhance even more the process of land denudation and desertification and may thereby cause the advancement of semiarid and semidesertic formations far into the heart of the Mediterranean biomes.

As documented by Naveh and Dan (1973), the harsher and more fragile the environment, the more far-reaching and irreversible will be the impacts of climatic and man-induced changes and the slower and more difficult will be the recovery processes. Thus, the complete destruction of the vegetation canopy in the semiarid zone at a very early date by grazing and cultivation exposed the soil to the direct impact of the harsh climate and to impeded rainfall and created conditions similar to those of the more arid regions with a scarce plant cover, climatic extremes, and flash floods. However, already at that time, they predicted that if these threatening trends of neotechnological landscape despoilation proceed unhampered, then even without climatic changes a similar fate will befall the more mesic Mediterranean mountains and hill regions, and the few remaining nature islands will be turned into overcrowded recreational slums, like most of the shores and inland waters of the Mediterranean Sea.

Effect of Climatic Variability on Herbaceous Plants

Let us now consider some of the ecological consequences of such climatic fluctuations on plant production and diversity, which seems to be presently the most crucial feature of global changes in the Mediterranean. In spite of the great progress achieved in our knowledge of the physiological basis of climatic stress responses in some important woody Mediterranean plants, we have still very little reliable information on these responses of woody plants and communities in the field.

For herbaceous plants we can utilize the information gathered in our earlier, long-term study in the deciduous open woodlands of Tabor oak (*Quercus ithaburense*) in the foothills of the lower Galilee. With 550 to 600 mm average annual rainfall, the annual rainfall variability was 20–30%, as measured by the mean deviation from the long-term average. But the monthly variability increases from 46% in the highest rainfall month of January to 77% in November, 88% in April, and 113% in October at the beginning of the rainly season. Like all other Mediterranean oak savannas, these serve as important cattle and sheep pastures. They are distinguished by their very species-rich and diverse herbaceous understory and are dominated chiefly by annual plants (Naveh and Whittaker, 1979).

As reported elsewhere in detail (Naveh, 1982, 1988), the great dependence of their productivity–and consequently also of livestock production–on annual and seasonal climatic fluctuations was demonstrated convincingly during a 6-year study of pasture and livestock production. In this period, the rainfall amplitude ranged from 380 to 708 mm between the driest and wettest year with a coefficient of annual rainfall variation of 23.3%. However, the amplitude of phytomass production was four times greater, ranging from 110 to 448 g/m^2, and its coefficient of variation was 64.6%. The correlation coefficient between total annual rainfall and phytomass production was 0.47. But a much greater climatic dependence of herbaceous phytomass production (as well as livestock production) could be detected by taking into account not only the total annual rainfall but also bimonthly rainfall distribution and minimum winter temperatures. These climatic variables could explain 95% of the variance in pasture and

Table 24-1. Rainfall and Temperatures in a Dry Year (1954–55), a Wet Year (1955–56), and an "Average" Rainfall Year (1956–57), but with Delayed Early Rains and a Cold January

a. Mean Temperatures in December–January (°C)

	December			January		
	Minimum	Maximum	Daily Average	Minimum	Maximum	Daily Average
1954–55	8.5	18.9	13.7	7.9	19.7	13.8
1955–56	9.0	17.7	13.4	7.1	18.4	12.8
1956–57	8.3	18.9	13.4	4.6	16.6	10.4

b. Monthly Precipitation (mm)

	1954–55	1955–56	1956–57	Average 1931–60
September				1.7
October		7.0		16.8
November	91.8	125.5	17.2	80.9
December	176.8	247.2	153.9	128.9
January	2.2	164.3	132.5	157.3
February	38.1	44.0	109.0	108.2
March	51.7	108.7	124.6	60.7
April	11.7	12.1	22.7	17.9
May	7.8		7.8	5.8
Total	379.6	708.8	567.7	578.0

livestock production. Intensive improvements by fertilizing and selective chemical weed control almost doubled average annual meat production from 105 to 194 kg/ha, but they exacerbated these interseasonal fluctuations even more and thereby demonstrated climatic dependence.

Of special interest were the effects of extreme climatic events, observed in this study during three consecutive, contrasting rainy seasons. As shown in Table 24-1, 1954–55 had the lowest annual rainfall of 380 mm and only 281 mm after December; 1955–56 had the highest and most well-distributed rainfall of 709 mm; and 1956–57 had about "average" rainfall of 568 mm, but with delayed early rains and January temperatures 2°–3°C lower. In Figure 24-1. the actual above-ground phytomass production during these three seasons is presented, as determined before each grazing cycle.

These not only show the striking differences between a very favorable and a very dry year, but also demonstrate that the total annual rainfall is not the only—and in the case of 1956–57—even not the most important independent climatic variable.

The striking rise in species richness from the driest to the following wettest year and again a drastic drop in species numbers, especially of legumes and other broad-leaved forbs is shown in Figure 24-2. These

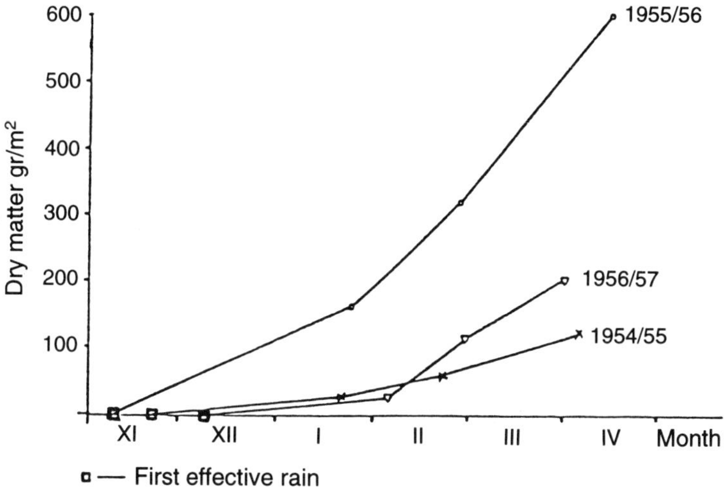

Figure 24-1. Herbage production in a dry year (1954–55), a wet year (1955–56), and an "average" year, but with delayed early rains and a cold January.

results show that a great number of the less abundant species, which contribute the most to species diversity, are the most sensitive to climatic fluctuations.

Favorable growth seasons with well-distributed rainfall from November through March, such as 1955–56, have apparently ensured the main-

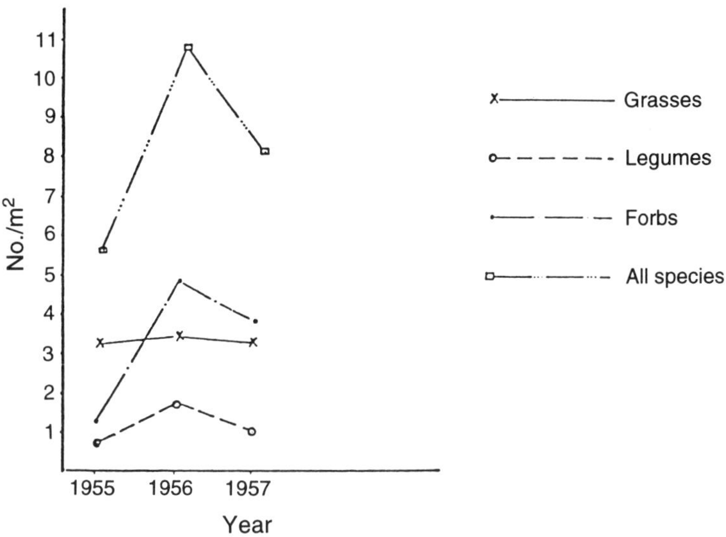

Figure 24-2. Changes in the number of species in eight permanent quadrats on April 1, in dry, wet, and "average" rain seasons.

tenance of seedbanks in the soil, providing rich reservoirs for floristic and genetic diversity and for natural selection to operate on. If such seed pools are available, they will greatly reduce the fitness uncertainty. This was apparently the case during millennia of traditional agro-pastoral land uses. Our studies showed that under moderate, rotational grazing management, such a species-rich, diverse, and productive herbaceous pasture vegetation can be maintained on a sustained basis in the woodland understory, in open, grassy patches in forests and maquis-shrublands, and in open grasslands. However, after short periods of too light grazing or complete protection, aggressive tall grasses and hemicryptophytic thistles will take over. Under heavy, uncontrolled grazing pressures, the least palatable and the most undesirable weedy species will dominate. Thereby both biological diversity and highly valuable economic productivity will be reduced greatly (Naveh, 1957; Naveh and Whittaker, 1979).

It is also obvious that the chances for such rich seed pools will be reduced greatly by an increase in climatic unpredictability. Therefore, climatic instability and stress, in combination with too light or too heavy grazing pressures, could lead to synergistic interactions, severely endangering the biological diversity, productivity, and stability of all Mediterranean subhumid and semiarid woodlands and grasslands in Israel and elsewhere and accelerating the rates of desertification processes in space and time. In view of the above-described climatic trends, it is very possible that such an acceleration is already happening.

Adaptive Genetic Diversity and Resilience

Of great relevance for our discussion are the important genetic-ecological studies carried out by Professor E. Nevo and his collaborators from the Institute of Evolution, Haifa University, shedding new light on the evolutionary basis of the Mediterranean biota and its adaptive responses to climatic stress. Their genetic-ecological studies of 21 species of plants and animals of the Mediterranean and steppe region range from the molecular to the landscape scales and include the two most important progenitors of our domesticated cereals, the wild emmer tetraploid wheat *Triticum dicoccoides* and the wild barley *Hordeum spontaneoum*. These studies revealed allozyme diversity and DNA polymorphisms which are clearly correlated with ecological factors. All these factors share a common denominator of increasing aridity stress as a major selective force for short- and long-term survival, and on the transition to the xeric region, increasing climatic fluctuations, both eastward and southward. Nevo (1988) and Nevo and Beiles (1989) concluded that this was the result of long-term evolution in the widely diverse climatic and edaphic macro- and microsite conditions of Israel.

On landscape scales, the striking spatial topoclimatic and ecological heterogeneity of these upland and mountain landscapes also seems to have induced a great biogeographic, taxonomic, and genetic diversity, even within short distances. Thus, for instance, even in the heart of the Mediterranean territory in Israel, on Mt. Carmel, Nevo et al. (1982), found in opposing north and south slopes, separated by only a few hundred meters, dramatically different populations of European and African floral and faunal elements. With distinctly higher genetic and species diversity on the drier and much more exposed and open southern slope, which is exposed to more than 100% higher radiation (Nevo, personal communication). These differences involve unrelated taxa such as plants, insects, snails, reptiles, birds, and small mammals. These ecotones are presently being studied closely in all their genetic, biological, ecological, climatological, and geographical aspects by an interdisciplinary team led by Prof. Nevo. Long-term monitoring will undoubtely reveal important information on species, population, community, and ecosystem dynamics under increasing climatic stress.

In a recent extensive study of the photosynthetic performance and ecological and genetic predictability in 107 genotypes from 27 wild emmer wheat populations in the Mediterranean territory of Israel, Nevo et al. (1991) found that the highest photosynthetic efficiency was displayed by populations of xeric marginal populations from the Samarian and Judean hills and from Mt. Gilboa. Genotypes and populations with high photosynthetic capacity can be identified by climatic factors and isozyme markers. Climatic variables alone significantly explained a substantial amount of all the variance of all photosynthetic variables, and especially water use efficiency. The main climatic variables were water availability factors, radiation, and temperature variables.

The results of these studies have far-reaching implications for our decisions on research and management priorities. They can serve as important guidelines for in-depth analysis of genetic diversity from the DNA and protein level to the landscape level. Such an analysis is crucial for the understanding of the evolutionary process of adaptation, speciation, and resilience under increasing climatic stress. They will provide the scientific basis for the prevention of the extinction of major herbaceous and woody keystone species, which are essential for the maintenance of ecosystem processes and landscape stability and integrity.

Nevo's studies also point out the great importance of the drier Mediterranean biomes, and especially their semiarid fringes along the 400-mm isohyet. It can be envisaged that under increasing climatic stresses, these most xeric and resilient vegetation types of the drier Mediterranean biomes may become the last refuges of Mediterranean plant and animal diversity. Here the growth season is not only shorter, but inter- and intraseasonal rainfall predictabilities are much lower, and moisture stress for summer-growing woody plants is more severe. These harsh conditions favored the development of an inherently more drought-tolerant, xeric, and thermophyllous flora which will most probably be better adapted to increasing climatic stresses and therefore could also be more resilient and elastic than those from more mesic habitats. If not abused by overgrazing and cutting or planted with dense monospecies pine plantations, they can be very rich in drought-evading shrubs and dwarf shrubs, grasses, and herbs, including the progenitors of cultivated cereals and other plants with great genetic potential for pharmaceutical, agricultural, and other future economic uses. They also have many sclerophyll woody and other taxa in common with the above-mentioned, more mesic biomes.

In such a semiopen, mosaiclike, low shrub-grass community on Mt. Gilboa, with widely scattered *Ceratonia siliqua* trees and denser stands of *Pistacia lentiscus* shrubs, we encountered the highest vascular plant species richness and diversity of all our study sites in northern Israel: 180 different vascular plant species in 1/10 hectare, including 147 herbs with 102 annuals and many geophytes, including the highly ornamental "Iris of Mt. Gilboa" (*Iris haynei*) and smaller shrubs and dwarf shrubs. It also ranked highest in animal species richness, with 14 reptile, 7 rodent, 31 bird, and 13 isopod species (Naveh and Whittaker, 1979).

With only 400 mm annual rainfall, 60 very dry and hot "sharav" days, and mean temperatures of 33°C at noon in June, this was the driest site of all our study plots. We explained its great biodiversity by its unique nature as a biological, ecological, and geographical ecotone, serving as a meeting point between the Mediterranean territory of the subhumid Samaria mountains (as their isolated outpost) and the semiarid Mt. Gilboa proper, to which even more xeric species have invaded from the lower-altitude semidesert Jordan plains. Having been exposed to traditional human and grazing pressure for many centuries, it is now grazed chiefly by gazelles. However, it is not included in the Gilboa Nature Reserve, and its fate is not assured at all.

From Mt. Gilboa, where Nevo and his coworkers found genotypes with the highest adaptive survival potentials and photosynthetic capacities, we also obtained the first proof that sclerophyllous woody plants may have undergone similar climatically induced intraspecific ecotypic differentiation. This is indicated by the results of a study on ecophysiological intraspecific variation in the sclerophyll evergreen shrub *Pistacia lentiscus* in northern Israel by Shaviv (1978). This shrub is one of the most important Mediterranean "keystone" species, because of its wide abundance, resilience, and effective soil improving and protecting features. In comparison with other populations of this shrub from much more mesic sites, the Gilboa population is distinguished by the most efficient active and passive water saving economy and the capability not only to survive during the hot summer months, but even to continue active photosynthesis and carbon fixation. It reached maximum photosynthesis at 35°C, as compared to 25°C for the Mt. Carmel population, and five times higher absolute photosynthetic values. This was the only ecotype to continue photosynthesis at high temperatures of 45°C, and at the same time it exhibited the lowest increase of respiration rate with increasing temperature, and had still positive values for photosynthesis/respiration at 40°C. It had the largest number of stomata per unit leaf area, the most rapid decrease in respiration rate when air humidity was lowered, and the lowest transpiration rates even when water was available, with low water potentials and a high albedo in the 500–700 mm region. In *Pistacia lentiscus* we also found striking ecotypic variation in root growth strategies, and those growing in poor calcareous soils were found most promising for soil erosion control and improvement and for the restoration of denuded slopes (Yogev and Naveh, 1986).

In conclusion, we should regard these xeric Mediterranean ecotones as a major genetic and ecological bank and climatic stress refuge, deserving urgent protection of their genetic and biological diversity.

Conclusions

The effective prevention of further disastrous increases in greenhouse gases level in the atmosphere, as well as of other detrimental and irreversible human-caused impacts on the ecosphere, will be possible, in my opinion, only by a substantial reduction of wasteful production and consumption in the industrial societies and by drastically curbing the population explosion in developing countries. The outcome of the recent United Nations Conference on Environment and Development in Rio de Janeiro in June 1992 has not given much hope for what Di Castri and Hansen (1992) called "a worldwide geopolitical willingness and a kind of new international economic order." This could have become a major driving force for such a cultural and socioeconomic environmental revolution. At the same time, however, we as ecologists and land managers may at least be able to help in reducing and mitigating the damaging effects on our few remaining less despoiled natural and seminatural systems. This can be achieved in the Mediterranean–as well as elsewhere–by a significant improvement in land use management, aimed at the protection of biodiversity and of other unique natural and cultural landscape assets. Any measure taken now to reduce and counter land despoilation and denudation and to minimize the loss of plant and animal species, especially in the most vulnerable sites in the mesic and drier ecotones, will increase the resilience of these landscapes and have far-reaching beneficial implications.

However, if the heavy and uncontrolled human and livestock pressures prevailing in most of the drier critical Mediterranean regions in southern Europe, the Levant, and North Africa continue unhampered, their destabilizing, mutual causal feedback loops with increasing climatic stress will accelerate the process of desertification. This could lead to an even more rapid shift of the subdeserts and desert biomes far into the Mediterranean territory in Israel and elsewhere. This will cause catastrophic destabilization of the drier mountain and hill landscapes, similar to those which have already occurred since the neolithic revolution during prehistoric and historic times, but within much shorter time spans.

We may conclude, therefore, that the threat of climatic change is a compelling additional reason to conserve and restore the health, integrity, and diversity of all natural and seminatural landscapes that are not yet irreversibly despoiled. This would be the best insurance policy against climatic uncertainties and environmental perils to rescue their vital ecological, socioeconomic, and evolutionary functions and their intrinsic natural and cultural values:

1. It would ensure sufficient landscape connectivity to allow free plant and animal migrations and nutrient movements across the landscape and to counter further fragmentization, urbanization, agroindustrial pollution, and scenic despoilation of the open landscape.
2. It would provide the greatest options for the sustainable functions of live-supporting ecosystems and for the redundancy of their functional groups and keystone species ensuring all vital ecological processes, sensu Chapin et al. (1992).
3. It would afford the best chances for the survival of rich biotic communities and their free and spontaneously developing and reproducing populations, and thereby ensure also further evolution of plant and animal species.

For this purpose we have to distinguish between two major ecosystem and landscape types:

1. Natural and close-to-natural ecotopes, in which flora, fauna, and ecosystem structure and function have undergone very little human modification. These are very rare in the Mediterranean and have remained only in inaccessible sites, such as rock cliffs and steep mountain slopes. Their major conservation object should be the complete protection from any interference and disturbance in order to safeguard not only their biota but also their natural patterns and cycles of self-regulating mechanisms and ecological processes within strict nature reserves.
2. Seminatural ecotopes, in which the largely spontaneous flora and fauna and their self-regulatory mechanisms have undergone the above-described far-reaching modifications through intensive and lasting anthropogenic perturbations. These include the few remaining wetlands, riverside ecotones, and other coupling ecotones, along with aquatic ecosystems in the lowlands and the sclerophyllous, coniferous, and mixed forests and the much more extensive woodlands, shrublands,

dwarf shrublands, and derived grasslands. These contain most of the natural Mediterranean plant and animal communities in different stages of degradation and regeneration.

Here our primary conservation object is twofold:

1. The prevention of undesirable, destructive processes, including too heavy mass recreation and livestock pressures and hot wildfires, caused by the accumulation of highly inflammable fuel, as well as water, air, and soil pollution.
2. The reestablishment and perpetuation of those multifactorial, homeorhetic flow processes that ensured the fine-grained spatial heterogeneity, the biological and cultural diversity and metastability, and the attractiveness of the Mediterranean landscapes. This can be achieved by developing detailed long-term but flexible management plans in space and time according to local site conditions and requirements. It will include different defoliation pressures of grazing, browsing, cutting, coppicing, pruning, and prescribed burning, rotated in space and time, as well a completely protected "control" treatment. A clear distinction in conservation strategies also has to be made here:
   - In nature reserves the main aim should be the highest attainable biological productivity, diversity, and stability by restoring all ecological processes and their protection, regulation, and carrier functions through scientific conservation management.
   - In nature recreation parks, this aim should be optimized with recreation amenities to allow maximum enjoyment with minimum damage to natural and cultural resources.

This will require, above all, much more efficient protection from external pressure and damage to nature reserves and parks by uncontrolled human interventions and recreation pressure. But it would be a grave mistake to assume that by complete and prolonged protection from any human interference the entirely hypothetical and mostly illusory homeostatic "climax" state could be attained. As has been explained above, most of these protected areas have evolved as human perturbation-dependent ecosystems. Therefore, we can ensure the continuation of all vital ecological processes and their homeorhetic flow equilibrium and evolutionary metastability only by dynamic conservation and restoration management for biodiversity, fuel management, and the maintenance of grassy patches and parklike mosaics. In view of the threat of climatic change, the prevention of dense, monotonous, species-poor, and highly flammable tree and tall shrub thickets becomes one of the most urgent management goals, also in nature reserves. It will open maximum options for woody and herbaceous plants and their structural community diversity, reduce their competition for light, space, and water, and provide suitable habitats for wildlife, even under increasing climatic stress.

Nature reserves and parks, and especially their ecotones, should also serve as the major permanent study and monitoring sites for biological and physical changes through time.

We should, however, also be aware of the possibility that the present locations and sizes of these protected areas may not be adequate to cope with impending climate changes. This implies the establishment of new systems of reserves and especially biosphere reserves to provide more diversified conditions, ranging from networks of small refugia with adequate corridors to larger preservation areas (Di Castri and Hansen, 1992).

As emphasized above, highest priority should be given to the protection of the last, richest plant refuges of the semiarid ecotones in Israel and elsewhere. The crucial question is whether there will be a sufficient stock of resilient species and subspecies left from their great genetic and floristic wealth which could serve as such genetic banks for restocking more mesic sites which may undergo a creeping process of desiccation.

In addition, sufficently large areas with highly natural, cultural, and scenic values, and especially those serving as corridors and/or buffering zones with nature reserves and parks, should be declared as managed landscape areas and/or biosphere reserves. In these, nondestructive and nonpolluting economic activities for the utilization of "hard" economic benefits should be continued, and, as far as possible, traditional agropastoral land uses should be restored. In highly disturbed sites, multipurpose restoration and afforestation should be carried out, using chiefly indigenous plants.

Recent helpful guidelines on protected landscapes for policy makers and planners have been presented by Lucas (1992).

In the remaining open uplands, used primarly for socioeconomic benefits, two major ecotechniques should be applied:

1. Controlled manipulation of the soil–plant–animal complex by controlled grazing, thinning, coppicing, and pruning of trees and controlled fire and pastoral agrotechnical improvements, according to site potentials and socioeconomic and other requirements and conditions.
2. Multipurpose afforestation, restoration, and reclamation wherever land denudation has reached an advanced stage, including roadsides, camping grounds, etc. The aim is the creation of seminatural, multilayered, and stable plant communities and ecosystems with multiple ecological, economic, and social benefits.

In recent years it has become more and more evident that the establishment of fast-growing pine trees in dense monospecies plantations on shallow, rocky, dry, and steep slopes has neither ecological nor economic justification. The main reasons are their high flammability, their great susceptibility to air pollution and especially to photochemical oxidants, and their synergistic effects with pests. These undesirable features could be further exacerbated with increasing climatic stress, air pollution, and fire hazards.

There is, therefore, urgent need for the diversion of afforestation on nontillable Mediterranean uplands from commercial coniferous and eucalyptus plantations to the establishment of open and parklike multiple-layer forests. These should be derived chiefly from local stock and based on the selection and multiplication of woody, multibeneficial keystone species from the drier Mediterranean biomes and ecotones. This should prepare the ground for the planting of special drought-resistant forest reserves in which the biological diversity and productivity of the Mediterranean arboreal vegetation and its rich herbaceous understory and the scenic attractiveness of these landscapes could be safeguarded.

For this purpose, special "climate stress refuges" should be established in which promising drought-resistant ecotypes could be multiplied and stored in long-term seedbanks for restoration projects.

Research Priorities

It is obvious that all these conservation and restoration challenges cannot be carried out without full support by a comprehensive research program. This should be inspired by a holistic systems approach, bridging the gaps between the manager and researcher, the academicians and the professionals, and between the different subdisciplines and subcultures into which contemporary ecology has been fragmented. This will require a new way of thinking about ecological problems with a transdiciplinary hierarchical view, as presented recently in a very lucid way by Allen and Hoekstra (1992). It should be based on long-term, well-integrated, problem-solving-oriented studies embracing all relevant biological and ecological complexity levels and scales from the DNA and subspecies, to ecosystem and landscape-terrain levels. If such a research program is of any practical value, it must be, of course, closely related to the above-described pressing problems of conservative and sustainable land use planning and management, facing uncertainties of climatic change. Its field work should therefore be carried out in representative sites of critical areas, in combination with experimental and larger-scale management and restoration treatments and monitoring.

Highest priority should be given to genetic and ecophysiological research for the study of the genetic-ecological basis and the inter- and intraspecific adaptive plant behavior under multiple stresses in the field, determining the reproduction and competition success and survival potentials of genotypes for their selection as promising woody and herbaceous keystone species. As a model for this research, the above-described studies by Nevo and his coworkers should be used. These should be expanded also to the community and ecosystem level. Greatest attention should be paid also to the edaphic and biotic factors, the root systems, and their rhizosphere and microflora.

There is also urgent need for the systematic study of the potential effects of the rise in greenhouse gases concentration on the Mediterranean vegetation. Not only CO_2 should be considered, but also photochemical oxidants and especially ozone, which is already causing widespread damage by chlorotic decline to *Pinus halepensis*, *Pinus pinea*, and broad-leaved maquis trees (Naveh et al., 1980; Busotti et al., 1991).

Combined field and fumigation studies in comparable conditions in coastal southern California indicated an aggravation of ozone formation under higher summer temperatures, with far-reaching effects on the natural vegetation (Malanson and Westman, 1991). Such studies could serve as a model for similar research in the Mediterranean proper.

It should be realized, however, that in reality it will be very difficult to separate the effects of climatic stress, as defined by the adaptive tolerance of the different species, and chronic and acute disturbances by grazing, cutting, fire, pests, and air pollution. These are coupled by deviation-amplifying (positive) feedback loops, and their interaction determines not only the fate of the individual plant and its community, but that of the landscape as a whole. Their understanding and study requires innovative, interdisciplinary approaches and methods. The same is also true for landscape ecological studies on the spatial and functional organization and the dynamics of landscape patterns and processes and on production, regulation, and carrier functions, as affected by different land use regimes. Such studies have to take into consideration different scenarios of climatic changes.

Such interdisciplinary, long-term multilevel and multidimensional research, combining field and laboratory investigation with integrative ecosystem and landscape studies, are presently completely lacking in the Mediterranean. The creation of special research funds, earmarked for these purposes, from international, European, and national sources will be necessary, in addition to the already existing funds, such as those of the European Union programs.

But even if sufficient funds were available today, there are almost no scientists in the Mediterranean who are acquainted with these methods and there is no proper interdisciplinary organization available for their training. These problems could be solved best by the creation of a special, independent organization or infrasturcture that could initiate and coordinate such studies and train young, promising scientists in these fields. Earlier efforts in this direction have unfortunately failed. However, in view of the combined threats of climatic change and accelerated deterioration of the open Mediterranean landscape, the need for such an International Mediterranean Ecosystem and Landscape Research Institute becomes even more urgent.

The Need for New Approaches and Tools

From our discussion emerges the need for an entirely new, holistic land use policy for the conservation and restoration of the Mediterranean landscapes as a whole. It should be based on well-coordinated, simultaneous efforts for better public and professional conservation education, more comprehensive planning and management, and more interdisciplinary and integrated research. For this purpose, neither one-sided agriculture, livestock, or forestry production nor nature conservation or recreation amenity-oriented approaches alone can provide a satisfactory solution. As described in detail elsewhere (Naveh and Lieberman, 1993), this requires a holistic systems approach of landscape-ecological determinism to be applied with innovative methods. These should aim at reconciling the conflicting demands for sustaining and improving the biological productivity, diversity, and scenic attractiveness of the open Mediterranean landscape with the needs of the people and the national economy. It should enable the optimization of all hard and soft landscape values and their long-term multiple benefit functions, by creating productive and attractive multiple use landscape mosaics in comprehensive master plans. In such master plans, all hard and soft values and their marketable and noneconomic richnesses could be expressed as functional parameters of production, regulation, protection, carrier, and information functions. To the latter, instead of monetary values, only relative values should be attached in a matrix of decision making, as suggested by de Groot (1992). For this purpose, innovative expert systems using fuzzy set theory are very suitable. Expert systems are computer programs based on "artificial intelligence" that can pose and answer questions relating to knowledge and information derived from human experts also in linguistic form. Fuzzy set theory enables the combination of such disparate landscape variables and subjective evaluations, including intangible values and functions, which cannot be quantified in a conventional way in numbers, but only in words. As shown by Negotia (1985), such verbal models can be used in fuzzy linear programming for the automatic transfer of knowledge from linguistic values in optimization functions.

Green Books for Threatened Landscapes

The realization of these plans will require a change in attitudes and concepts which cannot be initiated merely by interdisciplinary scientific research and by appropriate technological means, but only by a coordinated effort for public education at all levels. It requires the cooperation of all those who care for these uplands and their natural and cultural assets, those who study, plan, and manage them, and those who live from their resources. There is also urgent need to provide better tools for the decision making process in land use and conservation, taking into consideration the above-described threats.

As a first practical attempt to fulfill these functions, the preparation of Green Books for the conservation of threatened landscapes has been initiated by a special working group of the International Union for the Conservation of Nature and Natural Resources (IUCN); World Conservation Union Commission on Environmental Strategies, in cooperation with the International Association of Landscape Ecology (IALE). The first Green Book case study has been carried out it western Crete by multidisciplinary and multinational teams from the University of Thessaloniki and the University of Cambridge (Grove et al., 1993). Further case studies are planned in which a uniform methodology of such Green Books will be further developed. At the same time, the proposal for the preparation of worldwide red lists of valuable endangered landscapes has been adopted by the IUCN.

These green books should serve not only as a monitoring and prediction tool regarding the fate of open landscapes facing climatic change and increasing land pressure. They should also provide guidelines on their conservation, restoration, and sustainable development for decision making. The collection, analysis, synthesis, and interpretation of this information will be carried out with the help of advanced landscape-ecological methods, including integrated field surveys, appraisals, and landscape unit classifications, combined with remote sensing and dynamic Geographical Information Systems (GIS), and with multivariate statistical analysis and clustering methods. Their results, however, will be presented in these Green Books in nontechnical terms with ample illustrations and maps in an appealing and convincing way. Thereby they could also become an important tool for public education and be made available to the public, schools, universities, industries, business, politicians, and officials, and even to foreign visitors and tourists. Their object is not to produce another scientific report to be published and filed away sooner or later, but to become a usable document for professionals and the citizens. In contrast to Environmental Impact Statements, these Green Books should be active and positive iniatives to change already prevailing trends with the help of innovative, holistic, long-term, and multibeneficial planning and management strategies on regional landscape scales. Thereby they would become also a major tool to rescue endangered valuable landscapes included in the Red Lists.

Our hope is that in this way their semantic information will be transformed into pragmatic information, which becomes meaningful by its effects on the receivers and is expressed in their actions.

References

Allen TFH, Hoekstra TW. 1992. Towards a Unified Ecology. Columbia University Press, New York.

Boer MM, De Groot RS (eds). 1990. Landscape-Ecological Impact of Climatic Change. Proc Europ Conf Lunteren, The Netherlands 3–7 December 1989.

Bolin B, Doos BR, Jaeger J. 1986. The Greenhouse Effect, Climate Change and Ecosystems. Report No 29 SCOPE/ICSU. John Wiley & Sons, New York.

Bussotti F, Ferreti M, Gellini S, Barbolani E. 1991. New type forest damage to Mediterranean vegetation in Southern Sardinia forests. Eur J For Path 21:290–300.

di Castri F, Hansen AJ. 1992. The environment and development crises as determinants of landscape dynamics. In: Landscape Boundaries, F di Castri and AJ Hansen (eds). Springer-Verlag, New York, pp. 3–18.

Chapin FS, Schulze ED, Mooney HA. 1992. Biodiversity and ecosystem processes. Trends Ecol Evol 7:107–108.

De Groot R. 1992. Functions of Nature in Environmental Planning, Management and Decision Making. Wolters-Nordhoff, Amsterdam.

Farina A, Naveh Z (eds). 1993. Landscape Approach to Regional Planning: The Future of the Mediterranean Landscapes. Special Issue July 1993. Landscape and Urban Planning 24:1–4.

Gat Z, Rubin S. 1993. Analysis of Weather Conditions During the Winter Season of 1919/92. Israel Meteorological Service, Climatic Service. Bet Dagaan, Israel (In Hebrew).

Graber M, Cohen A. 1993. Climate change and air pollution: a review. In: Regional Implications of Future Climate Change, M Graber, A Cohen, and M Magaritz (eds). Israel Academy of Sciences and Humanities, Jerusalem, pp. 6–25.

GRIP (Greenland Ice-Core Project Members). 1993. Climate instability during the last interglacial period recored in the GRIP ice core. Nature 364:203–207.

Grove AT, Ispikoudis J, Kazaklis A, Moody JA, Papanastasis V, Rackham O. 1993. Threatened Mediterranean Landscapes: West Crete. Final Report to EEC. Department of Geography, University of Cambridge, Cambridge, UK.

Harte J, Torn M, Jensen D. 1992. The nature and consequences of indirect linkages between climate change and biological diversity. In: Global Warming and Biological Diversity, RL Peters and TE Lovejoy (eds). Yale University Press, New Haven, CT, pp. 325–343.

Hulme M, Wigley TML, Jones PD. 1990. Limitations of regional climate scenarios for impact analysis. In: Landscape-Ecological Impact of Climate Change. Proc Europ Conf Lunteren, The Netherlands 3–7 December 1989, MM Boer and RS de Groot (eds). IOS Press, Amsterdam, pp. 11–132.

IGBP. 1990. The Initial Core Project. Report No 12. The International Geosphere-Biosphere Program–A Study of Global Change. (IGBP) International Council of Scientific Unions (ICSU) Stockholm.

IPCC. 1990. Scientific Assessment of Climate Change. Intergovernmental Panel on Climate Change, Cambridge University Press, Cambridge, CA.

Jaeger J, Ferguson HL. 1991. Climate Change–Science, Impacts and Policy. Proc Second World Climate Conf (SWCC), Cambridge University Press, Cambridge, UK.

Jaffe S. 1992. Was there warming in Israel? Meterology Israel 1:92.

Jeftic L. 1993. Implications of expected climate change in the Mediterranean region. In: Regional Implications of Future Climate Change, M Graber, A Cohen, and M Magaritz (eds). Israel Academy of Sciences and Humanities, Jerusalem, pp. 278–302.

Katsoulis BD. 1987. Indications of change of climate from analysis of air temperature time series in Athens, Greece. Climate Change 10:67–79.

Kutiel H. 1985. The multimodality of the rainfall in Israel as reflected by the distribution of dry spells. Arch Meteorologie Geophysik Bioklimatologie 36B:15–27.

Kutiel H. 1992. Recent climatic variations in the Mediterranean Region. Proc Pan-Hallenic Symp Meteorology, Climatology and Atmospheric Physics, Thessaloniki, 21 May, 1992.

Kutiel H, Bar-Tuv A. 1992. Recent trends of decreasing autumnal sea surface temperatures (SST) in the Eastern Mediterranean Basin. Israel J Earth Sci 41:51–53.

Lascaratos, A, Gacic M, Oguz T. 1991. Recent atmospheric and oceanic climatic oscillations in the Northeastern Mediterranean and Black Sea. Proc Intern Workshop Regional Implication of Future Climate Change, April 28, 1991, Rehovot, Israel.

Le Houérou HN. 1988. Vegetation and land-use in the Mediterranean basin by the year 2050: a prospective Study. UNEP (OCA)/WG. 2/15.

Lucas PHC 1992. Protected Landscapes. A Guide for Policy Makers and Planners. Chapmann & Hall, London.

Malanson, GP, Westman WE. 1991. Modeling interactive effects of climate change, air pollution, and fire on a California shrubland. Climatic Change 18:363–376.

Mandel M. 1986. Water consumption of Israel according to ratio between the real and the potential evaporation. In: Quantitative and Qualitative Problems in the Present Water Budget in Israel. Proc Annual Meeting Israel Hydrol Assoc, Jerusalem, 29 October 1986 (in Hebrew).

Myers N. 1992. Synergism: joint effects of climate change and other forms of habitat destruction. In: Global Warming and Biological Diversity, RL Peters and TE Lovejoy (eds). Yale University Press, New Haven, CT, pp. 344–354.

Naveh Z. 1957. Report on Range Research at the Neve Yaar Experimental Station. FAO Working Party on Mediterranean Pasture and Fodder Development, 5th Meeting, Tel Aviv.

Naveh Z. 1982. The dependence of the productivity of a semi-arid Mediterranean hill pasture ecosystem on climatic fluctuations. Agric Environ 7:47–61.

Naveh Z. 1987. Biocybernetic and thermodynamic perspectives of landscape funtions and land use patterns. Landscape Ecol 1:75–83.

Naveh Z. 1988. The response of the herbaceous vegetation of a semiarid Mediterranean oak savanna to climatic fluctuations. In: Time Scales and Water Stress. 5th Int Conf on Mediterranean Ecosystems, F di Castri and F Floret (eds). UBS, Paris, pp. 361–371.

Naveh Z. 1991. Mediterranean uplands as anthropogenic perturbation dependent systems and their dynamic conservation management. In: Terrestrial and Aquatic Ecosystems–Perturbation and Recovery, O Ravera (ed). Ellis Horwood, New York, pp. 545–556.

Naveh Z, Dan J. 1973. The human degradation of Mediterranean landscapes in Israel. In: Mediterranean-Type Ecosystems. Origin and Structure, F di Castri and HA Mooney (eds). Springer-Verlag, Heidelberg, pp. 370–390.

Naveh Z, Kutiel P. 1990. Changes in the Mediterranean vegetation in Israel in response to human habitation and land uses. In: The Earth in Transition. Patterns and Processes of Biotic Impoverishment, GM Woodwell (ed). Cambridge University Press, New York, pp. 259–299.

Naveh Z, Lieberman AS. 1993. Landscape Ecology, Theory and Applications. Springer-Verlag, New York, 2nd edition.

Naveh Z, Whittaker R. 1979. Structural and floristic diversity of shrublands and woodlands in Northern Israel and other Mediterranean areas. Vegetatio 41:171–190.

Naveh Z, Steinberger EH, Chaim S, Rotman A. 1980. Photochemical air-pollutants–a threat to Mediterranean coniferous forests and upland ecosystems. Environ Cons 7:301–308.

Negotia. 1985. Expert Systems and Fuzzy Systems. Benjamin/Cummings Publishing Company, Menlo Park, California.
Nevo E. 1988. Genetic diversity in nature. Patterns and theory. In: Evolutionary Biology, Vol. 23, MK Hecht and B Wallace (eds). Plenum Publishing Corporation, New York, pp. 217–246.
Nevo E, Beiles A. 1989. Genetic diversity of wild emmer wheat in Israel and Turkey. Theor Appl Genet 77:421–455.
Nevo E, Carver BF, Beiles A. 1991. Photosynthetic performance in wild emmer wheat, *Triticum dicoccoides*: ecological and genetic predictability. Theor Appl Genet 8:334–460.
Nevo E, Bar-EI C, Beiles A, Yom-Tov Y. 1982. Adaptive microgeographic differentiations in allozyme polymorphism in land snalis. Genetica 59:61–67.
Prigogine I. 1976. Order through fluctuations: self-organization and social systems. In: Evolution and Consciousness: Human Systems in Transition, E Jantsch and CW Wadington (eds). Addison-Wesley, Reading, MA, pp. 93–130.
Schneider SH. 1989. The greehouse effect–science and policy. Science 243: 771–781.
Sestini G, Jeftic L, Milliman JD. 1989 Implications of expected climate changes in the Mediterranean region: An overview. MAP Technical Reports Series No 27. UNEP, Athens.
Shaviv I. 1978. The autecology of *Pistacia lentiscus* L. in Israel. PhD Thesis, Fac Agr Eng Technion, Haifa, Israel (Hebrew, English summary).
WCED. 1987. Our Common Future. World Commission on Environment and Development. Oxford University Press, New York.
Wigley TML, Farmer G. 1982. Climate of the Eastern Mediterranean and the Near East. In: Palaeclimates Paleoenvironments and Human Communities in the Eastern Mediterranean Region in later Prehistory, BAR International series 133, JC Bintliff and W Van Zeist (eds). British Archaeological Reports, Oxford, pp. 3–37.
Williamson P, Oeschger H. 1993. Climate instability in a warmer world. Climate Change 16:6–7.
Woodwell GM. 1992. How does the world work? Great issues of life and government hinge on the answer. In: Global Warming and Biological Diversity, RL Peters and TE Lovejoy (eds). Yale University Press, New Haven, CT, pp. 31–37.
Yogev I, Naveh Z. 1986. Effects of fertilizing and mulching on the establishment of *Pistacia lentiscus* in limestone debris for landscape reclamation. Proc 3rd Intern Conf, Israel Ecol Soc, pp. 403–412.

Index

N

O

P

Ecological Studies

Volume 68
Stable Isotopes in Ecological Research (1989)
P.W. Rundel, J.R. Ehleringer, and K.A. Nagy (Eds.)

Volume 69
Vertebrates in Complex Tropical Systems (1989)
M.L. Harmelin-Vivien and F. Bourlière (Eds.)

Volume 70
The Northern Forest Border in Canada and Alaska (1989)
J.A. Larsen

Volume 71
Tidal Flat Estuaries: Simulation and Analysis of the Ems Estuary (1988)
J. Baretta and P. Ruardij (Eds.)

Volume 72
Acidic Deposition and Forest Soils (1989)
D. Binkley, C.T. Driscoll, H.L. Allen, P. Schoeneberger, and D. McAvoy

Volume 73
Toxic Organic Chemicals in Porous Media (1989)
Z. Gerstl, Y. Chen, U. Mingelgrin, and B. Yaron (Eds.)

Volume 74
Inorganic Contaminants in the Vadose Zone (1989)
B. Bar-Yosef, N.J. Barrow, and J. Goldshmid (Eds.)

Volume 75
The Grazing Land Ecosystems of the African Sahel (1989)
H.N. Le Houérou

Volume 76
Vascular Plants as Epiphytes: Evolution and Ecophysiology (1989)
U. Lüttge (Ed.)

Volume 77
Air Pollution and Forest Decline: A Study of Spruce (*Picea abies*) on Acid Soils (1989)
E.-D. Schulze, O.L. Lange, and R. Oren (Eds.)

Volume 78
Agroecology: Researching the Ecological Basis for Sustainable Agriculture (1990)
S.R. Gliessman (Ed.)

Volume 79
Remote Sensing of Biosphere Functioning (1990)
R.J. Hobbs and H.A. Mooney (Eds.)

Volume 80
Plant Biology of the Basin and Range (1990)
C.B. Osmond, G.M. Hidy, and L.F. Pitelka (Eds.)

Volume 81
Nitrogen in Terrestrial Ecosystem: Questions of Productivity, Vegetational Changes, and Ecosystem Stability (1990)
C.O. Tamm

Volume 82
Quantitative Methods in Landscape Ecology: The Analysis and Interpretation of Landscape Heterogeneity (1990)
M.G. Turner and R.H. Gardner (Eds.)

Volume 83
The Rivers of Florida (1990)
R.J. Livingston (Ed.)

Volume 84
Fire in the Tropical Biota: Ecosystem Processes and Global Challenges (1990)
J.G. Goldammer (Ed.)

Volume 85
The Mosaic-Cycle Concept of Ecosystems (1991)
H. Remmert (Ed.)

Volume 86
Ecological Heterogeneity (1991)
J. Kolasa and S.T.A. Pickett (Eds.)

Volume 87
Horses and Grasses: The Nutritional Ecology of Equids and Their Impact on the Camargue (1992)
P. Duncan

Volume 88
Pinnipeds and El Niño: Responses to Environmental Stress (1992)
F. Trillmich and K.A. Ono (Eds.)

Ecological Studies

Volume 89
Plantago: A Multidisciplinary Study (1992)
P.J.C. Kuiper and M. Bos (Eds.)

Volume 90
Biogeochemistry of a Subalpine Ecosystem: Loch Vale Watershed (1992)
J. Baron (Ed.)

Volume 91
Atmospheric Deposition and Forest Nutrient Cycling (1992)
D.W. Johnson and S.E. Lindberg (Eds.)

Volume 92
Landscape Boundaries: Consequences for Biotic Diversity and Ecological Flows (1992)
A.J. Hansen and F. di Castri (Eds.)

Volume 93
Fire in South African Mountain Fynbos: Ecosystem, Community, and Species Response at Swartboskloof (1992)
B.W. van Wilgen et al. (Eds.)

Volume 94
The Ecology of Aquatic Hyphomycetes (1992)
F. Bärlocher (Ed.)

Volume 95
Palms in Forest-Ecosystems of Amazonia (1992)
F. Kahn and J.-J. DeGranville

Volume 96
Ecology and Decline of Red Spruce in the Eastern United States (1992)
C. Eagar and M.B. Adams (Eds.)

Volume 97
The Response of Western Forests to Air Pollution (1992)
R.K. Olson, D. Binkley, and M. Böhm (Eds.)

Volume 98
Plankton Regulation Dynamics (1993)
N. Walz (Ed.)

Volume 99
Biodiversity and Ecosystem Function (1993)
E.-D. Schulze and H.A. Mooney (Eds.)

Volume 100
Ecophysiology of Photosynthesis (1994)
E.-D. Schulze and M.M. Caldwell (Eds.)

Volume 101
Effects of Land-Use Change on Atmospheric CO_2 Concentrations: South and South East Asia as a Case Study (1993)
V.H. Dale (Ed.)

Volume 102
Coral Reef Ecology (1993)
Y.I. Sorokin (Ed.)

Volume 103
Rocky Shores: Exploitation in Chile and South Africa (1993)
W.R. Siegfried (Ed.)

Volume 104
Long-Term Experiments With Acid Rain in Norwegian Forest Ecosystems (1993)
G. Abrahamsen et al. (Eds.)

Volume 105
Microbial Ecology of Lake Plußsee (1993)
J. Overbeck and R.J. Chrost (Eds.)

Volume 106
Minimum Animal Populations (1994)
H. Remmert (Ed.)

Volume 107
The Role of Fire in Mediterranean-Type Ecosystems (1994)
J.M. Moreno and W.C. Oechel (Eds.)

Volume 108
Ecology and Biogeography of Mediterranean Ecosystems in Chile, California, and Australia (1994)
M.T.K. Arroyo, P.H. Zedler, and M.D. Fox (Eds.)

Ecological Studies

Volume 109
Mediterranean Type Ecosystems: The Function of Biodiversity (1994)
G.W. Davis and D.M. Richardson (Eds.)

Volume 110
Tropical Montane Cloud Forests (1994)
L.S. Hamilton, J.O. Juvik, and F.N. Scatena (Eds.)

Volume 111
Peatland Forestry (1995)
E. Paavilainen and J. Päivänen

Volume 112
Tropical Forests: Management and Ecology (1995)
A.E. Lugo and C. Lowe (Eds.)

Volume 113
Arctic and Alpine Biodiversity: Patterns, Causes and Ecosystem Consequences (1995)
F.S. Chapin III and C. Körner (Eds.)

Volume 114
Crassulacean Acid Metabolism: Biochemistry, Ecophysiology and Evolution (1995)
K. Winter and J.A.C. Smith (Eds.)

Volume 115
Islands: Biological Diversity and Ecosystem Function (1995)
P.M. Vitousek, H. Andersen, and L. Loope (Eds.)

Volume 116
High Latitude Rain Forests and Associate Ecosystems of the West Coast of the Americas (1995)
R.G. Lawford, P. Alaback, and E. Fuentes (Eds.)

Volume 117
Global Change and Mediterranean-Type Ecosystems (1995)
J.M. Moreno and W.C. Oechel (Eds.)